EDUCATION AND HERITAGE IN THE ERA OF BIG DATA
IN ASTRONOMY:
THE FIRST STEPS ON THE IAU 2020–2030 STRATEGIC PLAN

IAU SYMPOSIUM 367

COVER ILLUSTRATION:

'Eclipse entre Pehuénes'. Used with permission of Matias M. Cordero.

IAU SYMPOSIUM PROCEEDINGS SERIES

Chief Editor
MARIA TERESA LAGO, IAU General Secretariat
IAU-UAI Secretariat
98-bis Blvd Arago
F-75014 Paris
France
mtlago@astro.up.pt

Editor
JOSÉ MIGUEL RODRÍGUEZ ESPINOSA, IAU Assistant General Secretary
IAU-UAI Secretariat
98-bis Blvd Arago
F-75014 Paris
France
IAU_AGS@iap.fr

INTERNATIONAL ASTRONOMICAL UNION

UNION ASTRONOMIQUE INTERNATIONALE

EDUCATION AND HERITAGE IN THE ERA OF BIG DATA IN ASTRONOMY: THE FIRST STEPS ON THE IAU 2020–2030 STRATEGIC PLAN

PROCEEDINGS OF THE 367th SYMPOSIUM OF THE INTERNATIONAL ASTRONOMICAL UNION VIRTUAL MEETING INITIALLY PLANNED FOR BARILOCHE, ARGENTINA 8–12 DECEMBER, 2020

Edited by

ROSA M. ROS
Polytechnic University of Catalonia, Spain

BEATRIZ GARCÍA
National Technological University, Mendoza Regional Faculty, Argentina

STEVEN R. GULLBERG
University of Oklahoma, USA

JAVIER MOLDÓN
Institute of Astrophysics of Andalusia - CSIC, Spain

and

PATRICIO ROJO
National University of Chile, Chile

CAMBRIDGE UNIVERSITY PRESS
University Printing House, Cambridge CB2 8BS, United Kingdom
1 Liberty Plaza, Floor 20, New York, NY 10006, USA
10 Stamford Road, Oakleigh, Melbourne 3166, Australia

First published 2021

Printed in the UK by Bell & Bain, Glasgow, UK

Typeset in System LaTeX 2ε

A catalogue record for this book is available from the British Library Library of Congress Cataloguing in Publication data

This journal issue has been printed on FSC^{TM}-certified paper and cover board. FSC is an independent, non-governmental, not-for-profit organization established to promote the responsible management of the world's forests. Please see www.fsc.org for information.

ISBN 9781108490801 hardback
ISSN 1743-9213

Table of Contents

Session 1: State of the Art in Astronomy Education

Session 2: Citizen Science and Solar Eclipses

Session 3: Astronomy Education Research

Session 4: Astronomy in other disciplines to promote science vocations

Session 5: Innovation in Education

Session 6: Literacy in Astronomy

Session 7: Big Data in Astronomy

Session 8: Cultural Astronomy and Heritage

Session 9: Astronomy and Inclusion

Session 10: Informal Education in Astronomy

Posters

Session 1

Session 2

Session 3

Session 9

Session 10

Preface

This special volume includes contributions from the IAUS 367 Symposium, *Education and Heritage in the Era of Big Data in Astronomy*. It was intended to hold the symposium in Bariloche, Argentina at the time of the total solar eclipse in December 2020. Unfortunately, due to the Covid-19 global pandemic those plans had to be changed and the symposium was instead conducted virtually.

The new format operations were moved to Mendoza in Argentina, from where transmission, recording, and management of the symposium were each controlled using a platform provided by the IAU. When the "new reality" caused us to change the style of the meeting, a new and great opportunity presented itself: more connected people from around the globe, more contributions presented from countries not always represented at international symposiums, more parallel activities, and more channels for communicating astronomy to all, not only to professionals (astronomers, educators, communicators) but also to interested others through an inclusive and diverse framework. The impact of the symposium proved that this type of meeting is not only possible but can work very well and that this likely will be the starting point for a new way to organize such encounters, where both presenciality and virtuality will coexist and, from now on, the hybrid format will become the rule rather than the exception.

Astronomy education has become a major topic for the IAU's goals. Scientific results from this field have great potential to enhance the teaching and learning of astronomy for learners of many ages. Taking into account two of the goals of the IAU Strategic Plan for 2020-2030:

1) The IAU leads the worldwide coordination of astronomy and the fostering of communication and dissemination of astronomical knowledge among professional astronomers
2) The IAU stimulates the use of astronomy for teaching and education at school level

We consider that in this framework, new results and research methodologies from the cognition and learning science domains are now able to influence the work of astronomy educators, enabling them to make informed innovations for the teaching of astronomy and are part of the bases to reach the IAU objectives.

The primary goal of this symposium is to give perhaps for the first time a global vision of education and heritage in the frame of the goals of the IAU, taking into account the Plan 2020-2030 and to propose an eventual 'next steps' road map and a global astronomy education agenda for the next decade, while honoring the education from the past.

In this sense, we would foster inclusiveness in the advancement of astronomy and facilitate the advancement of the next generation of astronomers and scientists, through encouragement of the use of new methods of learning and best practices (including distance education: MOOCS) in pedagogy at university level, as well as the use of astronomy for teaching and education at school level, which are part of the definitions of the proposed Office of Astronomy for Education (OAE), which also pursuits the establishment of a Network of Astronomy Education Contacts (NAECs) to provide accessible materials and astronomy literacy guidelines globally. The invited speakers are not only international leaders in discipline-based education in astronomy and planetary science, but also in communication, history, inclusion, and protection of world's heritage, including the dark sky.

As with most other sciences, astronomy is being fundamentally transformed by the Information and Computation Technology (ICT) revolution. The data volume is growing exponentially, can be accessed remotely, and the observations can be performed even without a real knowledge of a telescope. The new approaches to the data permit the development of new tools, techniques, and resources for data analysis and produce discoveries which probably never would be reached with only traditional data analysis.

There were 10 main sessions with the following themes and articles from each are included in these proceedings:

1) State of the Art in Astronomy Education
2) Citizen Science and Solar Eclipses
3) Astronomy Education Research
4) Astronomy in Other Disciplines to Promote Science Vocations
5) Innovation in Education
6) Literacy in Astronomy
7) Big Data in Education
8) Cultural Astronomy and Heritage
9) Astronomy and Inclusion; The Role of Women and Girls in Astronomy
10) Informal Education: Museums, Planetariums, etc.

The symposium's educational program also included the following: A Galaxy Forum South America 2020; a Discussion and Round table: Research, innovation, literacy and inclusion in astronomy (in Spanish); three public conferences on Solar Eclipses (in English), Astrobiology (in English), Landscape and World Heritage (in Spanish) and Light Pollution (in Spanish); three workshops on Eclipses (in English and in Spanish) and Didactic Devices (in Spanish); an Associated Event on Quiet Sky Protection and Sustainable Development (in English).

We wish to extend our gratitude to our organizing institutions: Universidad Nacional de Río Negro – Sede Andina – San Carlos de Bariloche, Instituto de Tecnologías en Detección y Astropartículas (ITeDA, CNEA-CONICET-UNSAM), Consejo Nacional de Investigaciones Científicas y Técnicas (CONICET), Sociedad Uruguaya de Astronomía (SUA), Asociación Chilena de Astronomía (SOCHIAS), Fundación Balseiro (Bariloche), Research Infrastructure FOR Citizens in Europe (REINFORCE), European Gravitational Wave Observatory (EGO), Comisión Nacional de Actividades Espaciales (CONAE) and Asociación Argentina de Astronomía (AAA).

Rosa M. Ros
Beatriz García
Steven R. Gullberg
Javier Moldon
Patricio Rojo

Editors

Rosa M. Ros
Polytechnic University of Catalonia, Spain

Beatriz García
National Technological University, Mendoza Regional Faculty, Argentina

Steven R. Gullberg
University of Oklahoma, USA

Javier Moldon
Institute of Astrophysics of Andalusia - CSIC, Spain

Patricio Rojo
University of Chile, Chile

Organising Committee
Scientific Organising Committee

Beatriz García (Co-Chair)	ITeDA (CNEA, CONICET, UNSAM) – UTN FRM, Argentina.
Rosa M. Ros (Co-Chair)	Polytechnical University of Catalonia, Spain.
Akihiko Tomita	Wakayama Univ., Faculty of Education, Japan.
Andrea Sosa Oyarzabal	Univ. de la Rep. – Centro Univ. Reg. del Este, Uruguay.
Boonrucksar Soonthornthum	Nat. Astron. Research Institute of Thailand, Thailand.
Chenzhou Cui	Nat. Astron. Observatories, CAS, China.
Dongni Chen	Beijing Planetarium, China.
Hidehiko Agata	Nat. Astron. Observatory of Japan, Japan
Jay Pasachoff	Williams College-Hopkins Observatory, USA.
John Hearnshaw	Univ. of Canterbury Physics-Astronomy Department, New Zealand.
Julieta Fierro	UNAM, México.
Julio Fernández	Dept. Astron., Facultad de Ciencias, UdelaR, Uruguay.
Katrien Kolenberg	KU Leuven & Univ. of Antwerp, Physics Dep., Belgium.
Leonardo Pellizza	Inst. Arg. de Radioastronomía, CONICET, AAA, Argentina.
Néstor Camino	Univ. Nacional San Juan Bosco, CONICET, Argentina.
Nicoletta Lanciano	Univ. La Sapienza, Italia.
Paulo S. Bretones	Univ. Federal de São Carlos, Dep. de Metodologia de Ensino, Brazil.

Local Organising Committee

Mariana Orellana (Chair)	Nat. Univ. of Río Negro, Andean headquarters, CONICET, Argentina.
Elise Servajean	Univ. de Los Andes – Univ. del Desarrollo, Chile
Anahí Granada	Nat. Univ. of Río Negro, Andean headquarters, CONICET, Argentina.
Daniel Carpintero	FCAGLP – Univ. Nac.de La Plata, IALP (CONICET – UNLP), Argentina.
José Luis Hormaechea	Estación Astron. Río Grande (FCAGLP – UNLP, CONICET), Argentina.

List of participants in the Symposium IAU S367

Abreu Oliveira Vinicius de
Agata Hidehiko
Alba Martínez Durruty Jesús de
Alves-Brito Alan
Antonio Narino
Arifyanto Mochamad Ikbal
Barres de Almeida Ulisses
Bartus Paul
Bayo Amelia
Beamin Juan Carlos
Belmonte Juan
Bhakare Nikita
Bieryla Allyson
Bretones Paulo
Buckner Anne
Burton Michael
Caballero Mora Karen Salome
Caldú Primo Anahí
Camino Néstor
Canas Lina
Carreto Parra Francisco
Casado Johanna
Casu Silvia
Chatterjee Somenath
Chattopadhyay Soham
Chattopadhyay Sutapa
Cheung Sze-leung
Chinnici Ileana
Chis Paula
Choudhuri Arnab Rai
Clavijo Carolina
Colantonio Arturo
Colazo Marcelo
Cooper Sally
Coronel Salvador Sofia
Cremades Hebe
Cui Chenzhou
Cunnama Daniel
Curir Anna
Dalgleish Hannah
Borrayos Monica
Delgado Inglada Gloria
Deustua Susana
Devarapalli Shanti Priya
Deverakonda Ram Prasad
Doran Rosa
Duran Cintia
Durst Steve

Eff-Darwich Antonio
Ellegaard Ole
Elmegreen Debra
Espino Verónica
Falcon Nelson
Farmanyan Sona
Ferreira Rafael Ramón
Fierro Julieta
Figueiredo Newton
Fitzgerald Michael
Galperin Diego
García Beatriz
García Guillermo
Goez Theran Cristian
González-García César
Guevara Day Walter
Guevara Natalia
Gullberg Steven
Gutti Jogesh Babu
Hallberg Karen
Hammer Edith
Hanaoka Yoichiro
Hasan Priya
Hearnshaw John
Hemming Gary
Hoffmann Suzzane
Holbrook Jarita
Horaguchi Toshihiro
Iles Elizabeth J.
Impey Chris
Isidro Villamizar Gloria
Jiwaji Noorali
Jokin Ivo
Jones Graham
Kapoor Ramesh
Karaseur Fernando Ariel
Kerschbaum Franz
Kolenberg Katrien
Kołomański Sylwester
Komonjinda Siramas
Korhonen Heidi
Koribalski Baerbel
Kuznetsov Eduard
Labade Rupesh
Lanciano Nicoletta
Lanza de Oliveira Letícia
Lazzaro Daniela
Lebron Mayra

Lewis Fraser
Li Geng
Li Shanshan
López Alejandro
Maffione Nicolás
Malasan Hakim Luthfi
Martínez Usó María José
Martins Carlos
Metaxa Margarita
Mickaelian Areg
Mikayelyan Gor
Molenda-Zakowicz Joanna
Monteiro Martín
Morillo Acosta Marcela Janine
Nahar Sultana
Ödman Carolina
Ohnishi Kouji
Orellana Mariana
Ortega Alcides
Ortíz Amelia
Pantoja Pantoja Carmen A.
Paolantonio Santiago
Pasachoff Jay
Pastrana Claudio
Peña Ramírez Karla
Peralta Raphaël
Perkins Kala
Pernicone Verónica Leonor
Pössel Markus
Pović Mirjana
Prantzos Nikos
Pulatova Nadiia
Quinteros Cynthia P.
Quispe Adita
Randriamanakoto Zara
Rector Travis
Renchin Tsolmon
Ricciardi Sara
Rodrigues Lara
Rodríguez Espinosa José Miguel
Rogerson Jesse
Rokni Mahdi
Rollinde Emmanuel
Romagnoli Claudia
Ros Rosa María
Safonova Margarita

Sánchez González Javier
Sanhueza Pedro
Sebben Viviana
Sit Exodus Chun-Long
Sobreira Paulo
Solís-Castillo Basilio
Soonthornthum Boonrucksar
Sosa Andrea
Spasova Mina
Spathopoulos Vassilios
Stasinska Grazyna
Stavinschi Magda
Stoev Alexey
Stoeva Penka
SubbaRao Mark
Sundin Maria
Sylla Salma
Tamazawa Harufumi
Tampo Yusuke
Tancredi Gonzalo
Taylor Bernie
Testa Italo
Thorve Sonal
Toma Corina Lavinia
Tomita Akihiko
Torres Silvia
Torres Campos Ana
Troncoso Iribarren Paulina
Unda-Sanzana Eduardo
Usuda-Sato Kumiko
van Dishoeck Ewine
Vaquero José Manuel
Vasylenko Maksym
Vavilova Irina
Vera Victor
Villicana Pedraza Ilhuiyolitzin
Viñuales-Gavín Ederlinda
Voelker Anna
Walker Constance
Walker-Holmes Mizzani
Waller William
Wan Mohd Kamil Wan Mohd Aimran
Yakubu Muallim
Young Alex
Zotti Georg
Zou Siwei

Map of geographical distribution

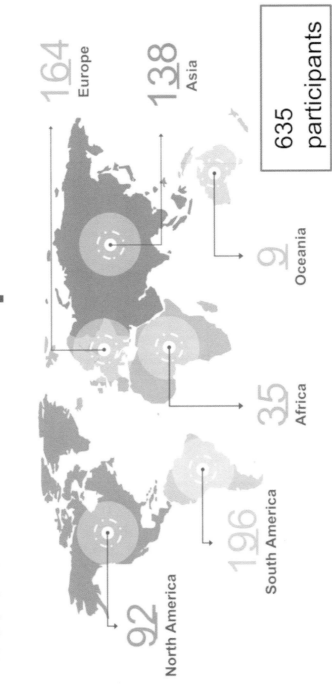

Figure 2. Geographical distribution of participants in IAU S367.

Group Photo

Attendees photographed during the meeting online
(Photo by Beatriz García)

Education and Heritage in the era of Big Data in Astronomy
Proceedings IAU Symposium No. 367, 2020
R. M. Ros, B. Garcia, S. R. Gullberg, J. Moldon & P. Rojo, eds.
doi:

Opening Session

Introduction

IAU President, Ewine van Dishoeck, gave enlightening remarks to open IAUS 367 Education and Heritage in the Era of Big Data in Astronomy. She began by recognizing this as the first IAU symposium dedicated to education and also that it is the first to be fully virtual. She said that while far away, we were more connected than ever. She discussed the IAU and its organization, including the roles played by its offices, divisions, commissions, and working groups. Ewine continued with outlining worldwide IAU networks and programs such as the Office of Astronomy for Development (OAD), Office for Astronomy Outreach (OAO), National Astronomy Education Coordinator (NAEC) teams, and the Network for Astronomy School Education (NASE). She emphasized that these groups help to educate the next generation of astronomers. She provided a most fitting opening to a groundbreaking symposium.

Education and Heritage in the era of Big Data in Astronomy
Proceedings IAU Symposium No. 367, 2020
R. M. Ros, B. Garcia, S. R. Gullberg, J. Moldon & P. Rojo, eds.
doi:10.1017/S1743921321001095

The IAU and Education: Introduction

Ewine F. van Dishoeck⬛, IAU President

Leiden Observatory, Leiden University,
P.O. Box 9513 NL-2300 RA, Leiden, the Netherlands
email: ewine@strw.leidenuniv.nl

Abstract. This paper provides a brief overview of the many facets of astronomy education and heritage, and how the IAU stimulates them. Activities range from training in astronomy through scientific meetings and schools for young astronomers, to using astronomy as a tool for development and for stimulating science education at school level. Communicating astronomy with the public and engaging in outreach activities with children to inspire curiosity is yet another way of how astronomy can help build a literate society and install a sense of global citizenship. The involvement of many people at all levels is key to success.

Keywords. Astronomy education, Outreach, Development, IAU Strategic Plan

1. Introduction

The mission of the IAU is to "promote and safeguard astronomy in all its aspects (including research, communication, education, and development) through international cooperation," as approved the XXXth General Assembly of the IAU in Vienna (Elmegreen & van Dishoeck 2018). The text in parenthesis is new and makes it explicit that the IAU has branched out beyond its original purpose of sharing astronomical knowledge and fostering communication among professional astronomers. Education is at the heart of many of the IAU activities.

The IAU is organized in Divisions, Commissions and Working Groups. Of particular relevance to this symposium is Division C on *Education, Outreach and Heritage* with 2385 members of the total ~12000 IAU members. Specifically, Commission C1 on "Astronomy Development and Education"; Commission C2 on "Communicating Astronomy with the Public"; and Commission C4 on "World Heritage of Astronomy" are active on the topics of this symposium. Big Data is part of Commission B2 of Division B on "Facilities, Technologies and Data Science".

Over the last decade, the IAU has created four Offices that enable the expanded vision (Fig. 1):

− *OYA:* The *Office of Young Astronomers*, hosted by the Norwegian Academy of Science and Letters, focuses on the training of young astronomers at University level, and organizes the International School for Young Astronomers (*ISYA*).

− *OAD:* The *Office of Astronomy for Development*, hosted in Cape Town South Africa in partnership with the NRF, focuses on the use of astronomy for development by capitalizing on the field's scientific, technological and cultural links and its impacts on society.

− *OAO:* The *Office for Astronomy Outreach* is a partnership with the National Astronomical Observatory of Japan, and focuses on providing access to astronomical information and astronomy communication with the general public.

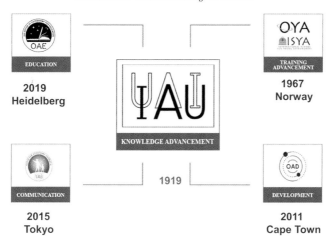

Figure 1. Offices of the IAU and their main missions. All of them have connections to education and there are many interactions between the Offices.

    — *OAE:* The *Office of Astronomy for Education*, hosted at the Haus der Astronomie in Heidelberg Germany, focuses on providing resources and training for using astronomy as a stimulus for teaching and education from elementary to high school level (astronomy and science education).

    Each of these Offices also have their own networks of coordinators worldwide, such as the National Outreach Coordinators (NOC) and the National Astronomy Education Coordinators (NAEC) networks and the regional nodes of the OAD (ROADs). These networks bring in huge additional manpower and resources to carry out the mission of the IAU. The IAU Strategic Plan is described in more detail in the paper by Hearnshaw (this volume). See also Andersen *et al.* (2019) for the first 100 years of the IAU and van Dishoeck (2019) for a vision of the IAU as it enters the second century of its existence.

2. The IAU and Heritage

    The night sky is available to everyone anywhere in the world, and forms an integral part of the environment that is experienced by humankind. Cultural heritage related to the night sky is therefore a vital component of cultural heritage in general. Across the world, there are an impressive number of astronomical records, sites and landscapes providing testimony to the diverse ways in which humans have viewed their connection with the universe from the earliest times to the present day (Ruggles 2015). The IAU through its Commission C4 works with UNESCO to recognize and raise awareness of the importance of astronomical heritage worldwide and to facilitate efforts to identify, protect and preserve such heritage for the benefit of humankind†. This includes protection of the visibility of the night sky itself. Heritage and education go hand in hand through the fascinating stories that are associated with historical records and sites.

3. The IAU and Education: astronomy training

    Education is embedded in the IAU at many different levels. Starting at the top, most IAU members are also educators at their universities by teaching classes at undergraduate and graduate level and training students in astronomy research. This includes teaching astronomy to non-science majors, providing them with a formal exposure to science.

† https://www3.astronomicalheritage.net/

Many excellent textbooks are available, some of them now also open source or accessible through on-line websites and videos. A role for the IAU could be to curate and raise awareness of the available material, provide access to reviews of best practices in college education, and help getting material translated into other languages. Some of this work is being undertaken by Commission C1.

PhD students and young astronomers further benefit from international meetings, including the IAU Symposia, Regional Meetings and General Assemblies which stimulate worldwide attendance and enable participation through travel grants. The IAU Junior Members (at postdoc level) have an active working group that organizes activities to stimulate personal development. The annual IAU PhD Prizes honor the best PhD theses in each Division, whereas the annual Gruber fellowships support high quality research of young astronomers, notably from countries in difficult economic conditions.

At MSc and beginning PhD level, the ISYA schools have played a key role in the development of astronomy worldwide. These three week schools, about 3 every 2 years, take place in developing countries and provide lectures and hands-on training at telescopes to about 30–50 students each. Since their start in 1967, more than 1600 students have been trained in total, many of which have gone on to become professional astronomers and leaders in their countries. They also build lifelong connections during the schools.

Progress in astronomy goes hand in hand with new large facilities. A key advantage of astronomy compared with other disciplines is that the bulk of the astronomical data are publically available, with major organizations having invested billions of dollars, Euros or Yens to make this possible. This means that anyone in the world can carry out frontline astronomy provided she/he knows how to access and work with the data. Starting late 2022, the IAU plans to organize annually two workshops to provide hands-on training through the I-HOW (IAU Hand-On Workshops). These workshops are aimed at PhD students, postdocs and young staff members in developing countries.

The OAD also regularly supports regional schools for BSc and MSc students on specific scientific topics, which often include a component of hands-on training in observations or working with Big Data sets. Astronomy for development can only happen if there is also development of astronomy in a country. Examples include the West African International Summer School for Young Astronomers, the Guatemalan School of Astrophysics, ArAS school for astrophysics in the Arab world, and the Joint Exchange Development Initiative for Africa (JEDI).

4. The IAU and Education: astronomy for development

Training in astronomy prepares students for much more than a career in astronomical research or instrumentation. The skills acquired in analyzing big data sets, performing large scale simulations, building and working with front-line technology or solving inherently complex and multidisciplinary problems makes astronomy students attractive to a wide range of sectors in society. Such a flow of talented people from astronomy to society is healthy and should be stimulated.

The OAD was established to promote the use of astronomy as a tool for development in every country. The OAD uses the UN Sustainable Development Goals (SDGs) as the global definition of development, for example in its annual call for proposals. The Regional Offices (ROADs) form the global core structure of the OAD, recently strengthened by the establishment of ROADs in Europe and the US. Interactions between them stimulate synergies, educate each other on tools and activities, but also raise awareness about cultural differences. Collaborations with fields other than astronomy are also included, such as the space sector, ICT, branches of social sciences and humanities, relevant industries and NGOs, art and cultural organisations. In this regard, the OAD serves as a 'hub' for collaborations across disciplines.

The OAD is currently developing a number of flagship projects that can serve as "signature dishes" and which have educational and/or heritage aspects. The first flagship centers on Astronomy for Socio-economic Development through Astrotourism: the use of small telescopes or mobile planetaria to stimulate the development of eco-tourism and create livelihoods in remote villages, with the dark night sky serving as the natural asset. The second one highlights Astronomy for Peace, Diplomacy and Global Citizenship with a particular focus on the Pale Blue Dot program (see below). Flagship 3 uses Astronomy Knowledge and Skills for Development, most notably using Big Data. This includes a fruitful partnership with the DARA Big Data project (Scaife & Cooper 2020).

5. The IAU and Education: astronomy for education at schools

The OAE provides a natural platform for promoting the use of astronomy as a gateway to STEM (science, technology, engineering, and mathematics) fields for young people. The OAE supports the astronomy community and educators in bringing the fascination of astronomy into schools at primary and secondary level. A worldwide network of National Astronomy Education Coordinators (NAECs) and Regional Education Offices are being established. Together, they promote astronomy in national curricula, identify accessible materials and set astronomy literacy guidelines (and liaise with education offices at ministries and with curriculum experts), support teachers with evidence-based education research and material (e.g., the "Big Ideas in Astronomy" booklet), encourage standards for teacher training activities, and help the community with its professional development (see Pompea & Russo 2020 for review).

The IAU is active in various other educational activities at school level. For example, Commission C1 has started the bi-yearly AstroEDU conferences and supports the AstroEDU website with peer-reviewed educational activities. In the context of its 100yr celebration†, initiatives included the Einstein schools, Astronomy Day in Schools, and support for teacher trainings (inherited from IYA2009) such as the Galileo teacher training, Network for Astronomy School Education (NASE), and Open Astronomy schools. Some of these activities will be embedded in the OAE in the future.

6. The IAU and Education: communicating with the public

Education also takes place more informally by communicating results to the general public and through engaging in outreach activities with them. This is the mandate of the OAO, and its long term vision is that that all people throughout the world will have access to knowledge of frontline astronomy; that all countries will have good access to astronomical research, culture and experiences to help build a literate society; and that astronomers are a strong part of the global citizenship.

The OAO generally does not create material itself, but works with the IAU and other organisations to increase the impact of its activities in education and public outreach. As such, it reaches the general public from the youngest children in kindergarten to families and senior citizens. Outreach is a strong component of many different units within the IAU, and the OAO interfaces with all of them. For example, the OAO works with Commission C2 to edit the "Communicating Astronomy with the Public" (CAP) journal as well as an Astronomy Outreach Newsletter, and co-organizes the biennial CAP conference. The OAO also provides easily accessible public-friendly information on astronomical terminologies and objects in the universe, and coordinates some of the worldwide citizen-science projects and campaigns, such as the public exoplanet naming IAU100 competition.

† www.iau-100.org,www.iau.org/static/archives/announcements/pdf/iau100-final-report-ann20019.pdf

The OAO NOCs are also the formal point of contact for engagement with amateur astronomy groups within each country. Amateurs, in turn, educate the general public in astronomy through the many events at their public observatories. Science musea and planetaria can also play a major role in this. Furthermore, the NOCs play a role in educating the public about the importance of the dark and quiet skies, including the recent threats that increased light pollution and the launch of satellite constellations pose not just for scientific exploration but also for public health and as cultural heritage†.

The OAD and OAO collaborate on the Pale Blue Dot Flagship, an education project that uses astronomy to promote a sense of global citizenship in young children. It is targeted at disadvantaged children aged 5 to 8, and builds on the "Universe Awareness" program led by Leiden Observatory. The fragility of our planet Earth as seen from space also provides an entry point to educate children about sustainability and climate change. As highlighted by Minister Pandor at the United Nations GA 75 Dialogue on "Astronomy: A unique educational tool for furthering the SDGs and stimulating a global perspective" (September 2020): *"Astronomy, more than any other science, inpires curiosity, optimism and hope in children"* (paraphrased).

7. The IAU and EID

Equity, inclusion and diversity (EID) are embedded in all IAU activities and are stimulated through the promotion and implementation of policies, structures, and programs such as those described above. The IAU has two working groups, on Women in Astronomy and on EID, that report directly to the Executive Committee. Following IAU Symposium 358 on "Astronomy for Equity, Diversity and Inclusion - a roadmap to action within the framework of the IAU centennial anniversary" held in Tokyo in 2019, a Springboard to Action‡ was published in early 2021. This booklet sets out recommendations to IAU National and Individual Members to work towards the more diverse and representative global astronomical community it envisions.

Recent highlights of the IAU's efforts on EID include "Inspiring Stars", an itinerant exhibit on astronomy for inclusion, "Hands in the Stars", the first international comparative list of astronomical words in sign languages, and several projects on astronomy for the visually impaired. As part of IAU100, global projects were carried out in 143 countries under the specific IAU100 theme "Inclusive Astronomy", which supported events like the IAU100 "Women and Girls in Astronomy", "NameExoWorlds", and "Astronomy Day in Schools", and more recently the OAO "Telescopes for All" projects.

8. Concluding remarks

This brief introduction has highlighted the many facets of education and heritage, and the role that astronomy and the IAU play in them. Education continues throughout people's lives and requires at each stage not just good teachers and high quality materials but also the skills to engage with students. The words of Benjamin Franklin resonate with what the IAU aims to establish: *"Tell me and I forget; teach me and I may remember; involve me and I learn."*

To make the IAU dream into a reality requires involvement from everyone, from research professionals to educators and students. The IAU is therefore grateful to its Offices, their Networks and Regional Offices, its membership and the many other people who have contributed as volunteers to *"use astronomy to make the world a better place"*

† www.iau.org/news/announcements/detail/ann21002/
‡ www.iau.org/static/publications/springboard-booklet-150dpi-2page-view.pdf

(Govender, priv. comm.). This includes the organizers of this symposium! Let's make sure to harness the inspiration of astronomy for science education: together, we can make it work.

References

Andersen, J., Baneke, D. & Madsen, C. 2019, *The International Astronomical Union* (Springer)

Elmegreen, D., & van Dishoeck E.F. 2018, *IAU Strategic Plan 2020–2030* www.iau.org/administration/about/strategic-plan

Pompea, S.M., & Russo, P. 2020, ARA&A, 58, 313

Ruggles, C.L.N. 2015, *Handbook of Archeoastronomy and Ethoastronomy* (Spinger)

Scaife, A.M.M., & Cooper, S.E. 2020, Astronomy in Focus XXX, Proceedings of IAU GA, 569 www.darabigdata.com

van Dishoeck, E.F. 2019, in *Under One Sky, IAU Centenary Symposium 349*, eds. C. Sterken, J. Hearnshaw, D. Valls-Gabaud (Cambridge University Press), 523

Session 1: State of the Art in Astronomy Education

Education and Heritage in the era of Big Data in Astronomy
Proceedings IAU Symposium No. 367, 2020
R. M. Ros, B. Garcia, S. R. Gullberg, J. Moldon & P. Rojo, eds.
doi:

Introduction

Each day of the symposium included two themed sessions. Session 1 on the opening day was themed "State of the Art in Astronomy Education."

These Proceedings begin with a talk by John Hernshaw discussing "The IAU Strategic Plan 2020–2030, a blueprint for forging a new social revolution in astronomy and for using astronomy as a tool for building a progressive society." John began with some background information by relating some of the history and evolution of the IAU and he continued with a description of its structure. He then included some of the ways in which the IAU has transformed how it operates including its evolution for better diversity and inclusion. John next went into a discussion of the IAU's Strategic Plan for 2020–2030 that first emerged at the 2018 IAU GA in Vienna. He brought out the five main goals for the new decade and gave insightful commentary for each of them. John continued with the need for fundraising to support these initiatives and included suggestions as to where the IAU can expand future activity. In his conclusion he stated "...we can expect the social revolution to continue, with the aims not only to foster the science of astronomy, but also to help build a more progressive society using astronomy as a tool." In consideration of all that was outlined by John, the IAU can anticipate a great decade ahead.

Following the talk Claudio Pastrana asked:

It is not necessarily a question, but there was a time when it was evident that the IAU had become an institution with testosterone excess. There was a conscious effort for a change? Or, simply, Women, were gaining their spaces as they usually do? With courage and intelligence?

and John Replied:

I think initially the early IAU pioneers that promoted the social revolution made a conscious effort – people like Jean-Claude Pecker and Edith Muller. But the drive to promote women in astronomy has really come from women themselves in recent decades. The need to promote gender equality is now widely recognized as a goal for the IAU and indeed for all sciences.

The next article is by Mirjana Povic and is called "Development of astronomy research and education in Africa and Ethiopia." In it Mirjana talks about the great potential in Africa and in particular in Ethiopia. She describes recent advances for astronomy research and education and also what needs to be done for the future. She cites the Square Kilometer Array (SKA) as an example of a contributor to Sustainable Development Goals. She gives other examples associated with such as the South African Astronomical Observatory (SAAO) and the Ethiopian Space Science Society (ESSS) that show the future to be bright for astronomy on the African continent.

A paper by Sally Cooper follows and is called "The National School's Observatory: Access to the Universe for all." Sally describes how the National School's Observatory

(NSO) makes access to the robotic Liverpool Telescope available to all schools in the United Kingdom and Ireland. Students are afforded the same access as are professional astronomers. The paper goes on to describe the value of the program and the importance of accessible data driven astronomy in education. Access is free and activities are available for students from seven to 18 years old, as well as for adults. She discusses the potential of the New Robotic Telescope (NRT) on La Palma and stresses the importance of the next generation of robotic telescopes for data driven astronomy education.

After the talk Amelia Bayo asked:

Hi Sally, such a great project!!! In Chile (within NPF) we are producing astronomy mirrors / segments out of carbon fiber, therefore much lighter, would that be something of interest for the new telescope?

and Sally replied:

Hi Amelia, I don't know much about the technical side of the new telescope. You can find out more about the telescope in the link and feel free to contact the team – they are very welcoming of interest. http://www.robotictelescope.org/team/

Then De Lara Rodrigues asked:

Dear Sally, connecting with Cintia's talk, I would like to know if NSO is interested in collaborating with non-English-speaking countries to translate at least some part of the website to other languages. It isn't easy to find high-quality and updated astronomy educational resources in other languages (and for the Southern hemisphere, as Cintia pointed out)

and Sally replied:

Please feel free to contact me at sally@schoolsobservatory.org The SchoolsObs is very open to working with countries to translate our resources. Please get in contact if you want to know more.

After Cooper's article you will find Sze-leung Cheung describing "A preliminary study of the impact of high school astronomy research-based learning in Thailand." In this he talks about student opportunities in Thailand with such as presenting in the student session of the Thai Astronomical Conference and mentored student research activities in the Advanced Teacher Training scheme. Preliminary results were said to support the valuable roles that planetariums can play in motivating students. It also was brought out that lecture settings are the norm in Thailand with few experimental activities, but that adding research in schools helps to develop thinking and produce new ideas. Results supported better personal development and self-confidence. He suggests that these findings might be useful in helping IAU Commission C1 to promote global research-based learning activities.

Following the talk Claudio Pastrana asked:

I can take PISA very seriously, because the monstrous inequality in OECD countries isn't parametrically corrected. Countries with many troubles are measured with the same rule. Let me say this in a nice form: This people are mistaken. How much are you, personally, concerned about PISA index? Isn't it a better course of action to take the problem and adjust the best possible solution, disregarding PISA?

and Sze – leung replied:

This is a good question, and I think it deserve a separate study. In my understanding, PISA are conducted differently in different countries, I have spoken to our NARIT director who also sits on the PISA board of Thailand, he told me that my countries play tricks in the scoring, so yes it is difficult to compare, but if you read carefully on the PISA report it actually gives many detailed information such as the distribution of the student performance within the same country, and these I think are very useful. It's difficult to find a single universal way of understanding everyone, but I think it's a good starting point.

Finally, within State of the Art in Astronomy Education Hakim Malasan and co-authors Rosa M. Ros, Chatief Kunjaya, Endang Soegiartini, Aprilia, and Riska Romadhonia talk about "Empowering science teachers in Indonesia through NASE workshops." Here they provide a four-year review of NASE workshops in Indonesia. They begin that their first Network of Astronomy for School Education Network (NASE) was held at Machung University in East Java in 2017. In 2018 and 2019 NASE workshops were held at the Institut Teknologi Sumatera in Bandar Lampung, Lampung Province, and in 2020 virtually hosted by Institut Teknologi Bandung in Bandung, West Java Province. They outline the great value that NASE workshops provide and the significant impact that each had with teachers in Indonesia.

Following the talk Muhammed Hafez bin Ahamat Murtza asked:

What is your opinion teaching astronomy using local folklore?

and Hakim replied:

Yes, that is what we try to do in Indonesia, especially for the local teachers. By bringing them to visit and investigate local archeological site and stimulate them with many star lores would in fact gain their enthusiasm and pride.

Poster presentations were included for each session as well. A listing of them by session has been added as an appendix.

Education and Heritage in the era of Big Data in Astronomy
Proceedings IAU Symposium No. 367, 2020
R. M. Ros, B. Garcia, S. R. Gullberg, J. Moldon & P. Rojo, eds.
doi:10.1017/S1743921321001101

The IAU Strategic Plan 2020–2030, a blueprint for forging a new social revolution in astronomy and for using astronomy as a tool for building a progressive society

John B. Hearnshaw[1]

[1]University of Canterbury, Christchurch, New Zealand
and IAU Vice-President on Executive Committee
email: `john.hearnshaw@canterbury.ac.nz`

Abstract. I discuss the second IAU Strategic Plan for the decade 2020–30 in the context of the overall evolution of the IAU in recent past decades. This article shows how the IAU has evolved dramatically since WW2. It is hardly recognizable in terms of its original organization and goals of a century ago. What was once an inward-looking body engaged purely with the procedures of astronomical research is now a dynamic and outward-looking organization, interacting with people, especially students and the public.

A large part of this success must be attributed to the IAU's unique body of individual members, whose number has grown strongly in recent decades. It is the individual members, especially through the Commissions and Working Groups, who have promoted these enormous changes in the outlook of the Union. This is a model for other scientific unions to follow, and especially for the work to promote the careers of women in science, for promoting the careers of young astronomers, for bringing students into astronomy or into science in general, for helping people with disabilities to have careers in astronomy, for engaging with the public, and for helping to develop astronomy and science in developing countries.

Looking to the future, the IAU's new Strategic Plan for the years 2020 to 2030 has five major goals for the coming decade: 1. The IAU leads the worldwide coordination of astronomy and the fostering of communication and dissemination of astronomical knowledge among professional astronomers. 2. The IAU promotes the inclusive advancement of the field of astronomy in every country. 3. The IAU promotes the use of astronomy as a tool for development in every country. 4. The IAU engages the public in astronomy through access to astronomical information and communication of the science of astronomy. 5. The IAU stimulates the use of astronomy for teaching and education at school level.

Future developments will also be engaging with the large number of amateur astronomers and helping to promote astro-tourism, which is perhaps the new frontier now growing rapidly around the world. The Strategic Plan is a blueprint for forging a social revolution in astronomy and for using astronomy as a tool for building a progressive society.

Keywords. IAU Strategic Plan, IAU individual members, a social revolution in astronomy, amateur astronomers, astro-tourism

---

1. Introduction

The International Astronomical Union (IAU) was founded on 28 July 1919 in Brussels by the International Research Council (IRC). It was one of the first of the new scientific unions to be formed in the era immediately following the Great War. However, the International Solar Union had preceded it (formed by Hale in 1904), as had also the

Figure 1. The first IAU General Assembly, Rome 1922. Note the paucity of female astronomers and the preponderance of old men.

Astrographic Congress, which dated back to 1887. These last two organizations were examples of international co-operation in astronomy from earlier decades.

In 1919 the IAU operated like a club for about 200 astronomers to discuss classification schemes, standards for measurement and collaborations. In this sense it was a closed and rather esoteric society for its members, who were a small fraction of scientists and an even smaller and largely invisible niche in society as a whole. The original IAU was neither fully international nor global, as the original members came from the USA, and Europe but excluded Germany, Austria, Hungary, Bulgaria and Turkey (the Central Powers of WW1). West Germany only became a national member in 1951 and most of the other Central Powers did so over the ensuing decade (though Hungary became a member in 1947)

The IAU was formed 'to promote and safeguard the science of astronomy in all its aspects through international cooperation'. From the birth of the organization, a triennial General Assembly (GA) was seen as the best way to promote these aims. The first GA was held in Rome in 1922, when just 83 astronomers participated (Fig. 1). The most recent was in Vienna in 2018 (the 30th GA) with about 3000. Currently, the IAU has 82 national members and over 11 000 individual members who are professional astronomers (generally with a PhD and employed in astronomy).

The structure of the IAU with both national and individual members made the Union fairly unqiue amongst scientific unions. But the presence of individual members, in due course, was found to be one of the key elements of the IAU's success. The individual members represent a task force of dedicated professionals who interact with each other globally to promote the aims of the Union. Because they feel a strong sense of belonging and of personal engagement in the affairs of the Union, this has proved to be a potent force for promoting the science of astronomy.

More information on the IAU and its history can be found in two recent publications (Sterken *et al.* 2019, Andersen *et al.* 2019).

2. Evolution and transformation of the IAU over 100 years

By the end of WW2, it was realized that an important component in promoting astronomy was to facilitate international collaboration. The triennial GA was not enough to meet these goals, and hence a new IAU commission for the Exchange of Astronomers was proposed. Commission 38 thus came into being in 1948 at the GA in Zürich. It helped fund exchange visits of astronomers between member countries, by supporting travel costs, but not living expenses or salaries. The typical duration of an exchange was three months or more. Travel grants were awarded for over six decades by the exchange of astronomers programme, from 1947 to 2009 and a total of 558 astronomers benefitted from these grants. Many came from developing countries (Hearnshaw 2019).

Commision 38 represents the first time the IAU started promoting people rather than just the modalities of how to carry out research. It was also notable as being the only one of two commissions to have funding and a budget (the other was C46 for the teaching of astronomy).

Commision 46 for the Teaching of Astronomy was established in 1964, largely on the initiative of Jean-Claude Pecker (France) and Marcel Minnaert (Netherlands). This in turn led to the first International School for Young Astronomers (ISYA) being held in Manchester in 1967. Graduate students at PhD and MSc level were the participants for these schools. For the first time the IAU was focussing attention on students. This represents a further important step in the Union's development as being more of a people-focussed organization rather than purely research-focussed.

It is notable that at the present time 42 ISYA schools have been held, and about 1500 students have participated in these ISYA. Many have gone on to distinguished professional careers in astronomy.

In 1994 the IAU began to be interested in promoting astronomy in developing countries. This was the year that the Working Group for the World-wide Development of Astronomy (WGWWDA) was established, with Alan Batten (Canada) as chair. The WGWWDA organized visits to developing countries, to give lectures, encourage research collaboratioins and promote IAU membership.

3. IAU structure

In 1919 the IAU was founded with an Executive Committee and 32 commissions covering most branches of astronomical research. Although some commissions were disbanded and new ones created, this structure proved to be inflexible as the size of the individual membership grew rapidly after WW2.

At the 1994 GA at the Hague, a structure with 11 divisions was adopted with commisions being attached to a division. By 2009 this structure had evolved to nine divisions and the 35 commissions and 53 workinmg groups that we have today. Notably, Division C covers the non-research activities of the Union, and in particular C1 is for astronomy education and devlopment, C2 for communicating astronomy with the public, C3 for the history of astronomy and C4 for world heriatge and astronomy. It is this Division for Education, Outreach and Heritage that has driven the transformation of the IAU and represents the face of its new-found social awareness.

A major new initiaitive came in 2009 at the time of the GA in Rio da Janeiro. It was here that the first IAU Strategic Plan 2010-20 was approved, which included establishing an Office of Astronomy for Development, which would have a small number of professionals who aim to use astronomy to promote economic development in developing countries. The OAD office was established in 2010 in Cape Town with Kevin Govender as its director, and is co-sponsored by South Africa's National Research Foundation, NRF.

OAD took over many of the functions of the former WGWWDA and another education commission working group, Teaching Astronomy for Development (TAD).

The success of OAD has led to three further offices being established. The IAU Office for Astronomy Outreach (OAO) was opened in 2012, being hosted and supported in Tokyo at the National Astronomical Observatory of Japan. Dr Hidehiko Agata is the supervising director and Lina Canas is the international outreach coordinator at OAO. The IAU Office for Young Astronomers (OYA) was established in Oslo in 2015. It is funded by the Norwegian Academy of Science and Letters and in turn supports the ISYA schools. OYA has no professional staff. On the other hand, ISYA has appointed a director and deputy director to coordinate up to two schools per year. They are Drs Itziar Aretxaga in Mexico and David Mota, a Portuguese astronomer based in Oslo. The IAU Office for Astronomy Education was established in 2019 in Heidelberg, Germany. It is hosted by the Haus der Astronomie of the Max Planck Gesellschaft and is funded in part by the Klaus Tschira Stiftung and the Carl-Zeiss-Stiftung. Dr Markus Pössel is the inaugural director.

The IAU thus has now a considerably more complex and multi-layered structure than was the case a century ago. The Executive Committee (EC) comprises twelve members, including the president, president-elect, general secretary and assistant general secretary, and six vice-presidents and two advisors. The secretariat, based permanently in Paris at the Institut d'Astrophysique, has two staff members dealing with finance and administration. Under the EC are nine divisions, 35 commissions, 53 working groups and four offices. The membership of the Union comprises 82 national members, 11801 individual members of whom 716 are junior members. Junior membership is a new membership category introduced in 2018 for those astronomers who have completed a doctoral thesis within the last six years.

4. A social revolution has been launched

The last fifty (almost sixty) years have seen a remarkable transformation in the way the IAU operates. It has become very much a people-focussed organization that not only promotes the science of astronomy with these new ways of operating, but also aspires to promote social change and development in society using astronomical advances as a tool to promote science, equity and inclusion more generally.

Several key examples can be cited for how the IAU now operates to achieve these new goals.

(a) The IAU has established several working groups that promote social change. These include the WG for Women in Astronomy (2003), the WG for Equity, Diversity and Inclusion (2015) and the WG for Junior Members (2018). All these are WGs directly under the Executive Committee. We can also cite the WG for Dark and Quiet Skies Protection, which has been proactive in 2020 in producing a report for the UN Committee on the Peaceful Uses of Outer Space (COPUOS). This report has widespread implications for the growth of astro-tourism and people's appreciation of an unpolluted starry night sky, as well as for the progress of astronomical research.

(b) The IAU has established itself as a role-model for promoting women in astronomy by appointing four women to serve as IAU president since 2006, and the Union will have three consecutive female presidents from 2015 to 2024. The female presidents of the IAU are
- Catherine Cesarsky (France) 2006–09
- Silvia Torres (Mexico) 2015–18

- Ewine van Dishoeck (Netherlands) 2018–21
- Debra Elmegreen (USA) 2021–24

(c) The IAU in 2009 organized the International Year of Astronomy, a world-wide celebration of the 400th anniversary of the first observations of the night sky with a telescope by Galileo. The IYA2009 was endorsed by the United Nations General Assembly. Hundreds of public events were organized in 148 different countries, and there were 12 Global Cornerstone Projects that involved the whole world. It is estimated that 815 million people world-wide were exposed to some aspect of astronomy by IYA activities during 2009. This is more than ten per cent of the entire world population (Cesarsky 2019).

(d) Strong support is given by the IAU to early-career astronomers, through the ISYA schools and the Junior Members programme. In 2015 the IAU also initiated a programme of prizes for the best PhD theses which were submitted world-wide. Up to ten prizes annually are offered.

(e) Support is given to developing countries to encourage the world-wide growth of astronomy education and research, through OAD. Promoting astronomy in Africa has been an especial area of interest. However there are also now ten regional offices of astronomy for development (ROADs) and two offices which are language centres that can translate stories and news in English into respectively Portuguese and into Chinese (LOADs).

(f) The IAU gives support to the hearing impaired with sign language for astronomical terms, and to people with other disabilities to enable them to embark on astronomical careers or assimilate astronomical news. The recent IAU symposium in Tokyo, IAUS 358: Astronomy for Equity, Diversity and Inclusion - a roadmap to action within the framework of the IAU 100th Anniversary highlighted the progress in these fields.

(g) In 2020 the IAU published a new Code of Conduct for all its members (see IAU Code of Conduct 2020). This included Anti-harassment Guidelines and also a Code of Ethics. The former article is designed to eliminate sexual harassment at IAU meetings, and also incidents such as verbal abuse. The second is designed to eliminate plagiarism and give due recognition to junior authors in multi-author publications.

(h) The IAU has actively promoted several presigious prizes offered in astronomy and cosmology. These are of course conferred on the elite few with outstanding achievements in their professional careers. However the prizes also promote astronomy in the eyes of the public, given the widespread media attention the recipients receive. The prizes include the Gruber Cosmology Prize, the Kavli Prize, the Shaw Prize and of course Nobel prizes when the prize in physics goes to an astronomer. It is notable that 24 Nobel Laureates have been awarded in astronomy since 1967, half of them in the last decade. If Boyle and Smith are included for their pioneering work on the CCD which has had such an infleunce on observational optical astronomy, then the number is 26 Nobel laureates since 1967.

We can reflect on why and how the IAU has been so successful in being socially proactive and a role-model for other scientific disciplines and other scientific unions. Probably three key factors have facilitated these changes:

- The IAU has a large number of individual members who are actively engaged in the affairs of the Union, especially working in the Commisions and Working Groups.
- The science of astronomy is visually very rich in images, and many discoveries therefore receive attention in the world's media with images of the latest discoveries, including those from current space missions. This has resulted in a large following by the educated

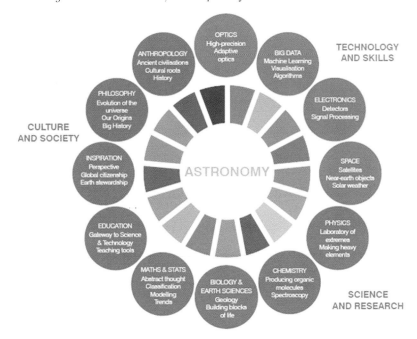

Connections of astronomy with other disciplines. The colours in the inner ring represent the
colours of the UN Sustainable Development Goals (SDGs) to which astronomy can contribute
(this "wheel" is a re-envisioning of the wheel on the cover of the Strategic Plan 2010–2020).

Figure 2. Connections of astronomy with other disciplines. From the IAU Strategic Plan
2020–2030.

public on the progress of astronomical research, especially discoveries such as black holes,
the big bang, cosmic expansion at one end of the distance scale, and also solar system
exploration of objects closer to Earth.

• Astronomy has hundreds of thousands of amateur astronomers world-wide, many of
whom own their own small telescopes and some of whom are actively engaged in useful
research, generally in stellar astronomy such as variable star observations. This fact is
rather unqiue amongst the sciences, and certainly has resulted in a widespread interest
in astronomy from the general public.

5. The IAU Strategic Plan 2020–2030

At the 30th GA in Vienna in 2018, the second IAU Strategic Plan for the decade
2020 to 2030 was approved (Elmegreen & van Dishoeck 2018). It was largely written by
president-elect Debra Elmegreen and by IAU president Ewine van Dishoeck. The new
Strategic Plan follows on from the success of the plan for the previous decade and builds
on the progress of the IAU over the previous century, especially in the transformation in
the previous half century to become more people-focussed and socially proactive.

Astronomy has links to many other disciplines and branches, including to culture and
society. Fig 2 is from the IAU Strategic Plan, 2020–30. These links to other disciplines of
scientific and social endeavour highlight the strength of astronomy as a multi-disciplinary
science, which is a key element of its success.

The new IAU Strategic Plan for the coming decade has five main goals as shown in
Fig. 3. The first concerns the core function of the Union, namely astronomical research,
but also the dissemination of the scientific knowledge acquired to the wider community.

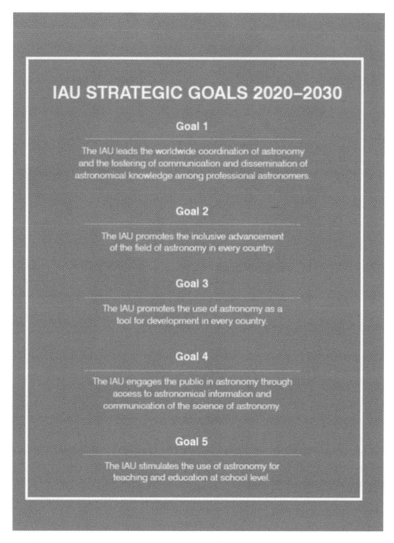

Figure 3. Five goals of the IAU for the coming decade. From the IAU Strategic Plan 2020–2030.

The following goals concern the world-wide development of astronomy as a science in all countries, especially the inclusive advancement of women in astronomy in all countries, and the use of astronomy for economic development, for outreach and communication to the public, and for teaching and education at school level. These last four goals are the socially proactive ones that characterize the present ambitions of the Union.

The following comments highlight some of the achievements already made on each of these goals, and also on future prospects.

(a) For goal #1, there is an extraordinary range of new investments underway to explore the entire electromagnetic spectrum coming from the universe and observed either from the ground or from space. In addition, the new science of gravitational wave astronomy and the relatively new field of neutrino astrophysics will complement electromagnetic observations. Deep space mssions will continue the

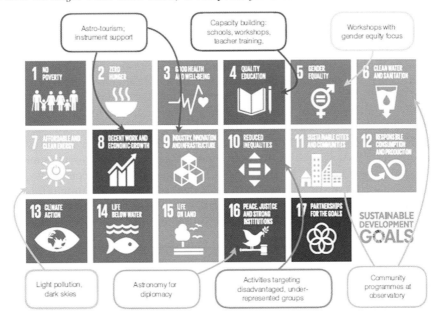

Some potential contributions of astronomy to the Sustainable Development Goals.

Figure 4. Over half of the UN's Sustainable Development Goals are supported by OAD projects From the IAU Strategic Plan 2020–2030.

exploration of solar system bodies. The projects include three extremely large telescopes, the James Webb space telescope, the Square Kilometre Array (SKA), the Vera Rubin telescope and advanced LIGO and Virgo gravitational wave detectors.

(*b*) For goal #2, it is noted that women comprise 11 to 20 per cent of individual IAU members in many developed countries. A handful of countries including France, Italy, Romania, Bulgaria, Brazil and South Africa have 21 to 30 per cent women. Very few national members have more than 30 per cent women astronomers, but they include Argentina and Thailand. As for the geographical distribution of astronomers, Africa is still very much under-represented when it comes to IAU members. Overall, there has been much progress, but clearly there is much still to do.

(*c*) For goal #3, the establishment of the OAD in 2012 in Cape Town represents one of the great successes of the IAU. Since 2013, 200 development projects have been funded in many countries which span the entire world (though as it happens not yet in Australia or New Zealand). This work has been under OAD director Kevin Govender. The work of OAD has been aligned with the seventeen UN Sustainable Development Goals and the claim in the Strategic Plan is that the OAD impacts on at least eight of the 17 UN goals. This is illustrated in Fig. 4.

(*d*) For goal #4, the OAO office in Tokyo has appointed National Outreach Coordinators (NOCs) to represent 127 countries (far more than the number of national IAU members). The NOCs form a supportive network of astronomers and outreach professionals taking astronomy to their local communities. This has been one of the principal successes of the Office for Astronomy Outreach. We can also mention the IAU/UNOOSA/IAC conference on Dark and Quiet Skies for Science and Society, to discuss light pollution's impact on astronomy, and also the adverse impact of radio noise and of satellite constellations. This was an on-line conference in October 2020 with nearly 1000 registrations. A 270-page report has been prepared for the UN committee COPUOS. Other work of the OAO, under Hidehiko Agata and Lina Canas is the CAP Journal, a peer-reviewed journal published twice a year, and the CAP Conference, held approximately every second year. (CAP is Communicating Astronomy to the Public).

(*e*) For goal #5, the IAU established the Office for Astronomy Education (OAE) in Heidelberg, Germany in 2019. It is hosted in the Haus der Astronomie of the Max Planck Institute for Astronomy with Markus Pössel as director. The appointment of National Astronomy Education Coordinators (NAECs) in many countries is underway. The OAE will analyse astronomy teaching in IAU countries and identify accessible materials and astronomy literacy guidelines. It will encourage standards for teacher training activities and it is planned to organize an annual International School for Astronomy Education (ISAE). The annual Shaw Prize-IAU Workshops on Astronomy Education that OAE will organize are funded by the Shaw Prize Foundation.

6. IAU International Fundraiser

With so many projects and new developments in the IAU's future plans, it is not surprising that these will require new funding. The IAU will expend about 4.2 million euros over the triennium 2022 to 2024, of which some three quarters comes from national member dues. The remainder comes from foundations, donations and royalties.

In April 2020, the IAU appointed a professional fundraiser to raise additional funds for the projects planned. She is Genevieve Marshall from the UK. The initial plan was to raise at least 100 000 euro per annum, perhaps more in later years.

7. Areas where the IAU needs to expand its activities in the future

The IAU has made huge progress in transforming itself into a modern and socially proactive organization supporting the use of astronomy for social change and serving as a role model for other scientific unions and organizations. However, there is still much work to do, and there are some areas where the Union has so far only made some initial steps.

Here are some of the things the Executive Committee could consider for the near future:

• More engagement with the community of amateur astronomers (especially those who want to contribute to research collaborations with professionals).

• Forging strategic alliances for outreach with planetariums, public outreach observatories and museums (with astronomy or space science divisions).

• Promoting astro-tourism in dark sky places. (See Fig. 5 for an example of thriving astro-tourism in New Zealand).

• Support for the Starlight Declaration (2007) on the right of all people to have access to a pristine, dark and starry night sky.

8. Conclusion

The International Astronomical Union has seen a transformation in its modus operandi during its first 100 years, especially so in its second half century since the late 1960s. It is now a much more people-focussed organization, serving the interests of astronomers of all nationalities and both genders.

With its second Strategic Plan 2020–2030, we can expect the social revolution to continue, with the aims not only to foster the science of astronomy, but also to help build a more progressive society using astronomy as a tool.

Figure 5. Astro-tourists at Mt John Observatory, New Zealand. Astro-tourism is a possible future area for more IAU engagement. Photo courtesy Fraser Gunn.

There is still much work to do in realizing existing goals, and there are new areas of engagement which the Union should consider tackling in the years to come.

More evolution of the IAU can be expected in the coming decades.

References

Andersen, J., Baneke, D. & Madsen, C. 2019. *The International Astronomical Union: Uniting the community for 100 years* Publ. Springer Nature, Cham, Switzerland

Cesarsky, C. 2019. Reflections on 100 years of the IAU. In *IAU Symposium 349*, 25

Elmegreen, D. & van Dishoeck, E. 2018. *IAU Strategic Plan 2020–2030*. https://www.iau.org/administration/about/strategic_plan/. Accessed 18 Jan 2021.

Hearnshaw, J.B. 2019. Commision 38 (Exchange of astronomers) and Commision 46 (Teaching of astronomy): two commissions that played a unique role in the history and development of the IAU. In *IAU Symposium 349*, 374

IAU Code of Conduct 2020. https://www.iau.org/news/announcements/detail/ann20013/. Accessed 18 Jan 2021.

Starlight Declaration, 2007. https://www.starlight2007.net/index_option_com_content_view_article_id_185_starlight-declaration_catid_62_the-initiative_itemid_80_lang_en.html. Accessed 18 Jan 2021.

Sterken, C., Hearnshaw, J. & Valls-Gabaud, D. 2019. *Under One Sky: The IAU Centenary Symposium*. Proceedings of IAU Symposium no. 349 held in Vienna, August 2018. Publ. Cambridge Univ. Press, Cambridge, England

Education and Heritage in the era of Big Data in Astronomy
Proceedings IAU Symposium No. 367, 2020
R. M. Ros, B. Garcia, S. R. Gullberg, J. Moldon & P. Rojo, eds.
doi:10.1017/S1743921321001009

Development of astronomy research and education in Africa and Ethiopia

Mirjana Pović[1,2]🄳

[1]Ethiopian Space Science and Technology Institute, Ethiopia.

[2]Instituto de Astrofísica de Andalucía (CSIC), Spain.
email: mpovic@iaa.es

Abstract. Africa has amazing potential due to natural (such as dark sky) and human resources for scientific research in astronomy and space science. At the same time, the continent is still facing many difficulties, and its countries are now recognising the importance of astronomy, space science and satellite technologies for improving some of their principal socio-economic challenges. The development of astronomy in Africa (including Ethiopia) has grown significantly over the past few years, and never before it was more possible to use astronomy for education, outreach, and development as it is now. However, much still remains to be done. This paper will summarise the recent developments in astronomy research and education in Africa and Ethiopia and will focus on how working together on the development of science and education can we fight poverty in the long term and increase our possibilities of attaining the United Nations Sustainable Development Goals in future for benefit of all.

Keywords. Astronomy education and research. Astronomy in Africa.

1. Introduction

Astronomy can be used as an important tool for development and for achieving the United Nations (UN) Sustainable Development Goals (SDGs)1. Being one of the most multidisciplinary sciences2, it can be used efficiently to promote education and to inspire-for and promote science, contributing directly to SDGs 4 and 8 (e.g., Govender 2009; Dalgleish 2020). The Network for Astronomy School Education (NASE)3, NUCLIO4, Galileo Teacher Training Program5, and Universe Awareness (UNAWE)6 are only some of the examples where through astronomy tens of thousands of children, students, and teachers have been involved in science. Astronomy can also effectively contribute to socio-economical growth of countries and therefore to SDGs 8 and 10 (e.g., McBride et al. 2018). We have great examples such as South Africa, Chile, or Canary Islands in Spain and Hawaii in US, where governments selected astronomy and space science to be priority fields for their socio-economic development (e.g., see South African Ten-Year Plan for Science and Technology7, or the astronomy program of the Government of Chile8). Astronomy is also one of the most cutting-edge sciences and one of the principal

1 https://sdgs.un.org/
2 https://www.iau.org/static/education/strategicplan_2010-2020.pdf
3 http://sac.csic.es/astrosecundaria/en/Presentacion.php
4 https://nuclio.org/
5 http://galileoteachers.org/
6 https://www.unawe.org/
7 https://www.gov.za/documents/ten-year-plan-science-and-technology#
8 https://www.conicyt.cl/astronomia/files/2013/11/Roadmap_Astronomia_v3.pdf

contributers to technological development and innovation over the last decades (e.g., Vigroux 2009), contributing directly to different SDGs including SDG 9. Astronomy also showed over the past decades to be an important tool for promoting peace and diplomacy through large international collaborations. The Square Kilometer Area (SKA) is a great example in this aspect, and direct contributer to SDGs 16 and 17. In the digital revolution that we are leaving now we can see direct contributions of astronomy, through wifi discovery, development of computing (e.g., through grid computing), communication (e.g., satellite communication), global positioning system (GPS; e.g., through atomic clocks based on distant quasars observations), and imaging (e.g., through charged coupled devices (CCDs)), and in the future through big data and related technologies (e.g., through SKA revolution) (e.g., Rosenberg et al. 2013). Benefits of astronomy for socioeconomical development in Africa have been recognised by the African Union (AU) and highlighted in the Common African Position on the Post-2015 Development Agenda9. As a result of that the very first African Space Strategy has been developed under the AU, recognising space science and astronomy as important for achieving SDGs. In addition, many African countries are putting important efforts in developing astronomy and space science in terms of research, institutional development, infrastructure, human capacity development (HCD), and policies (e.g., Pović et al. 2018).

2. Astronomy developments in Africa

Over the past years significant improvements have been done regarding astronomy development across Africa. Development of radio astronomy is one of the principal priorities through the SKA and African Very Long Baseline Interferometry Network (AVN). Through these big international collaborations, beside South Africa, different African countries are involved including: Botswana, Ghana, Kenya, Madagascar, Mauritius, Mozambique, Namibia, and Zambia. All of these countries signed for the first time in 2017 the memorandum of understanding to collaborate on development in radio astronomy. Ghana was the first country among all African SKA partners to build the radio observatory in Kuntunse, by converting the 32m telecommunication antenna into radio telescope (e.g., Wild 2017). Ghana became the third African country with professional radio astronomy facilities after South Africa and Mauritius. Starting from 2025, during the SKA phase 2, thousands of dishes are planned to be built in South Africa and other African partner countries. Recently, South African Radio Astronomy Observatory (SARAO) has been established by joining the South African SKA and HartRAO. The first phase of MeerKAT has been completed in 2018 with establishment of 64 radio array dishes of 13.5m each (e.g., Camilo 2018). Thanks to this, we are now able to obtain from Africa some of the most detailed radio images of the Universe. Mauritius, in collaboration with India, is operating a radio telescope since 1992. Recently, Nigeria successfully assembled and installed 3m radio telescope mainly for HCD purposes (e.g., Pović et al. 2018). Namibia is now planning to build the first millimeter-wave radio telescope in Africa together with Netherlands (e.g., Backes et al. 2017). Hydrogen Intensity and Real-time Analysis eXperiment (HIRAX) is under development in South Africa with plans to build 1000 6m dishes that will operate in 400-800 MHz window (e.g., Newburgh et al. 2016).

Besides radio astronomy, development in optical astronomy also experienced a lot of progress over the past years. In South Africa the world largest 11m SALT telescope (e.g., Buckley et al. 2006) plus some another 20 South African and international telescopes are operating under the South African Astronomical Observatory (SAAO). New MeerLICHT robotic 0.65m optical telescope has been recently built in South Africa to be synchronized with MeerKAT and its radio observations. In Morocco, Oukaimeden Observatory

9 https://sustainabledevelopment.un.org/content/documents/1329africaposition.pdf

experienced strong development since its inauguration in 2007, including the establishment of 60cm TRAPPIST-North telescope dedicated to the search of extrasolar planets
(e.g., Benkhaldoun 2018). Ethiopia built in 2014 Entoto Observatory with two twin 1m
telescopes (e.g., Tessema 2012). 1m telescope has been moved from La Silla in Chile to
Burkina Faso who is now trying to build its first astronomical observatory (e.g., Carignan
2012). Different countries are now working on site testing for development of new optical
observatories, including Algeria, Egypt, Ethiopia, Kenya, and Tanzania (e.g., Pović et al.
2018). Regarding development of gamma-ray astronomy in Africa, H.E.S.S observatory in
Namibia with its five Cherenkov telescopes is still one of the best ground-based facilities
in the field of high-energy astronomy.

Regarding HCD in astronomy, remarkable progress has been made over the past years.
New post-graduate programs have been established across the continent (e.g., in Egypt,
Ethiopia, Kenya, Morocco, Namibia, Nigeria, Rwanda, South Africa, Sudan, Uganda,
etc.). Office of Astronomy Development (OAD-IAU) with its constant support contributed significantly to HCD. Hundreds of African MSc students have been trained
through the National Astrophysics and Space Science Programme (NASSP)1 in South
Africa. Many other students benefited through trainings and their studies supported by
the AVN programme, Development in Africa with Radio Astronomy (DARA; e.g., Hoare
2018)2, SKA-HCD programme, International Science Program (ISP)3, or The African
Initiative for Planetary and Space Science (AFIPS)4. Public awareness has been improved
across the whole continent. In many countries primary and secondary school education
benefited through different teachers training programs such as NASE, Galileo Teachers
Training, or NUCLIO. The African Astronomical Society (AfAS)5 has been established
recently in 2019, with aim to be the voice of astronomy in Africa and to contribute to
address the challenges faced by the continent through the promotion and advancement of
astronomy, and through the activities under some of its main committees such as Science
Committee, Outreach Committee, and recently established African Network of Women in
Astronomy (AfNWA). Finally, the recently launched African Strategy for Fundamental
and Applied Physics (ASFAP)6 aims to strengthen in future the development of physics,
including astronomy, across the continent.

3. Astronomy developments in Ethiopia

The Ethiopian Space Science Society (ESSS)7 was established in 2004 as a civic society
whose aim was to promote the development of astronomy and space science in Ethiopia
and its benefits for the society (e.g., Tessema 2012). ESSS activities were fundamental
for later development of Entoto Observatory in 2014 and the Ethiopian Space Science
and Technology Institute (ESSTI)8 in 2016, the first research center of such kind in
Ethiopia and all East-African region. Under the ESSTI Astronomy and Astrophysics
(A&A) Research and Development Department different activities have been carried out
over the past years. These include:

• Running of MSc and PhD post-graduate program in A&A, forming some of the very
first MSc and PhD students in the country in the field. Currently, 5 PhD and 11 MSc

1 http://www.nassp.uct.ac.za/
2 https://www.dara-project.org/
3 https://www.isp.uu.se/
4 https://africapss.org/
5 https://www.africanastronomicalsociety.org/
6 https://africanphysicsstrategy.org/
7 https://www.ethiosss.org/
8 https://etssti.org/

students graduated under the department, and another 3 PhD and 4 MSc students are about to graduate in 2021.

- HCD of young staff members has been an important task since ESSTI establishment, through organisation of different trainings, workshops, seminars, etc.
- Research in astronomy is one of the main activities of department, with currently three research groups being established in extragalactic astronomy, stellar astronomy, and cultural astronomy, and with a small cosmology group being under the development. Research contributed to astronomy and science developments in Ethiopia, through some of the very first publications, new international collaborations, and visibility given to the ESSTI and Ethiopia (e.g., Pović 2019).
- A lot of work has been done over the past years on the institutional development of the ESSTI, starting from scratch with development of all its structure, departments, different guidelines, establishment of different committees, etc.
- Significant work has been done in organising international meetings, including the IAU symposium 356 on 'Nuclear Activity in Galaxies Across Cosmic Time' in 2019 that was the third IAU symposium organised in Africa in the last 100 years, 2017 IAU Middle-East and Africa Regional Meeting, 2017 IAU International School of Young Astronomers, and 8th African Space Leadership Congress in 2019. These events were fundamental for strengthening international collaborations, research, and HCD.
- Significant work has been also carried out over the past years in putting the two Entoto telescopes in an operational mode, and supporting the site testings at the north of Ethiopia, close to Lalibela.
- Department took also an active participation in development of some of the very first policies and road maps, such as the Ethiopian Space Science Policy (green and white papers), and the Ethiopian Space Science and Technology Road Map (in progress).
- Close collaboration has been established with different schools and universities, and continues outreach programmes have been carried out. The department also carried out different activities with school teachers, including two NASE workshops.
- Finally, through the initiative STEM for GIRLS in Ethiopia and the ESSTI Gender Office, different activities have been carried out over the last 2 years for promoting STEM fields among secondary school girls and their teachers.

International collaborations and knowledge transfer are fundamental in supporting the further developments in astronomy in Africa and Ethiopia for the benefit of our all society.

References

Backes, M., et al., 2017, Proceedings of the 4th Annual Conference on High Energy Astrophysics in Southern Africa, Cape Town, South Africa, PoS(HEASA2016)029
Benkhaldoun, Z., 2018, Nature Astronomy, 2, 352
Buckley, D., et al., 2006, IAUS, 232, 1 AfS, 2, 352
Camilo, F., 2018, Nature Astronomy, 2, 594
Carignan, C., 2012, AfrSk, 16, 18
Dalgleish, H., 2020, Astronomy & Geophysics, 61, 6.18
Govender, K., 2009, IAUS 260, Valls-Gabaud, D. & Boksenberg, A., eds.
Hoare, M., 2018, Nature Astronomy, 2, 505
McBride, V., et al., 2018, Nature Astronomy, 2, 511
Newburgh, L. B., et al., 2016, SPIE, 9906, 5
Pović, M., et al., 2018, Nature Astronomy, 2, 507
Pović, M., 2019, IAUS 365, 15, 3
Rosenberg, M., et al., 2013, arXiv:1311.0508
Tessema, S. B., 2012, AfrSk, 16, 41
Vigroux, L., 2009, IAUS 260, Valls-Gabaud, D. & Boksenberg, A., eds.
Wild, S., 2017, Nature, 545, 144

Education and Heritage in the era of Big Data in Astronomy
Proceedings IAU Symposium No. 367, 2020
R. M. Ros, B. Garcia, S. R. Gullberg, J. Moldon & P. Rojo, eds.
doi:10.1017/S1743921321001186

The National Schools' Observatory: Access to the Universe for All

Sally E. Cooper[1] ⓘD

[1]Astrophysics Research Institute, Liverpool John Moores University,
IC2 Liverpool Science Park, 146 Brownlow Hill, Liverpool, L3 5RF, UK
email: nso@ljmu.ac.uk

Abstract. The National Schools' Observatory is an educational platform that offers free access to all schools in the UK and Ireland to the world's largest robotic telescope, the Liverpool Telescope. The website offers activities, resources for teaching and importantly Go Observing, the telescope interface. The website receives 1.5 million visitors a year and has registered users in 80 countries. The next generation of robotic telescopes offer a unique opportunity to build in education, that is open and accessible to all.

Keywords. Astrophysics – Instrumentation and Methods for Astrophysics, Astrophysics – High Energy Astrophysical Phenomena, Physics – Physics Education

---

1. The National Schools' Observatory

The National Schools' Observatory (NSO)† (Newsam, A. (2007)) offers free access to all schools in the UK and Ireland to the world's largest robotic telescope, the Liverpool Telescope (LT). The NSO is a well established astronomy platform that has over 1.5 million visits per year. It offers access to learning, activities, teacher resources and importantly Go Observing, the telescope interface. To date the NSO has received over 200,000 observing requests; providing schools the same access to the skies as professional astronomers. The NSO updated its website in 2017 and it offers general users access to all data and learning materials.

2. The Liverpool Telescope

The NSO provides opportunity for observing on the 2.0 metre Liverpool Telescope (Steele *et al.* (2000), Steele, I. (2004)). The LT is located at the Roque de los Muchachos Observatory on La Palma, 2396 metres above sea level, home to an array of telescopes. The LT is a professional telescope with a broad range of instrumentation and observing capabilities. It was built by Telescope Technologies Ltd and is run by the Astrophysics Research Institute at Liverpool John Moores University, in the UK. Approximately 10 percent of observing time on the LT is provided to the NSO. The telescope is fully robotic and autonomous so that there is no need for human intervention for normal operation. This makes it ideal for schools and users as they do not need to concern themselves with the technical side of observing. Even for professional astronomers, the telescope handles constraints on scheduling such as the weather, scientific priority and whether an object is observable. Its robotic nature and autonomous scheduling is fundamental to the simplicity of Go Observing.

† www.schoolsobservatory.org

3. Go Observing

Modern telescopes are complicated. Go Observing is the interface between the telescope the NSO website where users can simply click to observe (4 clicks to observe a section of the Moon). It is designed to be dynamic and flexible and offers different observing programmes for teachers and their students. However, it is not possible to observe everything in the Universe! A limited number of objects can be selected but these include: stars, planets, galaxies, nebulae and much more. There are advanced options for those that want more control such as the choice of filter and exposure time. For those that want a 'live' experience of observing, see Lewis, F. (in these proceedings).

4. Activities and Access

Go Observing is the tool for observing but it is only one step in the astronomical process. LTImage is the software that is provided by the NSO to enable an enthusiastic new astronomer to explore the Cosmos. At a fundamental level it is an image processing tool but it can be used to do 'real' science; explore dynamic spectrum using scaling and false colour, measure the size and brightness of an object and create 3 colour images. Alongside these tools, the NSO platform offers a range of classroom and home based activities that lead the user to make observations and engage with the data in different ways rather than just point and see. The activities are suitable for ages 7 to 18 and for any keen adult. Access is completely free for teachers in the UK and Ireland to register. Other teachers and students and keen astronomers can register for free as a general user which provides the user with full access to the website learning materials and data archive of all observations taken to date.

5. The New Robotic Telescope

The NSOs early involvement in the design phase of the Liverpool Telescope was fundamental to its success. Liverpool John Moores University is leading the design of the New Robotic Telescope (NRT) (Copperwheat *et al.* (2015), a 4.0 metre robotic telescope in a similar location on La Palma. The NRT provides new challenges, both in its physical engineering and also in data transfer. For the NSO, this provides a chance to widen its educational activities, instead of being limited to traditional astronomy education methods, and engage students in wider STEM areas of engineering and computing. Our experiences with data in the world are more important than ever, with teaching being intermittently delivered remotely and online over the 2020/2021 school years in the UK and in many countries around the world. Data driven astronomy education is important to provide access to the universe for all, not just for future astronomers. The next generation of robotic telescopes planned for the future offer a unique opportunity to build in education from the outset that is open and accessible to all.

References

Copperwheat, C. M. *et al.*, 2015, *Experimental Astronomy*, 39, 1, 119–165

Newsam, A., 2007. The National Schools' Observatory. *A&G*, 48(4), 4–22

Steele, I. A., Newsam, A. M., Mottram, C., and McNerney, P., 2000. Enabling schools and public access to the Liverpool Robotic telescope. *Adv. Global Communications Technologies for Astronomy*, 4011, 133–145.

Steele, I. A, 2004, The Liverpool telescope. *Astronomische Nachrichten*, 325(6–8), 519–521.

Education and Heritage in the era of Big Data in Astronomy
Proceedings IAU Symposium No. 367, 2020
R.M. Ros, B. Garcia, S. Gullberg, J. Moldon & P. Rojo, eds.
doi:10.1017/S1743921321000582

Empowering Science Teachers in Indonesia through NASE Workshops

Hakim L. Malasan[1,2]**, Rosa M. Ros**[3]**, Chatief Kunjaya**[1]**,
Endang Soegiartini**[1]**, Aprilia**[1]**, and Riska W. Romadhonia**[2]

[1]Astronomy Division, Faculty of Mathematics & Natural Sciences,
Institut Teknologi Bandung, Bandung, Indonesia
email: hakim@as.itb.ac.id

[2]ITERA Astronomical Observatory in Lampung,
Institut Teknologi Sumatera, Lampung, Indonesia
email: hakim.malasan@itera.ac.id

[3]IAU-NASE, Spain

Abstract. Youths and kids in Indonesia since almost two decades ago have been showing significant increase of interest in space sciences, especially astronomy. One of the main factors is due to the annual event of National Science Olympiad which includes Astronomy as the subject. The increasing level of public interest, especially younger generation on astronomical events, such as eclipses, moon sightings, meteor showers has been constantly observed from time to time. Being aware that Astronomy course does not included in primary and secondary education level's curricula, teachers are somewhat desperate and are not capable to play role as clearing house in science related to space. The IAU Network of Astronomy for School Education Network (IAU-NASE) course was started in 2016 in Machung University, East Java as the pilot project in Indonesia. The course has attracted significant interest from teachers and university staff, especially in East and Middle Java Provinces. Being confident with the enthusiasm of teachers who expressed that NASE course could fulfil their needs to teach and instruct students in a very efficient way, it was organized consecutively at Bandar Lampung, Lampung Province in 2018 and 2019 (hosted by Institut Teknologi Sumatera) and in 2020 at Bandung, West Java Province (hosted by Institut Teknologi Bandung). The most recent NASE course on 21–23 August 2020, conducted in on-line mode, was attended by 74 participants, although primarily aimed at 15 School teachers, and was quite successful. The on-line observational activity turned out to be the most impressive session for the participants. We report and review four years of IAU NASE courses in Indonesia, with various documentation and brief analysis of the positive impact to the teachers and instructors attitude in teaching astronomy at secondary level of education.

Keywords. Astronomy, Education, NASE

1. Background

In 2004 the annual astronomy olympiad was introduced in Indonesia. This event was held to encourage young teenagers in Indonesia, especially high school students, to deepen their knowledge and skill in space sciences. Despite lack of astrophysics contents in the curriculum, students have been showing great interest in the competition. This national astronomical olympiads form a structured framework of event capable of reaching out to and motivating high-school students to study astronomy and astrophysics, rewarding theme at each step (Stachowski & Sule 2019)

The proliferation of astronomical clubs and communities in Java Island with addition of some clubs in Sumatra Island and South Sulawesi stimulated young pupils to expose themselves onto astronomical problem solving oriented activities. Astronomical events such as lunar crescent, solar and lunar eclipses, meteor showers are among the most popular and very much awaited by public, especially, students. Astronomy learning, thus, is not concentrated in a certain area but spread across the country. The need of distance learning for astronomy development to overcome barriers in delivering education to the grassroot is needed since 2015 (Yamani & Malasan 2015)

There is a growing number of science teachers who demand practical science trainings to equip them better in the classrooms. Nevertheless, since science teachers in the secondary level education for the whole country outnumber professional instructors available to train astronomy, besides lack of vision from most of school principals and education board in the province, this demand can not be fulfil immediately and timely, to catch the constant growing of interested students and to match timeline of annual science competition. The Ministry of Education & Culture has spent considerable length of time to adopt astronomy as one of the branches in the National Science Olympiad. In 2007 Astronomy became part of this national event. It is not surprising that in the district level, more than 10,000 students participated in the contest (Wiramiradja & Kunjaya 2006)

In this regard, the IAU-Network for Astronomy School Education (IAU-NASE) whose goal is to train science teachers in the secondary level of education is needed.

2. Network of Astronomy for School Education (NASE) course in Indonesia

NASE course, comprises of series of lectures, workshops and practical observations preferably in mother language of hosting country, provides effective means for teachers to make simple practical tools to support teaching in the class. NASE course promotes active learning process in astronomy by doing real activities such as observation of astronomical objects and phenomena.

The 1st NASE course in Indonesia (or the 82nd International NASE course) was held in Malang, East Java. Hosted by MaChung University whose Rector is an astronomer, it was participated by 34 participants mostly from East and Central Java provinces on the period 25–28 July 2016. Four instructors: from Argentine (Beatrize Garcia), Indonesia (CK), Japan (Akihiko Tomita) and Spain (RMR) were fully involved in the activities, including translation of original materials into Indonesian language and composing structure of course. The structure of NASE course has been specified by identifying local requirement prospect of development by also refering to the existing educational curricula in secondary level of education in Indonesia. It was thought that issues like local wisdoms which take advantages from astronomical phenomena and archeo-astronomy are suitable to be incorporated in the course, especially in Indonesia. An excursion to Badut Temple (*Candi Badut*), the oldest Hindu temple in East Jave was arranged. As stated in the local newspaper, Astronomy is an effective bridge for pupil to gain passion on general science. Science teachers should prepare and equip themselves better to deliver contents of astronomy to pupils since early stage of education. Participants in this first course were appointed as local instructors for Indonesia.

After being absence for a year, the NASE course was resumed in 2018 right after the occasion of the 10th Southeast Asia Astronomy Network (SEAAN) meeting, and in 2019. It was fully hosted by Institut Teknologi Sumatera (ITERA) who administer the ITERA Astronomical Observatory in Lampung. Similar to first NASE course, a preparation stage consisting of updating translated materials, training for six local instructors who are lecturers in ITERA. Contents of workshops, lectures are all follow closely to

(Ros & Hemenway 2015). A wider spectrum of audience were interested in joining the NASE course in Lampung. It consisted of undergraduate students, member of astronomical communities, beside science teachers in high schools. Most of participants came from Sumatra Island, with a few from Java Island.

Structure of NASE course adopted for activities in Indonesia in 2018 and 2019 comprises of Four Lectures delivered by professional astronomers, i.e. Stellar Evolution, Cosmology, History of Astronomy and Solar System; Ten Workshops led by instructors, i.e. Local horizon and watches of sun, Movement of the stars, the sun and the moon, Stages and Eclipses, Briefcase from the young astronomers, Solar spectrum and sunspots, Life of stars, Astronomy beyond the visible, Expansion of the Unvierse, Planets and exoplanets, and Preparation of observations; Excursion to nearby potential archaeoastronomy sites, museum, old library. Participants and instructors investigate and discuss astronomical ingredient of the visited site. In 2018 a visit to Lampung Regional Province Museum was arranged, while in 2019 participants were taken to *Pugung Raharjo*, a compound of relics from the megalithic period such as statues, punden terrace, corpses stones, altars, menhir, hollow stones and dolmen.

In mid-2020, the world is facing Covid19 pandemy. Therefore the conventional NASE course must be reshaped to match the situation and to adopt the tight health protocol. Visiting a venue for workshop was impossible. The NASE course 2020 hosted by Institut Teknologi Bandung (ITB) and supported by instructors from ITERA was then decided to be conducted by means of on-line platform. We composed a 32-hours course in full on-line mode by means of Zoom plaftorm. The first plan of course was aimed at 15 teachers in Bandung city, but later we decided to broaden the audience. As a result, a total of 74 participants ranging from high school pupils up to university lecturers were enrolled in the course. It is interesting to learn that college student significantly occupies 50% of total participants. These college students are mostly active in astronomical clubs and communities. Two weeks prior to the course, Local Organizing Committe distributed materials and moduls packed in a kit, put it on google drive and delivered it through post office to registered participants. All participants then prepared materials for workshop two days before the course is commenced. For the entire course, participants created ten teaching aids they learnt from the activity in NASE workshop. These include a Sundial, a Star demostrator, a solar demonstrator, set of ruler, simplified quadrant, horizontal goniometer, and planisphere, a Solar structure, a Spectrometer, and a Rocket. Furthermore, activities that need to be carried out outdoor or need interactions between instructors and participants were all displayed by means of demo videos provided by NASE.

The most attractive activity which is a new feature in NASE course conducted in online mode, is the real-time observations. An instructor situated in Lampung Province, set up a portable 20 cm telescope equipped with a ZWO ASI 178 mm camera and allsky camera. Supported by a modest internet connection, ZWO software, PC and Zoom platform, instructor spent about two and a half hours showing Jupiter and Saturn in real-time to participants who are distributed throughout the country (see Fig. 1).

3. Implications

Network of Astronomy for School Education (NASE) course comprises of lectures, workshops and practical observation has been providing the teachers of how to teach astronomical theory easier and more attractive to pupils regardless its level of education. In particular the workshops have been helpful for the teachers to device their own teaching aids. These teaching aid can be made easily, and the theory behind it is easy to understand.

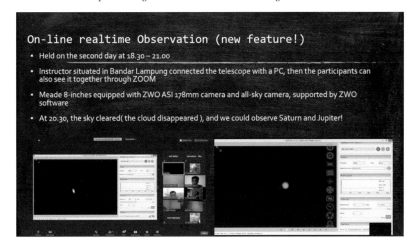

Figure 1. Real-time observation in on-line mode for participants in NASE course.

The amateur astronomers along with astronomical communities who usually held public events for kids sometime run out of ideas on how to make practical astronomy more interesting and easy to understand by children as young as pre-school ages (even to toddlers!). Thanks to NASE course, several workshops have been inspiring and successfully gave fresh ideas to deliver contents of astronomy by asking children to take part in the activities.

Many of the participants in previous NASE course in Malang, East Java (2016) are in fact consistently participate in the following course. It is interesting that they mostly came to the following course at their own expense. The major change in NASE course conducted in Bandung (2020) through distant learning by common on-line platform was positively responded by potential participants. The large number of participants in the on-line NASE course shows large interest. It is recommended that in the future on-line NASE course not only be aimed to registered participants but could also be opened to general public.

To conclude, it is proven through the NASE courses in Indonesia since 2016, what NASE has provided is beyond participants expectation. They have real experience and adventure on education, engagement and learning.

Acknowledgments

We thank ITERA for continuous encouragements & P3MI ITB for providing fund for the on-line NASE courses. Local instructors consist of Robiatul Muztaba, M. Isnaenda Ichsan, Dear Michiko M. Noor, Nova Respita, Elisa Kencana, Andi Fitriawati (ITERA), Elisa Fitri, Muthia Dewi (ITB), Yudhiakto Pramudita (UAD). Nindhita Pratiwi was the coordinator of NASE course in Lampung on the period 2018–2019. We are indebted to Dr. Patricio Rojo (IAU) for suggestions that have improved significantly the manuscripts.

References

Stachowski, G. & Sule, A 2019, *EPJ Web Conferences*, 200, 01011
Yamani, A., & Malasan, H.L. 2015, *Pub. Korean Astron. Soc.*, 30,715
Wiramihardja, S.D. & Kunjaya, C. 2006, *Proc. The 9th Asian-Pacific Regional IAU Meeting*, Ed. Sutantyo,W et al., 311
Ros, Rosa M. & Hemenway, M. (editors) 2015, *14 Steps to the Universe: Astronomy course for teachers and science graduates (2nd ed*, Network for Astronomy School Education NASE, International Astronomical Union IAU, Albdeo-Fulldome

Education and Heritage in the era of Big Data in Astronomy
Proceedings IAU Symposium No. 367, 2020
R. M. Ros, B. Garcia, S. R. Gullberg, J. Moldon & P. Rojo, eds.
doi:10.1017/S1743921321001113

A preliminary study of the impact of high school astronomy research-based learning in Thailand

Sze-leung Cheung[1]🆔 and Matipon Tangmatitham[1]

[1]National Astronomical Research Institute of Thailand,
260 Moo 4, T. Donkaew, A. Maerim, Chiangmai, 50180, Thailand
email: cheungszeleung@narit.or.th

Abstract. In Thailand, annually there are more than 50 high school students presenting in the Thai Astronomical Conference (Student Session) and more than 20 high school students joining research activities mentored by their teachers and NARIT staff through the "Advanced Teacher Training" scheme. These opportunities offer a unique experience for students to learn various skills through proposing a research question, design research methodologies, acquire different knowledge conducting research, present and communicate their results and response to criticism. Data collection for this qualitative research study is conducted through interviews with the senior high school students who completed their research presentations, with a control group of students who did not have research-based learning experience but had other informal learning experiences such as planetarium visit, or after school astronomy activities. The study looks into students' learning behaviour, attitude towards science, skills acquired for other subjects, interest in science careers.

Keywords. research-based learning, impact study

1. Background

In Thailand, there are a couple of educational problems. Firstly, the classes are traditional lecture-based with a very large class size - usually more than 50, and also with a classroom culture that teachers discourage students to ask questions during classes, which makes student lack the ability to think and questions. More, the teaching is very heavily orientated to teach examinations techniques (which is mainly multiple choices, and the process thinking is very much lack in classroom settings.

The PISA (Programme for International Student Assessment) scheme is a standard assessment that runs in multiple countries to understand (and compare) student performance in reading, mathematics, and science. Take the 2018 assessment as an example, Thailand is way below the mean score, in all aspects of reading, mathematics, and science, which has reflected the shortcoming of the Thai education system. People may argue that the universal test such as PISA is not suitable nor meaning to represent multicultural aspects of education, however, it still serves as a good starting point for referencing baseline.

On the other hand, Thailand has successfully included astronomy in its national science curriculum since 2001. This important move has resulted in a series of teacher training offered by the National Astronomical Research Institute of Thailand (NARIT) annually, they come in three different levels - beginning, intermediate and advanced level. In the advanced level teacher training, teachers will guide 20 high school students (grade 10-12) to conduct research projects with support from NARIT staff. And learning from Japan,

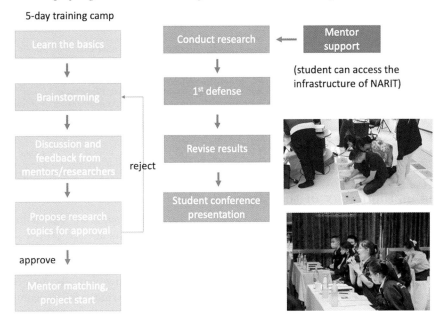

Figure 1. Flow chart and some photos of the research learning activities.

Thailand organized "The Thai Astronomical Conference Junior session (TACS)" every year, around 50 high school students joined the conference every year. TACS is described more in detail in Sappankum *et al.* (2018).

There have been many studies showed that active learning increases student performance in science, engineering, and mathematics, such as by Freeman *et al.* (2014). Teaching methods can also be classified into a spectrum of methods, range from the more teacher-directed methods to more student-directed methods, such as lectures, worked examples, interactive lecture, flipped classroom, questioning, discussion-based, scenario-based, case-based, collaborative learning, inquiry-based, problem-based, or project-based methods. Research-based learning is also project-based learning with added values from research exercises.

In our case, the flow of the research-based learning is outlined in Fig 1. Assuming the students have zero background in astronomy, they are asked to join a 5-day training camp to learn basic astronomy and research skills. Within the camp, a lot of time is devoted to brainstorming sessions in which the students, teachers, staff, and researchers will discuss and come up with a research project. The topic proposed will be subject to an approval process, rejected ones have to repeat the steps and come up with a new topic. This process played a very important role to allow the students to think logically and know what they are going to do, and also how to ask good questions. Then later the support staff or researchers will teach the students specific skills they needed for the project they proposed. After the topics were confirmed, mentor matching takes place, and students will learn many new things in a short time and then conduct their researches. Like proper research, the students are asked to present their results, both in the written and oral format in a defense with all research students, teachers, staff, and researchers present. They have to face criticism, they have to revise their results if necessary, and will be finally presented at the TACS.

Table 1. Control set questions asked to both the planetarium visitors and research students.

learning behaviour	• I often read about astronomy
	• I can access astronomy information from other channels
Take away from the activity	• I learn something new about astronomy in the activity
	• I came away with a stronger interest in astronomy
	• I want to join the activity again in the next 12 months
	• I want to look up more information about certain topics of astronomy
Attitude towards science	• Learning about science and technology is useful for my life
	• Science and technology can shape a better society
Experience	• If I did not join the activity, I will miss an important experience in my life
	• It is more fun to learn in the activity than in school

2. The methodology of the impact study

The research learning activities have been organized for years, we see an increased interest in the schools wants to join the activities, but there was never a proper study to understand the impact of these activities, which is the main motivation for conducting this study to look into the details of its impact.

The study is conducted both in a quantitative and qualitative way, through comparing the data with the same set of questions given to planetarium visitors compare to research students, and also through interview with the research student. Table 1 listed the questions we presented to understand the learning behaviour, take away from the activity, attitude towards science and experience.

3. Preliminary results

In the prototype run, due to COVID-19 pandemic we were only able to collect very small sample (n=6 for planetarium visitors with age 11-20 and high school education level within a two-month interval; n=1 (out of the 20) research student who conducted the 1-hour in person interview) and therefore not conclusive. Therefore we only see this as a prototype run and will continue the study in the coming year activities with a new round of students and the students are well informed for the study, and also collect data from 3 more planetarium. Here are some observations are found for further testing when we have a larger sample size.

• Planetarium visitors has stronger motivation than research student to join the activities

• Research student become more resourceful than planetarium visitors

• Research student does not necessarily finish the project with stronger interest in astronomy than planetarium visitors

• Research student has a stronger "take home" impact than planetarium visitors (want to join the program again, look up additional information)

• Research student has shown a stronger believe in science and tech is good for the society, and usefulness of science in daily life

• Planetarium visitors and research students both enjoy the valuable experience of joining the activities

In addition, 10% of planetarium visitors (n=35) disagree ; 10% feel neutral towards science and technology can develop a better society. From the qualitative study through the interview with the student, we learn the following reflections:

• in Thailand, schools are usually conducted in lecture settings, students has little experience doing experiments;

• joining the research project helps to present experiments in school, write science report and also the skills and precaution in collecting data, interpretation, develop thinking and produce new ideas

- the learning was also helpful for personal development on interpersonal skills, be responsible, build self confidence
- students considering if they will study astrophysics in university

4. Conclusion and next steps

The result presented here is still very preliminary, and further tests have to be continued, especially to increase sample size. More importantly, we are looking for other methods to understand the critical thinking aspect of the students, and also compare the results with students in different school environments, and use a different control group in classroom learning setting. Once we have enough data over years, we can look into the study path and career path of the participants.

Apart from the study, NARIT is planning to launch a scaled-down but international version of the program in 2021 via the NARIT operated UNESCO International Training Center in Astronomy. And we are also looking into the possibility to export this experience to IAU Commission C1 to promote research-based learning activities globally.

References

Freeman, S., Eddy, S., McDonough, M., Smith, M., Okoroafor, N., Jordt, H., & Wenderoth, M. 2014, *PNAS*, 23, 8401

Sappankum, P., & Tangmatitham, M. 2018, *Communicating Astronomy with the Public Conference*, 2018, 156

Session 2: Citizen Science and Solar Eclipses

Section 2: Culture, Structure, and Social Life

Education and Heritage in the era of Big Data in Astronomy
Proceedings IAU Symposium No. 367, 2020
R. M. Ros, B. Garcia, S. R. Gullberg, J. Moldon & P. Rojo, eds.
doi:

Introduction

Session 2 was for "Citizen Science and Solar Eclipses" and the section of the Proceedings opens accordingly with "Preliminary Report on the 14 December 2020 Total Solar Eclipse Observations," a paper by Jay Pasachoff that was presented at a public talk during the symposium. The symposium was timed to coincide with this solar eclipse to include an excursion for viewing. This unfortunately did not happen due to the global Covid-19 pandemic. Jay and his research team were able to secure permission to enter Chile and traveled to Temuco for their observations in study of the solar corona. Jay outlines their research objectives and comparisons of predictions and observations were posted by NASA's Goddard Space Flight Center. Regrettably clouds obscured the viewing of totality from many locations.

After the presentation Margarita Metaxa asked:

What is the difference between the inner and outer corona?

and Jay answered:

In the lower solar corona, there are helmet streamers. They consist of closed magnetic loop-like arcades connecting to the solar surface. Farther out in the solar corona, the streamers extend to a radial stalk connecting to the out-flowing solar wind. These structures, with photospheric light scattered off coronal electrons, are part of the K-corona (kontinuierlich-corona); they have polarization and no absorption lines. Even farther out is the F-corona (Fraunhofer corona). with photospheric light scattered by dust largely near the orbit of Mercury. The F-corona shows absorption lines and is basically unpolarized. The standard "separation of the F- and K-coronas" uses the depth of Fraunhofer lines or the percentage of continuum polarization.

Nestor Camino asked:

In the eclipse of 2020, will any group reproduce Edington's experience?

Jay replied:

No group carried out the Eddington experiment-to test Einstein's general theory of relativity by detecting displacement of star images near the eclipsed Sun-at the 2020 eclipse, but there is already at least one group preparing such measurements for 2024. The Eddington experiment, once a professional type of observation but now able to be performed by advanced amateur astronomers, was carried out most recently at the 2017 eclipse in the United States.

The following article is one by Alcides Ortega and Nelson Falcón that is called "Astronomical Society of the University of Carabobo: Scientific enculturation agent in Valencia City (Venezuela)." Alcides give a great overview of the Astronomical Society of the University of Carabobo (SAUC) and the value it has added by motivating cultural appropriation and scientific enculturation while embracing aspects of diversity. The many activities and goals of SAUC are described and the description highlights the motivating power of astronomy to the public.

The next article is one by Emmanuel Rollinde called "Modeling astronomy education, the case of F-HOU tools: SalsaJ and Human Orrery." Here the author introduces these two means of using astronomy to motivate interdisciplinary science education. Such A Lovely Software for Astronomy in Java (SalsaJ) is software that allows students to replicate data analysis made by astronomers and the author gives good example of its use and value. Next, he shows the effectiveness of using a "human Orrery" that lets students enact planetary movements as a group in a manner that helps them experience the proper relative speeds.

Following the talk Breezy Ocana Flaquer asked:

You showed the students can calculate the velocity of the stars in the center of the milky way, for that do you need to download some data, i assume, if so, from where?

and Emmanuel replied:

The data are the images of the stars around the center of the galaxy. Students have to pick up one of the stars that makes a full loop (this is the tricky part...). Then they plot the location of this star and recover the ellipse. They do not calculate the velocity of the star. Data are available on the GHOU web site handsonuniverse.org/france (it is under construction and should be ready early 2021...) but you may send us a mail if you need data soon...

Next, Cynthia Quinteros said:

May I ask you the link to the publication on the use of SalsaJ? Is there, by any chance, any partial reproduction of that work in English? Many thanks in advance!

and Emmanuel responded with:

SalsaJ publication : https://arxiv.org/pdf/1202.2764.pdf in French : https://hal.archives-ouvertes.fr/hal-02303711/document the francophone conference : https://astroedu-fr.sciencesconf.org/

The final article presented with "Citizen Science and Solar Eclipses" is "A dialogue between Vygotsky's learning theory and peer instruction in astronomy classes," written by Jamili De Paula, Denise Ferraz, and Newton Figueiredo. The authors discuss what was learned from the analysis of polls about peer instruction following a 2018 Brazilian undergraduate astronomy course regarding astrometry and celestial mechanics. The peer instruction methodology used was explained and the questionnaire/polls were described. The authors conclude by highlighting the observed value that was added to the learning of the students using this method of instruction.

After the talk Rade Marjanović asked:

In your opinion is it possible to use the peer instruction method in senior years of elementary school, in particular in classes of physics (with a dash of astronomy, since I am astronomer ?

and Newton answered:

Yes, it is possible and some of my students are actually doing that with very good results.

As evidenced by these opening-day articles presented in the proceedings, IAUS 367 Education and Heritage in the Era of Big Data in Astronomy was off to a great start.

Education and Heritage in the era of Big Data in Astronomy
Proceedings IAU Symposium No. 367, 2020
R. M. Ros, B. Garcia, S. R. Gullberg, J. Moldon & P. Rojo, eds.
doi:10.1017/S1743921321000843

Preliminary Report on the 14 December 2020 Total Solar Eclipse Observations

Jay M. Pasachoff[ID]

Williams College–Hopkins Observatory, Williamstown, MA 01267, USA
Chair, International Astronomical Union Working Group on Solar Eclipses
email: `eclipse@williams.edu`

Abstract. This paper summarizes preliminary scientific observations from sites in Chile and Argentina from which the totality was observed on 14 December 2020 at the minimum of the solar-activity/sunspot cycle.

Keywords. Sun eclipse

1. Introduction

The prediction of the 14 December 2020 total solar eclipse's path across Patagonia dovetailed nicely with the plans for IAU Symposium 367 in San Carlos de Bariloche, with the participants being bused on eclipse day north for an hour or so to totality. Though the worldwide COVID-19 protocols led to the symposium being changed to virtual, the location where the meeting participants would have gone was near where cleared skies would have led to observational sucess. These included Piedra del Águila, north of the original meeting site at Bariloche.

My team's planning for the years before the eclipse was based on the Atlantic Coast of Argentina, in Las Grutas, but COVID-related travel shutdowns led to the cancellation of many groups' tours and of travel plans even for scientific teams. The presentation of lists of research groups to the Argentinian government some two months before totality did not promptly allow our admission to Argentina, and my own team wound up working through the American Embassy in Chile to get permission to enter. The existence of a major airport, Temuco (ZCO), within totality in Chile was a major advantage to deciding where my scientific team should go. Our main goals were related to the study of the solar corona.

2. Observational Goals

- Study how the magnetic field, changing over the 11-year sunspot cycle, constrains the coronal streamers and polar plumes
- Assess how coronal mass ejections (different at each eclipse) propagates through space, making "space weather" that impinges on Earth
- Measure the intensities and distributions of hot coronal gas at different temperatures (highly ionized iron: Fe XIV, Fe X, new: Ar X=argon^{+9})
- Carry out >1 Hz measurements to assess one of the proposed methods of coronal heating
- Measure velocities of moving gas and changing coronal magnetic field

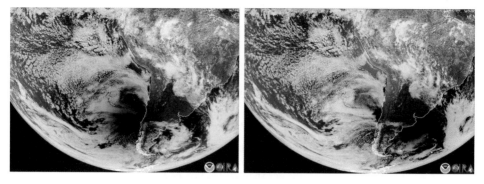

Figure 1. Two GOES-16 Advanced Baseline Imager (ABI) composite images showing the umbra as it approaches and covers part of Chile and Argentina. The Rayleigh-corrected imagery combines the two visible bands (centered at 0.47 and 0.62 μm), along with information from the 'vegetation' band (centered at 0.86 μm). (Tim J. Schmit, NOAA/NESDIS Center for SaTellite Applications and Research (STAR)).

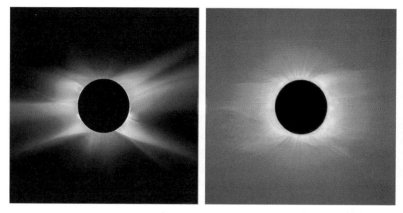

Figure 2. (left) Predictive Science Inc's posted computer model as of days before the eclipse. (right) A composite image based on observations from Andreas Möller of Germany from Piedra del Águila, Argentina, and composited within hours after the eclipse by Wendy Carlos working with me.

- With Marcos Peñaloza-Murillo, again measure the effect of the abrupt extinguishing of solar heating on terrestrial atmospheric temperature and pressure, potentially causing atmospheric gravity waves

3. Contents

As we had done at the preceding eclipses, including 2018 in the United States (Mikić, *et al.* 2018) and 2019 in Chile (Pasachoff, *et al.* 2020), we planned to compare our observations of the coronal streamers and other aspects of coronal configurations with the shapes predicted by a group from Predictive Science Inc of San Diego, California, based on the preceding months' of observations of the photospheric magnetic field with NASA's Solar Dynamics Observatory. Of course, the far side of the Sun hadn't been observed from two to four weeks before totality, so deviations between the predictions and the observations (Fig. 2) could stem from such time delays in magnetic-field measurements in addition to needed improvements in the computer model.

Though most sites in Chile were under clouds, including the one where my team had set up, near Pucón and the Villarrica volcano's ski area, there were a few regions at

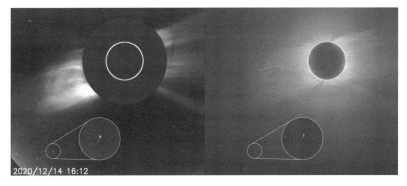

Figure 3. Two comets show on the images made through a hole in the widespread clouds at Piedra del Águila by Andreas Möller, with the composite image made in my collaboration with Vojtech Rušin and Roman Vaňúr of the Slovak Astronomical Institute, Tatranská Lomnica, Slovakia. In the SoHO image, colored red, a white circle shows the actual size of the solar disk, which is hidden behind the NRL C2 coronagraph's occulting disk.

Figure 4. (left) The corona from Gorbea, Chile. (Patricio Rojo, U. Chile, Santiago), (right) Prominences from El Condor near Las Grutas, Argentina (Verónica Espino, Planetario Galileo Galilei, Buenos Aires).

which clouds parted for at least a fraction of totality, including Piedra del Águila, south of the original meeting site at Bariloche and therefore close to where the Symposium participants might have been bused on eclipse day.

Our comparison of the predictions and our observations were posted the next day by NASA's Goddard Space Flight Center: https://www.nasa.gov/feature/goddard/2020/scientists-use-nasa-data-predict-appearance-corona-dec-14-total-solar-eclipse.

A citizen-scientist the day before the eclipse had located a sungrazer comet on an image from the European Space Agency's Solar and Heliospheric Observatory (SoHO); the U.S. Naval Research Laboratory's "C2" coronagraph observed it, and with prodding from me Karl Battams there gave it number SOHO-4108, jumping ahead of a couple of thousand unnumbered SoHO comets. It was reported officially on Minor Planet Electronic Circular 2020-Y19, 2020, COMET C/2020 X3 (SOHO), https://minorplanetcenter.net/mpec/K20/K20Y19.html. We discussed the comet images from SOHO and on a composite of several dozen of Andreas Möller's images from Piedra del Águila made for Vojtech Rušin and me by Roman Vaňúr (Fig. 3). https://www.nasa.gov/feature/goddard/2020/recently-discovered-comet-seen-during-2020-total-solar-eclipse-SOHO

Figure 5. A slitless spectrum, showing coronal lines from highly ionized iron in the red and in the green with brighter chromospheric and prominence lines, including H-alpha in the red and the yellow line from which helium was discovered about 150 years ago. (credit: Diego Guerrero (Córdoba, Argentina) and Robert B. Slobins (flashspectrum.com), Tallahassee, Florida).

Across the band of totality, several other scientists were able to image the corona through clouds. Patricio Rojo of the University of Chile imaged the corona from Gorbea in western Chile (Fig. 4a). Verónica Espino of the Planetario Galileo Galilei imaged the prominences from El Condor near a cloudier Las Grutas from eastern Argentina (Fig. 4b).

Coronal spectra show especially two emission lines in the visible range from highly ionized iron, revealing that the corona is millions of kelvins hot. U.S. amateur astronomer Robert Slobins, who has been observing and recording "slitless spectra" –with the narrow band of chromosphere or corona near the limb acting as its own slit– since 1984, coached Córdoba, Argentina, amateur astronomer Diego Guerrero on setting up the equipment and even directed him, 200 km long-distance, from Las Grutas to a favorable location at Ministro Ramos Mexía. The coronal red line from 9-times-ionized iron, barely visible in this reproduction (Fig. 5), should be revealed after calibration to be stronger than the green line from 13-times-ionized iron at this minimum stage of the sunspot cycle.

In spite of the clouds, our meteorological station at Pucón, run there by Michael Roman (U. Leicester, UK) and Theo Boris (Collegiate School, New York City) provided temperature, pressure, wind, and insolation measurements for the eclipse day ±1. A team including a dozen undergraduates and faculty sponsored by the Montana Space Grant Consortium, led by Jennifer Fowler, launched weather balloons as high as 32 km on eclipse day ±1 and detected eclipse-driven gravity waves emanating from the path of totality. They flew radiosondes measuring temperature, pressure, relative humidity, and wind speed and direction.

A team led by dual-citizen Demián Gómez of The Ohio State University made ionospheric measurements from Argentina.

Though COVID-related restrictions led to difficulties in location and kept most tourists away from the eclipse path, and kept us from bringing our coronal-oscillation experiment meant to verify specific theories of coronal heating (with institutional restrictions on my own travel and that of my astrophysics-major students as well as that of Michael Person from MIT), enough observations were made to provide scientific advancement as a result of the 14 December total solar eclipse.

4. Acknowledgments

Our expedition to Chile and subsequent data reduction received major support from grant AGS-903500 from the Solar Terrestrial Program, Atmospheric and Geospace Sciences Division, U.S. National Science Foundation. Christian Lockwood (Williams College '20) was supported by a grant from the Sigma Xi Honorary Scientific Society. Predictive Science Inc. was supported by the US Air Force Office of Scientific Research, NASA, and NSF. V. Rušin is supported by the project VEGA 2/0048/20 (Slovak Academy of Sciences).

References

Mikic, Z., Cooper, D., Jon, A. L., Ronald, M. C., Duncan, M., Lisa, U., Pete, R., Roberto, L., Tibor, T., Viacheslav, T., Janvier, W., Miloslav, D., Jay, M. P., & Wendy, C., 2018, "Predicting the Corona for the 21 August 2017 Total Solar Eclipse," *Nature Astronomy* 2, pp. 913–921, https://doi.org/10.1038/s41550-018-0562-5; https://rdcu.be/5iVS

Pasachoff, J.M. et al. 2020, *Early Results from the Solar-Minimum 2019 Total Solar Eclipse*, for IAU Symposium 354, "Solar and Stellar Magnetic Fields: Origins and Manifestations", Copiapo, Chile, July 2019, pp. 3-14 (Cambridge University Press).

Peñaloza-Murillo, M.A. et al. 2020, Air temperature and humidity during the solar eclipses of 26 December 2019 and of 21 June 2020 in Saudi Arabia and in other eclipses with similar environments, in preparation Arxiv, http://arxiv.org/abs/2011.11460

Education and Heritage in the era of Big Data in Astronomy
Proceedings IAU Symposium No. 367, 2020
R. M. Ros, B. Garcia, S. R. Gullberg, J. Moldon & P. Rojo, eds.
doi:10.1017/S1743921321000673

Astronomical Society of the University of Carabobo: scientific enculturation agent in Valencia city (Venezuela)

Alcides Ortega[1]ⓘ and Nelson Falcón[2]ⓘ

[1]Dirección de Informática, University of Carabobo
BOX 129 Avda Bolivar norte, Valencia, Venezuela
email: `aortega@uc.edu.ve`

[2]Dept. of Physics FACYT, University the Carabobo,
Box 129 Avda Bolivar norte, Valencia, Venezuela
email: `nelsonfalconv@gmail.com`

Abstract. The Astronomical Society of the University of Carabobo (SAUC) is an activity of permanent scientific dissemination, which uses Astronomy as a tool for the scientific enculturation of the local population and non-formal teaching of Science and Technology. The SAUC base their Learning activities through the Bachelard epistemology and Morin pedagogy. Furthermore the focus of the dissemination and popularization of Astronomy must focus on knowledge for life and on overcoming epistemic obstacles between knowledge expert and knowledge taught. The SAUC's activities are focused on holding Master Class, seminars, astronomy courses for amateurs; development of multimedia materials and the national astronomical ephemeris. The qualitative evaluation, after two decades of activities of the SAUC allows us to conclude that astronomy can affect as a motivating axis for cultural appropriation and scientific enculturation by broad sectors of the local community, regardless of age, gender, race, socio-economic activity, or ideological-cultural diversity.

Keywords. scientific enculturation, epistemology, history of science

---

1. Introduction

The regularities of the movement of the stars in the celestical sphere, demonstrated the awareness that the natural world could be rationally understood, that there is an "order" and "rules" that govern it, and that knowledge of them to our most ancient ancestors, allowed the satisfaction of specific needs such as the farming and fishing days. This cultural heritage on the role of astronomy to encourage the study of nature; It is usually used for non-formal education and the dissemination of science and technology. The Astronomical Society of the University of Carabobo (SAUC) is a permanent scientific dissemination activity, which uses Astronomy as a tool for the scientific enculturation of the population, and for the non-formal teaching of Sciences and Technology, in the City of Valencia Venezuela .The SAUC is a permanent scientific dissemination activity, which uses Astronomy as a tool for the scientific enculturation of the population, and for the non-formal teaching of Sciences and Technology, in the City of Valencia Venezuela. The activities of the SAUC consist of the holding of periodic events (scientific fairs, courses and masterclasses, stars parties, convention of amateur astronomers, etc.) and permanent information about news topics of astronomy and Space Sciences (on Radio, TV, newspapers and social networks). The Astronomical Society of the University of

Carabobo began as a permanent activity for the study and dissemination of astronomy at the end of the last century, coinciding with two notable astronomical events in the region. Although there were already groups of astronomy fans in the region, they did not carry out permanent outreach programs. The transit through the perihelion of Comet Hale-Bopp, in April 1997, dazzled the local population, showing itself as the brightest comet of the century and remaining visible to the naked eye for a long time. It gave rise to several days of astronomical observation and continuous interaction with the community. The total solar eclipse February 1998 was, a phenomenon of singular majesty that led us to plan activities outside our geographic environment distant more than 300 kilometers from our city. The logistics carried out for the trip to the zone of totality, allowed us to bring together local astronomy fans. In addition, the interaction with astronomers from other regions, promoted the realization of group activities and use of common resources for the divulgation of astronomy in our country. For more than two decades, this activity has brought together the University Community, the Carabobo State Engineers Center, the Venezuelan Association for the Advancement of Sciences and the Associations of Astronomy amateurs, with the purpose of disseminating science and technology. The non-formal education in astronomy, as a tool for the scientific enculturation of the local population and non-formal teaching of Science and Technology is the propose of SAUC; and the conceptual systematization was present several years ago Falcon & et al (2002) Falcon (2009). Now we present the conceptual connection of the astronomical heritage and popularization of Astronomy with the Bacherlard Epistemology and the Morin pedagogy.

2. Overview

The Astronomical Society of the University of Carabobo, following the premise formulated by UNESCO, within the framework of the "Education for a sustainable futurproject, launched in 1996, bases its Learning activities on the cognitive theories of Gastón Bachelard and Edgar Morín, according to which the approach to the dissemination and popularization of Astronomy should focus on knowledge for life and on overcoming epistemic obstacles between expert knowledge and knowledge taught. Bachelard Bachelard (2000), in his work contributes his conception of epistemological obstacle. And he poses it as the psychological difficulties that do not allow the appropriation of knowledge, minimizing the transition from a pre-scientific state to a scientific state, through the observation of phenomena. The Epistemological obstacles , as Bacherlard says, can be reduce to the list: (a)The first obstacle: primary experience, (b) General knowledge as an obstacle to scientific knowledge, (c) An example of a verbal obstacle: sponge. On the over-extension of familiar images (d) Unitary and pragmatic knowledge as an obstacle to scientific knowledge, (e)The substantialist obstacle (f) Psychoanalysing realists, (g) The animist obstacle, (h) The myth of digestion, (i) The libido and objective knowledge, (j) The obstacles to quantitative knowledge, and (k) Scientific objectivity and psychoanalysis. In the Scientific Spirit, Bacherlard Bachelard (2000) advocates the imitation of the rationalism and values of Science, rather than its methods

Scientific culture must be in a state of permanent mobilization, replace closed and static knowledge with open and dynamic knowledge, dialectize all experimental variables, favoring the evolution of reason. The verbal obstacle prevents the advancement of science. Language plays a preponderant role in the recognition of phenomena, and must be used in a responsible way, the abusing explanations, associations, linguistic subjectivities, puts the act of understanding and the act of communication at risk. The approach to the natural world through phenomenology opens the way to the formation of scientific knowledge in a continuous search for knowledge. Scientific discursive rationalization is opposed to basic convictions such as light certainties. One should not start

Bacherlard Epistemology	Movement of the "stars" in the Celestial sphere	Astronomy examples	Morin pedagogy in XXI century
• Foundation	• Spatial universality • Temporal regularity • Changes in cycles	• Planetary apparent movement • Celestial dynamics in the solar system • Stellar evolutions • Solar cycles activity	• Blindness: error and illusion • Teach understanding
• Coherence	• Rational understand • Transcendence over the local environment	• Solar System: structure and evolutions • Structure of the Milky Way • Local Groups of galaxies • Our place in the Universe	• Relevant knowledge • ("for the here and now") • Ethics of the human specie
• Dialectic	• Laws of the natural world • News observations and phenomenology	• Exoplanets exploration • Galaxies and cosmology • Large scales structure	• Teach the human condition • Coping with uncertainty
• Problems	• Adaptations to the changes reality • New conceptual reality	• Relativity and gravitations • Climate change controversy • Exobiology and conservations • Future of the human race	• Earthly identity

Figure 1. Heuristic comparison between the non-formal education of astronomy and the Bacherlard and Morin theses.

from the truth; the school knowledge learned is in permanent doubt, and must be continuously validated. The obstacle of pragmatic knowledge, utility constitutes a principle of exclusion and an epistemological obstacle to knowledge. The obstacle of quantitative knowledge, also known as false rigor, blocks thought, it is a primary symbolic system that sometimes prevents the understanding of new knowledge. For other hands, Edgar Morín, proposes restructuring education in the twenty-first century; through what he called" the seven knowledge for the education of the future" Morin (1999): (1) Recognize the blind spots of knowledge: the error and the illusion, (2) The principles of the relevant knowledge, (3) Teaching the human condition, (4) Teach the identity planetary (5) To face the uncertainties, (6)The teaching of the understanding and (7) Ethics of the human race. The astronomy as a common scientific culture It postulates that education should be for life in the here and now. In addition, it is necessary to reorganize, not only the act of teaching, but also the fight against the defects of the system is important because it responds to Morin's principle of rational uncertainty, which invites us to ask ourselves about nature and the concept of humanity. It allows to inquire about the place of humanity in the universe and the finitude of life. This is where astronomy reveals itself to us as the science of interconnection between different disciplines and addresses fundamental problems. The regularities of the movement of the stars in the celestial spheres, showed that the natural world could be rationally understood. This cultural heritage on the role of astronomy to promote the study of nature; It is useful for non-formal collective education, and for the dissemination of science and technology among young people. The conceptual implications of Bacherlard's epistemological and Morin's cognitive theses, in popularization of the astronomy can be summarized in Fig. 1. The conception about a dynamic rationalism, in the formation of scientific knowledge (Bacherlkard's foundation, column 1), remains manifested by the astronomy historical development ; and by the systematic observation of the phenomena that take place in the apparent movement in the celestial sphere. (Column 2). These can be exemplified with relevant topics of current astronomy (column 3) that come to raise the thesis of the "Seven knowledge for the

Figure 2. The non-formal education in astronomy, as a tool for the scientific enculturation of the local population and non-formal teaching of Sciences and Technology.

Education of the Future" in the pedagogy of E. Morin (Column 4). It is clear that the list in column 2 is neither complete nor restrictive, but only indicated as relevant examples.

3. Implications

The activities of the SAUC focus on the didactic transposition of knowledge, avoiding the details of the rigor of metalanguage; in the use of the scientific spirit in the sense of Bacherlard's Philosophy. The astronomical observations of the present time, is concordant with the idea of education for the here and now, according to the Philosophy of Edgar Morin. The Astronomical Society of the University of Carabobo, develops outreach activities with a non-formal teaching approach of Science and Technology of a ludic nature Falcon (2009), such as seminars, amateur astronomy courses; traveling exhibitions, etc. In addition SAUC produces of astronomy multimedia materials (videos, software, books, etc.). These activities and resources would generate the enculturation of the local population in science and technology. The qualitative evaluation of the process of scientific enculturation of the general population, can be evidenced by the heterogeneity and massive attendance of the public to the activities carried out (Fig. 2).

Part of the activities of the Astronomical Society is to collaborate with the youth sciences festivals, organized by the Venezuelan Association for the Advancement of Science, and also in the professional improvement activities of the Carabobo State Engineers Center. Also the SAUC produces, every years, the astronomical ephemeris of Venezuela for free, a work that was previously carried out by the Cajigal Naval Observatory until 2006. The qualitative evaluation, after two decades of activities of the Astronomical Society of the University of Carabobo, allows us to conclude that astronomy can affect as a motivating axis for cultural appropriation and scientific enculturation by broad sectors of the local community: Regardless of age, gender, schooling, socio-economic activity or ideological-cultural diversity.

References

Bachelard, G. 2000, *La formación del espíritu científico*, 30, 490.

Falcón, N; Muñoz, R. & Vegas-Castejon, J. 2002, *Astronomy Communication*, Mahoney, T. Ed. La Laguna, Spain, 159.

Falcón. N. 2009, *The Role of Astronomy in Society and Culture*, Valls-Gabaud, D. & Boksenberg, A. Eds, IAU Symposium. 260,710.

Morin, E. 1999, *Los siete saberes necesarios para la educación del futuro*, UNESCO, Paris.

Education and Heritage in the era of Big Data in Astronomy
Proceedings IAU Symposium No. 367, 2020
R. M. Ros, B. Garcia, S. R. Gullberg, J. Moldon & P. Rojo, eds.
doi:10.1017/S1743921321000466

Modeling astronomy education, the case of F-HOU tools : SalsaJ and Human Orrery

Emmanuel Rollinde[1] ⓘ

[1]CY Cergy Paris Université, LDAR, F-95000 Cergy, France
Universités de Paris, Artois, Paris Est Creteil, Rouen
email: emmanuel.rollinde@cyu.fr

Abstract. This communication introduces two cases of the use of astronomy as a motivating context to interdisciplinary science education with emphasis on modeling activities. Firstly, a dedicated software, called SalsaJ, allows students to reproduce the same data analysis as made by astronomers. The case of exoplanet detection will be used as an exemple. Secondly, bodies of learners are considered to model movements of planets with a Human Orrery (a Spatio-Temporal Map of the Solar System), connecting thus mathematics, physics, geography and arts.

Keywords. methods: data analysis, solar system: general, astronomy education

1. Introduction

The association F-HOU is part of a worldwide project, Global Hands-On Universe†, that aims to promote a method of learning sciences through astronomy. We are a community of science teachers, astronomy researchers and science education researchers, whose main goal is to provide science and mathematics teachers from middle school to high school with interactive and collaborative tools (software, instruments, methodology, documents shared on the Internet). Based on an investigative pedagogy that emphasizes observation, the global objective is to develop students' curiosity, to maintain their taste for science and technology, and to lead them to argue and reason (Rollinde *et al.* 2020). This approach requires interaction between physical sciences and mathematics, and can also be easily integrated into multidisciplinary teaching (combining literature, philosophy, technology, language teaching and sports).

2. Modeling activities and astronomy

Our pedagogical approach is based on the principle of modeling and the resulting link between mathematics and physical sciences (Hestenes 2006; Helding *et al.* 2013). To address and reflect on a problem, an expert or student will use two registers as described in Tiberghien (1994) and Sensevy *et al.* (2008), see Fig. 1 (right): the empirical register which consists of objects, phenomena and actions on objects and phenomena; the theoretical register which corresponds to an abstraction applicable to any real object. As for students, the theoretical register corresponds to a set of statements that the student holds to be true and that allow him or her to make sense of the real. A model can then be considered as a link between these two registers. Modeling activities make students work from one register to another, or within one register. Modeling in the context of mathematics education makes a specific emphasis on horizontal and vertical mathematization (Blum & Leiss 2007; Yvain-Prébiski & Modeste 2020). The first is the

† http://handsonuniverse.org/france

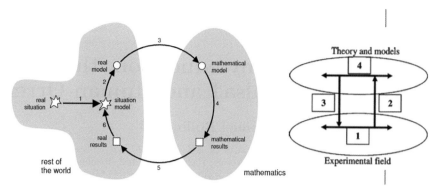

Figure 1. Two illustrations of modeling process. *left:* the modeling cycle presented in Blum & Leiss (2007). *right:* the theory of the two worlds presented in Sensevy *et al.* (2008). See main text for application to astronomy in education.

transformation of the real model into a mathematical model, which requires to choose a set of physical parameters. The second is the obtaining of a mathematical result that can be later on translated into real results, or a description of the real situation.

Astronomy provides teachers and students with real situations that are very specific. First, the real situation in astronomy is a 'single case' situation, meaning that one cannot modify the observed celestial objects, or move around to see them from different point of view. In most cases (at least in the educational context), few objects will be observed to infer general laws. This was the case for Kepler and Newton who had to infer classical mechanics relations and laws from the observation of one single Solar System. This is less the case today with the era of big data even in astronomy. Second, real situation is observed through the light send or reflected by the objects. Hence, the situation given to the learner is already a model of the real situation provided by either an image or a spectrum. Making sense of this modeled situation is not straightforward and imposes the use of modeling cycle. Based on astronomy, and compared to 'everyday-situations' (Yvain-Prébiski & Modeste 2020) students' activities will enhance two aspects of modelisation: inferring the right model that should relate best to the real situation; choosing physical parameters that can be measured in the modeled situation (the image) and that can later on make sense of the real situation (after the two mathematizations).

3. Modeling activities with data analysis

To make data manipulation and measurement feasible in the class, a specific software, SalsaJ, has been developed in Java (Doran *et al.* 2012; Rollinde *et al.* 2016). Its acronym stands for "Such A Lovely Software for Astronomy in Java". Downloaded free of charge from the project site, it works on all platforms and can be used in different European languages as well as in Arabic and Chinese. The pedagogical sequence we present here is based on the detection of an exoplanet through its transit in front of its host star. The students have at their disposal a time series of images (Fig. 2, left) which accounts for their empirical register. The objective is to find clues revealing the presence of a planet in orbit around a star.

A first modeled situation of a star and an exoplanet must be derived through the understanding that the light received by the observer (empirical register) is emitted by point stars (theoretical register). The learner must then use this model for predicting that the planet blocks part of the light when it is located between the observer and the star (theoretical register). Through this first cognitive task that connects two registers, the photometry emerges as a good candidate to use the model for making sense of the

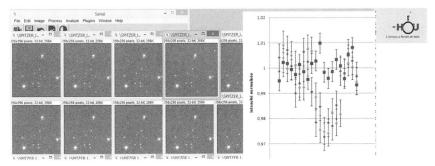

Figure 2. Modeling activities on the detection of an exoplanet. The real problem is provided by a series of image of the same field taken at regular time interval (*left*). The model used is that of a star and an exoplanet orbiting around it. The mathematical prediction of the model is a graph showing the evolution with time of the luminosity of different stars in the field (*right*).

Figure 3. Modeling activities on the Solar System. Students enact the movements of planets and comets around the Sun on a 2D representation of the orbits, a so called Human Orrery. *Left:* Students enact and observe movements as seen from different referential frames. *Right:* Primary school children in front of the Human Orrery they have build in their courtyard.

observed images and to make measures. Once the students proceed and make the measure using the tools provided by the software, they end up with a mathematical representation of the evolution of the luminosity of three different stars with time (Fig. 2, right) that may be compared to the prediction. The mathematical resolution consists in a statistical analysis of the significance of the departure from a constant luminosity for one of the series of data (or one of the star). The modeling cycle ends by connecting the amplitude of the observed decrease to the ratio of the surface of the star versus that of the planet.

4. Modeling with the learning body

A "human Orrery" (Fig. 3) allows the learners to enact the planets' movement with correct relative speed. An Orrery is a mechanical or digital device designed to model the motion of the planets around the Sun and their changing positions in the sky. On a human Orrery, the orbits of planets and comets are drawn at a human scale allowing movements in the Solar System to be enacted by the learners (Rollinde 2016; Rollinde & Decamp 2019). We give here one example of how this tool may be used to help students learn practice modelisation (Rollinde, Decamp & Derniaux 2021).

All grade 10 students of a French high school were involved in two successive teaching-learning sessions. The learning material consisted of two versions of the Human Orrery. One was an A4 model used individually on a table, with reference frames and trajectories drawn on a translucid paper; the other was a Human size version, with reference frames and celestial bodies enacted by students' bodies. Students enacted movements with their fingers and with their bodies respectively. Activities illustrated the movements of Earth, Mars, and Sun during 24 hours and one year, observed from different reference frames (terrestrial, geocentric or heliocentric). Students had to use the model for predicting trajectories in each referential frame. To do so, they enacted the trajectories, observed them, and discussed whether length and speed of the planets were different when observed in each referential frame. By doing so, they also proceeded through the modeling cycle, had to make predictions and test them. Hence, the model of the Human Orrery helps to make sense of what happens to a trajectory during a change of referential frame.

5. Conclusion

This communication intends to motivate and enhance the explicit reference and use of modeling activities in the field of astronomy education. The set up and evaluation of such activities requires to combine expertises from professional astronomers, didactics researchers (in mathematics and science) and teachers. In January 2021, a francophone conference† has gathered 200 people coming from those three communities. We are confident that the international effort led by UAI on astronomy education will be fruitful in the set up of this community.

References

Blum, W., & Leiss, D. 2007, *Mathematical Modelling. Education, Engineering and Economics-ICTMA*, 12, 222–231.

Doran, R., Melchior, A. L., Boudier, T., Ferlet, R., Almeida, M. L., Barbosa, D., & Roberts, S. 2012, *arXiv preprint* arXiv:1202.2764.

Helding, B., Megowan-Romanowicz, C., Ganesh, T., & Fang, S. 2013, *in Modeling Students' Mathematical Modeling Competencies*, 327–339, Springer, Dordrecht.

Hestenes, D. 2006, *Proceedings of the 2006 GIREP conference: Modeling in physics and physics education*, 31, 27. Amsterdam: University of Amsterdam.

Rollinde, E. 2016, *International Journal of Science and Mathematics Education*, 17(2), 237–252.

Rollinde, E., Ferlet, R., Melchior, A. L., Delva, P., Chagnon, G., & Salomé, P. 2016, *Le Bulletin de l'Union des Professeurs de Physique et de Chimie*, 4(983), 469–496.

Rollinde, E., & Decamp, N. 2019, *Journal of Physics: Conference Series, IOP Publishing*, 1287(1), 012011.

Rollinde, E., Pennypacker, C., Doran, R., Darhmanoui, H., Handa, T., Kothari, K., Lewis, F., Robberstad, J., & Megowan, C. 2020, *Futures of Education - GHOU 2020, Intern report*, hal-03089436

Rollinde, E., Decamp, N., & Derniaux, C. 2021, *Physical Review Physics Education Research*, submitted

Sensevy, G., Tiberghien, A., Santini, J., Laubé, S., & Griggs, P. 2008, *Science education*, 92(3), 424–446.

Tiberghien, A. 1994, *Learning and instruction*, 4(1), 71–87.

Yvain-Prébiski, S., & Modeste, S. 2020, *Proceedings of the CIEAEM conference*, 71, 139–150.

† AstroEdu-FR, https://astroedu-fr.sciencesconf.org/

Education and Heritage in the era of Big Data in Astronomy
Proceedings IAU Symposium No. 367, 2020
R. M. Ros, B. Garcia, S. R. Gullberg, J. Moldon & P. Rojo, eds.
doi:10.1017/S1743921321000855

A dialogue between Vygotsky's learning theory and peer instruction in Astronomy classes

Jamili De Paula[1], Denise Pereira de Alcântara Ferraz[2] and Newton Figueiredo[3] (ORCID)

[1]Programa de Pós-graduação em Educação em Ciências, Universidade Federal de Itajubá
Caixa Postal 50, 37500-903, Itajubá, MG, Brazil
email: `jamilidepaula@gmail.com`

[2]Instituto de Física e Química, Universidade Federal de Itajubá
Caixa Postal 50, 37500-903, Itajubá, MG, Brazil
email: `deferraz@unifei.edu.br`

[3]Instituto de Física e Química, Universidade Federal de Itajubá
Caixa Postal 50, 37500-903, Itajubá, MG, Brazil
email: `newton@unifei.edu.br`

Abstract. Active learning methodologies have been used to teach science, technology, engineering, arts and mathematics at higher education institutions in several countries. We report the results of using peer instruction in an Astronomy undergraduate course taught at a research university in Brazil. The course syllabus covered topics on astrometry and celestial mechanics at an introductory level and was offered in the second semester of 2018. In order to better investigate the effect of the interaction among students, we have asked them to talk to their peers after the first poll regardless of the outcome. We have then analyzed the outcomes of all peer instruction polls, before and after student interaction, as well as the course evaluation questionnaires answered by the students at the end of the semester. From these analyses we were able to establish an approximation between peer instruction and some key elements of Vygotsky's social interactionist theory.

Keywords. Active learning, Astronomy education, Peer instruction, Vygotsky's social interactionist theory.

---

1. Introduction

Peer instruction is a student-centered active learning methodology developed by Eric Mazur at Harvard University in the early 1990s (Mazur 1997) that has been successfully used in science, technology, engineering, arts and mathematics (STEAM) classes since then (e.g. Crouch & Mazur 2001, Lenaerts *et al.* 2002, Lucas 2009, Wood 2009, Zingaro & Porter 2014). There is robust evidence that this methodology enhances students' understanding (e.g. Smith *et al.* 2009) and reduces dropout rates (e.g. Watkins & Mazur 2013).

Turpen & Finkelstein (2009) showed that the way peer instruction is implemented by instructors is not unique. Despite of these variations, a typical peer instruction session usually comprises of seven steps (Mazur 1997, Vickrey *et al.* 2015):

(*a*) Question posed
(*b*) Students given time to think

(*c*) Students record individual answers
(*d*) Students convince their neighbors (peer discussion)
(*e*) Students record revised answers
(*f*) Feedback to instructor: tally of answers
(*g*) Instructor's explanation of correct answer

Since the interaction among the students plays a pivotal role in this methodology, the main objective of this research was to perform an approximation between peer instruction and Vygotsky's social interactionist theory (Vygotsky 1980) using data collected in an Astronomy course taught at Universidade Federal de Itajubá, a research university in Brazil.

In order to perform this semester-long investigation, a new design for the course was implemented in 2018 so that peer instruction could be used throughout the semester. Given the encouraging results we have found, we intend to keep this approach for future offerings.

In the next session we provide a brief overview of the course, after which we present and discuss the implications of this research.

2. Overview

The syllabus of the Astronomy course covers topics of astrometry and celestial mechanics at an introductory level. It was taught along 16 weeks in the second semester of 2018 for undergraduate students, most of them majoring in meteorology, physics or chemistry. Forty-seven students enrolled in the course, but nine of them withdrew before the end of the semester, so the dropout rate was 19%.

Each week the students had a pre-class session on a virtual learning environment, followed by two 110-minute long sessions on campus. During on-campus sessions the students had hands-on activities, problem solving group tasks and peer instruction classes, but no traditional lectures were given.

In peer instruction sessions, a multiple-choice conceptual question was posed to the students, who were instructed to write their answers in individual paper cards, after having some time to think. Then the instructor asked them to discuss their choice with a peer who had chosen a different alternative. After a few minutes, they were asked to answer again the same question in another paper card. The instructor followed the strategy described by Smith *et al.* (2011), so that in each class the students were asked to answer six multiple-choice questions about the same subject and they voted in both polls, regardless of the score of the first poll.

At the end of the semester, the students answered a questionnaire designed to provide a feedback on their perception about the course. The questionnaire had ten multiple-choice and five open-ended questions. We have analyzed not only the quantitative results from the polls, but also the qualitative results found in the students' answers to the questionnaire. In this paper we present some of these results and a full description of the data analysis can be found in De Paula *et al.* (2020).

3. Implications

Figure 1 shows, for each question in which the students interacted with their peers, the overall score of the class on the second poll as a function of the score on the first poll. In 17 cases the score of the class increased after the interaction among the students. In two cases (shown in red) both scores were the same, while in one case (shown in green) the score on the second poll was lower than the first score. These results are consistent with the literature, as shown, for example, in the review paper by Vickrey *et al.* (2015).

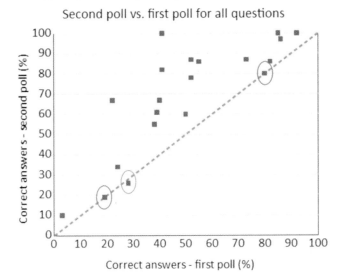

Figure 1. Second poll vs. first poll overall score for all peer instruction questions. Adapted from De Paula *et al.* (2020).

These results show that a large number of students who chose a wrong alternative in the first poll switched to the correct answer in the second poll. Under the framework of Vygotsky's social interactionist theory, we may consider that, for these students, the concept assessed by the question was within their *zone of proximal development*. Thus, in the first poll they could not figure out the right answer by themselves, but after interacting with their peers, they were able to answer the question correctly.

We should also note that peer instruction classes encourage the students to express their thoughts in words when they interact with their peers (Mazur 1997), which reminds us of Vygotsky's concepts of *language* and *cultural mediation* (Vygotsky 1986). Language plays a key role in his theory, since by verbalizing a conflict, the students find a way to solve the conflict under the mediation of their peers, who are the *more knowledgeable others* - another important concept of the theory.

After performing this quantitative analysis, we focused our attention on the questionnaire the students answered at the end of the semester. By analyzing the answers to the open-ended questions, we were able to identify other elements of Vygotsky's theory.

One student wrote:

> "This methodology, which requires us to answer the questions and then have a discussion with our peers, was essential for a better learning of the subject. When we talk to our colleagues, those who explain and those who listen to the explanation reach a better conclusion."

Yet another student wrote:

> "The multiple-choice questions were very important to our learning because we could check if we had really understood the subject, as well as being able to agree or disagree with our classmates."

Two key concepts of the social interactionist theory can be pointed out from these statements: the *social interaction* among the students and the use of *language* while interacting with their peers.

References

Crouch, C.H., & Mazur, E. 2001, *Am. J. Phys.*, 69, 970

De Paula, J., Figueiredo, N., & Ferraz, D.P.A. 2020, *Cad. Bras. Ens. Fis.*, 37, 127

Lenaerts, J., Wieme, W., & Van Zele, E. 2002, *Eur. J. Phys.*, 24, 7

Lucas, A. 2009, *PRIMUS*, 19, 219

Mazur, E. 1997, *Peer Instruction: A User's Manual* (Upper Saddle River: Prentice-Hall)

Smith, M.K., Wood, W.B., Adams, W.K., Wieman, C., Knight, J.K., Guild, N., & Su, T.T. 2009, *Science*, 323, 122

Smith, M.K., Wood, W.B., Krauter, K., & Knight, J.K. 2011, *CBE-Life Sci. Educ.*, 10, 55

Turpen, C., & Finkelstein, N. 2009, *Phys. Rev. Spec. Top. Phys. Educ. Res.*, 5, 20101

Vickrey, T., Rosploch, K., Rahmanian, R., Pilarz, M., & Stains, M. 2015, *CBE-Life Sci. Educ.*, 14, 1

Vygotsky, L. 1980, *Mind in Society* (Cambridge: Harvard University Press)

Vygotsky, L. 1986, *Thought and Language* (Cambridge: MIT Press)

Watkins, J., & Mazur, E. 2013, *J. Coll. Sci. Teach.*, 42, 36

Wood, W.B. 2009, *Annu. Rev. Cell Dev. Bi.*, 25, 93

Zingaro, D., & Porter, L. 2014, *Computers & Education*, 71, 87

Session 3: Astronomy Education Research

Education and Heritage in the era of Big Data in Astronomy
Proceedings IAU Symposium No. 367, 2020
R. M. Ros, B. Garcia, S. R. Gullberg, J. Moldon & P. Rojo, eds.
doi:

Introduction

The first article in this session is about an invited talk by Markus Pössel, Carolin Liefke, Niall Deacon, Natalie Fischer, Juan Carlos Muñoz, Markus Nielbock, Saeed Salimpour, and Gwen Sanderson that is titled The IAU Office of Astronomy for Education. In it the authors describe the new IAU Office of Astronomy Education (OAE). The OAE was first approved at the 2018 GA in Vienna and came into being in December 2019. Its organization, goals, and activities are described and the value it provides for astronomy education is emphasized.

Following the talk questioning was active and Mirjana Povic asked:

I can see that many African countries still do not have established NAEC team. Can from the African astronomical society we help you with that?

The response was:

While three of the officers of the African Astronomical Society have been involved in the NAEC selection process in different roles, I believe we have not formally asked the AfAS for support in this matter. Thanks for the suggestion, we will follow up on that!

Noorali Jiwaji asked:

In the world map can you explain what the yellow means

Markus Pössel replied:

The yellow areas in the NAEC world map are those countries where we have a nomination, but will still need to confirm the NAEC candidates. There is a distinction here between countries who have an official IAU National Committee for Astronomy as their official national IAU body, and countries who have not; since we want the NAECs to be anchored in their national communities, for countries with an NCA, the NCA confirms the NAEC Teams. Where there is no NCA, the OAE itself makes the confirmation, based on the candidates' qualifications and recommendations from the community.

After the talk Amelia Bayo asked:

Hi Sally, such a great project!!! In Chile (within NPF) we are producing astronomy mirrors / segments out of carbon fiber, therefore much lighter, would that be something of interest for the new telescope?

Salma Sylla asked:

In how many languages, the resources are available?

Markus answered:

All of our translation projects are currently in the future; based on a first analysis of the prevalence of languages spoken by school-age children/young adults, we are looking at about 30 to 40 languages, though – which will be considerable work, as we are well aware!

Lara Rodrigues asked:

In there a link for the astronomy education survey worldwide you mentioned?

Markus answered:

You can find the NAEC summaries of their country's educational systems, and the role of astronomy within those systems, here: https://www.haus-der-astronomie.de/oae/worldwide

Mayra Lebron asked:

How do the OAE Nodes and Centers relates with the NAECs?

Markus answered:

Those are separate structures, although of course we expect there to be overlap. The main role of the OAE Nodes and OAE Centers is to be part of the OAE overall – they are *not* regional offices (such as the ROADS of the OADs), but all of them have an international perspective and reach – they support the overall mission of the OAE, not the activities of the OAE regarding a specific country (although there might be some overlap, e.g. when it comes to supporting certain translations, or helping organise regional events). The NAECs, in contrast, are country-specific.

Finally, within State of the Art in Astronomy Education Hakim Malasan and co-authors Rosa M. Ros, Chatief Ku Next Susana Duestua writes about IAU Strategic Plan 2020-2030 and Division C Education, Outreach and Heritage. She discusses how Division C fits into and supports the goals of the Strategic Plan. She outlines how the division stays dynamic and finishes with plans for great work in the future.

Following this is an article called "Research, Innovation, Scientific Literacy and Inclusion in Astronomy" written by Nicoletta Lanciano, Amelia Ortiz-Gil, and Néstor Camino. This was written about a roundtable session that addressed issues of research, innovation, scientific literacy, and inclusion. Details were introduced about each and contributions of the participants in these areas were recognized.

Students' Preconceptions in Astrophysics – How to break them down? was written by Corina Toma. In it she talks about the most common preconceptions of students entering into the study of astrophysics and how to break them down. She emphasizes the use of active teaching methods that include experiments and models.

After the talk Kelly Lepo asked:

Thanks for the great talk Corina. Are there any resources that NASA or other organizations could make that would help help you in the classroom to overcome your student's preconceptions?

And Corina answered:

Of course. I use many times NASA, ESA, ESO resources

The final paper of this section called "Meet the skies of world : First Intecontinental-experimental course of teacher training in Astronomy, of a cooperative and participatory type, in a time of physical distancing" was written by Nicoletta Lanciano and Nestor Camino. The authors describe virtual course development necessitated by the global pandemic. Meet the Skies of the World is the first with a second to follow. Key items are outlined and they finish with what will be accomplished by the next course.

Following the talk Lara Rodrigues asked:

Nicoletta and Néstor, congratulation on your work. It is very inspiring! Have you evaluated it with some method? And also, do you plan to follow-up with the participant teachers' and see how they apply what they learn with their students? Thank you!

And Néstor replied:

Hola Lara. Thank you for your greetings. We had an evaluation by the comments of the more than 100 participants, but we don't have any "tracing" of the way each teacher worked with their students. We will develope another course, level II, next year, with the same teachers, that will be another way of evaluation. thank you. yours. Néstor.

Education and Heritage in the era of Big Data in Astronomy
Proceedings IAU Symposium No. 367, 2020
R. M. Ros, B. Garcia, S. R. Gullberg, J. Moldon & P. Rojo, eds.
doi:10.1017/S1743921321000934

The IAU Office of Astronomy for Education

Markus Pössel[1]ⓘ**, Carolin Liefke[1], Niall Deacon[1], Natalie Fischer[1,2],**
Juan Carlos Muñoz[1], Markus Nielbock[1], Saeed Salimpour[1]
and Gwen Sanderson[1]

[1]IAU Office of Astronomy for Education and Haus der Astronomie,
Königstuhl 17, 69117 Heidelberg, Germany
Contact email: oae@astro4edu.org

[2]Forscherstation – Klaus-Tschira-Kompetenzzentrum
für frühe naturwissenschaftliche Bildung gGmbH,
Speyerer Straße 6, 69115 Heidelberg, Germany

Abstract. Since January 2020, the International Astronomical Union has an Office of Astronomy for Education (OAE). The OAE, which joins the previously existing IAU Offices for Astronomy for Development (OAD), Astronomy Outreach (OAO) and Young Astronomers (OYA) is hosted at Haus der Astronomie, a center for astronomy education and outreach operated by the Max Planck Society in Heidelberg, Germany. This contribution outlines the mission of the OAE, the current state of the office, its background, mission and collaborative structure, as well as the activities that have already started or are planned for the future.

Keywords. astronomy education, IAU Strategic Plan 2020–2030, International Astronomical Union

The IAU Office of Astronomy for Education is the newest office of the International Astronomical Union, and a key element of the Union's Strategic Plan 2020–2030. As such, we are grateful to the organisers for allowing us to present both our plans for and the current state of the IAU OAE at this symposium.

1. The Office of Astronomy for Education's role within the IAU

The IAU is the worldwide umbrella organisation of and for professional astronomers. Traditionally, its role has been two-fold: Since its foundation in 1919, the Union has organised scientific meetings, including General Assemblies every three years and IAU Symposia such as IAUS 367. Also, the IAU has been instrumental for the establishment, via its resolutions, of formal community consensus on issues where such consensus is deemed useful — such as the definition of specific coordinate systems, of the astronomical unit, or, famously, of what a planet is. The IAU has also traditionally been responsible for naming celestial objects, or features on the surfaces of such objects (such as craters of the Moon).

But over the past decade, the IAU has also become much more active beyond those traditional areas of activity. Notably, at the 2009 General Assembly, the IAU adapted its Strategic Plan 2010–2020, "Astronomy for Development — Building from the IYA 2009" (Miley 2009), which grew out of activities in the framework of the International Year of Astronomy 2009 (Russo & Christensen 2009). In the plan's successor, the extended Strategic Plan 2020–2030 (van Dishoeck & Elmegreen 2019), the perspective has broadened: In that plan, education and outreach come into their own, on par with the continuing goal of leveraging astronomy for development.

Figure 1. The OAE logo, designed by Juan Carlos Muñoz and Gwen Sanderson, combines representations of classic and modern tools of education.

The organisational structure of the IAU has two main strands. On the one hand, the IAU provides a framework for astronomers to self-organise, bringing together those who are experts on, and interested in engaging with, specific astronomical topics. For that purpose, there are within the IAU nine Divisions, most of them defined along astronomical sub-disciplines (such as Planetary Systems and Astrobiology in Division F, or Sun and Heliosphere in Division E), with an additional layer of organisation in the shape of 35 Commissions, each associated with one or more divisions. For our purposes, the relevant division is Division C, "Education, Outreach and Heritage," whose president, Susana E. Deustua, is also on the steering committee of the IAU Office of Astronomy for Education. Specifically, Commission C1 (President 2018–2021: Paulo S. Bretones) is concerned with "Astronomy Education and Development."

In order to accomplish specific tasks, divisions or commissions can institute Working Groups. Such Working Groups are by default term-limited, namely instituted at an IAU General Assembly for the period of time until the next such assembly three years later. There are exceptions in the shape of so-called Functional Working Groups that fullfill more long-term tasks (such as the naming of minor bodies), and thus are exempt from the three-year limit.

It is at this point, namely when considering alternatives to getting things done via temporary Working Groups, that the idea of IAU Offices comes into play. Prior to 2019, the three offices were the Office of Astronomy for Development (OAD; McBride & Venugopal 2020) in Cape Town, South Africa; the IAU Office for Astronomy Outreach in Tokyo, Japan (OAO; (van Dishoeck & Elmegreen 2019) and the (virtual) Office for Young Astronomers (OYA; Gerbaldi *et al.* 2011) tasked with organising the annual International School for Young Astronomers (ISYA).

This, then, was the context for the establishment of the IAU Office of Astronomy for Education — completing the quartet of IAU offices with a mission that is centered around the leveraging of astronomy for the benefit of education in primary and secondary schools (Fig. 1). Both the overall aim and five specific goals were specified within the IAU's Strategic Plan 2020–2030, which was approved in August 2018 at the XXXth IAU General Assembly, in Vienna, Austria. The plan lists the overall mission of the IAU as "The mission of the International Astronomical Union is to promote and safe-guard astronomy in all its aspects (including research, communication, education and development) through international cooperation." Based on this mission, the plan spells out five specific goals, including the following as Goal 5: "The IAU stimulates the use of astronomy for teaching and education at school level."

After the Strategic Plan had been approved, newly-elected IAU General Secretary Teresa Lago swung into action: On October 31, 2018, the IAU launched an international

Figure 2. Establishment of the IAU Office of Astronomy for Education at the 1st Shaw-IAU Workshop on Astronomy for Education at IAU headquarters in Paris on 17 December 2019. Center group left to right: OAE Deputy Director Carolin Liefke, OAE Director Markus Pössel, IAU General Secretary Teresa Lago signing the Memorandum of Understanding, IAU Division C President Susana Deustua and IAU President-Elect Debra Elmegreen. Image: IAU/A. Gustin.

call for Letters of Intent to institutions willing to host the new IAU Office of Astronomy for Education, to be submitted by December 31, 2018. From the 23 Letters of Intent received in response, an ad-hoc selection committee made a short-list. On 28 February, 2019, the nine short-listed applicants, namely institutions from Australia, China, France, Germany, India, Italy, Netherlands, and the USA, were asked to submit a full application. By the 30 June, 2019 deadline, six proposals had been received. Haus der Astronomie, Heidelberg (Pössel 2011), emerged as the clear winner once the proposals had been evaluated and ranked, and also received the unanimous approval of the IAU Executive Committee. The decision was communicated to Haus der Astronomie on 13 September, and a Memorandum of Understanding establishing the OAE came into force on 17 December, 2019, signed by Hans-Walter Rix as managing director of the Max Planck Institute for Astronomy representing the host institution, and IAU General Secretary Teresa Lago for the IAU (Fig. 2).

2. OAE: Planned and current activities

From the IAU mission, and the goals formulated in the IAU Strategic Plan 2020–2030, it is clear that setting up the OAE's activities must begin with the question: Which actions aimed at fostering the leveraging of astronomy to benefit primary and secondary education are appropriate and effective for an IAU office to take, at an international level? There are several aspects to this. First of all, what actions are appropriate at an international level? After all, the most important interactions in education happen locally, between teachers and learners, and between the learners themselves. Also, even within the IAU community and the wider astronomy community, there are numerous actors supporting astronomy education with individual actions, but also within the framework of large-scale projects, such as Universe Awareness (UNAWE; Miley, Ödman & Russo 2020), EU Space Awareness (Russo 2015), the Galileo Teacher Training Program (GTTP; Doran 2012) or the Network for Astronomy School Education (NASE; Ros 2012). It is clear that any additional IAU effort needs to make sure not to re-invent any wheels.

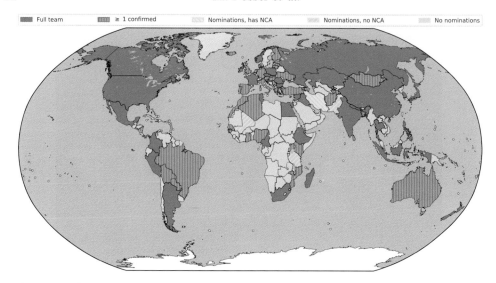

Figure 3. Overview of the current reach of the network of National Astronomy Education Coordinator Teams (NAEC Teams). Figure: N. Deacon and G. Sanderson.

The strategy that we formulated as part of the Haus der Astronomie application to host the OAE takes into account those boundary conditions. Our conclusion from those boundary conditions is that it is the task of the OAE to foster professionalisation in astronomy education both for educators and for professional astronomers involved in education, to create, curate, translate and help disseminate fundamental resources, to foster the creation of community-wide standards both for astronomical resources and for teacher training events, and to foster and support suitable infrastructure for astronomy education. Involvement of the various national and regional astronomy education communities requires active networking.

2.1. *Networking*

Liaising between the OAE and the various national astronomy education communities are the National Astronomy Education Coordinator Teams (NAEC Teams). The worldwide NAEC network comprises astronomers and educators with expertise in primary and secondary education. NAECs interface between the OAE and the educational community in their respective countries; they are tasked with identifying local needs in astronomy education, promoting astronomy in national curricula, and developing teaching resources and training events.

Each NAEC team is composed of up to 5 members, covering their country's diversity in terms of gender, geographical distribution, ethnicities and, where applicable, languages. In countries that have a National Committee for Astronomy, the NCA can directly nominate their NAEC team. Individuals can also submit self-nominations; if their country does not have an NCA, the OAE team evaluates their application. As of the time of this writing in January 2021, the tally of NAECs is the following:
- 318 Nominations / Self-Nominations from 87 countries
- 300 confirmed NAECs from 82 countries
- 230 NCA- Nominations from 60 countries
- 86 Self-Nominations from 34 countries (of which 9 are awaiting approval)

19 countries have not sent any nominations. The current reach of the NAEC network is shown in Fig. 3.

In addition to the NAECs, there are plans for more specific extensions of the OAE, in the shape of what we call OAE Centers and OAE nodes — hosted by institutions who are willing to commit specific funds and FTE resources to the OAE, with the goal of solidifying the long-term international reach of the office. Nodes and Centers differ both in scope and amount of resources provided by the host institution, but all of them are active internationally: their scope is not primarily regional; instead, all of them are meant to have international reach. Centers are expected to have similar dedicated staffing and funds as the OAE itself, whereas Nodes are more limited in scope and resources.

Of course, our networking activities do not end with the NAECs, the OAE Centers and OAE Nodes. In addition, the OAE will aim to keep in touch with astronomy education stakeholders within and without the IAU. One eventual aim is to create a data base of professional astronomers who are interested in astronomy education, and in furthering the mission of the OAE. Last but not least, given that the goals of the various IAU offices have some overlap, the OAE participates in regular coordination meetings with the other offices.

2.2. *OAE Reviews*

OAE Reviews are meant to be a key tool for the office's professionalisation of astronomy education. They are meant to provide professional astronomers and astronomy education pracitioners with a ready means of getting up to speed in areas pertinent to astronomy education — from teaching methods to background information in astronomy education to astronomy-specific topics, for example daylight observations, or indigenous astronomy. By default, each OAE Review is written by an OAE Review Panel consisting of educators and researchers, and the creation process always includes the solicitation of community feedback.

Currently under way are three OAE Reviews: The OAE Review on Equity, Diversity and Inclusion has constituted its OAE Review Panel, which has met several times remotely in order to organise work on the review. The OAE Review on Digital Teaching and Learning, intended to summarize best practices on how to efficiently use digital technology for educational purposes, including (but not limited to) interactive simulations, e-assessment tools, and online teaching, is currently recruiting its Panel.

The OAE Review on Astronomy Education around the World, which is led by OAE staff (Saeed Salimpour and Niall Deacon), will be based on, and analyze in a larger context, descriptions of various countries' educational systems by the respective NAEC teams. The descriptions are meant to cover the role that astronomy plays in primary and secondary education in each country. To date, first versions of 64 of these country-specific documents have been published on the OAE website.

The establishment of additional OAE Review panels is planned for 2021. Once we are joined by OAE Centers and OAE Nodes, the supporting roles ("OAE support scientist" for an OAE Review) will be filed by OAE Center and OAE Nodes personnel. The OAE Astronomy Education Seminar Series (see below) is also meant to aid us in finding prospective OAE Panel members, as well as resources useful for inclusion in OAE Reviews.

3. Astronomy Education Infrastructure

An important part of the OAE's mission of fostering astronomy education is the creation and operation of suitable infrastructure.

3.1. *astroEDU*

astroEDU (https://astroedu.iau.org) is an open-access online resource of peer-reviewed astronomy education activities (Russo *et al.* 2015) which has since its inception been a collaborative effort between various institutions. It is a project of the International Astronomical Union under the framework of the IAU Office of Astronomy for Development. The project is supported by the IAU OAD, Universe Awareness, Leiden University, LCO, Space Awareness, Europlanet and European Union. IAU astroEDU is part of Europlanet 2020 RI and EU Space Awareness projects and it has received funding from the European Union's Horizon 2020 research and innovation programme under grant agreements No 654208.

astroEDU has now been transferred to the stewardship of IAU OAE as part of the core astronomy education resources. OAE will provide/fund the framework both for the public-facing astroEDU website and for the underlying journal management system, as well as organisational support. The scientific responsibility will remain with the scientific community, and notably with astroEDU editor-in-chief Michael Fitzgerald of Edith Cowan University, Perth.

3.2. *Astronomy education resources data base*

A key part of OAE planning is the maxim not to re-invent wheels. The number one go-to resource for research astronomers in search of specific bits of literature is the NASA Astrophysics Data Service (ADS; Accomazzi *et al.* 2015). OAE is currently in negotiations with the NASA ADS team to expand the ADS data base to include astronomy education resources, in English as well as in other languages, with the OAE acting as curator for the pertinent ADS collection.

3.3. *Big Ideas in Astronomy*

Big Ideas in Astronomy (Retrê *et al.* 2019) is a project by the Leiden Observatory, Leiden University (the Netherlands) and Institute of Astrophysics and Space Sciences (Portugal) in the framework of the IAU Commission C1: Working Group on Research and Methods. The aim of this project was to create a document that constitutes a proposal for a definition of astronomy literacy: a list of "big ideas," complete with descriptions in terms of sub-ideas, that any astronomy-literate person should be familiar with. The Big Ideas project having reached a state where the main initial work is completed, extensive feedback from the community has been obtained, and a new version that incorporates that feedback created, its focus has now moved to creating faithful translations of the Big Ideas document, as well as occasional revisions at regular intervals.

With a memorandum of understanding signed in October 2020, OAE has taken over stewardship for the Big Ideas in Astronomy project, with the previous IAU Commission C1 working group becoming Big Ideas Advisory Panel. The rollout of version 2 is currently underway. The website is already online although not yet ready for public distribution. Work is ongoing to put the existing translations into the Big Ideas booklet design. The full rollout of version 2 is envisioned by the end of the first quarter of 2021.

4. Astronomy Education Resources

Part of our mission is to provide astronomy educators worldwide with a basic collection of open, high-quality educational resources, either curated or newly created, translated wherever possible into the learners' native languages, and available under licenses that allow for free use (namely suitable Creative Commons licenses). Full strategic planning will only become possible once the OAE Centers and Nodes are in place, and we have a more reliable idea about the personnel and funds we can bring to bear on the various projects; the following is an indication of where we are starting:

4.1. *Multilingual Glossary*

We expect translation of various resources into dozens of languages to be an important part of OAE activity over the years to come. As a solid basis for such translations, we have begun creating a multilingual glossary, as a joint project with OAO (Lina Canas, Hidehiko Agata). On the OAE side, the project lead is Niall Deacon. The project aims to produce a glossary of a few hundred astronomical terms that will commonly come up in primary or secondary school lessons, with translations both of the terms itself and brief definitions (providing context) into as many languages as we can manage. The initial batch of terms will be the result of an iteration projects: Starting out with a list drawn from the Japanese glossary (Agata) as well as the Big Ideas document, we have asked the NAECs for additions, and are currently in a stage where expert panels are meant to select the most pertinent terms from the current pool, separately for primary and secondary school teaching. To ensure the quality of the translations, we will institute a review process where, ideally, each translation will have been checked by a professional astronomer who is also a native speaker.

4.2. *Astrophotography Contest*

Suitable images are an important teaching tool. In a number of areas of astronomy, numerous excellent images are available, thanks to the image publications notably from the NASA/ESA Hubble Space Telescope and the ESO telescopes. In other areas, notably in situations where high-end and/or space telescopes are unsuitable, there are some gaps when it comes to images accessible freely under open licenses. The aim of our Educational Astrophotography Contest, our astrophotography competition launched in early 2021, is to fill some of the gaps. Winning images will receive cash prizes, and will be released as Open Educational Resources under a suitable Creative Commons license. Depending on the success of the initial contest, we expect to follow up with more targeted contests over the following years.

4.3. *Visualisation resources*

In order to kick-start the OAE resource pool for images and animations, we are currently working on diagrams/animations/visualisations specifically related to the topics of the "Big Ideas in Astronomy," both with HdA's Thomas Müller and with the visualisation specialist Stefan Payne-Wardenaar.

4.4. *Standards for astronomy education resources*

Part of the mission of the OAE is to establish community-consensus standards for what makes for high-quality education resources. To start the process, we will solicit ideas about such standards from astronomy educators as well as external experts in early 2021.

5. Events

5.1. *Shaw-IAU Workshops*

The Shaw Prize Foundation has generously provided funding to the IAU for an annual Shaw-IAU Workshop on Astronomy for education. The First Shaw-IAU Workshop was held in Paris in December 2019, and served to provide the nascent OAE with feedback from the astronomy education community regarding plans and strategy.

The Second Shaw-IAU Workshop on Astronomy for Education was a fully online event held on Oct 6-9 2020. It brought together the NAECs and other key actors in the field of astronomy education. The Opening Session also served as the official launch event of the OAE, featuring Ewine van Dishoeck as President of the International Astronomical Union (IAU); Theresia Bauer, MdL, Chair of the Zeiss Foundation Administration as well as Minister for Science, Research and Art of the German State of Baden-Württemberg; Beate Spiegel, Managing Director of the Klaus Tschira Stiftung; Kenneth Young, Chairman of the Shaw Prize Council and Vice Chair of its Board of Adjudicators, representing the Shaw Prize Foundation; Teresa Lago as General Secretary of the IAU; with a keynote address by Svein Sjøberg, University of Oslo, on the ROSE study (Relevance Of Science Education).

Over the following days, 347 participants from 82 countries were able to attend 31 talks in sessions dedicated to Making astronomy education equitable, diverse and inclusive; Astronomy education within the IAU; Astronomy education in low-tech environments; Astronomy education resources; and Astronomy education around the world. Participants were also able to visit poster displays and "NAEC booths," in which a total of 51 NAEC Teams presented themselves and their countries.

5.2. *OAE Seminar Series*

We are currently making plans for an OAE Online Seminar Series. In the series, we want to provide information about the same topics that are also covered in the OAE Reviews — with the series effectively acting as a "trial run" for the different reviews. We also want to introduce astronomers and astronomy education practitioners to the basics of astronomy education research. The overall aim is to bring together experts and members of the astronomy, astronomy education and general education research communities to instigate fruitful discussions about how to engage in astronomy education. Our target group consists of anyone interested in understanding the fundamentals of astronomy education and astronomy education research, whether a beginner or a seasoned expert.

5.3. *Schools for Astronomy Education (SAE) and standards for teacher training events*

As per the Strategic Plan 2020–2030, the OAE will eventually organise an Annual International School for Astronomy Education (ISAE) for teachers, as well as Regional Schools for Astronomy Education. While the current pandemic makes such endeavours impossible at this time, for us, it in fact makes sense for the ISAEs, and Regional SAEs to commence at a later stage: Another of our goals, after all, is to establish community-consensus standards for teacher training events, and the proper order is for our own activities in that field to commence after at least a preliminary version of the standards has been established. As in the case of astronomy education research standards, we will start the process by soliciting ideas about, and views on, such standards from astronomy educators as well as external experts in early 2021.

6. Who does the work?

So who are the actors — who does the work? The IAU OAE is operated by Haus der Astronomie (HdA), a center for astronomy education and outreach in Heidelberg, Germany, which is administered by the Max Planck Institute for Astronomy (MPIA) in Heidelberg, which in its turn is part of the Max Planck Society for the Advancement of Science, Germany's largest society for fundamental research. As such, OAE is based, in

Figure 4. Aerial view of the MPIA Campus, showing the main building of the Max Planck Institute for Astronomy (right), MPIA's Elsässer laboratory (top), and Haus der Astronomie as seat of the IAU Office of Astronomy for Education (left). Image: Dominik Elsässer.

style, in a spiral-galaxy-shaped building on Königstuhl mountain (cf. Fig. 4) — although for its first year, OAE staff operated almost exclusively from home, meeting up virtually at regular intervals.

Funding for OAE operations is provided by the Klaus Tschira Foundation, the Carl Zeiss Foundation, and the Shaw Prize Foundation, in addition to regular funding that comes from the budget of the IAU itself.

The management of the OAE consists of Markus Pössel as OAE Director, who is also the Managing Scientist of Haus der Astronomie, and Carolin Liefke as Deputy Director, who is a Haus der Astronomie staff member with a dual appointment at Heidelberg University. Staff funded by the OAE budget includes Markus Nielbock, Juan Carlos Muñoz and Niall Deacon as OAE Coordinators, Saeed Salimpour as OAE Astronomy Education Research Coordinator, and Gwen Sanderson as Organizational Assistant. In the area of primary education, we are supported by Natalie Fischer (HdA) as a consultant. OAE work profits greatly from collaboration with additional Haus der Astronomie staff.

At the point of this writing, namely in January 2021, we are about to sign Memoranda of Understanding that will enlarge the OAE beyond the central Heidelberg office: Following our 2020 call for OAE Centers and Nodes, we are now looking forward to establishing OAE Centers Italy, India, China, Cyprus, and Egypt, as well as OAE Nodes in France, South Korea, and Nepal.

OAE strategic planning and operations are supported by the OAE Steering Committee, which consists of Susana Deustua (Space Telescope Science Institute; chair), Coryn Bailer-Jones (Max Planck Institute for Astronomy), Matthias Bartelmann (Heidelberg University), and Teresa Lago (IAU General Secretary). In 2020, the Steering Committee met five times, namely in January, April, July, September and December.

Up-to-date information about OAE activities is available on the office's website at [http://www.astro4edu.org].

7. Conclusion

With the IAU Office of Astronomy for Education, the IAU has taken a decisive step towards the goal of leveraging astronomy for education, in particular for education in the STEM subjects (science, technology, engineering and mathematics). Over the coming years, the OAE aims at fostering astronomy education world-wide, following the strat-egy outlined in this contribution, while remaining flexible when it comes to reacting to specific community issues. Both in the framework of the IAU and beyond, there have been numerous powerful initiatives, on different scales, to support astronomy education, and the OAE is happy to be able to take its place in the existing vibrant astronomy education international community.

References

Accomazzi, A., Kurtz, M. J., Henneken, E. A., Chyla, R., Luker, J., Grant, C. S., Thompson, D. M., Holachek, A., Dave, R. & Murray, S. S. 2015 in A. Holl, S. Lesteven, D. Dietrich & A. Gasperini (Eds.), *Library and Information Services in Astronomy VII, ASP Conference Series* 492, pp. 189–197. arXiv:1503.04194

van Dishoeck, E. & Elmegreen, D.M. 2018, *IAU Strategic Plan 2020–2030*, www.iau.or/tati/ducatio/trategicplan-2020-2030.pdf

van Dishoeck, E. & Elmegreen, D.M. 2020, *Proceedings of the International Astronomical Union* 14(A30), pp. 546–548. DOI: 10.1017/S1743921319005337

Doran, R. 2012, *Proceedings of the International Astronomical Union* 10(H16), p. 547. DOI: 10.1017/S1743921314012046

Gerbaldi, M., DeGreve, J. P. & Guinan, E. 2011, *Proceedings of the International Astronomical Union* 5(S260), pp. 642–649. DOI: 10.1017/S174392131100295X

McBride, V. & Venugopal, R. 2020, *Proceedings of the International Astronomical Union* 14(A30), pp. 553–554. DOI: 10.1017/S1743921319005350

Miley, G. 2009, *Astronomy for Development: Strategic Plan 2010–2020*, www.iau.or/tati/ducatio/trategicplan-2020-2030.pdf

Miley, G., Ödman, C. & Russo, P. 2020, in Visser, J., & Visser, M. (Eds.), *Seeking Understanding* (Brill — Sense), pp. 119–135, DOI: 10.1163/9789004416802, arXiv:2001.00456

Pössel, M. 2011, *Astronomie und Raumfahrt im Unterricht* 48(3–4), pp. 31–34.

Retrê, J., Russo, P., Lee, H., Penteado, E., Salimpour, S., Fitzgerald, M., Ramchandani, J., Pössel, M., Scorza, C., Christensen, L. L., Arends, E., Pompea, S. & Schrier, W. 2019, *Big Ideas in Astronomy: A Proposed Definition of Astronomical Literacy.* www.iau.org/static/archives/ announcements/pdf/ann19029a.pdf

Ros, R. 2012, *Physics Education* 47(1), pp. 112–119. DOI: 10.1088/0031-9120/47/1/112

Russo, P. 2015, *Proceedings of the International Astronomical Union* 11(A29A), p. 398. DOI: 10.1017/S1743921316003409

Russo, P., Gomez, E., Heenatigala, T. & Strubbe, L. 2015, arXiv:1501.07116

Russo, P, & Christensen, L. L. 2009, *International Year of Astronomy 2009: Final Report*, www.astronomy2009.org/static/archives/documents/pdf/iya2009_final_report.pdf

Education and Heritage in the era of Big Data in Astronomy
Proceedings IAU Symposium No. 367, 2020
R. M. Ros, B. Garcia, S. R. Gullberg, J. Moldon & P. Rojo, eds.
doi:10.1017/S1743921321001083

IAU Strategic Plan 2020–2030 and Division C Education, Outreach and Heritage

Susana Deustua[iD]

STSCI, 4700 San Martin Drive, Baltimore, MD USA
email: deustua@stsci.edu

Abstract. The IAU Strategic Plan 2020–2030 identifies five goals to guide the Union's activities through this decade: to lead the global coordination of professional astronomy, to promote inclusive advancement of astronomy, to promote astronomy as a tool for development, to promote public engagement in astronomy, and, to stimulate using astronomy for primary and secondary education. Education permeates through all these goals – professional development, informal education and formal education for the very young through to the PhD level. For the IAU's Division C, Education, Outreach and Heritage, the plan provides the foundation for supporting teachers worldwide who use astronomy as a tool for teaching science, the humanities and the arts, to promote astronomy education research, to engage life-long learners, and to engage everyone with our collective history and heritage to understand the Universe. I describe Division C's remit, where we are now in this 3rd decade of the 21st century.

Keywords. Keyword1, keyword2, keyword3, etc.

1. Introduction

In 2018, the IAU Strategic Plan 2020–2030 was released. Its goals are summarized here:

- To lead the global coordination of professional astronomy,
- To promote the inclusive advancement of astronomy,
- To promote astronomy as a tool for development,
- To promote public engagement in astronomy, and,
- To stimulate using astronomy for primary and secondary education.

The IAU has nine divisions, of which six are focussed on specific science research areas. Three divisions, A, B and C, might be considered supporting or service divisions as their purview is broad, and necessary for the promotion of the astronomical sciences.

Division C Education, Outreach and Heritage objectives are:

1. to support teachers worldwide who use astronomy as a tool for teaching science, the humanities and the arts,
2. to promote astronomy education research
3. to engage life-long learners
4. to engage everyone with our collective history and heritage to understand the Universe.

2. Division C education, outreach and heritage

This Division's activities are broad and wide ranging, with its members active in:

Education – from primary to tertiary (university) levels and include formal (in school) and informal (out of school) education, informed by astronomy education research.

Outreach – communicating astronomy with the public

Heritage – the legacy of tangible astronomical artefacts and structures and the intangible attributes of astronomy in our cultural histories (past and present), as well as the history of astronomy.

Division C promotes astronomy in developing countries and has an interest in maintaining dark skies the absence of which affects education, outreach and cultural practices as well as research.

The Division's governing body consists of a President, Vice President, and a Steering Committee of six members plus the four commission presidents and an Advisor, who is often the past president. All of these individuals are elected by the Division members.

Commissions focus their activities within an area; thus each of the four commissions within Division C have their own remits: Astronomy Education and Development (C1), Communicating Astronomy with the Public (C2), History of Astronomy(C3) and World Heritage and Astronomy (C4). Additionally, the IAU allows interdivision commissions, and Divisions C and B support Commission B-C Protection of Existing and Potential Observatory Sites, which spans the interests of the two divisions.

Commissions have an Organizing Committee, a President and Vice President, who are elected by the Division Members. Working groups are normally part of a commission and are organized by to carry out one or more programs and projects; they have a chair or co-chairs. Additionally, working groups may find that their programs span more than one commission, and thus form intercommission working groups. The parent commissions may belong to different divisions.The working groups are the heart of the Division; members are encouraged to actively participate.

In response to the Strategic Plan 2020-2030, in 2018 the Division restructured its working groups to support the goals of the new strategic plan and in anticipation of the establishment of the Office of Astronomy for Education. We established an open proposal process with defined critera, and Working Group proposals were reviewed by the Steering Committee. One of the happy consequences of this process was the elevation of the Working Group Astronomy for Diversity and Inclusion from Comission C1 to become a working group of the IAU Executive Committee.

The opportunity to propose for new working groups or propose to continue existing working groups occurs every three years, normally occurring in the few months before the General Assembly. Although the 2021 General Assembly was postponed to 2022 due to the COVID-19 pandemic, the business year cycle was still maintained.

2.1. *Commissions and working groups*

The final complement of commissions and their working groups for the 2018-2021 triennium are listed here. Please visit the webpages of Division C, its commissions and working groups for more information.

1) Astronomy Education and Development (C1)
   - Astronomy Education Research and Methods
   - Astronomy Competitions for Secondary School Students
   - Astronomy Education Resources (AstroEdu)
   - Network for Astronomy School Education (NASE)
2) Communicating Astronomy with the Public (C2),
   - CAP Conferences
4) History of Astronomy(C3)

5) World Heritage and Astronomy (C4).
   - Astronomical Heritage in Danger
   - Windows to the Universe: Classical and Modern Observatories

2.1.1. Intercommission working groups

Division C supports the following intercommission working groups
- C1-C3-C4 WG Archaeoastronomy and Astronomy in Culture
- C1-C3-C4 WG Ethnoastronomy and Intangible Astronomical Heritage
- C1-F2-F3-H2 WG Education and Training in Astrobiology
- B7-C1 WG Achieving Sustainable Development within a Quality Lighting Framework
- B4-C3 WG Historical Radio Astronomy (Inter-Union IAU-URSI

2.1.2. Division and inter-division working groups

Some working groups have projects that are within the overall remit of a division or overlap with another division's interests. Within Divison C these are:
- WG Star Names
- C-E WG Solar Eclipses

3. Keeping the division Dynamic

Key to keeping the division dynamic are the Division's individual and associated members who number 2384 active members, of which 182 are junior members. All members are encouraged to actively participate in the working groups, especially junior members and early career astronomers. The division's members include astronomers, historians, professional educators and communicators, embracing research and practice in education, outreach and tangible and intangible astronomical heritage.

The Division coordinates activities with the four IAU Offices: Astronomy of Astronomy for Development (OAD), Office for Astronomy Outreach (OAO), Office for Young Astronomers (OYA) and the recently established Office of Astronomy for Education (OAE). The origin of these offices are partially in the precursor to the Division; connections between the Division and the Offices continue. In particular, OAO and Commission C2 work closely together on the CAP (Communicating Astronomy to the Public) conferences, which are held every two to three years, and Commission C1 has a role supporting the OAE through its emphasis on education research and the newly formed online Astronomy Education Journal

Rewarding Excellence: Ph.D. Prize

Rewarding research is vital for encouraging junior scholars, and one way is recognize outstanding achievement. Nominations and self-nominations for the Division C Ph.D. prize are accepted annually from any country for thesis topics in:
- Thesis topics in the remit of the Division
- Astronomy Education Research
- Astronomy Education Practice
- Outreach and Informal Education
- History of Astronomy
- Cultural Astronomy (including ethnoastronomy and archaeoastronomy)
- Diversity, Equity, and Inclusion (DEI) in astronomy

Details of the prize are found at https://www.iau.org/science/grants_prizes/phd_prize/

Communicating: Conferences

Promoting astronomy education, outreach, history and heritage also involves communicating and sharing results. To this end, the Division supports international symposia as

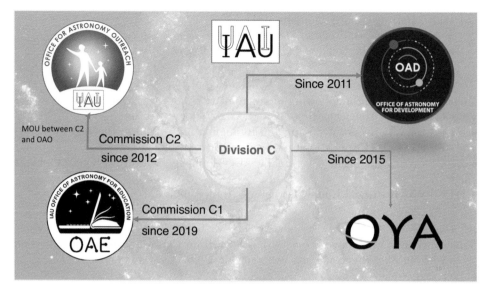

Figure 1. The four IAU Offices and Division C.

well as ancillary activities (e.g. outreach events, education workshops) at IAU scientific
conferences. Recent IAU Symposia supported by Division C include:

• IAUS 367 Education and Heritage in the Era of Big Data in Astronomy – the first
steps on the IAU 2020–2030 Strategic Plan, Argentina (this symposium)

• IAUS 358 "Astronomy for Equity, Diversity and Inclusion – a roadmap to action
within the framework of the IAU centennial anniversary, Japan

• International Symposium on Education in Astronomy and Astrobiology, Netherlands
Joint IAU and European Astrobiology Campus Ancillary activities at IAU science
conferences

• Division Days at IAU General Assemblies

4. The future

In the few years since the publication of the IAU Strategic Plan 2020-2030, the
Division's activities have been focused in support of its goals, and in preparing for the
next decade. I anticipate the Division will expand into scholarly research in the theory
and practice of education (formal and informal) and outreach, and continue collaborations with the OAE, OAO, OAD and OYA to promote astronomy in all countries. While
the Division attracts a broad cross-section of astronomers, we still need to encourage
early career as well as professionals in related fields to engage in the activities of, and,
with the Division. This is important for developing new ideas, methods and approaches
to the diverse projects in the Division's portfolio. all continents

These are exciting times!

Dr. Susana Deustua is a scientist at the Space Telescope Science Institute in Baltimore,
MD USA, working on the Nancy Grace Roman Space Telescope. She is currently the
president of the IAU Division C, Education, Outreach and Heritage. Her research interests
are supernova cosmology, absolute calibration, and solar system studies. She has a long-
term interest in astronomy education, and served for six years as Director of Education
of the AAS. As a member of the Supernova Cosmology Project, she shared in the 2015
Breakthrough Prize in Fundamental Physics and the 2007 Gruber Cosmology Prize.

Education and Heritage in the era of Big Data in Astronomy
Proceedings IAU Symposium No. 367, 2020
R. M. Ros, B. Garcia, S. R. Gullberg, J. Moldon & P. Rojo, eds.
doi:10.1017/S1743921321000417

Research, innovation, scientific literacy and inclusion in astronomy

Nicoletta Lanciano[1], Amelia Ortiz-Gil[2] and Néstor Camino[3]

[1]National Coordinator of the Research Group on Sky Pedagogy of the MCE (Movimento di Cooperazione Educativa), University La Sapienza, Roma, Italia
email: `nicoletta.lanciano@uniroma1.it`

[2]Astronomical Observatory, University of Valencia, Spain
email: `amelia.ortiz@uv.es`

[3]National Coordinator of NAEC Argentina – OAE IAU
Complejo Plaza del Cielo, CONICET-FHCS UNPSJB
Esquel, Patagonia, Argentina
email: `nestor.camino.esquel@gmail.com`

Abstract. With this roundtable we wished to discuss about different topics related to research, innovation, scientific literacy and inclusion in Astronomy education, putting together the expertise of educators from all around the world. The present paper includes the introduction to the roundtable prepared by the chairs of the session as well as the contributions from several of the participants.

Keywords. research, innovation, scientific literacy, inclusion, Astronomy education

---

1. Introduction to the roundtable

To start the discussion and exchange of ideas, the chairs of the roundtable proposed some initial thoughts that we describe in this section.

1.1. *Research*

We wish to tackle the following questions: (1) Who conducts the investigation. (2) On what topics is research conducted in Education/Didactics of Astronomy. These questions are being addressed to:

- Teachers and professors in their own professional activity.
- Teachers and professors who work with researchers in didactics.
- Researchers in didactics who do experiments in critical mode.
- Astronomers who work with researchers in didactics to transform the knowledge they produce in Astronomy (new discoveries, formulation of new hypotheses, questions and doubts about previous theories) into knowledge for the teaching of Astronomy.

Research on Didactics/ Education is different from the outreach activities addressed to the general public, and dissemination. These researchers are interested in the thoughts, misconceptions, difficulties, and challenges that the learners face. This also includes their initial hypothesis, which can be locally true only, their questions and their partial answers.

But above all we need to focus on the processes that lead to the construction of knowledge: the tools, words, models and activities that help, or those that confound them or prevent them from learning. And this is not something that worries professional astronomers, disseminators or amateurs.

As researchers in didactics we have learnt that a crucial problem in Astronomy is to reconcile what we see, what we perceive and what we think is the truth.

We have learnt, from the research on Didactics of all the scientific disciplines conducted during the decades from 1970 to 1990 that the pre conceptions of the learners are crucial, be them children or adults. We study how to detect them but not only through tests or the use of words (oral registry) but also by means of iconic, gestural and symbolic registries, while remaining open to the ones we are not yet aware of and are brought in by the students. There is research about the conceptions that are already built but find obstacles, difficulties that can be epistemological, perceptual, conceptual, and research about the learning progression at different ages and at different levels of knowledge acquisition.

There is also research related to the history of sciences and to the different cultures in our planet, in different epochs and different places: the mistakes, hypothesis, discoveries, formulation of "laws"; the presence/absence of men and women; the existence sin antiquity of scientific communities – Alexandria's library and museum in Egypt, Bagdad's library-.

Other research focuses on the differences related to gender, social background – in particular between countries from the North and South hemispheres. One example is the Globolocal Project (www.globolocal.net).

1.2. *Innovation*

In the literature, most of the definitions of "Innovation" include two keywords: change and improvement through something new. But new does not mean necessarily better than what there was before, nor does it mean good in itself. New can be new locally, for specific people at a given time. Perhaps what one (teacher, researcher) thinks is new (because it is new for him, or because he has thought about it and found it) is not new for everyone, in all cultures and in all periods: but it is still an enrichment for your own activity.

Innovation is not the indiscriminate use of technological tools, either. as it can be done while still being attached to an obsolete pedagogical framework. This is particularly true during the current pandemic times when everybody is using IT resources (meeting platforms like ZOOM or MEET, electronic devices, virtual tools) at all education levels. Their use does not ensure an improvement from a pedagogical and didactical point of view, it depends on the use the teachers make of them. In fact, it might well be more innovative to learn by observing directly the natural world, our own surroundings or the night sky.

This leads us to consider which are the desirable characteristics that a particular educational innovation action must comply with. For example, it has to be sustainable: it should be applicable even after the initial funding has been spent, the experience published in a professional journal or the initial time scheduled for it. It should be transferrable: the action should be possible to use outside the context in which it has been developed. Finally, it must take into account the efficiency with which the learning is attained, like whether the learning goals are reached, and whether the students get better results thank previously. The goal of any innovation in education is to facilitate the adaptation to the societal changes regarding the knowledge, technology, information, new languages, communication and scientific research. But is should also help to include everybody: it must serve as an inclusion tool.

1.3. *Scientific literacy*

Astronomy is a scientific discipline whose activity is not directed nor its specificity is directed at education, or learning or teaching, or to link the construction of knowledge that it generates in relation with the sky with the daily lives of people. It is a professional activity that does not require that connection. With regard to teaching-learning, there are two international currents, very different at the deepest level:

• A current based on information and content: a relationship is not sought between what is read, heard, known, and what is seen and experienced directly by looking at the sky, but using books or optical instruments and models instead. This modality of Astronomy teaching is widely spread, especially in the formal context of education, at all levels, including university and the training of specialists in Astronomy.

• By contrast, the other current is essentially concerned with the connection of people with the sky, mainly through observation from their own place, with strong ties to social, affective, cultural aspects (different according to latitude and longitude, the situation of the country, among other aspects), on what can be observed in a given moment, putting in relation the Earth and the sky. All this helps the autonomous and open construction of questions of those who learn, in many and diverse levels of understanding, and with great depth in the space-time dimension of phenomena, using especially the eyes and the body to observe and learn directly, in addition to the use of recording and measurement instruments, generally simple and inexpensive.

What makes the difference between both currents is how and for what purposes it is observed, since astronomy essentially requires observation. It is not the observation activity itself that characterizes each one, but the function it fulfils for the construction of knowledge in both streams.

1.4. *Inclusion*

Inclusion, in a broad sense, is a concern that is present at the IAU in all aspects of research, innovation and scientific literacy. With this theme, we intend to reflect on social inclusion of different people regardless of their wealth, the regions of the world where they live, the type of work they do, their mother tongue (specially in the case of non-English speakers who work in science), their gender, the technology to which they have access, their use of the physical senses, the possibility of using their body, their health, among many other possible factors that generate differences, exclusion, violence and discrimination in the world.

Therefore, research, innovation and scientific literature must be centred on each individual student to allow them to reveal their potentialities. One way to achieve this is through the Universal Design of Learning (UDL) framework which is based on the premise that everyone learns in a different way, has a specific personal style of learning. UDL proposes strategies and technologies that take into account these multiple styles.

2. Contributions from the participants in the round table

2.1. *Identifying the accessibility of educational resources, by Stefania Varano (INAF)*

We present here a study by the Inclusion Group of the National Institute for Astrophysics in Italy on a list of tags and keywords for the definition of specific inclusive features of educational activities, in order to identify the inclusive potential of activities available on several platforms for astronomy education. Tags, as icons, represent in brief the accessibility of specific resources. Such awareness allows to select what is most suitable to oneself and to build ones own customized path and experience.

These tags are indented to highlight characteristics of activities that are efficient in some contexts and with some specific needs. They have been identified on the one hand on the basis of the self-evaluation of our experience; on the other hand this list results from the advice of experts working with (or representing) disadvantaged groups and/or Special Educational Needs (SEN). Some examples:

- "Short and simple texts": activities that make use of little text and simple language, also accessible to non-native speakers and people with some speech disorders.
- ÒSafe Materialsó: activities that make exclusive use of non-sharp, non-hazardous materials, which can also be used in prisons and with young children and some people with cognitive disabilities.

2.2. *Astronomy Heritage in Education, by Alessandra Zanazzi (INAF)*

This is a project that plans to involve students in the discovery of astronomical sites and the history of astronomy in the cities. It combines astronomy, history, art, history of science, philosophy, development of thought and technologies in a truly transversal way. It shows how astronomy and the relationship with the sky have always been part of culture, traditions and development of people being very much integrated in monuments, ancient churches and works of art in every historic city.

The project starts from the realization of astronomical walks in the city with students and from the detection of the success, collected even with young people, of initiatives such as the observation of the passage of the sun on monumental sundials at solstices or equinoxes, or the use of ancient instruments, armils and Zodiacs. We have also written a tourist-astronomical guide (for Padua, Florence and Palermo), easy to use, non-specialistic, with an attractive graphic; also the guide-book has a map and itineraries on different themes, which helps to explore the places with an astronomical perspective.

2.3. *Use of technology in teaching, by Sara Ricciardi (INAF – OAS)*

The idea I'd like to bring to this roundtable is about the use of technology in teaching. It is often superficially thought that bringing a robot or a device into the classroom is equivalent to doing an innovative teaching, as if bringing a pen into the classroom would make children become writers. Technology is a powerful, interesting and natural instrument to learn about science first of all because it is an instrument of science itself. Nonetheless the use of an instrument does not warrant a personal and significant learning. Technology could be a means of expression enabling creativity in the classroom but could be used as a dull and unexpressive tool: the same difference you can see between a free writing and a dictation. Pushing the example even further, should you keep from teaching children to write because you think they won't be writers or journalists when they grow up? That is why it is important to recognize technology as a means of expression able to transform children from users to creators in our societies where children are already soaked in technology.

References

Lanciano, N. 2020, *Avec curiosité, s'interroger sur les erreurs des élèves. P 57-64 ; In : Apprendre par lerreur, direction de M.Graner, A.Giordan, Chronique sociale Ed. Strumenti per i giardini del cielo*

D'Addio. A.C., & April, D. 2020, *GEM report UNESCO: teachers need training on inclusion*, https://gemreportunesco.wordpress.com/2020/10/05/teachers-need-training-on-inclusion/ (retrieved 11 January 2021)

Globo Local Project oficial web site: www.globolocal.net

Education and Heritage in the era of Big Data in Astronomy
Proceedings IAU Symposium No. 367, 2020
R. M. Ros, B. Garcia, S. R. Gullberg, J. Moldon & P. Rojo, eds.
doi:10.1017/S1743921321000491

Students' Preconceptions in Astrophysics – How to break them down?

Corina Lavinia Toma[iD]

Computer Science High School «Tiberiu Popoviciu», Cluj-Napoca, Romania
email: corlavtoma@gmail.com

Abstract. At any stage of education, the students begin to study astrophysics with previous incorrect preconceptions that impedes the understanding of new scientific notions. It is a waste of time to add new concepts on a "weak basis". In this case the duty of any teacher is to spend more time to carefully remove preconceptions from students' minds. In the present article there are reviewed some of the most common preconceptions of the students when they are learning astrophysics with the right solutions to break them down using some active teaching methods, especially experiments and models.

Keywords. Education, gravitational lensing, solar wind, stellar color

1. Introduction

The term preconception refers to the previous knowledge that the student has gained from his own experience, before starting to study a new subject. This includes the ideas and concepts that each student has reach from the environment. Some of these preconceptions are often incorrect and even illogical that create a real problem for science teachers in their didactic approach. Usually when we say preconception we mean a wrong idea about something. "Science teachers, even more than others if that is possible, don't understand that one doesn't understand", Bachelard (1967). When the students begin to study astrophysics they have previous preconceptions that impedes the understanding of new scientific notions. Too much effort is spend to add new concepts on a "weak basis", so it is the duty of any teacher to spend more time to carefully remove preconceptions from students' minds.

2. Content

In the present article there are reviewed some of the most common preconceptions of the students when they are learning astrophysics with the right solutions to break them down. There are indicated some active teaching methods, especially experiments and models, as provided K. Tobin in his book about constructivism, Tobin (1993).

First preconception: How does the light propagate?
Usual answer: The light propagates rectilinearly. The student, even the one in the high school has this opinion after his direct observations and also from the experiments and theory of geometrical optics. Scientific answer: The light propagates curvilinearly. When the student studies the electromagnetic waves and he finds out about the momentum of the photon, he can understand that the photon has a kinetic mass and only in this case he realizes the notion of the curved trajectory of the light. The scientific answer comes from the quantum optics.

Figure 1. The model for Einstein cross.

To reinforce this new concept of propagation of the light on a curved trajectory we propose to our students a contest: which group of three students makes the most interesting and correct presentation of the famous Edington's experiment in 1919, as the first prove of Einstein's theory of general relativity. We give them the bibliography and the next hour, when they present their works we can discuss also about Sun total eclipse and the space curvature. To motivate the students to be curious and to study the astrophysics with pleasure, we introduce a tricky concept the gravitational lensing, where a massive cosmic object focuses light from another object beyond it to produce a distorted or magnified image. This is the same effect studied by Edington, but in this case the deflection of the light is made by any kind of star, a constellation or a galaxy. The astronomical discovery of gravitational lenses in 1979 gave additional support for general relativity.

To explain better this concept we use 3D videos simulations or drawings, e.g. ESA simulation [https://www.spacetelescope.org/videos/heic1106a] and then we made with the students simple experiments, we learned during the courses of the Network for Astronomy School Education (NASE). These experiments are explained in the book "14 Steps to the Universe", Moreno *et al.* (2017). Every teacher can reproduce the models for the Einstein's cross (Figure 1) or the Einstein's ring after watching the corresponding videos [https://www.youtube.com/watch?v=v1SdQE1pZZo], using very common materials.

Second preconception: What is between the Sun and the Earth?

Usual answer: ... there is vacuum or nothing. If the light can come so big distance from stars to us, the students believe that it must be nothing in space between stars and the Earth. The scientific answer: there are clouds of interstellar dust and the solar wind (electrons, protons and few ions).

The solar wind is invisible and the students doubt it is something real after all explanations about the Earth magnetic field and the charged particles of this wind and the solar composition [http://solar-center.stanford.edu/FAQ/Qsolwindcomp.html]. In the moment of the discussion about the northern light or aurora borealis their ideas begin to change. There are so many great photos, videos with this very nice phenomenon so the students are very motivate to find out that the aurora is the result of collisions between gaseous particles in atmosphere with charged particles from the solar wind. The students' question is: why the huge curtains in the sky have different colors and the they have to find the answer on internet, as homework. Another question for a project is: do other planets have auroras and if yes, what are the differences between them?

Third preconception: Does the Sun move?

Usual answer: NO! The students consider that the planets move around the Sun, but the Sun is at rest; even when the students know that there is no celestial body at rest, they cannot imagine the entire Solar System in movement. This error comes from the fact that the representations of the Solar System are drawn in plan not in space. The scientific answer: Our whole Solar System orbits around the centre of Milky Way and in the same time the Sun rotates around its axis.

There are videos with models in 3D with the trajectories of the planets around the Sun, meanwhile our star is moving in the galaxy [https://www.youtube.com/watch?v=rQJDEhlE-DY]. In order to emphasize the real rotation movement of the Sun around its axis there is a NASA simulation [https://solarsystem.nasa.gov/solar-system/sun/in-depth/] and a NASE experiment with the solar sunspots in the same book "14 Steps to the Universe", Costa *et al.* (2017). The students are very astonished when they realize that Galileo Galilei has calculated with centuries ago that the Sun rotation period decreases from the Equator to the Poles, using only the observation of these sunspots. To be sure that the students will retain the concept of the movement of the Sun in space it's a good exercise for them to compare the values of the revolution velocities for the Earth and the Sun.

Fourth preconception: What percentage of the entire Solar System mass is the Sun mass?

Usual answer: The Sun mass represents any value between 10% to 70% of the entire Solar System mass, but more often the students choose the value of 50%. Students believe that the Sun has a mass about as large as the entire mass of the rest of the Solar System because the figures about the Solar System can't be made to scale. The real value is 99.8%. To remove this preconception we use a "sweet" experiment with 100 pieces of cubic sugar on a tray. We ask the students to put aside the sugar that represents the mass of the Sun. It is an experiment very interesting for young students. We take care to ask those who know the answer to come into play. Students train and their curiosity grows as the answers are more and more varied. When they find out that only a small part of a sugar cube remains for the entire mass of the Solar System, most exclaim: it is not possible, how can it be so? Then it follows the explanation of the law of universal gravitational attraction between any two bodies.

There are many other students' preconceptions in astrophysics learning. Here we will review shortly only two: First, the hotter stars are the red ones: the students associate the red color with the wood fire and they deduce that the hottest stars are the red ones. They don't think that the flame of methane gas is blue and they don't know that it's temperature is higher than that of the wood fire. They understand everything after we teach them the Wien law, the stars life and the interpretation of the Hertzprung-Russel diagram. Many years, our students achieved a model of HR diagram we learned in 2007 at the First ESO-EAAE Astronomy Summer School, Radeva (2007). Second, the galaxies are moving through the Universe: the students believe that the galaxies move through the space but it is the space which expands, dragging the galaxies. A very simple experiment with a model to "see" this expansion using a balloon and some glued "paper galaxies" on it (Figure 2), is explained in the same NASE book "14 Steps to the Universe", see Figures 8a and 8b in Moreno *et al.* (2017).

Figure 2. The "paper galaxies" are not moving in the space.

3. Conclusion

Preconceptions are very deeply imprinted in the student's mind. Every teacher must identify the preconceptions, to break them down and then to reconstruct the concepts using especially active teaching methods: experiments and models. This educational approach means many hours of lesson preparation and a great teaching experience.

References

Bachelard, G. 1967, *La formation de l'esprit scientifique, Librairie philosophique*, J. VRIN, 5e edition

Tobin K. 1993, *A Practice of Constructivism in Science Teaching*, Lawrence Eribaum Associates, Inc. Publishers, page XV, ISBN 0-8058-1878-2

Moreno, R., Deustua, S. & Ros, R.M. 2017, *Expansion of the Universe* in: Ros, R.M. & Hememway, M.K. (eds.), "14 Steps to the Universe", Second Edition, p. 130–141 , ISBN: 978-84-15771-46-3

Costa, A., García, B. & Moreno, R. 2017, *Solar Spectrum and Sunspots* in: Ros, R.M. & Hememway, M.K. (eds.), "14 Steps to the Universe", Second Edition, p. 98–107 , ISBN: 978-84-15771-46-3

Radeva, V. 2007, *What Makes a Star so Special*, 1st ESO-EAAE Astronomy Summer School, Garching 2007, page 103

Education and Heritage in the era of Big Data in Astronomy
Proceedings IAU Symposium No. 367, 2020
R. M. Ros, B. Garcia, S. R. Gullberg, J. Moldon & P. Rojo, eds.
doi:10.1017/S174392132100065X

Meet the skies of the world first intercontinental-experimental course of teacher training in astronomy, of a cooperative and participatory type, in a time of physical distancing

Nicoletta Lanciano[1] and Néstor Camino[2]

[1]National Coordinator of the Research Group on Sky Pedagogy of the MCE (Movimento di Cooperazione Educativa), University "La Sapienza", Roma, Italia.
email: nicoletta.lanciano@uniroma1.it

[2]National Coordinator of NAEC Argentina – OAE IAU Complejo Plaza del Cielo.
CONICET-FHCS UNPSJB., Esquel, Patagonia, Argentina.
email: nestor.camino.esquel@gmail.com

Abstract. For over 30 years, the MCE Sky Pedagogy Research Group (Italy) and Plaza del Cielo Complex (Argentina) had been offering teachers and educators many projects and activities related to training and teaching/learning processes having the sky and the study of the multiple relationships of humans with the sky as a focus of research. We have developed didactical methods based on direct experience, naked-eye activities, cooperation and exchange of experiences. The pandemic forced us to suspend the in-person meetings, which are at the center of our research in the Teaching of Astronomy, so we decided to react constructively exploring virtuality designing the course "Meet the skies of the world". Two courses where developed from June to September 2020, each one having 4 virtual meetings of 2,5 hours each, with more than a hundred participants from different countries. A Level II course will be developed during the first semester of 2021.

Keywords. Naked eye observation. Collaboration. Teachers training course. Didactics of Astronomy. International.

1. A new way to teach Astronomy to overcome the COVID-19 pandemic

Over the years, as the result of many researches on the Teaching of Astronomy, we have developed didactical methods based on direct experience, naked-eye activities, cooperation and exchange of experiences, and we have proposed, in the city or in nature, residential training workshops in which many teachers and school educators, of all levels and various disciplines from different continents and countries of the world, had participated (Lanciano 2019; Lanciano et al, 2008; Camino 2012).

The suspension of all meeting activities during the months of the pandemic forced us to suspend the in-person meetings and in particular the residential courses. So, we asked ourselves what positive answer we could try to give. We reacted constructively and creatively, helping us to "travel and meet" under the skies of different countries of the world in a virtual mode, a resource that seemed to be the exact opposite of our

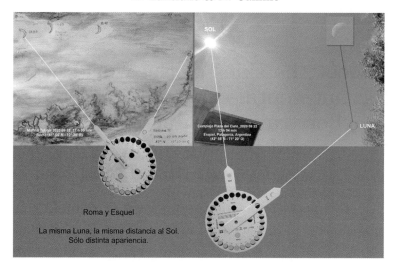

Figure 1. Same Earth, different locations, different skies.Comparison of observations of the
Sun and the Moon from North-South and East-West locations.

customary practice, which was based on in-person contact, on direct experience, on acting
and reasoning together.

2. Meet the skies of the world: a virtual, collaborative, international course of Didactics of Astronomy

We decided to explore the virtual possibility to meet teachers from distant countries
and cultures. So, the course "Meet the skies of the world" was born, proposed by the
Italian Group and organized in collaboration with friends and colleagues from Argentina,
Colombia and Brazil, open to participants from other countries in Europe and America.

This "experiment" was possible, and had a great pedagogical coherence, thanks to an
active collaboration that spans many years through the common International "Globo
Local" Project (www.globolocal.net, Lanciano 2012; Lanciano et al, 2019; Rossi et al,
2015). The course had seven coordinators/teachers and researchers: Nicoletta Lanciano
(Italia), Néstor Camino (Argentina), Rita Montinaro (Italia), Marina Tutino (Italia),
Elisa De Sanctis (Italia), Telma Cristina Dias Fernandes (Brazil), and Liliana Piragua
(Colombia).

3. Meet the skies... Level I

Two courses where developed, having 4 virtual meetings of 2,5 hours each, during
two months. The first one was during June-July (including a solstice); the second during August-September (including an equinox). Between the meetings, an equal number
of hours were planned in order to develop individual work to observe the sky (observing, registering and making drawings and photographs of the position of the Moon and
the Sun in the local sky, stars and planets, among others, see Fig. 1), to construct the
proposed didactical tools (local horizon representation, Sun-Moon goniometer, celestial
planispheres, Earth Parallel Globe), and to study, discuss and exchange their observations, questions, and didactical reflections. For instance, one teacher finds, among the
strengths of the course that "insisting on the use of physical tools and objects, which in
high school are always too denigrated and considered childish" was very important during
the course. Comparing observations of the same phenomena (Moon phase, for instance)
from different latitudes and longitudes, led us, each time, to question ourselves about

variables and constants in the sky: what is part of our identity, what is a common feature for everyone because we share the planet, and its impact on Culture and Education. The 4th meeting was dedicated to discussing and evaluating the process developed, and to share future experiences according with each participant interests, student's groups, locations, etc. More than a hundred teachers from Italia, Brazil, Colombia, Argentina, many also living in USA, Sweden, Germany and Portugal, participated in the courses: feeling differences in the sky because our longitudes (the Moon culminating for ones, just rising for others) and latitudes (winter for ones, summer for others).

4. The key items of "Meet the skies of the world" course

Our peculiar and essential activities of observation of the sky by naked eye. Also in a physical distancing course, we proposed to observe the sky, to look through the window to see if the Moon rises during the meeting time, etc. Everything in the course tries to lend support for real observation in the best possible way by means of sharing registers and discussions in plenaries. As stated by Ovide Décroly "Observing is more than perceiving ... means making comparisons, noting global or individual differences and similarities; observing means building a bridge between the world and thought". In the words of a participant "I am passionate about using educational science experiments as a means of integration and interaction. I realize that we have the best experiment available for this: our sky". The sky is beautiful and free, for everyone, and a significative tool to comprehend space and time from any place and in any date.

High presence of cultural links. We propose a high cultural level but not a difficult one, and with clear references to various cultures, languages and historical periods: tales of myths and traditions linked to the sky, with references to the history of science, anthropology, general culture, etymology of words related to the sky, and to figurative and poetic art. Furthermore, we have songs that accompany observation, the expectation of a phenomenon, that support the memory of what is being known and that express the meanings in poetic form. They are songs, for example, with words by the Italian poetry Gianni Rodari on the Moon, or the ancient Greek Empedocles on the elements of the Universe, or composed by some of us in the Group on Sky Pedagogy and others by some participants, or from popular traditions of countries.

To practice a multicultural perspective: how different it is to live in the South or in the North of the world! For example, only between the Tropics there are two days in the year with the Sun at the Zenith; only in some region of the world there are four "astronomical" seasons, and in others there are only two seasons like someone from Tanzania told us: "we do not experience the four seasons taught in books, so learning is just by cramming from books without understanding because we have only three seasons -hot, rain, cold- or in some region only two -hot and cool- ". Every aspect linked to the sky -time zones, seasons, phases, the length of the day- are well known to all "in theory", but experiencing them and meeting them in the words, drawings and photos of someone with whom you speak directly, virtually this time, causes a different emotional attention that strengthens awareness and radically imposes itself on memory. We have added to this the exchange of traditions linked to anthropological aspects that characterize the periods of the year and the deadlines of the calendars: the group thus experienced how, for example, the June Solstice is Summer for the Northern hemisphere but could bring snow to the Southern hemisphere.

We work in contrast to the standardization (of a unique language imposed on everyone in the world and of unique methods of explanation based on texts, without objects, with the word given only to those who "teach") for a conscious scientific citizenship. So, all the meetings and the chats were developed in Latin languages: Italian, Spanish and Portuguese (as is the tradition of the international meetings of teachers of Freinet pedagogy from all over the world, in which everyone speaks their own language and everyone collaborates by speaking slowly to help each other understand and communicate).

5. WHAT IF..., IN 2021? WE ARE PREPARING "MEET THE SKIES... LEVEL II"

We've begun to organize a Level II course, inviting those teachers who participated in both Level I courses in 2020; it will have more complex observational activities with specific problems to solve, making explicit what is constant and what changes because of the different topocentric positions and times of observation.

Furthermore, it is in process one of the outcomes of the two courses: the organization of some exchanges of Scholastic Correspondence, a Freinet technique, already widely experimented between Italian and Argentine teachers on sky observation, and between Italian and Brazilian teachers on the use of the Diary of the Sky (Nardi et al, 2020), taking into account the related school levels and the differences in the organization of the school year, in addition to remote face-to-face activities centered to share observations of the Moon, night sky and of diurnal phenomena such as shadows, Earth Parallel Globe and solstices and equinoxes.

We all want to overcome the global situation caused by the COVID-19 pandemic, but anyway we will continue working to share our skies, and to learn and teach, in peace, for a better future world. The Teaching of Astronomy from a multicultural and collaborative perspective is a wonderful way to do this!!

References

Camino, N. (2012). "La Didáctica de la Astronomía como campo de investigación e innovación educativas". En Bretones, Paulo (compilador), Actas electrónicas del I Simpósio Nacional de Educação em Astronomia (SNEA I). Rio de Janeiro, Brasil.

Globo Local Project oficial web site: www.globolocal.net

Lanciano, N and Camino, N. (2008). "Del ángulo de la geometría a los ángulos en el cielo. Obstáculos para la conceptualización de las coordenadas astronómicas". Enseñanza de las Ciencias, 26 1, Pp. 69 a 82. Barcelona, España.

Lanciano N., Giordano E., Rossi S., Berardo M., (2012) "La sphére de la Terre entre astronomie et géométrie – Le Projet Int. Globo Local, Poster, Actes EMF2012, Enseig. des mathématiques et contrat social Enjeux et défis pour le 21°s., Ginevra (Svizzera).

Lanciano, N., (2019), Strumenti per i giardino del cielo, IV ed., Ed. Asterios, Trieste.

Lanciano, N., Mautone, O., Montinaro, R. (2019). Laboratori sulla Luna a 50 anni dal primo sbarco umano, L'insegnamento della matematica e delle scienze integrate. Vol.42 A n.2, p 109–138.

Nardi, R., T. C. Fernandes, T. C., Lanciano, N. (2020). "Research about the adaptation process of Astronomy didactic material – the Diary of Sky – from the context of the Northern Hemisphere to the Southern Hemisphere", International Conference on Physics Education (ICPE) Co-editors Deena Naidoo and Douglas Clerk, Johannesburg, South Africa 2018 Journal of Physics: Conference Series 1512.

Rossi, S., Giordano, E., Lanciano, N., (2015), "The Parallel Globe: a powerful instrument to perform investigation on Earth's illumination", http://stacks.iop.org/0031-9120/50/32 Phys. Educ. 50 (2015) 32-41.

Session 4: Astronomy in other disciplines to promote science vocations

Education and Heritage in the era of Big Data in Astronomy
Proceedings IAU Symposium No. 367, 2020
R. M. Ros, B. Garcia, S. R. Gullberg, J. Moldon & P. Rojo, eds.
doi:

Introduction

This section begins with the invited talk "Stellarium: Simulation for Research and Outreach" by Georg Zotti. In his presentation Georg describes the history of Stellarium's development and the value that it provides. Here you will see him discuss capabilities of the open-source program and he provides good example. He talks about its planetarium-like utility and work for the future.

Following the presentation John Briggs asked:

The latest versions of Stellarium seem to have more demanding requirements for graphics and will not run, for example, on some older laptops. Can you explain about the advantages of the newer graphics requirements, as I am sure exist? (Stellarium is one of our very most wonderful current tools for astronomy education!)

Georg responded:

Stellarium is currently based on Qt5 which requires OpenGL 2.1. Some graphics simply requires support for OpenGL shading language. If you want a nice and modern simulation, old computers are just that, sorry: too old. But the current Stellarium runs on a tiny Raspberry Pi, so I think the requirements are not high. Any computer with Intel Core-i 2xxx or later should work well, or any earlier with a Geforce 8000 and later (2008?). For older laptops, you can still run version 0.12.9 or so. It is based on Qt4. However, the latest developments in historical accuracy are not included. But I used such an Atom netbook (built 2010) to track my telescope in a stay in Namibia

Anahí Caldú asked:

Georg, thank you very much for the nice talk and to all the Stellarium team for such a great work! I imagine everything is done in a voluntary way? Is it possible to raise funds somehow to hire more developers?

Georg answered:

Unfortunately... I could use full time work hours while developing this exhibition to advance precession/nutation and some other details. But most hours are our weekends and evenings. My observing time suffers...

Stellarium can be supported via OpenCollective donations (link on stellarium.org). But it's not enough to make a living from that. We use that for e.g. hardware replacements, Apple publishing licenses (!), open-access fees for publishing in qualified journals or as incentive for students working on some future additions.

I would like to be involved in a R & D program that would allow some months of concentrated development to bring this program further. We are still on version 0.* until we have fixed aberration and the Lunar axis orientation I mentioned. And then the Qt framework on which Stellarium is based just has moved on to their next decade with Qt6 (released this week), which will require considerable effort adapting, maybe until 2022, esp. w.r.t. scripting support and maybe even graphics.

Cristian Goez Theran asked:

Is there any way to indicate the vernal points (without having to indicate the ecliptic and celestial equator), as well as indicate the galactic center in Stellarium?

and Georg answered:

Equinox and solstice points: yes: Sky & Viewing options (F4), Markings tab, find Equinoxes/Solstices.

Galactic center: Not yet, but you are the first to ask for it. Should not be hard to do. :-)

Next is Chris Impey's invited "Online Resources for Astronomy Education and Outreach" where he talks about the many astronomy education resources now available on the internet. Chris describes the global role that the IAU plays in astronomy education with initiatives such as the International Schools for Young Astronomers (ISYA). He describes how access to resources such as outreach, textbooks, and videos has been digitally enhanced and expresses the value added to their utility through internet availability. Chris cites examples such as Stellarium and other platforms as well.

"Galaxy Forum South America-Argentina" by Steve Durst, Margarita Safonova, Santiago Paolantonio, Marcelo Colazo, and Geng Li. The authors outline activity of the Galaxy Forum South America 2020 that was held on December 8th, 2020 and offered by the International Lunar Observatory Association (ILOA) with support from the Instituto de Tecnologías en Detección y Astropartículas and the IAU. Forum topics presented here regard astronomy from the Moon, the history of constellations, UV observations from the Moon, CONAE deep space station activity, and the skies of ancient China.

Johanna Casado, Gonzalo de la Vega, Wanda Díaz-Merced, Poshak Gandhi, & Beatriz García contributed "SonoUno: a user-centred approach to sonification." The authors describe sonoUno for human-computer interface for astrophysical data access, collection, sonification, and analysis. A strongpoint of this software is that it was developed from the start to be centered on the user.

After the talk Tim Spuck asked:

SonoUno question: I'm PI on Innovators Developing Accessible Tools for Astronomy where we have developed the Afterglow Access software making image analysis accessible to BVI. SonoUno sounds very interesting. Are you doing any work with whole image sonification, or is it limited to sonification of data plots? When will SonoUno be available? Is it Web-based software or does it require an install on a computer? Would love to talk more sometime.

Johanna Casado answered:

SonoUno is available at github and it is not sonorixe images but data...I will send a more detailed answer this nigh, sorry...

The last paper in this section is called "SciAccess: Making Space for All" and was written by Anna Voelker, Caitlin O'Brien, and Michaela Deming. Anna tells about the SciAccess Initiative regarding the need for greater diversity and inclusion in STEM for scientists with disabilities. She gives an overview of the initiative's projects and outlines some of the inclusive practices that can help. She concludes by stressing the importance of engaging more of those who previously had been excluded from STEM.

Education and Heritage in the era of Big Data in Astronomy
Proceedings IAU Symposium No. 367, 2020
R. M. Ros, B. Garcia, S. R. Gullberg, J. Moldon & P. Rojo, eds.
doi:10.1017/S1743921321000752

Stellarium: Simulation for Research and Outreach

Georg Zotti[ID]

Ludwig Boltzmann Institute for Archaeological Prospection and Virtual Archaeology,
Hohe Warte 38, A-1190
Vienna, Austria
email: `Georg.Zotti@univie.ac.at`

Abstract. Over the past decade the free and open-source cross-platform desktop planetarium program Stellarium has gained not only most of the computational accuracy requirements for today's amateur astronomers, but also unique capabilities for specialized applications in cultural astronomy research and astronomical outreach. A 3D rendering module can put virtual reconstructions of human-made monuments in their surrounding landscape under the day and night skies of their respective epochs, so that the user can investigate and experience the potential connection of architecture, landscape, light and shadow, and the sky. It also played a key role in an exhibition about Stonehenge in Austria.

Exchangeable "skycultures" allow the presentation of constellation patterns and mythological figures of non-Western cultures. Stellarium's multi-language support allows community-driven translation of the whole program, which predestines its use in education also in minority languages.

Stellarium is developed by a very small core team, but is open to external contributions.

Keywords. Stellarium, virtual observatory, desktop planetarium, historical astronomy simulation, virtual archaeoastronomy, cultural astronomy, outreach

1. Introduction

In summer of the year 2000, Fabien Chéreau created a student project. His aim was to use modern 3D computer graphics to create a fast and realistic realtime simulation of the night sky. He decided to make this project and its source code available on the Internet, thereby attracting a handful of collaborators. By around 2006 they had raised the attention of a growing user community which earned them the title of "project of the month May 2006" on the SourceForge open-source code repository. Shortly after, Stellarium was used by ESO to easily access and browse VLT images (Kapadia et al. 2008).

The open-source nature of Stellarium also attracted the present author to use Stellarium for research and simulation in historical and archaeoastronomy Zotti & Neubauer 2011, 2015. However, the program at that time was not well suited for historical simulation, given some trade-offs in implementing simplified astronomical models in favour of high execution speed. Over the recent years, several of these issues have been solved, and several features unique to Stellarium now invite users and also researchers in the fields of cultural astronomy (archaeo- and ethnoastronomy) and history of astronomy to use it as tool for research and outreach activities.

2. Motivations for simulating past skies

There is countless evidence that the sky and its phenomena have inspired humans since earliest times (Ruggles 2015). Many earlier cultures have left traces in form of built monuments, and frequently those buildings have been erected with axes or other viewports aligned with simple but impressive celestial phenomena, like solstice sunrises or sunsets. Where those monuments are still at least partially preserved in ruins, we could visit those sites and aim at recreating the observations which caused the orientation or illumination effect in question, when sunlight may fall into a natural or artificial (or reshaped) cavity on particular dates of the year. However, there are several reasons why such observations nowadays are not really possible.

First, many archaeological sites are too fragile to be visited extensively, and well-known sites actually are in danger of destruction by overtourism. In many cases the best way to preserve an archaeological site from the elements after excavation is to carefully fill up the site again after all relevant data have been retrieved. Nowadays, documentation should include repeated 3D recording by laser scanning or photogrammetric modelling to allow recreating the excavation process (Filzwieser et al. 2016). In other cases, surface features have even vanished, and only subsoil features have been detected by aerial imaging and geophysical archaeological prospection methods like magnetometer or ground-penetrating radar surveys (Trinks et al. 2018).

The more relevant factor are secular changes in the sky. The slow changes in earth's axis tilt slightly change the solstice sunrise and sunset positions along the horizon, and precessional movement shifts the starry sky along the ecliptic, likewise changing rising and setting points of bright stars suspected to have been the target of some orientations.

Therefore, and also to make research results more accessible to a wider audience, a contemporary approach should involve computer graphics simulation which must combine an accurate astronomical simulation engine with some elements that represent the foreground, be it a simple panoramic photograph or rendering of one particular observing location, or a full three-dimensional simulation of a landscape with accurate virtual reconstructions of buildings through which the user can walk in virtual space, similar to the experience in first-person computer adventure games.

Since the 1990s desktop planetarium programs have become popular, and with advances in computer graphics, a few titles have allowed adding landscape panoramas to provide a better feeling of immersion into the observing location. Given their visual appeal and practical use, the programs are highly popular in the amateur astronomy community, and they are also frequently used for research in topics of cultural astronomy.

3. Stellarium milestones for simulation of historical sky vistas

The first group of Stellarium's developers have created a very realistically looking sky simulation based on several relevant studies from astronomical and computer graphics literature (Schaefer 1993; Preetham et al. 1999; Jensen et al. 2001; Tumblin & Rushmeier 1993; Devlin et al. 2002; Jensen et al. 2000; Larson et al. 1997). The program soon covered a wide spectrum of applications. One of the early developers added the functionality to show constellations of other cultures. He later forked off a branch (spinoff project) from which he developed a digital planetarium project from which further spinoff planetarium projects grew worldwide. Unfortunately such forks usually prevent changes made to one subproject finding their way back to the original project. The desktop version of Stellarium can be used in a planetarium dome either with fisheye projector or built-in predistortion for projecting onto a curved mirror, and it can run automated presentations programmed in JavaScript, however it has no further planetarium show-oriented infrastructure like a dome video player. Other developers concentrated on the application

Figure 1. Comet C/1858 L1 Donati (left) in a contemporary illustration (Weiß (1892), Fig. XX) and (right) simulated in Stellarium 0.20.4. Further tweaking could improve the match of tail curvature.

for observation support by creating program extensions (plugins) for simulated ocular and sensor views, and even for driving GOTO telescopes. Stellarium was then based on the Qt4 C++ toolkit, allowing the development of the program with an unconventional user interface which is however easy to operate and works on all three major desktop platforms (MS Windows, Linux/X11, Apple MacOS X). It also allows translation of the program into dozens of user languages, which is performed by several hundred voluntary collaborators utilizing a simple web interface. These factors, and obviously the free availability, have attracted the attention of a large community of users. Currently Stellarium is regularly released around equinoxes and solstices, and each version has seen several hundred thousand downloads.

Around 2010 I have joined the team of developers and since then added a few elements which improved its applicability for research in historical sky simulation. I have first participated in including atmospheric refraction and extinction, and while we started to develop a 3D foreground renderer (Zotti & Neubauer 2012a,b), others implemented the important correction for ΔT, the irregular slowdown of Earth's rotation, which had been missing from the program so far, but is essential especially for accurate eclipse simulation. Stellarium can meanwhile apply over 30 models for ΔT correction. However, users interested in historical eclipses must be aware that to our knowledge it is not useful to attempt eclipse simulation in the remote past and expect to derive solid conclusions, given that reliable observation reports only date back to about the 8th century BC (Stephenson et al. 2016), and simply extrapolating ΔT's parabolic trend may soon lead to many hours of error, which for Solar eclipses means that while the geographic latitude of the shadow's central axis may be correctly determined, the longitude can be wrong by tens of degrees. This is not a problem of Stellarium in particular, but an open question about Earth's rotation in general.

A major challenge was then posed by an upgrade in the underlying Qt C++ toolkit in 2013/14 after which Stellarium (version 0.13 and later) had to drop support for hardware not capable of providing a sufficient level of OpenGL or DirectX graphics functionality. Unfortunately, also most developers from the first group left the project shortly after these works, which has been maintained since around that time by Alexander Wolf.

This version also saw integration of a simple comet tail model mostly taken from previous work ((Zotti 2001; Zotti & Traxler 2003)) and based on own observations of C/1996 B2 Hyakutake and C/1995 O1 Hale-Bopp, consisting of two textured slim parabolas which model ion (straight, pointing away from the sun) and dust tails (curved, depending on solar distance and velocity), and the addition of an optional data file for more than 1000 historical comets (see Fig. 1) from several sources (Frommert 2014;

98 G. Zotti

Figure 2. An illustrative simulation of artificial light at night in Stellarium 0.20.4. The landscape panorama from Vienna's historical Kuffner observatory (late 19th century) was created by M. Prokosch and made available on the Stellarium repository of publicly available landscapes. It was modified by the author by adding a simple hand-painted layer of nocturnal illumination: street lights, bright windows and the light glow of the city of Vienna towards the east.

Yeomans & Kiang 1981; Mucke 1985). A week under the pristine skies of Namibia convinced me to implement a visualisation of the Zodiacal light for version 0.13.2 (Kwon et al. 2004).

In addition to the already existing simulation of global light pollution following the Bortle scale (Bortle 2001), an optional *localized light pollution layer* (see Fig. 2) was then added to the landscape foregrounds (Zotti & Wuchterl 2016).

The higher minimum level of graphics hardware required to run the program finally allowed us to integrate the Scenery3D landscape rendering plugin (Zotti 2015, 2016) and a further plugin which shows diurnal track (declination arc) visualisations for solstices, the "cross-quarter" days (between solstices and equinoxes), and the "lunistice" declinations relevant to many archaeoastronomical studies. Also azimuths to configurable locations like sacred mountains can be displayed.

A major leap forward in creating a reliable tool for historical simulation was the implementation of an accurate long-time model of precession (Vondrák et al. 2011, 2012) and IAU 2000B nutation (McCarthy & Luzum 2003) in version 0.14.0. (Unfortunately a sign error in the original formulation of the nutation matrix Hilton et al. 2006, eq.21) went undetected until reported by a user and finally fixed in version 0.20.2.)

Shortly after, the 3D visualisation found its first real applications by other researchers (Frischer et al. 2016, 2017), where we could identify the summer solstice sunrise orientation of the entrance axis in the so-called Antinoeion in Hadrian's villa in Tivoli near Rome.

Supported by ESA's *Summer of Code in Space* (SOCIS) programme, we could develop code to access the JPL DE430 and DE431 ephemerides (Folkner et al. 2014), the latter of which provides planetary positions into times as early as −13.000, which significantly extends the applicability over the time range recommended for the classic VSOP87 analytical solution (Bretagnon & Francou 1988) used by default.

In preparation for a major exhibition (below), we could put considerable effort into the program for some time. The scenery 3D plugin was made more efficient and meanwhile allows us to use Stellarium's time control to also make parts of the 3D foreground semi-transparent or invisible when they do not fit to the time currently set in Stellarium. This allows the simulation of evolving monuments, like phases of building and destruction (Zotti et al. 2018).

Many users appeared to have problems understanding the seasons beginnings and how calendar dates relate to them. The application of the Julian calendar for all dates before October of 1582 leads to a known drift of the dates for season beginnings against the "canonical" dates engraved in most people's minds (and in the Christian rules for Easter computation) of March 21st, June 21st etc. Of course, for dates in the 5th millennium BC, this calendar error pushes summer solstice deeply into "July", a named date that does not really make sense in the historical context. Other cultures also have developed their own calendars and have left observational reports recorded with them. The latest release (0.20.4) introduced the Calendars plugin which allows display and handling of various calendar systems in parallel. This plugin will be extended in future versions.

Two problems identified in the original implementation have then still persisted for a long time. The rotation and orientation of planet axes, best seen in the Moon showing part of its back face in early prehistory, will finally follow more accurate procedures (e.g. Urban & Seidelmann 2013) with release of version 0.21.0 in early 2021.

A final known accuracy issue that remains to be solved is the aberration of starlight, most noticeable in lunar occultation simulation. Only after solving this we will be able to assess overall acuracy of the simulation, and may have to find remaining errors.

Astronomical improvements not primarily aimed at historical applications included the optional rendering on non-spherical planetary bodies and the addition of planetary feature nomenclature labels, both of which were implemented by students again sponsored by ESA's SOCIS programme. Over the last years, Stellarium's maintainer Alexander Wolf has added a module named AstroCalc which provides a wide variety of ephemeris tabulation and visualisation options. We have also resurrected and extended the User Guide which meanwhile contains almost 400 pages (Zotti & Wolf 2020b).

A data set which has not been significantly changed over the last decade is the use of the HIPPARCOS (ESA 1997), Tycho 2 (Høg et al. 2000) and NOMAD (Zacharias et al. 2004) star data. While cross-identification with a few other catalogs was added and some data errors removed, Stellarium does still not compute stellar proper motion in 3 dimensions (only working with the linear proper motion components) and does also not simulate motion of binary star components around their common center of gravity. A handful of bright stars therefore deviate from applications which include those corrections (De Lorenzis & Orofino 2018). A future complete remodelling of the star catalog should of course be based on results from GAIA.

4. Application of 3D foregrounds

The visualisation of 3D models under the simulated sky allows an almost natural game-like experience while studying building axes, view corridors, or also the appearance of shadows (Fig. 3). A workflow involving GIS (Geographical Information System) software, a self-made data converter and the open-source Blender 3D modelling program has been presented (Zotti 2019; Zotti et al. 2019) which allows the creation of larger landscapes. The capabilities of the Scenery3D plugin have been tested with two datasets: a 3D laser scan from the Neolithic temple of Mnajdra in Malta (Hoskin 2001), and a LiDAR (airborne laser scan) based model of the Chankillo landscape in Peru (Ghezzi & Ruggles 2015). Both models are discussed in detail elsewhere (Zotti et al. 2020b).

The immobile rigid foreground rendered by the Stellarium Scenery3D plugin may not be enough in some situations when interaction with the scene is required, for example to study the operation of historical observational instruments. Therefore a connection

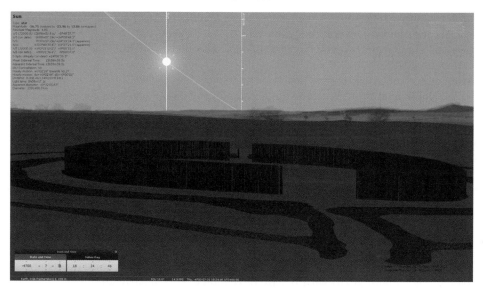

Figure 3. A simple virtual reconstruction of the Neolithic enclosure (*Kreisgrabenanlage*) Pranhartsberg 2 in Lower Austria shown with the Scenery3D plugin in Stellarium. Dozens of these monuments, consisting of circular deep ditches and palisade walls with usually 2 or 4 entrances, were erected in Neolithic Europe (ca. 4800-4500 BC), and archaeologists still discuss their purpose and use. Their traces can be usually detected only in aerial images and magnetometry surveys. One frequently discussed idea is the orientation of entrances towards solstices and other astronomically determined directions. The entrance on the far side shows a passage oriented towards summer solstice sunset, probably also visually augmented by an aligned pair of posts. The ArchaeoLines plugin indicates summer solstice declination and the Solar vertical. A vertical line indicates altitudes, and the Calendars plugin provides dates in proleptic Julian and Gregorian calendars, the latter giving a more intuitive date with respect to the seasons. However, this is the only such monument in over 30 studied in Lower Austria to show such orientation – the rest most commonly shows orientation following the terrain slope (Zotti & Neubauer 2015).

kit to the Unity game engine has been developed (Zotti et al. 2020a) which allows use of Stellarium as sky background renderer and provider of astronomical information for interactive, vivid and game-like applications.

The Stellarium team is willing to host interesting 3D models of astronomically relevant sites released under a permissive license, similar to the many conventional photographic landscapes contributed by users worldwide. However, while astronomical data usually can be freely accessed shortly after observation, keepers of accurate data of cultural heritage sites interesting to archaeoastronomical simulation unfortunately appear to generally not allow the release of such data to the public, even if that would certainly raise interest and awareness around such sites. We sincerely hope this statement can be disproved as soon as possible and robust, well-made, accurately georeferenced 3D sceneries made available.

5. Outreach: The skyscape planetarium

An exhibition project on Stonehenge in the MAMUZ museum for prehistory in Mistelbach, Austria, opened in 2016, provided developing time for some improvements. The exhibition showed replica of parts of Stonehenge's central "horseshoe" stones in original size. The surrounding circle of stones and the outer landscape was to be provided as seamless projection by 5 projectors on a 25m wide and 4m high canvas. For this screen form factor, Stellarium's capabilities of presenting a rendered sky view not only with the

Figure 4. The skyscape planetarium, a 25×4 m screen with 5 overlapping projectors providing the landscape backdrop for the central replica horseshoe of Stonehenge in an archaeological exhibition. The scripting functions of Stellarium were used for the narration in daylight hours, and a human operator could operate the program during special tours.

perspective or stereographic projections found elsewhere, but with several others, and especially cylindrical (or Mercator and Miller variants), proved essential.

An automated show should present relevant pieces of information and take visitors on a narrated tour through the history of Stonehenge and related sites. We did not use Stellarium's 3D capabilities but limited ourselves to pre-rendered artificial panoramas, with the sky background always provided by Stellarium. The changing hues of the sky behind seen through the stones still gave an almost three-dimensional impression. When fitting to the narration, solstice sunrises or diurnal tracks could be shown. Stellarium provides a scripting engine based on JavaScript for such shows, which can also play audio files or insert images onto the screen (Fig. 4).

The big advantage of using a "live" computer program over a pre-recorded movie however lies in the possibilities the program can provide for special occasions. On a few evenings, the museum offered flashlight tours for children who entered a dimly lit museum with a tour guide. At some point they woke up a "man from the past" who then joined the group and explained some exhibits in his own language, augmented by acting, music and dancing, all this while the stars of the unpolluted night sky twinkled through the Stonehenge panorama. At some point, twilight had to set in, and the group welcomed the famous summer solstice sunrise seen from within the central stones of Stonehenge with a little ceremony. The highly successful exhibition was extended for the 2017 season.

Operating Stellarium's user interface on the big canvas would have been distracting. We therefore developed the RemoteControl plugin which allows the control of the program with an HTTP control interface (Zotti et al. 2017). This means Stellarium can be controlled from a web browser, and enables a museum operator to control the program on a portable device like a simple tablet computer or even smartphone. Meanwhile several other projects are using this RemoteControl API for their own purposes to communicate with Stellarium.

6. Skycultures

Since early in its development Stellarium has allowed displaying various collections of non-Western constellations (termed "skycultures") and a text panel of background information. Constellations can be shown just as bordered regions in the sky (IAU borders), constellation stick figures, artistic renditions, or any combination. In addition, asterisms in the modern sense of non-official stick figures and "ray helpers", longer orientation lines, can be displayed.

Over the past years several users from non-Western cultures contributed new or improved existing skycultures, but it became clear that we will also need to extend the available functionalities, because several concepts like Lunar stations (highly important in many Asian cultures) or Dark Constellations (formed by dark clouds in the Milky Way) are currently not available. A more advanced solution for the skycultures which also must involve considerations about multilingual transliteration and translation is still an important area of future research and development (Zotti & Wolf submitted).

7. Other contemporary and scientific uses

Stellarium has a plugin for the display of meteor showers with data retrieved from the International Meteor Organisation (IMO). Another plugin can display artificial satellites. A temporary popular pastime was the observation of "Iridium flares", bright reflections of sunlight by flat antennas on the first generation of Iridium communication satellites, which could be predicted and impressively visualized until the satellites were taken down. Stellarium is capable of simulating lunar and solar eclipses with meanwhile pretty good accuracy, although a few more timestamped videos (from known sites) should be investigated before we can give a final assessment. Alexander Wolf has created several further plugins to show historical supernovae and novae, pulsars, quasars and exoplanet systems, and Guillaume Chéreau made the HiPS surveys (Fernique 2017) accessible for Stellarium. He also added an extension to show multi-resolution texture maps for the planets. A few examples for the simulation of transient astronomical phenomena were given earlier (Zotti & Wolf 2020a), and the benefits of showing historical (super)novae in the context of the reporters' skycultures is discussed in detail elsewhere (Zotti et al. 2020b).

8. Conclusion and further work

Stellarium provides several unique capabilities useful for astronomical teaching, research and outreach in the domains of astronomy basics, historical astronomy, archaeoastronomy and ethnoastronomy. Likewise it is widely appreciated in the amateur astronomy community as simple and accurate tool for presentations and preparation of own observations, and it can be used during observation for telescope control and ocular/sensor view simulation. It is used by many authors as de-facto standard tool for astronomical illustrations. The constellation artwork created by Johan Meuris for Stellarium has been seen in other places outside the program, just like a multitude of screenshots, unfortunately often without giving proper credits.

As open-source tool Stellarium grows mostly by the personal ambition, motivation and interests of its authors and contributors. Over the past decade, its accuracy has been significantly improved in some most relevant parts, but it is always advisable to compare critical results to those from other sources. Authors of scientific work are still advised to keep existing accuracy limitations in mind which we give in the User Guide.

The team accepts suggestions and code contributions on the project's Github website (Stellarium 2021), but pace of development depends on available resources. The team will soon have to face the next major upgrade of the Qt framework (Qt6) to keep Stellarium available for the next decade.

Acknowledgements. The Ludwig Boltzmann Institute for Archaeological Prospection and Virtual Archaeology (https://archpro.lbg.ac.at) is based on an international cooperation of the Ludwig Boltzmann Gesellschaft (A), Amt der Niederösterreichischen Landesregierung (A), University of Vienna (A), TU Wien (A), ZAMG-Central Institute for Meteorology and Geodynamics (A), 7reasons (A), ArcTron 3D (D), LWL-Federal state archaeology of Westphalia-Lippe (D), NIKU-Norwegian Institute for Cultural Heritage (N) and Vestfold fylkeskommune-Kulturarv (N).

References

Bortle, J.E., *S&T* Feb. 2001, 126

Bretagnon, P. & Francou, G., 1988, *A&A*, 202, 309

De Lorenzis, A. & Orofino, V., 2018, *Astronomy and Computing*, 25, 118

Devlin, K., Chalmers, A., Wilkie, A. & Purgathofer, W., 2002, in: *Proc. EUROGRAPHICS*

ESA (ed.), 1997, *The HIPPARCOS and TYCHO catalogues. Astrometric and photometric star catalogues derived from the ESA HIPPARCOS Space Astrometry Mission*, ESA Special Publication, vol. 1200

Fernique, P., 2017, *HiPS – Hierarchical Progressive Survey*, Tech. rep., IVOA

Filzwieser, R., Neubauer, W., Nau, E. & Toriser, L., 2016, *Archaeologia Austriaca*, 100, 199

Folkner, W.M., Williams, J.G., Boggs, D.H., Park, R.S. & Kuchynka, P., 2014, *The Planetary and Lunar Ephemerides DE430 and DE431*, IPN Progress Report 42-196, JPL/NASA

Frischer, B., Pollini, J., Cipolla, N., Capriotti, G., Murray, J., Swetnam-Burland, M., Galinsky, K., Häubner, C., Miller, J., Salzman, M.R., Fillwalk, J. & Brennan, M.R., 2017, *Studies in Digital Heritage*, 1, 1, 18

Frischer, B., Zotti, G., Mari, Z. & Vittozzi, G.C., 2016, *Digital Applications in Archaeology and Cultural Heritage (DAACH)*, 55–79

Frommert, H., 2014, Comet data of all comets up to 1994, including some predictions up to 2000, online at http://spider.seds.org/spider/Comets/comets.data.txt, retrieved 01/2014

Ghezzi, I. & Ruggles, C.L.N., 2015, in: C.L.N. Ruggles (ed.), *Handbook for Archaeoastronomy and Ethnoastronomy*, vol. 1, chap. 62, 807–820 (New York: Springer Reference)

Hilton, J.L., Capitaine, N., Chapront, J., Ferrandiz, J.M., Fienga, A., Fukushima, T., Getino, J., Mathews, P., Simon, J.L., Soffel, M., Vondrak, J., Wallace, P. & Williams, J., 2006, *Celestial Mechanics and Dynamical Astronomy*, 94, 351

Høg, E., Fabricius, C., Makarov, V.V., Urban, S., Corbin, T., Wycoff, G., Bastian, U., Schwekendiek, P. & Wicenec, A., 2000, *A&A*, 355, L27

Hoskin, M., 2001, *Tombs, Temples and their Orientations*, chap. 3: The Temples of Malta and Gozo, 23–36 (Bognor Regis: Ocarina Books)

Jensen, H.W., Premoze, S., Shirley, P., Thompson, W., Ferwerda, J. & Stark, M., 2000, *Night Rendering*, Tech. Rep. UUCS-00-016, Computer Science Department, University of Utah

Jensen, H.W., Stark, M.M., Premože, S., Shirley, P., Durand, F. & Dorsey, J., 2001, in: *Proc. SIGGRAPH 2001* (ACM)

Kapadia, A., Chéreau, F., Christensen, L.L., Nielsen, L.H., Gauthier, A., Hurt, R. & Wyatt, R., 2008, in: L.L. Christensen, M. Zoulias & I. Robson (eds.), *Communicating Astr. with the Public 2007: Proc. from the IAU/National Observatory of Athens/ESA/ESO Conf. 8-11 October 2007*, 220–224 (ESA/Hubble)

Kwon, S.M., Hong, S.S. & Weinberg, J.L., 2004, *New Astron.*, 10, 91

Larson, G.W., Rushmeier, H. & Piatko, C., 1997, *A Visibility Matching Tone Reproduction Operator for High Dynamic Range Scenes*, Tech. Rep. LBNL-39882, Ernest Orlando Lawrence Berkeley National Laboratory

McCarthy, D.D. & Luzum, B.J., 2003, *Celestial Mechanics and Dynamical Astronomy*, 85, 37

Mucke, H., 1985, *Die Sterne*, 61, 5/6, 276

Preetham, A.J., Shirley, P. & Smit, B., 1999, in: *Proc. SIGGRAPH 1999*, 90–100 (ACM)

Ruggles, C.L. (ed.), 2015, *Handbook for Archaeoastronomy and Ethnoastronomy* (New York: Springer Reference)

Schaefer, B.E., 1993, *Vistas in Astron.*, 36, 311

Stellarium, 2021, Github repository, online at https://github.com/Stellarium/stellarium

Stephenson, F.R., Morrison, L.V. & Hohenkerk, C.Y., 2016, *Proc. R. Soc. A*, 472, 20160404

Trinks, I., Hinterleitner, A., Neubauer, W., Nau, E., Löcker, K., Wallner, M. et al., 2018, *Archaeol. Prospection*, 25, 3, 171

Tumblin, J. & Rushmeier, H., 1993, *IEEE Computer Graphics & Application*, 13, 6, 42

Urban, S.E. & Seidelmann, P.K. (eds.), 2013, *Explanatory Supplement to the Astronomical Almanac*, 3rd ed. (Mill Valley, CA: University Science Books)

Vondrák, J., Capitaine, N. & Wallace, P., 2011, *A&A*, 534, A22, 1

Vondrák, J., Capitaine, N. & Wallace, P., 2012, *A&A*, 541, C1

Weiß, E., 1892, *Bilderatlas der Sternenwelt*, 2nd ed. (Eßlingen bei Stuttgart: J. F. Schreiber)

Yeomans, D.K. & Kiang, T., 1981, *MNRAS*, 197, 633

Zacharias, N., Monet, D.G., Levine, S.E., Urban, S.E., Gaume, R. & Wycoff, G.L., 2004, in: *AAS Meeting Abstracts*, BAAS, vol. 36, 1418

Zotti, G., 2001, *A Multi-Purpose Virtual Reality Model of the Solar System (VRMoSS)*, Master's thesis, TU Wien

Zotti, G., 2015, in: C.L. Ruggles (ed.), *Handbook for Archaeoastronomy and Ethnoastronomy*, vol. 1, chap. 29, 445–457 (New York: Springer Reference)

Zotti, G., 2016, *Mediterranean Archaeology and Archaeometry*, 16, 4, 17

Zotti, G., 2019, in: L. Henty & D. Brown (eds.), *Visualising Skyscapes: Material Forms of Cultural Engagement with the Heavens*, 35–54, Routledge Studies in Archaeology (Routledge)

Zotti, G., Frischer, B. & Fillwalk, J., 2020a, *Studies in Digital Heritage*, 4, 1, 51

Zotti, G., Frischer, B., Schaukowitsch, F., Wimmer, M. & Neubauer, W., 2019, in: G. Magli, A.C. González-García, J.B. Aviles & E. Antonello (eds.), *Archaeoastronomy in the Roman World*, chap. 12, 187–205, Historical & Cultural Astronomy (Springer International Publishing AG)

Zotti, G., Hoffmann, S., Wolf, A., Chéreau, F. & Chéreau, G., 2020b, *Journal for Skyscape Archaeology*, 6, 2

Zotti, G. & Neubauer, W., 2011, in: C.L.N. Ruggles (ed.), *Archaeoastronomy and Ethnoastronomy: Building Bridges between Cultures*, 349–356, IAU S278 (Cambridge U. Press)

Zotti, G. & Neubauer, W., 2012a, in: G. Guidi & A.C. Addison (eds.), *Proc. VSMM2012 (Virtual Systems in the Information Society)*, 33–40 (Milano: IEEE)

Zotti, G. & Neubauer, W., 2012b, in: M. Ioannides, D. Fritsch, J. Leissner, R. Davies, F. Remondino & R. Caffo (eds.), *Progress in Cultural Heritage Preservation*, LNCS, vol. 7616, 170–180 (Heidelberg: Springer)

Zotti, G. & Neubauer, W., 2015, in: F. Pimenta, N. Ribeiro, F. Silva, N. Campion, A. Joaquinito & L. Tirapicos (eds.), *SEAC2011 Stars and Stones: Voyages in Archaeoastronomy and Cultural Astronomy*, 188–193, no. 2720 in BAR International Series (Oxford: Archaeopress)

Zotti, G., Schaukowitsch, F. & Wimmer, M., 2017, *Culture and Cosmos*, 21, 1&2, 269

Zotti, G., Schaukowitsch, F. & Wimmer, M., 2018, *Mediterranean Archaeology and Archaeometry*, 18, 4, 523

Zotti, G. & Traxler, C., 2003, in: M.H. Hamza (ed.), *Proc. of the Third IASTED Int. Conf. on Visualization, Imaging, and Image Processing*, 964–969, IASTED (Benalmádena, Spain: ACTA Press)

Zotti, G. & Wolf, A., 2020a, in: M.T. Lago (ed.), *Astronomy in Focus: As presented at the IAU XXX General Assembly, 2018*, 184–186, IAU (Cambridge University Press)

Zotti, G. & Wolf, A., 2020b, *Stellarium 0.20.4 User Guide*, https://stellarium.org

Zotti, G. & Wolf, A., submitted, in: A.C. González-García, M. Rappenglück, G. Zotti et al. (eds.), *Beyond Paradigms (Proc. SEAC2019)*, BAR International

Zotti, G. & Wuchterl, G., 2016, in: F. Silva, K. Malville, T. Lomsdalen & F. Ventura (eds.), *The Materiality of the Sky: Proc. of the 22nd Annual SEAC Conf., 2014*, 197–203 (Sophia Centre Press)

Education and Heritage in the era of Big Data in Astronomy
Proceedings IAU Symposium No. 367, 2020
R. M. Ros, B. Garcia, S. R. Gullberg, J. Moldon & P. Rojo, eds.
doi:10.1017/S1743921321000399

Online Resources for Astronomy Education and Outreach

Chris Impey[iD]

Steward Observatory, University of Arizona, Tucson, AZ 85721, USA
email: `cimpey@as.arizona.edu`

Abstract. The growth of the Internet has facilitated the easy availability of resources for teaching astronomy and doing astronomy outreach. This overview concentrates on resources that are free or open access. Basic teaching materials like textbooks and lab activities can be found, along with higher level items such as concept inventories and interactive instructional tools. There is also a small but growing research literature on astronomy instruction to be found online. Astronomers engaged in outreach can have access to large image collections, tools for doing citizen science, and planetarium apps. These resources are of enormous value to both novice and seasoned instructors, and anyone conveying the excitement of astronomy to a public audience.

Keywords. Education, Outreach, Communication

1. Introduction

The Internet has transformed access to astronomy materials. Before 1995, instructors were mostly reliant on printed textbooks, 35-mm slides, and their own lecture notes. Depending on local resources, they might also be able to make use of labs and hands-on activities. Astronomers doing outreach could use small telescopes to show people the night sky, assuming they had access to a dark site, but otherwise the main way to communicate was a public lecture. Now, astronomers can choose from a wide variety of online resources to augment their teaching and inspire public audiences.

The utilization and availability of online resources for astronomy has been increasing steadily, but the trend accelerated as the world dealt with fallout from the COVID-19 pandemic. The impact of universities was profound, as the wholesale shift to online learning exacerbated inequalities of access (Marinoni *et al.* 2020). It also presented challenges for astronomy instructors, who had to sacrifice classroom interaction and hands-on activities for remote learning. The context for astronomy teaching is active learning, which has been definitively been shown to yield higher learning gains than passive methods like lecturing (Freeman *et al.* 2014). Astronomy instructors had to find modern teaching methods that can be implemented online. Outreach has also been affected by COVID-19, as star parties cannot be held safely, and science centers have had to shut their doors (Collins *et al.* 2020). A silver lining is that the pandemic has let astronomers address the carbon footprint of all their activities (Stevens *et al.* 2020).

2. The Role of the IAU

The International Astronomical Union has a unique role in astronomy education and outreach as the only organization that can harness networks and sponsor events with global reach. IAU Commission 46, Astronomy Education and Development, was set up in 1964, and in 1967, the IAU started its International Schools for Young Astronomers

Figure 1. The IAU Office of Astronomy for Education (OAE) operates the International School for Young Astronomers and has activities housed in the Haus der Astronomie in Heidelberg (top), while the Office for Astronomy Outreach runs the CAP Journal and the Communicating Astronomy with the Public conference series (bottom). Courtesy IAU.

(ISYA) program (Gerbaldi 2007). As of 2021, forty schools have been held around the world, targeting and reaching large percentages of women and students in developing countries. Apart from ISYA, for a long time, education did not have a major role in the organization. The first IAU Colloquium on astronomy education was not held until 1980, and by the time of the second, there had been a hundred IAU conferences on topics in research (Percy 1988). The Office for Astronomy Outreach (OAO) was established 20 years ago. Through the work of Commission 55, Communicating Astronomy with the Public, the first conference on Communicating Astronomy with the Public was held in 2005, and the Communicating Astronomy with the Public Journal started in 2007, with two issues each year since then (Fienberg *et al.* 2014).

The past decade has seen an acceleration of the IAU's commitment to education and outreach, with the impetus provided by widespread public interest in the International Year of Astronomy (IYA) in 2009. The IYA activities reached 815 million people in 148 countries (Russo & Christensen 2010). The Office of Astronomy for Development (OAD) was opened in South Africa in 2011, with a mandate to spur international development and expand astronomy education and outreach globally (Chapman *et al.* 2015). There have been an increasing number of IAU symposia and colloquia devoted to education and outreach, a new set of Shaw-IAU Workshops on "Astronomy for Education," and the launching of a biennial conference series on the theme of "Astronomy Education: Bridging Research and Practice." IAU education and outreach now has a physical home for the Office of Astronomy Education (OAE) at the Haus der Astronomie in Heidelberg, Germany. The IAU's new strategic plan (IAU 2020) invokes education and outreach in four of its five goals. Figure 1 shows some of the activities of the OAE and the OAO.

While most of the resources for astronomy education and outreach are in English, some are multi-lingual or address non-Western cultures. The IAU operates the Astronomy Translation Network, with 380 volunteers translating materials across 45 languages (Shibata & Canas 2019). Some of the popular Crash Course Astronomy videos have been translated into up to a dozen languages (Plait 2016). Also, Andrew Fraknoi has written and periodically updated a resource guide with hundreds of references and web links to astronomy in non-European cultures (Fraknoi 2019).

3. Teaching and Outreach

Almost every professional astronomer is engaged in some form of teaching and also does outreach. Astronomers who work in higher education teach as part of their job function, and there have always been good resources for helping them improve their classes and methods (Pasachoff & Percy 2005), often with a focus on introductory astronomy teaching in the United States (Prather *et al.* 2009; Waller & Slater 2011). Proceedings of IAU meetings contained 283 papers about astronomy education in an 18-year period (Bretones & Neto 2011), and an update was provided based using a recent survey of the members of IAU Commission C and its Working Group on Theory and Methods in Astronomy Education (Bretones & Neto 2011). From 2001 to 2013, research articles were published in Astronomy Education Review (Fraknoi 2014), and with the 2020 launch of the Astronomy Education Journal, astronomers once again have a vehicle for publishing research and sharing best practices. Recently, the Institute of Physics has published two ebooks that summarize online resources and methods for teaching introductory astronomy (Impey & Buxner 2019; Impey & Wenger 2019).

Astronomy is exceptionally well-suited to outreach, since there are many spectacular images of objects in the night sky, and the cosmos has universal appeal. The IAU has also recognized that outreach can facilitate sustainable development through the world (Guinan & Kolenberg 2016). It has evolved from just part of a scientist's duties to a distinct career path that is well-suited to astronomers (Cominsky 2018). Scientists do outreach even when their institutions do not reward or incentivize that work (Rose *et al.* 2020). Perversely, there was a sense that popularizing science might adversely affect a scientist's reputation, a phenomenon dubbed the "Sagan effect" (Joubert 2019). However, a large survey of IAU astronomers found outreach to be widespread worldwide, and largely immune from peer criticism (Entradas & Bauer 2019). Major observatories have multi-faceted outreach programs that can convey complex research to broad public audiences (Madsen & West 2020; Griffin 2003).

4. Textbooks and Instructional Materials

Introductory astronomy textbooks are mostly published to meet the demand of the roughly 250,000 students each year who take astronomy as an undergraduate science requirement in the United States (Fraknoi 2001). One free and open-source textbook for learning introductory astronomy is a project of the non-profit OpenStax program at Rice University. The book, which can be used online or downloaded in several formats, was written and vetted with the assistance of 70 astronomers (Fraknoi 2017a). There is an Open Educational Resources Hub associate with the book, where ancillary materials by the authors or adopters are made available free of charge as well. The textbook is updated annually, and students can download it or access it on their phones, tablets, and laptops. Over 100,000 students have used the book since it came out, distributed among 400 institutions. Another free textbook is at the Teach Astronomy web site, based on a printed book authored by Chris Impey and William Hartmann. It contains 520 articles, organized into 19 chapters, comprising over 600,000 words. With a click, Google Translate does a serviceable job of converting the articles into dozens of foreign languages. The site also has a curated set of 45,000 astronomy articles from Wikipedia, 1200 short video clips covering most astronomical topics, and a unique clustering tool applied to the textbook articles and thousands of images. Details about the site and technical background on the clustering technology and the content database have been published (Impey *et al.* 2016).

Andrew Fraknoi has provided a listing of both full collections of lab activities (mainly from university astronomy departments) and a selection of individual activities that are particularly useful (Fraknoi 2017b). The popular CLEA labs, Contemporary Laboratory

Exercises in Astronomy, are included in this list (Marschall 1998). ComPADRE is a digital library of educational resources in physics and astronomy intended for instructors and students (Deustua 2004). The project was sponsored by the American Association of Physics Teachers and the American Astronomical Society and it has been supported by the National Science Foundation. The collection is diverse, covering tutorials, activities, labs, simulations, animations, and papers and conference proceedings on physics and astronomy education research. There is more physics than astronomy content, but the physics collection includes topics relevant to teaching introductory astronomy, such as radiation, light, atomic structure, and gravity. The Center for Astronomy Education (CAE 2021) features an extensive collection of resources for both instruction and assessment of introductory astronomy. Based on twenty years of pedagogy research, the materials have all been tested and validated. The CAE web site hosts instructional strategy guides on using lecture tutorials (Prather *et al.* 2012), think-pair-share questions, a thousand assessment questions, ranking tasks, banks of multiple-choice exam questions, and a full set of lecture slides for an introductory astronomy course. An important ingredient for the universe of online instructional materials is astroEDU, a web platform for peer assessment of astronomy activities (Russo *et al.* 2015).

5. Concept Inventories

A concept inventory is a research-based assessment instrument that probes a student's understanding of key concepts in a subject. Typically, it is administered with a carefully defined curriculum, and student learning is measured before and after the concept has been covered in class. Concept inventories were pioneered in physics (Hestenes *et al.* 1992), but they have since spread to astronomy and other subjects in science and even beyond (Sands *et al.* 2018). Concept inventories have been developed on general space science and astronomy, positional astronomy, lunar phases, light and spectra, stellar properties, planet formation, and size, scale, and structure. Before attempting to use any concept inventory, it is advisable to get guidance (Madsen *et al.* 2017). Concept inventories must be used carefully, since they have to be grounded in item response theory (Wallace & Bailey 2010). The original concept inventory for astronomy was the Astronomy Diagnostic Test (Hufnagel 2001).

6. Interactive Tools

Kevin Lee at the University of Nebraska has created interactive materials on astronomy called ClassAction for use at the introductory college level or the high school level (Lee *et al.* 2006). They are dynamic think-pair-share questions, with 500 items in 22 topic modules. These resources can be imported into Powerpoint and there is a browser and module editor (for both PC and Mac) that allows instructors to customize the modules. The web site has instructions and comments about the pedagogy behind the materials. Figure 2 shows an example. The same web site has 15 lab modules built around a set of simulators of physics and astronomy phenomena, with students able to set up initial conditions and vary parameters, acting very much like scientists. Each lab has pretests and posttests for measuring student learning. Conversion of the interactive simulations from Flash to HTML5 is underway. Carl Weiman, with his PhET project at the University of Colorado, was the first to create simulations based on education research, and the first to provide students with a game-like interface that encourages exploration and discovery (Weiman 2010). App versions of the simulations work on both Android and iPhone devices, and they have been converted to HTML5 to work on all web platforms. The web site has information on how to use the simulations effectively and it provides the

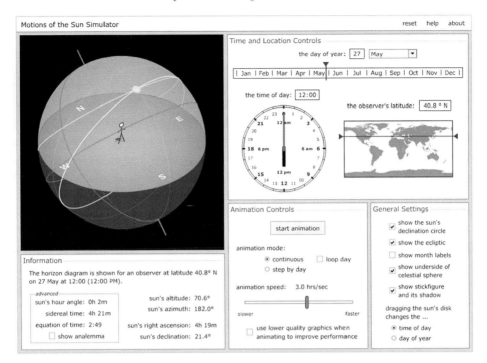

Figure 2. Example of an interactive simulation for conveying the motion of the Sun, part of a set of simulations where students can control the parameters, in this case location on the Earth and time of day and year. Hosted by the University of Nebraska. Courtesy Kevin Lee.

instructional scaffolding required to promote conceptual learning. Most of the PhET simulations are in physics, but there are several dozen simulations on gravity, radiation, and stellar properties.

7. Videos and Image Collections

Segueing from education to outreach, some categories of astronomy content can be used equally well for students and for public audiences. Videos can provide important augmentation or enrichment for an astronomy class. They can also be inserted into a talk given to the public. The subject of astronomy has long been well served by long format videos from national media producers in the United States such as PBS/NOVA and National Geographic. A newer phenomenon is short format video, often made by individual scientists, NASA or ESA missions, or educators, sometimes with inexpensive equipment (Roos & Van den Bulck 2019). YouTube hosts many excellent videos about astronomy. Launched in 2005 and operated by Google, YouTube is the second most popular web site in the world. Over 400 hours of content are uploaded every minute, and a billion hours of content get viewed every day (Zhou *et al.* 2016). A video web search for "astronomy" returns 3.5 million results, with 6000 new videos added daily. One of the best resources is the Astronomy Crash Course series, hosted by Phil Plait and created and distributed by PBS Digital Studios (Schmidt 2015). Short astronomy videos created by undergraduates are on YouTube's Active Galactic Videos channel (Impey *et al.* 2018), and two other popular YouTube channels are SciShow Space, hosted by Hank Green, and HubbleCast (Christensen *et al.* 2007).

Astronomy eclipses all other science fields with its diverse and spectacular images. The pre-eminent resource comes from the Hubble Space Telescope, particularly the 4000 plus

Figure 3. The Hubble Heritage Project released one spectacular image from the Hubble
Space Telescope every month from 1998 to 2016. Courtesy the Space Telescope Science
Institute.

images from the Hubble Heritage Project (Kessler 2012), see Figure 3. The European
Southern Observatory has a searchable archive with over 14,000 images (ESO 2021).
Other major observatories such as NOAO, NRAO, and AAO have image archives, with
few restrictions on their use. For the Solar System, NASA's Jet Propulsion Lab hosts the
Photojournal, with over 24,000 images from its missions over the past 50 years (NASA
2010). Astropix is a JPL archive of over 7000 images from the space-based HST, Spitzer,
Chandra, GALEX, Herschel, Planck, and WISE telescopes, and from ESO's ground-
based telescopes. Astropix adds value to images by aggregating all the relevant contextual
information, such as the source, sky position, field of view, wavelength of observation,
and a color map. This metadata is part of the worldwide standard called Astronomical
Visualization Metadata (Hurt *et al.* 2007). Finally, the iconic web site Astronomy Picture
of the Day (APOD) started at the same time as the Internet, in 1995. Its first post had
just 14 views and now it has close to two billion views. The over 9000 APOD images
have played a major role in anchoring astronomy in the consciousness of the public
(Bonnell & Nemiroff 2006).

8. Sky Viewing Tools

There are a number of free or open source planetarium software packages that can be
used in a lab or a classroom setting for introductory astronomy or projected live for a
public lecture. The most prominent is Stellarium, supported on all the major operating
systems. In addition to visible sky objects, and constellation maps from ten cultures,
Stellarium has 600,000 stars from the Hipparcos and Tycho-2 catalogs, and the ability to
import 200 million more. There is also a mobile version for Symbian, Android, and iOS
(Hughes 2008). Other highly rated and free programs are Celestia, SkyChart, and Aladin.

Examples have been published on using these tools in a classroom (Persson & Eriksson 2016). The most popular commercial planetarium software is Starry Night College, which has lesson plans, pre- and post-assessment resources, and interactive student exercises. There are many smartphone apps to view the night sky, some of which give an immersive experience and connect with articles from Wikipedia and object catalogs that reach far fainter than the naked eye sky. The apps work on both iOS and Android platforms and are free or available for modest cost (Young 2015). As counterpoint to its comprehensive maps of the Earth, Google made sky maps available. Zoomable maps of the Moon and Mars are also online, and the Moon resources include 3D models and 360-degree panoramas (Connolly *et al.* 2008).

Educators, their students, and members of the public can harness a network of robotic telescopes to take their own images of astronomical objects. The MicroObservatory is funded by NASA and was developed by scientists and educators at Harvard-Smithsonian Center for Astrophysics (Gould *et al.* 2006). Typically, about 50 bright objects can be observed by the small telescopes, including planets, star clusters, and a few nebulae and galaxies. Students select their object, the filter, and a field of view, and submit a request. Data is typically taken within a week by one of the telescopes in the network. The Internet enables views of the night sky to be integrated with images from ground- and space-based telescopes, capabilities with great potential for teaching astronomy. The exemplar of this is the WorldWide Telescope (WWT), an open source collection of applications and data, hosted on GitHub, with data available in the cloud (Goodman *et al.* 2011). Originally developed by Microsoft Research and available only as a Windows application, WorldWide Telescope now has a web client so it can be used in a browser on any desktop computer or handheld device. The project realizes the long-held vision of an open source, virtual observatory (Gray & Szalay 2002). WWT has had 10 million active users. The capabilities of WorldWide Telescope seem overwhelming at first, but the project provides over 50 examples of "tours" for instructors to use in the classroom to give students a sense of the richness of the night sky (Figure 4).

9. Citizen Science Projects

Research astronomy is fueled by enormous, multi-wavelength data sets, and the IAU has organized an effort to harness these resources for education and public outreach (Cui & Li 2018). Citizen science is a phenomenon that is a direct result and beneficiary of the spread of Internet access around the world. Millions of volunteers work on many thousands of projects across all fields of science, without any formal training. Some of them use research-level data sets, others crowd-source measurements of the natural world (Bonney *et al.* 2014; Marshall *et al.* 2015). Volunteers classified galaxies from the Sloan Digital Sky Survey into categories according to their morphology. It was strikingly successful; non-scientists classified 900,000 galaxies with a reliability not very different from that of trained professionals (Lintott *et al.* 2011). Galaxy Zoo grew a few years ago into Zooniverse, operated by the Citizen Science Alliance as an umbrella for many citizen science projects. Citizen scientists are working or art and archaeology, weather data and animal classification. The site has over a million registered volunteers, and their collective efforts have led to over a hundred research papers. The astronomical projects underway include looking for solar coronal mass ejections (Solar Stormwatch), detecting bubbles in the interstellar medium (Milky Way Project), using light curves to detect extrasolar planets (Planet Hunters), analyzing images of Mars (Planet Four), looking for stars where planets are forming (Disk Detective), and analyzing time-lapse images to find undiscovered asteroids (Asteroid Hunter).

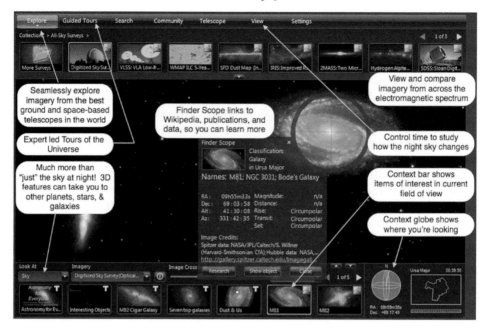

Figure 4. Screen shot from the WorldWide Telescope application and web tool. The night sky is represented and images from many observatories are included. Users can create scripted multimedia "tours" of the night sky. Courtesy Patricia Udomprasert.

10. Visualizations and Virtual Worlds

A final category of online resources for astronomy education and outreach involves the creation of immersive experiences. NASA's Eyes is an app for desktop PCs and Macs, and for mobile devices. It's the work of the Visualization Technology Applications and Development Team at the Jet Propulsion Laboratory. The app is sophisticated tool for embedding NASA's planetary data assets in an environment of data visualization where the user has flexibility in exploring the data. The "Eyes on the Earth" module lets a user monitor the planet's vital signs, trace the movement of water around the planet, and interact with a global temperature map. The "Eyes on the Solar System" module allows a user to inspect planet and moon surfaces with mapping data from NASA missions, and recreate Solar System exploration with missions from 1950 to 2050. The "Eyes on Exoplanets" module renders 3D space and populates it with 1000 confirmed exoplanets. Users can inspect physical properties of the exoplanets, make comparisons between the exoplanet systems and Solar System geometry, and overlay their habitable zones. For the more adventurous astronomy educator or outreach specialist, astronomy exhibitions and experiences can be created in virtual worlds (Minocha & Reeves 2010). The virtual world called Second Life was established in 2003 and reached a peak of a million users in 2013. The author has taught in Second Life, and students in his class did outreach by creating interactive and animated museum exhibits there (Gauthier 2007). As Internet bandwidth and computer power rise, more people will be explore Second Life and other richly realized virtual worlds (Crider 2020).

11. Conclusion

For anyone teaching astronomy or conveying the excitement of the subject to public audiences, the Internet provides a wonderful selection of resources. Materials for the classroom are often research-validated and well-tested, and most of the materials for outreach are free of copyright restrictions. Methods of active engagement not covered in this short article include messaging (i.e. Slack, Discord) and live audience polling (i.e. Poll Everywhere). Science literacy is a concern throughout the industrialized world (Liu 2009), and astronomy is an important vehicle for raising levels of awareness about the contents of the universe, space, deep time, and our place in the cosmos. The IAU has been proactive in raising awareness of education and outreach, and astronomers have many ways to contribute to the new astronomy education ecosystem (Pompea & Russo 2020). This article provides a starting point for engaging in this important work.

References

Bonnell, J.T., & Nemiroff, R.J. 2006, Astronomy: 365 Days, Harry N. Abrams, New York, NY

Bonney, R., Shirk, J.L., Phillips, T.B., Wiggins, A., Ballard, H.L., Miller-Rushing, A.J., & Parrish, J.K. 2014, *Science*, 343, 1436

Bretones, P.S., & Neto, J.M. 2011, *Astronomy Education Review*, 10, 1

Bretones, P.S. 2019, *EPJ Web of Conferences*, 200, 01022

CAE 2021, Center for Astronomy Education, https://astronomy101.jpl.nasa.gov/

Chapman, S., Catala, L., Mauduit, J.-C., Govender, K., & Louw-Potgieter, J. 2015, *South African Journal of Science*, 111, 1

Christensen, L.L., Kornmesser, M., Shida, R.Y., Gater, W. & Liske, J. 2007, *Communicating Astronomy with the Public*

Collins, M., Dorph, R., Foreman, J., Pande, A., Strang, C., & Young, A. 2020, A Field at Risk: The Impact of COVID-19 on Environmental and Outdoor Science Education, Policy Brief, Lawrence Hall of Science, Berkeley, California

Cominsky, L.R. 2018, *Nature Astronomy*, 2, 14

Connolly, A., Scranton, R., & Ornduff, T. 2008, Preparing for the 2009 International Year of Astronomy, ASP Conference Series 400, San Francisco, California, 96, https://www.google.com/sky/

Crider, A. 2020, Astronomy Education, Volume 2: Best Practices for Online Learning Environments, IoP Science, https://iopscience.iop.org/book/978-0-7503-1719-1/chapter/bk978-0-7503-1719-1ch4

Cui, C. & Li, S. 2018, Astronomical Data and Analysis Software and Systems XXVII Conference, Santiago, Chile, http://daepo.china-vo.org/

Deustua, S. 2004, Mercury, 33, 19, https://www.compadre.org/astronomy/

Entradas, M., & Bauer, M.W. 2019, *Nature Astronomy*, 3, 183

ESO 2021, European Southern Observatory, https://www.eso.org/public/images/

Fienberg, R.T., Christensen, L.L., & Russo, P. 2014, *Communicating Astronomy with the Public*, 14, 4

Fraknoi, A. 2001, *Astronomy Education Review*, 1, 121

Fraknoi, A. 2014, *Jounal of Astronomy and Earth Sciences Education*, 1, 37

Fraknoi, A. 2017a, *Physics Teacher*, 55, 502

Fraknoi, A. 2017b, http://www.fraknoi.com/wp-content/uploads/2017/12/Laboratory-Activities-for-Astro-101.pdf

Fraknoi, A. 2019, The Astronomy of Many Cultures: A Resource Guide, Astronomical Society of the Pacific, San Francisco, CA, https://astrosociety.org/education-outreach/resource-guides/multicultural-astronomy.html

Freeman, S., Eddy, S.L., McDonough, M., Smooth, M.K., Okoroafor, N., Wodt, H., & Wenderoth, M.P. 2014, *Proceedings of the National Academy of Science*, 111, 8410

Gauthier, A.J. 2007, *Communicating Astronomy with the Public*, 1, 32

Gerbaldi, M. 2007. Highlights of Astronomy, Volume 14, Cambridge University Press, Cambridge, England, 221

Goodman, A., Fay, J., Muench, A., Pepe, A., Udomprasert, P, & Wong, C. 2011, Astronomical Data Analysis Software and Systems. XXI. ASP Conference Series Vol. 461, 267, http://www.worldwidetelescope.org/webclient/

Gould, R., Dussault, M., & Sadler, P. 2006, *Astronomy Education Review*, 5, 127, http://mo-www.cfa.harvard.edu/MicroObservatory/

Gray, J., & Szalay, A. 2002, *Communications of the Association for Computing Machinery*, 45, 50

Griffin, I. 2003, Astronomy Communication, ASSL Volume 290, Springer, Dordrecht, Holland, 139

Guinan, E.F., & Kolenberg, K. 2016, Astronomy in Focus, Volume 1, International Astronomical Union, Cambridge University Press, Cambridge, England, 390

Hestenes, D. Wells, M., & Stackhamer, G. 1992, *The Physics Teacher*, 30, 141

Hufnagel, B. 2001, *Astronomy Education Review*, 1, 47

Hughes, S.W. 2008, *Science Education News*, 52, 83, http://stellarium.org/

Hurt, R.L., Gauthier, A.J., Christensen, L.L., & Wyatt, R. 2007, *Communicating Astronomy with the Public*, 1, 450, https://astropix.ipac.caltech.edu/

IAU 2020. IAU Strategic Plan 2020-2030, International Astronomical Union, https://www.iau.org/static/administration/about/strategic_plan/strategicplan-2020-2030.pdf

Impey, C.D., Hardegree-Ullman, K.K., Patikkal, A., & Austin, C.L. 2016, *International Journal for Innovation, Education, and Research*, 4, 117, http://www.teachastronomy.com

Impey, C.D., Wenger, M., Austin, C., Calahan, J. & Danehy, A. 2018, *Communicating Astronomy with the Public*, 24, 32

Impey, C.D., & Buxner, S.B. 2019, editors, Astronomy Education: Evidence-Based Instruction for Introductory Courses, Institute of Physics, Bristol, England

Impey, C.D., & Wenger, M. 2019, editors, Astronomy Education: Best Practices for Online Learning Environments, Institute of Physics, Bristol, England

Joubert, M. 2019, *Nature Astronomy*, 3, 131

Kessler, E.A. 2012, Picturing the Cosmos: Hubble Space Telescope Images and the Astronomical Sublime, University of Minnesota Press, Minneapolis, Minnesota, https://hubblesite.org/resource-gallery/learning-resources/hubble-heritage

Lee, K.M., Guidry, M., Schmidt, E.G., Slater, T.F., & Young, T.S. 2006, Proceedings of the National STEM Assessment Conference, National Science Foundation, Washington, DC, http://astro.unl.edu/classaction/

Lintott, C.J. *et al.* 2011, *Monthly Notices of the Royal Astronomical Society*, 410, 166, https://www.zooniverse.org/

Liu, X. 2009, *International Journal of Environmental and Science Education*, 4, 301

Madsen, C. & West, R.M. 2000, Information Handling in Astronomy, Kluwer, Dordrecht, Holland, 25

Madsen, A., McKagan, S.B., & Sayre, E.C. 2017. *The Physics Teacher*, 55, 530.

Marinoni, G., Van't Land, H., & Jensen, T. 2020, The Impact of COVID-19 on Higher Education Around the World. International Association of Universities, Paris, France

Marschall, L.A. 1998, IAU Colloquium Volume 162, New Trends in Astronomy Teaching, Cambridge University Press, Cambridge, England, 79

Marshall, P.J., Lintott, C.J., & Fletcher, J.N. 2015, *Annual Reviews of Astronomy and Astrophysics*, 53, 247

Minocha, S., & Reeves, A.J. 2010, *Learning, Media, and Technology*, 35, 111

NASA 2010, NASA Tech Brief, NPO-47264, https://photojournal.jpl.nasa.gov/

Pasachoff, J.M., & Percy, J.R. 2005. Teaching and Learning Astronomy: Effective Strategies for Educators Worldwide, Cambridge University Press, Cambridge, England

Percy, J.R 1988, IAU Colloquium Volume 162, New Trends in Astronomy Teaching, Cambridge University Press, Cambridge, England, 2

Persson, J.R., & Eriksson, U. 2016, *Physics Education*, 51, 25

Plait, P. 2016, *Slate Magazine*, https://slate.com/technology/2016/02/crash-course-astronomy-translated-into-different-languages.html

Pompea, S. & Russo, P. 2020, *Annual Reviews of Astronomy and Astrophysics*, 58, 313

Prather, E.E., Rudolph, A.L., & Brissenden, G. 2009, *Physics Today*, October Issue, 41

Prather, E.E., Slater, T.F., Adams, J.P., & Brissenden, G. 2012, Lecture-Tutorials for Introductory Astronomy, 3rd Edition, Pearson, New York, New York

Roos, M. & Van den Bulck, N. 2019, *EPJ Web of Conferences*, 200, 01004

Rose, K.M., Markowitz, E.M., & Brossard, D. 2020, *Proceedings of the National Academies of Science*, 117, 1274

Russo, P., & Christensen, L.L. 2010. International Year of Astronomy 2009 Final Report, https://www.astronomy2009.org/resources/documents/IYA2009_Final_Report/index.html

Russo, P., Heenatigala, T., Gomez, E., & Strubbe, L. 2015, eLearning Papers No. 40, http://www.openeducationeuropa.eu/en/elearning_papers

Sands, D., Parker, M., Hedgeland, H., Jordan, S., & Galloway, R. 2018, *Higher Education Pedagogies*, 3, 60

Schmidt, J.T. 2015, *The Journal of the Gilded Age and Progressive Era*, 14, 284, https://www.pbs.org/show/crash-course-astronomy/

Shibata, Y., & Canas, L. 2019, GAM 2019 Blog, Astronomers Without Borders, https://astronomerswithoutborders.org/gam2019-news/gam-2019-blog/4892-bringing-more-astronomy-into-your-language-the-iau-astronomy-translation-network.html

Stevens, A.R.H., Bellstedt, S., Elahi, P.J., & Murphy, M.T. 2020, *Nature Astronomy*, 4, 843

Wallace, C.S., & Bailey, J.M. 2010, *Astronomy Education Review*, 9, 010116-1

Waller, W.H., and Slater, T. 2011, *Journal of Geoscience Education*, 59, 176

Weiman, C.E. 2010, *The Physics Teacher*, 48, 225, https://phet.colorado.edu/

Young, M. 2015, *Sky and Telescope Magazine*, March 2015, 68

Zhou, R., Zhemmarat, S., Gao, L., Wan, J., & Zhang, J. 2016, *Multimedia Tools and Applications*, 75, 6035

Education and Heritage in the era of Big Data in Astronomy
Proceedings IAU Symposium No. 367, 2020
R. M. Ros, B. Garcia, S. R. Gullberg, J. Moldon & P. Rojo, eds.
doi:10.1017/S1743921321000831

Galaxy Forum South America-Argentina 2020

S. Durst[1] **, M. Safonova**[2]**, S. Paolantonio**[3]**, M. E. Colazo**[4] **and G. Li**[5,6]

[1]International Lunar Observatory Association, ILOA, USA.
email: info@iloa.org

[2]Indian Institute of Astrophysics (IIA), Bangalore, India.
email: margarita.safonova62.@gmail.com

[3]Córdoba Observatory, Argentina.
email: paolantoniosantiago@gmail.com

[4]The National Commission for Space Activities (CONAE), Argentina.
email: mcolazo@conae.gov.ar

[5]National Astronomical Observatories, Chinese Academy of Sciences, China.

[6]University of Chinese Academy of Sciences, China.
email: ligeng@bao.ac.cn

Abstract. Galaxy Forum (GF) South America 2020, was held virtually on December 8, 2020 on the opening eve of IAU 367 by the International Lunar Observatory Association (ILOA Hawai'i) with the support of the Instituto de Tecnologías en Detección y Astropartículas (ITeDA, CNEA-CONICET-UNSAM) and IAU. Galaxy Forum is an education and outreach program sponsored by ILOA, an interglobal enterprise incorporated in Hawaii as a non-profit organization to expand human knowledge of the Cosmos through observation from our Moon and to participate in internationally cooperative lunar base build-out.

As a IAU-367 associated event, Galaxy Forum featured comments by Dr. Beatriz Garcia and presentations by ILOA Director Steve Durst (ILOA Hawai'i, USA), Marcelo Colazo (CONAE, Argentina); César Gonzalez García (CSIC, Spain); Li Geng (NAOC, China); Santiago Paolantonio (Córdoba Observatory, Argentina) and Margarita Safonova (IIA, India). In this contribution, the overview of the contributions permits an approach to the GF interests.

Keywords. astronomy education, history of astronomy, astronomy from the Moon

1. Astronomy from the Moon, Precession, Epochs and 21st Century Astronomy, presented by Steve Durst

Earth axial precession, called "Earth's 3rd Motion", is the geo-dynamic process which results in the Sun appearing from Earth at any equinox to move counter-clockwise on the ecliptic through constellations of the zodiac: The Precession of the Equinoxes. Through evolving 21st Century techniques (such as VLBI and astronomy from the Moon), accurately observing the Earth's rotation/precession may help more precisely determine the apparent arrival of the Sun on the ecliptic in the constellation of Aquarius at the time of the vernal equinox, which is now calculated about 2597 AD using the IAU 1928/current map.

With the approach to J2000.0/New Millennium from the 1960s/1970s especially, astrophysics and astrometry scientists have been focusing intensely on Earth precession rates and expressions.

Work on Precession and Rotation of the Earth has accelerated since 1930, when the 88 constellations and their boundaries - fixed by Delporte along strict lines of declination and right ascension as they existed at epoch 1875.0 – finally became ratified and published by the International Astronomical Union (IAU). We will also discuss our efforts to form a Working Group with the focus on Earth Precession, Constellations and Epochs.

2. History of Constellations and the *Uranometría Argentina*, presented by Santiago Paolantonio

The current astronomical community unanimously accepts the division of the celestial sphere into 88 constellations, according to what was established by the International Astronomical Union. After the formation of the Union, in the first Assembly of 1922 the exclusive use of the Latin names for the constellations and their abbreviations was resolved with the three letter system. In the following meeting, the Belgian National Committee of Astronomy examined the pending issue of the limits of the constellations, presenting a motion to review them. The astronomer Eugne Delporte was given the responsibility of the complete theoretical demarcation. In his work presented in 1930, to define the constellations and their limits, Delporte took especially into account what was done half a century earlier in the famous work of the Uranometría Argentina (Paolantonio & Garcéa 2019) of the *Observatorio Nacional Argentino* (1877–1879).

The reason for using the *Uranometría Argentina* was based on the fact that in this work a detailed investigation was made of the current situation at that time, unifying the stellar denominations and solving a proposal of the limits of the constellations, which used arcs of right ascension circles and parallels of declination, choosing them in such a way that they did not deviate too much from those used in the most important atlas of the time, and minimizing the changes of belonging of the stars to the constellations.

Delporte opted to use the reference equinox of 1875.0, to coincide with the for the Southern Uranometry in order to form a set with it, although by that time the positions were already given referred to 1900.0. For the maps that accompanied the report, the projection chosen in the *Uranometría* were defined taking as reference the proposal made in the *Uranometría Argentina*.

3. Prospect for UV Observations from the Moon: the Journey of LUCI, presented by Margarita Safonova

For every human endeavor, there are always people asking 'WHY?' Why going to space, when we can observe safely from the ground? Why going to the Moon, when we can observe cheaper from the near-Earth orbit? It happened before: "Space travel is utter bilge", claimed Richard van der Riet Woolley on assuming the post of British Astronomer Royal in 1956 (Wolley 1956). On the eve of the Apollo 11 landing, he insisted that "from the point of view of astronomical discovery, it [the Moon landing] is not only bilge but a waste of money" (Wolley 1958). 50 years later, the referee of the LUCI paper asked why do we need to go to the Moon, why can't we launch on a CubeSat, or even suborbital flight? And the reason for going to the Moon was not only because observing from the ground, or from near-Earth orbit, is becoming more and more problematic: up to 25,000 small satellites may be launched by 2026, with additional 42,000 SpaceX Starlink satellites, making night sky opaque for astronomy; while the amount of man-made orbital debris becoming unmanageable. ESA Space Debris Office estimates (as of November 2020) more than a million debris objects larger than cm size and 128 million objects in 1 mm to 1 cm range (ESOC 2020).

Our main reason was that we had an opportunity to do astronomy from a frontier, due to the Team Indus, an Indian contestant for the Google Lunar X PRIZE competition MoonShot, who offered to mount LUCI on their lander as a transit telescope to perform a survey of the available sky from the surface of the Moon. Our choice was for the UV telescope, because the Earth's atmosphere absorbs and scatters UV photons preventing observations of the active Universe. UV-emitting phenomena are generally associated with high-energy activity: massive star formation, hot transients such as supernovae (SNe), which stay UV bright for hours to days, AGN flare M-dwarfs with UV-flaring activity, and flashes from cosmic collisions. The UV range is a critical tool for classifying and studying these hot transients, and the Moon presents unprecedented platform for UV astronomy especially, with essential absence of atmosphere and ionosphere offering an unobstructed view of the space, and low gravity as a stable platform for telescopes.

Lunar Ultraviolet Cosmic Imager (LUCI) is a near-ultraviolet (NUV, 200–320 nm) all-spherical mirrors imaging telescope for transit astrometry from the lunar surface. Though the launch was cancelled, there are currently several initiatives in place and LUCI is an innovative telescope designed to take advantage of these opportunities. LUCI 0.5° field of view (FOV) and a weight of only 1.2 kg makes it unique. No other UV space payloads have been previously reported with an all-spherical optical design for imaging in the NUV domain and a weight below 2 kg. Another unique feature is the high brightness limit – LUCI can observe bright UV sources not accessible by the more sensitive large UV missions. The processing and analysis of the data is intended to be performed by the students and to be open to the public as soon as the processing is done. Thus LUCI can be straight away engaged in the citizen science program that we plan to start at the Institute.

We are continuing with our space instruments development, and have designed and built a wide-field NUV Transient Surveyor (NUTS) that can be flown on a range of available platforms: CubeSats, larger space missions, or even go to the Moon (Mathew *et al.* 2019). NUTS is a Ritchey-Chrétien (RC) telescope with a solar-blind photon counting detector and an FPGA-based processing unit. The 3° FOV is especially intended for performing transient survey of the UV sky. NUTS is fully assembled and calibrated, and is stored along with LUCI in a class-100 facility. As LUCI and NUTS are fully developed and ready to fly, they can demonstrate the diverse science capabilities of Moon-based, small, low-cost UV payloads.

4. CONAE's Activities Related to Deep Space Stations, presented by Marcelo Colazo

The National Commission for Space Activities (CONAE, https://www.argentina. gob.ar/ciencia/conae) is the Argentine Space Agency with the capacity to act publicly and privately in scientific, technical, industrial, commercial, administrative and financial matters, as well with competence to propose policies for the promotion and execution of activities in the space area for peaceful purposes. With this objectives, CONAE must propose and execute a National Space Plan, considered as a State Policy, to use and take advantage of science and space technology for peaceful purposes and provide information to the country in order to collaborate in an effective government management.

Within the framework of the intergovernmental agreement signed between the People's Republic of China and the Argentine Republic, the subsequent Amendment to the aforementioned agreement signed, and the inter-institutional agreements signed between the National Commission for Space Activities (CONAE), China Launch and Tracking General Control (CLTC), and the Province of Neuquén, the CLTC – CONAE-NEUQUEN Station was established to provide support to the Chinese lunar exploration program. In the same way, the European Space Agency (ESA) and the government of the Argentine Republic through CONAE, signed in 2009 an agreement for the establishment of the station called ESA Deep Space 3 (DSA3,) in the city of Malargue, located in Mendoza province to support European interplanetary exploration space missions.

The agreements allow the use of 10% of the operational time of the stations for CONAE and its projects in cooperation with national and international partners. The installed technology allows not only monitoring and telecommunication with spacecraft, but also scientific research projects. Our country promoted the possibility of using these facilities for space and scientific activities. Several scientific institutions in the country have been working together in radio astronomical projects as continuum observations of radio sources and the Sun using the original backend instruments located at the stations since 2015. In order to make effective use of the time available, in 2019 CONAE opened a call for research opportunity addressed to the national scientific community. Seven projects were presented that cover the entire time available, in which the Argentine Institute of Radio Astronomy (IAR) and other institutions of the country's astronomical community participate, to begin working in the coming months.

By 2021 CONAE, in collaboration with IAR and the Institute of Technologies in Detection and Astroparticles (ITeDA), plans to finish manufacturing and putting into operation an Argentine instrument for scientific use that will take full advantage of the capabilities of the stations.

5. Picturing the Skies in Ancient China, presented by Geng Li

This contribution is a review of astronomy and culture in ancient China, as well as a prospect of historical astronomy research. It was commonly known by ancient Chinese people since very early ages that stars appeared in the sky can indicate seasons. Patterns of groups of stars from the Neolithic ages inferred that observations have been made at that time. The constellations in China were called "asterism", associated with royal governance and human society under the philosophical idea of "correspondence between heaven and man" (Xiaochun & Kistemaker 1997).

This idea also resulted in continuous celestial phenomenon records among the past two millenniums. More than 38,000 items concerning sunspots, eclipses, historical supernovae and novae, comets and meteor showers, auroras that could help us to track back to the historical skies. Despite the purpose of the ancient Chinese observers was not on scientific research, it has been an indispensable legacy for us to uncover the mystery of the universe, not only for astrometric but also astrophysical approach. From the star atlas and star catalogues, we could possibly extract useful astrometric information. An online database called "Ancient Chinese Astronomical Phenomenon Catalogue (ACAPC)" is under construction and will be launched in the near future. Hopefully, this effort can make better understanding beyond the boundary of science and humanities.

References

ESOC 2020, Space debris by the numbers, ESOC, Darmstadt, Germany. 18 November 2020. https://www.esa.int/Safety Security/Space Debris/Space debris by the numbers

Mathew, J., Nair, B.G., Safonova, M. et al. 2019, *ApSS*, 364, 53

Paolantonio, S. & García, B. 2018. Uranometria Argentina and the constellation boundaries. *Proc. of the IAU*, 13 (S349), 505-509. doi:10.1017/S1743921319000681

Wolley, R. 1956, quoted in the Daily Telegraph, 3 January 1956

Wolley, R. 1958, Interview in the Daily Express, 20 July 1958

Xiaochun, S. & Kistemaker, J. 1997, The Chinese Sky During the Han: Constellating Stars and Society. *Sinica Leidensia*, Vol. 38, Brill, February 1997.

Education and Heritage in the era of Big Data in Astronomy
Proceedings IAU Symposium No. 367, 2020
R. M. Ros, B. García, S. R. Gullberg, J. Moldon & P. Rojo, eds.
doi:10.1017/S174392132100079X

SonoUno: a user-centred approach to sonification

Johanna Casado[1,2] **, Gonzalo De La Vega, Wanda Díaz-Merced,**
Poshak Gandhi[3] **and Beatriz García**[1,4]

[1]Instituto de Tecnologías en Detección y Astropartículas (CNEA, CONICET, UNSAM),
Mendoza, Argentina.
email: `johanna.casado@iteda.cnea.gov.ar`

[2]Instituto de Bioingeniería, Facultad de Ingeniería, Universidad de Mendoza, Argentina.

[3]School of Physics & Astronomy, University of Southampton, SO17 1BJ, UK

[4]Universidad Tecnológica Nacional, Argentina

Abstract. Though there are a variety of astronomy sonification software packages, none of them shown high granularity evidence of having been designed with a user-centric focus. SonoUno is a sonification package created taking into account the user from the beginning, and incorporates end user contact feedback for continuous improvements to the software. In this contribution, SonoUno user cases are presented with the soft corresponding updates, as well as the first description of a recent web page development.

Keywords. methods: data analysis, instrumentation: miscellaneous

1. Introduction

Astrophysics tools are mostly visual, despite studies showing that multimodal approaches can enhance data analysis possibilities. In general, multimodal tools broaden accessibility to scientific data for those with varying performance, learning styles and disabilities. Several projects around the world attempt to make astronomy accessible only at educational aspect, motivating people to progress to aspects no related to the mainstream of astronomy research. These works present different sensorial inputs, like sense of touch, for example Planetariums for blind people, the vibrating universe (De Leo-Winkler *et al.* 2019), different 3D mockups and texts on braille.

Moreover, there are some sonification tools that allow to sonify specific data sets, as Sonification Sandbox (Davison and Walker 2007), MathTrax (https://prime.jsc.nasa. gov/mathtrax/), xSonify (Diaz-Merced *et al.* 2011), Sonifyer (Dombois *et al.* 2008), Sonipy (Worrall *et al.* 2007) (https://github.com/lockepatton/sonipy), Planethesizer (Riber 2018), StarSound (Cooke *et al.* 2017) and SonoUno (https://github.com/ sonoUnoTeam/sonoUno). First four of them are outdated. Particularly, Sonifyer was design to sonify electroencephalography data, MathTrax have educational purpose and the other software present astronomical applications. About the graphic user interface (GUI), Sonipy don't have one and the others shown complex GUI with a lot of elements and in some cases present pop-up windows, forcing end users to change between windows. In general, these software are centred on the data set, the common software frameworks or programmer experience, leading the end user needs out of the loop until the end of the development. In contrast, sonoUno has been designed to be User Centred (UC), its development has been motivated by the desire to bring multimodal access to astrophysical data to all people and aspects of expertise, irrespective of their performance styles or functional diversity. This digital interface may allow people with different learning styles and multiplicity of disabilities to succeed in their transitions through performance aspects. This tool allows to import data, plot, sonify and mark points of interest, with several options available on the sound parameters and plot styles. Moreover, the software allows to script mathematical states avoiding pop-ups. Starting from sonoUno software theoretical framework and ISO standard 9241-171:2008, we present in this contribution some user case studies and the consequent software updates.

2. Methodology

On April 2019 a focus group (FG) sessions were conducted at Southampton University; during that session participants made suggestions and comments about what they need and what they propose to change in the soft. Participants were contacted directly by the recruiter/facilitator through email or through Southampton Sight. Southampton Sight is an award-winning charity organisation in Hampshire County, UK, supporting blind and partially sighted people since its inception in 1899.

About the groups, Group A: four people, two visually impaired and two with low vision (no expertise); Group B: one person with no sensorial or physical disabilities (computing specialist); Group C: one professional astronomer with low vision; Group D: three professional astronomers with no disabilities. The iterations were voice recorded after previous consent approved by the ethics committee at Southampton University, and the audio recordings and consent forms were kept in a password protected file during a year at the Universidad de Mendoza in Argentina. Both audio recordings and consents were erased after a year.

In addition, a group of specialists related to astronomy in a diversity of aspects, were contacted by email to install different version of sonoUno. They used the soft on their own, with their own data, carrying data analysis activities characteristic of their daily practices; and shared with us feedback about its use and recommendations. The next section presents some end users recommendations, collected from all the contacts, and the analysis for the implementation of modifications, taking into account the sonoUno framework and main goals.

3. Result

3.1. End users recommendations

During the FG sessions, the participants make some suggestions to improve the software according to their needs and expectations, for example: "put the play and pause on the same button" (Group A and C); "some of the continuous decreasing data use the same note" (All groups); "one of the panel wasn't on the panel section" (Group B); "the slider should show the real data point, not the position" (Group D); "why do the software only shown the first ten rows of the data set?" (Group D); "to change the instrument they have to press stop and play again" (All groups). During the email exchange with specialist in astronomy and astrophysics some suggestions were focused on the time between notes, to match the points distance (for variability); the mark point button, to show the mark when it is pressed; a loop function; to plot and sonify two or three columns against x at same time; and a line command interaction. The discussion and implementation of changes according the suggestions, were based on accessibility. A screenshot of the software and all updates are available on GitHub web page (https://github.com/sonoUnoTeam/sonoUno).

3.2. *SonoUno updates*

As a first step, the recommendations were divided in 3 groups: "graphic", "sound" and "complex implementation". A complex implementation may be, data operations and graphic changes that would require to implement aspects in the internal structure of the algorithm that may affect several functionalities of the software. The graphic modifications selected for implementation lacked that complexity and were fully implemented. For example, play and pause is now on the same button, the panel 'Data Parameter' was changed to the panels section on the menu to follow a good tasks linearization, when the mark button is pressed the mark line appear above the position line and all the elements on the interface display a textual description when the mouse is hoover over them.

About the sound recommendations, the first was a bug discovered during the focus group session, now sound parameters can be changed without pressing any additional button. Another was based on the sound resolution, with a previous library using MIDI notes (in our context of use and when using python 2, MIDI restricts the number of notes, only around 70 notes on Piano instrument for example), as a preliminary approach and in parallel to Python 3 update the library Pygame was used to generate and play sound modifying sound parameters directly on its waveform. Temporarily the resolution problem was solved, but the way to include MIDI sounds stay on agenda. About the problem of the reproduction time, the limitation is based on the minimum tempo that the graphic user interface allows, this time can't be modified. Instead, is proposed to reduce the number of dataset elements averaging the closest values, this update has to be tested in the next contact with end users. The last one was that time between notes must match the points distance, according to do that the space between notes was measured and the program take in consideration each gap of silence between notes.

Ultimately, in the complex implementations group, the sliders bars were modified to show the real dataset value and not the position on the array. Then, the possibility of write the desired value above the slider was reached. About the data-grid that in the past only shown the first ten values, a way to show all the dataset without freezing the graphic user interface was addressed, right now the dataset was shown in pages of 100 values and it has to be tested in future end users exchanges.

Afterward, some recommendations present potential but require changes on the interface framework, instead of that was decided to address those by first implementing these updates with the command line element on the interface and test it before to include any change in the graphic user interface. This approach allows not only to ensure that the functionality is useful, but also perform user exchanges to decide the optimum way to include it in the graphic user interface maintaining the user centred concept. These final updates incorporate a loop function (because some datasets like light curves were best represented with that functionality) and to display and sonify more than one column in the same or different plots. The recommendation to generate several audio files with one command from different data sets (not related to data analysis) seems to be very useful, but didn't correspond with the sonoUno main goal: 'Creation of a human-computer interface suitable for the access, collection, sonification and analysis of astrophysical data centred on the user from the beginning'. This software is mainly centred on data analysis, the creation of audio files is a consequent for portability and exchange with colleagues.

3.3. *Web Page development*

Addressing other recommendation and a wide and global use of the software, a Web Page development was initiated at the beginning of 2020 (Developed under the Project REINFORCE-GA 872859 with the support of the EC Research Innovation Action under the H2020 Program SwafS-2019-1the REINFORCE). Installing a full application with all its dependencies may be too much work for some users that don't require complex functionalities or are occasional users of the software. It was decided that a web interface, with fewer functionalities that the Python base program, would help overcoming these limitations.

A web interface implies the use of HTML, JavaScript and CSS to do the programming, which results in different code bases for the native and web versions, which requires extra programming effort. Additionally, web access is also achieved through a web browser, which introduces an intermediate software layer, especially when it comes to user interface, with its own rules and standards. The implementation can be divided in front-end and back-end. The front-end is the most fundamental part, as it provides the actual data presentation and sound functions and is being currently implemented with many functions already working. This section provides the interactive part of the experience, which is also the most appealing for the users, has a high contrast design that can be customized through CSS and also takes advantage of the ARIA features of modern web browsers that allow to customize the screen reader workflow. Some of the limitations of the web interface will come from the slower processing speeds on the web browser and some limitations of producing sound. On the other hand, back-end part has been planned to take jobs that depend on more complex software capabilities so the user would not need to install such large and complex software packages locally, but probably transferring data from the local device to the back-end server makes it inconvenient for large data sets.

4. Conclusion

We can conclude that the proposed development has the potential to enhances the actual scientific work bringing other sensorial styles to data analysis. Also, sonoUno makes possible to analyse data by people with functional diversity, permitting to explore data and make decisions by themselves. This method shows the importance of a user centered approach influencing over computational science, engineering and software design, among others, to take into account the end user with high granularity from the beginning, making the tools more usable, efficient and useful, taking into account the accessibility.

Finally, it is very important to address the cause of inequity to remove it or find an alternative way. There are a lot of cases where remove barriers is easy if a user centred approach with a good design was made from the beginning.

References

Cooke, J., Díaz-Merced, W., Foran, G., Hannam, J., & Garcia, B. 2017, *Proc. IAU*, 14(S339), 251–256

Davison, B. K., & Walker, B. N. 2007, *Proc. ICAD*, 2007, 509–512

De Leo-Winkler, M. A., Wilson, G., Green, W., Chute, L., Henderson, E., & Mitchell, T. 2019, *Journal of Science Education and Technology*, 28(3), 222–230

Diaz-Merced, W., Candey, R. M., Brickhouse, N., Schneps, M., Mannone, J. C., Brewster, S., & Kolenberg, K. 2011, *Proc. IAU*, 7(S285), 133–136

Dombois, F., Brodwolf, O., Friedli, O., Rennert, I., & Koenig, T. 2008, *Proc. ICAD*, 2008

Riber, A. G. 2018, *Proc. ICAD*, 2018, 219–226

Worrall, D., Bylstra, M., Barrass, S., & Dean, R. 2007, *Proc. ICAD*, 2007

Education and Heritage in the era of Big Data in Astronomy
Proceedings IAU Symposium No. 367, 2020
R. M. Ros, B. Garcia, S. R. Gullberg, J. Moldon & P. Rojo, eds.
doi:10.1017/S1743921321000430

SciAccess: Making Space for All

Anna Voelker⑩, Caitlin O'Brien and Michaela Deming

Department of Astronomy, The Ohio State University
4055 McPherson Laboratory, 140 W 18th Ave, Columbus, OH, United States, 43210
emails: `voelker.30@osu.edu`, `obrien.847@osu.edu`, and `deming.32@osu.edu`

Abstract. The SciAccess Initiative ("SciAccess") is dedicated to advancing disability inclusion and diversity in STEM education, outreach, and research. In this paper, the authors present an overview of accessible STEM programs run by the SciAccess Initiative, including an annual conference, international working group, and space science mentorship program for blind youth. Recommendations for creating accessible mentorship programs and networking events, both virtually and in-person, are detailed so that these inclusion-focused efforts may be replicated by others.

Keywords. SciAccess, accessibility, disability, diversity, inclusion, astronomy outreach

1. Introduction

In 2018, the SciAccess Initiative was founded in response to an overwhelming need to address a lack of accessibility, diversity, and visibility for scientists with disabilities in the STEM community. Made possible due to The Ohio State University ("OSU") President's Prize, the SciAccess Initiative began as a one-time conference dedicated to promoting disability inclusion in STEM. From there, it grew into an international initiative that has since branched off into myriad programs working towards a more equitable future.

While it is estimated that 26% of Americans have disabilities Census (2012), people with disabilities represent only 8.3% of overall workers, with an estimated 1.6% employed in science and engineering in the U.S. NCSES (2019) Advancement in these fields is highly influenced by networking opportunities such as conferences, internships, career fairs, virtual seminars, and social events Mickey (2019). When these opportunities are inaccessible, disabled students and professionals are denied the same experiences as their nondisabled colleagues, which in turn can harm their employment prospects.

People with disabilities comprise the world's largest minority group U.S (0000), yet are severely underrepresented in the STEM fields. In response to these discrepancies in employment and education rates, SciAccess seeks to foster equitable STEM opportunities by providing a space in which disabled scientists, educators, students, and advocates can share their experiences with one another and with their nondisabled peers. SciAccess aims to advance the development and dissemination of best practices in accessible STEM research and education through a growing series of international programs, as outlined below.

2. Overview

A chronological overview of SciAccess Initiative projects is presented below.

SciAccess 2019 Conference. On June 28 and 29, 2019, OSU hosted SciAccess: The Science Accessibility Conference. This international event brought together 250 scientists, educators, students, and disability rights advocates to share best practices for

STEM accessibility. The SciAccess 2019 Conference featured over sixty speakers, including keynote presentations from Dr. Temple Grandin, professor of animal science and renowned autism advocate, and Anousheh Ansari, the first female private space explorer.

Making Space for All Webinar Series. Making Space for All was an educational webinar series hosted by the OSU Department of Astronomy and Center for Cosmology and AstroParticle Physics. It featured underrepresented researchers in space science and focused on providing accessible astronomy outreach content during the COVID-19 pandemic. This series culminated in the SciAccess 2020 Conference.

SciAccess 2020 Conference. On June 29, 2020, OSU hosted SciAccess 2020: The Virtual Science Accessibility Conference. With the worldwide transition to online learning, this event took place virtually and brought together speakers and attendees from around the world who share a dedication to inclusive science. The SciAccess 2020 Conference had over one thousand total registrants and 555 participants on the day of the event, with attendees joining from 46 nations and all seven continents, reaching as far as the South Pole. The conference culminated with a keynote presentation by Dr. Soyeon Yi, who shared her experiences as the first and only South Korean astronaut.

SciAccess Working Group. The SciAccess Working Group is a collective of professionals that meets virtually each month to discuss the latest developments in accessible STEM. The group publicizes STEM accessibility projects for use worldwide and primarily consists of researchers and educators working in the STEM fields, as well as professionals with disability and accessibility backgrounds. Individuals interested in joining can do so by filling out the form at go.osu.edu/wg.

SciAccess Zenith Mentorship Program. The SciAccess Zenith Mentorship Program ("Zenith") is a virtual program for blind and low vision (BLV) students interested in astronomy. Established in August 2020 in partnership with OSU and the Ohio State School for the Blind, Zenith connects 8-12th grade BLV students ("Zenith scholars") with OSU student mentors. Using multi-sensory resources such as 3D-printed astronomical models, provided by the nonprofit See3D (See3D.org), and data sonification, Zenith provides an accessible entry point into astronomy and scientific research. After a successful pilot program in autumn 2020, Zenith became a registered student organization at OSU and will continue hosting a new cohort of Zenith scholars from around the world each semester.

3. Implications

Best Practices for an Accessible In-Person Conference. SciAccess has employed a wide variety of methods for ensuring conference accessibility. Based on this experience and the corresponding attendee feedback, the following is recommended for in-person events:

• **Quiet room:** A quiet room, or sensory friendly room, provides attendees with a designated space for taking a break from socializing and from conference commotion. The SciAccess 2019 quiet room included service dogs, art supplies, and playdough. The benefits of quiet rooms have also recently been seen at select commuter airports David (2019).

• **Color communication badges:** Introduced by Autism Network International, color communication badges allow conference attendees to share their communication preferences nonverbally Autistic Self (2014). Red, yellow, and green slips of paper are inserted into the nametags of the attendees, who choose which color they wish to display at any given time. Green means an individual is looking to socialize and meet new people, yellow means they would only like to be approached by those they already know, and red means that they would not like anyone to initiate a conversation with them at this time. This system allows attendees to clearly communicate their preferences and eases the anxiety of networking for all participants. BLV SciAccess attendees were provided with braille color badges.

- **Pronouns on nametags as the default:** Displaying pronouns (such as he/him, they/them) on attendee name tags normalizes the practice of not assuming someone's gender. On the SciAccess 2019 registration page, attendees were asked to specify their pronouns and were informed that this selection would be displayed on their nametag. They were also given the option to skip this question, but by making it the default, nearly all conference attendees chose to display their pronouns on their nametag, creating a more inclusive environment. Because of the precedent set by SciAccess 2019, this practice was then adopted by the IAU 358 Symposium in Tokyo in November 2019.

- **Braille and large print materials:** During registration, ask attendees for their accommodation requests. Offer braille and large print event programs for BLV participants.

- **Tactile map:** SciAccess 2019 used a tactile, thermoform map of the conference venue in order to help BLV attendees navigate the building.

- **Guide volunteers:** Conference volunteers were on-call at all times at the central information desk. If a BLV attendee requested assistance locating a specific room, volunteers were trained to guide the attendee to their destination.

- **Making food accessible:** If offering a buffet, braille descriptions can be taped to the table to identify food. SciAccess 2019 also provided attendees with a list of nearby restaurants that were wheelchair-accessible. To help with food navigation, have additional guide volunteers available during meal breaks.

- **Sign language interpreting:** Sign language interpreting is essential for the inclusion of Deaf attendees. If a conference has concurrent sessions, organizers should provide sign language interpreters for each individual Deaf attendee so that they can go to the session of their choice, instead of being restricted to a single conference strand.

- **CART captioning:** Communication Access Real-time Translation (CART) uses a human captioner to provide a word-for-word transcription of an event. Captioning not only benefits Deaf and hard-of-hearing attendees, but also supports accessibility for second language learners Gernsbacher (2015). For the best quality, use human-transcribed CART captions instead of auto-generated captioning services.

- **Accessible seating:** In each conference room, comfortable armchairs were available for those with chronic pain and for anyone unable to sit in a rigid chair for long periods of time. Ensure that armchair placement does not impede wheelchair access. For social events, avoid high-top cocktail tables, which are inaccessible for people using wheelchairs and anyone who is unable to stand for long periods of time. Instead, use standard round tables for networking events.

- **Slide descriptions:** Train all speakers to describe their presentation slides. This means verbally describing all visual content so that BLV audience members are included.

Fostering Accessibility During Online Events. As virtual events increase in popularity in the wake of COVID-19, it is essential that accessibility remains at the forefront of program design. Based on the work of SciAccess 2020 and the Making Space for All webinar series, the following is recommended for ensuring the accessibility of online programs:

- **Pronouns:** While most online events do not have nametags, organizers can still encourage attendees to share their pronouns by adding them to their display name.

- **Increased breaks:** Increase the frequency of breaks in order to relieve "Zoom fatigue."

- **Sign language interpreting and CART Captioning:** Organizers using Zoom can follow these guidelines in order to successfully incorporate sign language interpreting and CART captioning within their virtual event: go.osu.edu/zoomaccess.

Developing an Accessible Mentorship Program. Zenith aims to model and propagate best practices for accessible mentorship programs. The first step to creating such a program is to build connections with the local community. If an organization is looking to work with blind students, they could begin by contacting their local school for BLV students (in the U.S., there is generally one per state). The next step is to reach out to their science teachers and see if the school would be interested in partnering on a science mentorship program. Such a partnership combines the organization's STEM expertise with the school's knowledge of student needs and accessibility. Organizations should also consider partnering with local universities to recruit student mentors and seek expert guidance from faculty advisors.

Zenith holds regular guest lectures with space scientists, including BLV astronomers, allowing the students to meet role models with similar life experiences. Each semester, Zenith scholars select a topic that they are passionate about and work with their mentors to develop a professional presentation on their subject. The students then present their work for family, friends, and university professors at a research symposium, the program's culminating event. Organizations interested in furthering this work by creating their own Zenith chapter can connect with the SciAccess Initiative to receive further guidance. Additional resources can be accessed at sciaccess.org.†

4. Conclusion

These accessibility initiatives, whether in-person or virtual, serve an essential function in connecting people who have been historically excluded from the STEM fields. In addition to fostering an exchange of ideas, resources, and best practices, they also serve as eye-opening experiences for younger generations of students with disabilities who are faced with a severe lack of representation in STEM. Accessibility is an active, ongoing, and intentional commitment to creating inclusive engagement opportunities. When the talents and perspectives of people with disabilities are neglected, science as a whole suffers. By creating accessible programs and events, organizations prove to disabled students and future scientists that not only is there space for them in STEM, but a profound need.

References

U.S. Census. *Nearly 1 in 5 people have a disability in the u.s., census bureau reports*, 2012.

National Center for Science and Engineering Statistics. *Women, minorities, and persons with disabilities in science and engineering*, 2019.

Ethel L. Mickey. *Stem faculty networks and gender: A meta-analysis, 2019.*

U.S. Dept. of Labor. *Diverse perspectives: People with disabilities fulfilling your business goals*, n.d.

Kathleen David. *Kids and adults with autism flying easier in pittsburgh, with airport's help*, 2019.

Autistic Self Advocacy Network. Color communication badges, 2014.

Morton Ann Gernsbacher. "Video Captions Benefit Everyone". In: *Policy Insights from the Behavioral and Brain Sciences*, 2(10), 2015.

† Similar content to that contained within this paper will be published in other conference proceedings as work regarding the SciAccess Initiative has been presented at multiple events.

Session 5: Innovation in Education

Education and Heritage in the era of Big Data in Astronomy
Proceedings IAU Symposium No. 367, 2020
R. M. Ros, B. Garcia, S. R. Gullberg, J. Moldon & P. Rojo, eds.
doi:

Introduction

Boonrucksar Soonthornthum was invited to discuss "Strategies on Astronomy Education in the Era of Digital Transformation" and opens this section. He stresses the uses of astronomy literacy and data intensive astronomy in the support of astronomy education. He continues about how big data astronomy plays an important role in school and higher education. A focus to increase astronomy literacy is to promote astronomy on an even wider scale with regional and inter-regional collaboration.

After the presentation Paulo Bretones asked:

Let me know about not only careers in astronomy research, but also about education studies. Do you have contact with people of the education field with whom you are promoting studies about curriculum, evaluation, learning, and the teaching process? Do you have any contact with colleagues in this area?

Boon answered:

Yes, as I show here is a slide on "A strategy to disseminate 'Astronomy Education' through 'School network' in Thailand". We have delivered 10-inch Dobsonian telescopes, which were built in Thailand, with intensive training, to 460 schools in 76 provinces in Thailand. So, now, we have a very strong school network, actually, in Thailand. We also organized teacher's training programs in 3 levels: basic, intermediate, and advanced. We provided teachers with astronomy education materials for schools all over Thailand. Since we are an "IAU Southeast Asia Regional Office for Development (IAU SEA ROAD), we extend teachers' training activities for teachers in the Southeast Asia. As Thailand also been appointed as an "International Training Centre on Astronomy (ITCA) under Auspices of UNESCO", we can extend these activities to more regions in the world, especially Africa and South America.

Paulo continued:

Yes, yes very interesting. I knew that you have offered Master's or Ph.D. degrees in astronomy and about the specific contents regarding astronomy research. I asked you that if you also have a postgraduate program in education or sociology or psychology. Do you have any connection or contact with these people in the universities where you combine astronomy with education, sociology, psychology etc. in order to be able to go in depth about teaching and learning and how teachers reflect about astronomy education.

and Boon replied:

Yes, we do, and we also know the importance of astronomy education for research and realize how important it is to use "STEAM education" to promote astronomy education. We now have several collaborations with faculties and universities in Thailand, especially Chiang Mai University, to promote astronomy education and astronomy education research. We hope to extend more collaborations on astronomy education and research to other universities internationally.

Rosa M. Ros, Beatriz García, Ricardo Moreno, Mahdi Rokni, and Noorali Jiwaji described "NASE workshop: Eclipses and Gravitational Lenses." Their article describes how astronomy can be used to inspire interest in science in general. They describe the nature of the IAU's Network of Astronomy School Education (NASE) program and its contributions, using as an example eclipses that have fascinated the public as well as that of gravitational lensing.

Following the presentation Devendra Bisht asked:

It's an extraordinary way to teach Astronomy. How to download the materials and where?

Rosa answered:

The materials presented here are in NASE website http:nasepreogram.com. At present NASE web page has 16 subpages with the information of the courses in the different languages in which the course is given. All materials are in English, Spanish, Portuguese, French, Romanian, Mandarin Chinese and Persian. We are working on the translation of the basic book and / or the ppt for presentation such as Russian, Indonesian, Japanese, Armenian, Mongolian, Kiswahili, Greek and Catalan. In addition, in this moment, we are beginning to translate to, Hungarian, Cantonese Chinese, Thai, South Korean and Arabic.

Anahí Caldú asked:

Georg, thank you very much for the nice talk and to all the Stellarium team for such a great work! I imagine everything is done in a voluntary way? Is it possible to raise funds somehow to hire more developers?

Rosa then invited Noorali to explain the interest to translate NASE materials:

Noorali continued:

Yes, we are translating to Kiswahili then we can open up to many more teachers in the country than just handling English. There are some difficulties of a lack of scientific vocabulary but it is an interesting help for teachers.

Paula Chis commented:

The great interest that NASE has in giving the courses in the languages of the country, that in Romania most teachers do not handle English easily and that by having the ppt translated and the books with all the NASE contents are very useful and teacher use in many secondary schools.

Beatriz Garcia replied :

There were lot of congratulations and greetings from different countries and people. She continued that it should be mentioned that the videos used within the training course are silent so that the instructors can explain them in the language of the country and adapt them to the corresponding latitude and cultural context. Thus in the experiments simple materials are used to get in that place, or with typical toys of the place ...

The article that follows is one called "A HOLOtta Fun: Explaining Astronomy with 3D Holograms" by Anne Buckner. Anne stresses the value of using 3D holograms instead of 2D images for outreach and describe how this can be done with anyone's virtual presentations. Such can be useful for explaining complex topics in an easy-to-understand manner in less formal public talks. Explanation and direction is given.

After Anne's presentation Mary Kay Hemenway asked:

Please share hologram link from end of talk in chat.

and Anne provided:

https://youtu.be/Xv7YH-y7UzQ

Raphaël Peralta asked:

Do you have instructions for building bigger holograms ?

Anne answered:

Yes for larger holograms you use the same videos but (1) choose a large screen and (2) scale up the perspex pyramid. For the 60inh screen the scaled up pyramid needed structural support as it was heavy but for unto 40 inch screens the pyramid doesn't need supports. Here is a template for smaller ones (just scale measurements with screen size): https://www.bealsscience.com/post/2016/02/15/3d-hologram-projector-for-you-phone-or -tablet

This link is better as it gives measurements! https://www.johnnosscience.com/ make-a-holo-projector.html

For very large screens using acrylic sheets rather than think plastic sheets would be better for structural integrity

For the 60@ my acrylic sheet was 6mm thick, for 32 inch it was 2mm thick. I think uptown about 20 inches thin acetate sheets should be fine to make it

Priya Hasan and Najam Hasan contributed "Astronomy Data, Virtual Observatory and Education." The authors begin by discussing that virtual learning is valuable in astronomy, both synchronously and asynchronously. They discuss their work and give as example several projects. They then present lessons learned from their project for others to use in the future.

"Real data for Astronomy class in the college and University" was written by Ilhuiyolitzin Villicana Pedraza, Francisco Carreto Parra, and Julio Saucedo Morales. Through this the authors talk about how astronomy education can be used to inspire people to study other sciences as well. They also relate that students with this exposure that do not enter STEM fields will often continue to support astronomy and space sciences no matter where they are. They provide examples in support.

"IBSE-Type Astronomy Projects Using Real Data" by Fraser Lewis talks about Inquiry-Based Science Education (IBSE) activities employing resources from the National School's Observatory and the Faulkes Telescope Project. Both provide free access to 2-meter robotic telescopes and the described activities are designed to be projects for students that are teacher-free.

Education and Heritage in the era of Big Data in Astronomy
Proceedings IAU Symposium No. 367, 2020
R. M. Ros, B. Garcia, S. R. Gullberg, J. Moldon & P. Rojo, eds.
doi:10.1017/S1743921321000922

Strategies on Astronomy Education in the Era of Digital Transformation

Boonrucksar Soonthornthum[1]

[1]National Astronomical Research Institute of Thailand (NARIT),
Princess Sirindhorn AstroPark, 260 Moo 4 Tambon Donkaew, Amphoe Mae Rim, Chiang Mai
50180, Thailand.
email: boonrucksar@narit.or.th

Abstract. Two important strategies on astronomy education in the era of digital transformation proposed on this presentation are the uses of "Astronomy Literacy" and a "Deep Learning" through "Data Intensive Astronomy" to support astronomy education.

Astronomy literacy can create several thinking skills to young generation and public for promoting human capacity buildings on science and technology.

Nowadays, the astronomical data archives are impressively large and the "digital age" has made it easy to make the data available to astronomers, researchers, under graduate and graduate students and even to publics. Big data in astronomy has then played an important role in astronomy education both in higher education and school education. Astronomers and researchers can access "big data" for the deep learning through data intensive astronomy on their research works and school students and publics.

We hope to extend these strategies through regional and inter-regional collaboration to promote astronomy education in wider scale.

Keywords. astronomy education, digital age, astronomy literacy, deep learning.

1. Introduction

To promote "Astronomy Education" thoroughly to all regions across the globe, the IAU has emphasized astronomy education as one of the major key roles in the IAU Strategic Plan 2020–2030. In 2019, IAU has established the "Office of Astronomy Education (OAE)" (Fig. 1) and the "IAU Commission C1: Education and Development" will serve as a driving force and a crucial mechanism of the IAU to support the plan and action of the OAE on astronomy Education.

At present, we have already been in the era of the "Digital Transformation". Most people learn how to utilize digital technologies on their livings to adapt themselves to a drastically change in their careers on the quick flow of the digital technologies (Fig. 2).

Astronomy activities have also to be modified and adapted from the traditional ways to the new normal ways through the applications of the digital technologies especially in the unprecedented difficult period of the "COVID-19" pandemic or even in the "Post-COVID" time. Digital technology will become a powerful tool to serve our regular activities eg: conferences, symposia, schools and workshops etc. and digital network solutions will enable new types of creativities and innovations.

Astronomy is one of the oldest sciences which has strongly driven curiosities, passions and inspirations of young generations and publics. Therefore, a strategy through astronomy education for promoting "Human Capacity Building (HCB)" and "Human Resource Development (HRD)" in science and technology can be driven through "**Astronomy**

Figure 1. IAU strategic plan and the Office of Astronomy for Education.

Figure 2. Digital transformation and global digital network (Zaki 2020).

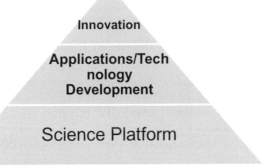

Figure 3. A strong science platform to support applied sciences and technology development to build innovation.

Literacy" eg: teacher trainings, school student trainings etc. eg: teacher trainings, school student trainings etc.

Astronomy requires students to learn several disciplines especaially Physics, Mathematics, Statistics, other branches of science and engineering to understand deeply on the mysteries of the universe. Therefore, astronomy has led to novel and innovative questions of the universe which requires a "**Deep Learning**" especially the use of "**Data Intensive Astronomy**" as a second strategy on promoting astronomy education. And finally, a strong "**Science Platform**" can be established to support another "**applied sciences**" and "**technology development**" which lead to the final goal of building "**Innovation**" (Fig. 3).

Figure 4. PISA 2018 worldwide ranking (OECD).

2. Astronomy Literacy as a driver of human capacity buildings

Astronomy education through the mechanism of astronomy literacy can be used as a powerful driver for human capacity buildings in science and technology.

Astronomy literacy is one of the powerful tool to promote several thinking skills eg: critical thinking, logical thinking, systematic thinking, analytical thinking etc. of the young students and publics.

The "Programme for International Student Assessment (PISA) worldwide ranking is one of a tool for international student assessment which is used widely in Thailand and several countries to help in evaluating and improving thinking skills on science, mathematics and reading for school's students. Although many people believe that this assessment would help to promote the development on science literacy skills in the country, but there are still some queries on how this pre-designed test would fit well with education culture in every country. Therefore, PISA worldwide ranking may still be insufficient for overviewing knowledges and thinking skills of young students in the country. Astronomy literacy can be one of a vital tool to support the development on knowledges and thinking skills for students.

Figures 4 shows a PISA 2018 worldwide ranking in 79 countries (en.wikipedia.org) in the world.

Teacher trainings and student trainings in astronomy are the activities used for human resource development and human capacity buildings through astronomy education. The "on-site communications" on trainings, workshops, schools can be effectively replaced by "on-line communications" during the COVID-19 pandemic (Fig. 5).

Figure 5. Transformation from teacher "on-site" training to "on-line" training.

3. An example in using a strategy to create a school network and disseminate "Astronomy Education" in Thailand

The National Astronomical Research Institute of Thailand (NARIT) is a national institute under the Ministry of Higher Education, Science, Research and Innovation with a main mission to promote research, education and outreach in science and technology using astronomy.

For driving a country-wide astronomy education, NARIT has initiated the "school network" since 2015 by delivering the home-made 10-inch Dobsonian telescopes and learning materials to selected schools throughout Thailand with the intensive trainings (Fig. 6). At present, NARIT has distributed 460 telescopes to 460 schools in 76 provinces in Thailand. The schools become a strong network to co-organize astronomy activities, either on-site or on-line, with NARIT.

NARIT has been endorsed by the IAU since 2012 for hosting a "Southeast Asia Regional Office of Astronomy for Development (SEA ROAD)" and also endorsed by the UNESCO to be an "International Training Centre in Astronomy(ITCA)". So, we can extend both regional and inter-regional collaborations to cooperate and support astronomy education in an international level (Fig. 7).

Commission C1 initiates several working groups which expected impact on astronomy education in a global scale. Thailand, through NARIT, has some opportunities to cooperate with Commission C1 organized astronomy education activities which gave a broad impact in enhancing astronomy literacy and human capacity buildings in Thailand. NARIT hosted a Network for Astronomy School Education (NASE) workshop during 21–24 May 2019 and there were 38 teachers from NARIT's schools network joined NASE courses and brought some projects and materials for teaching in schools.

NARIT had also opportunity to join a first "Astronomy Day in Schools" during 10–17 November 2019. There were132 schools with 45,970 participants involved in "Astronomy Days in Schools" program in Thailand. The Astronomy Day in Schools initiative is an

Figure 6. A home-made 10-inch Dobsonian telescope and the distribution to the school network with intensive training by NARIT.

IAU100 Global Project with the vision of mobilizing and disseminating astronomy activities in schools. The most popular astronomy education activities involved: 1) Lectures related to Astronomy Topics 2) Astronomy Exhibitions 3) Day and Night Observations. This will bring a good opportunity for students to directly interact and engage with astronomers in their communities, and learn about the important role of astronomy in our daily lives which is a very important strategy in promoting astronomy literacy for young students in schools.

During the COVID-19 pandemic, NARIT had joint with the IAU Commission C1 organized the "2nd NASE Workshop", on-line, in Thailand during October 17-18, 2020 with 130 teachers participated (Fig. 8).

NARIT also co-organized the "Astronomy Day in Schools", on-line, during the event of the "Great Total Solar Eclipse" in Argentina on December 14, 2020 and there were 180 schools participated on this activity. Moreover, a webinar on "A Dialogue on the Role of Astronomy in the COVID-19 Era" was organized with the IAU OAD on August 25, 2020 which the digital communication has played an important role in replacing the regular on-site activities.

4. Deep Learning: Data Intensive Astronomy for Education

The "Deep Learning" through "Data Intensive Astronomy" is one of the proposed strategic goals in astronomy education for human capacity buildings on Science and Technology. NARIT creates a "University Network" with many universities in Thailand and abroad to collaborate on astronomy research and education. NARIT's researchers has now worked closely with university academic staff to supervise both undergraduate and graduate students using observational data collected from the 2.4-m Thai National Telescope, the Thai Robotic Network Telescope and the 40-m Thai National Radio Telescope including the observational data from NARIT's international collaborative network projects such as Gravitational Wave Optical Transient Observers (GOTO) project, Jiangmen Underground Neutrino Observatory (JUNO) project and Cherenkov Telescope Array (CTA) project (Fig. 9).

The use of those massive observational data collected at the Data Archive Center at Princess Sirindhorn AstroPark (NARIT's Headquarter) under co-supervisions between NARIT's researchers and the university academic staff, the graduate and undergraduate students from many universities in Thailand and abroad have the opportunities to

Figure 7. MoU signings for SEA ROAD and ITCA of NARIT.

access through the varieties of massive observational data. Therefore, the data intensive learning through a High -Performance Computing (HPC) cluster at NARIT's headquarter (Fig. 10) can support vitally for the graduate and undergraduate student's projects. We also use data intensive astronomy as a tool for preparing the 21 Century skills for

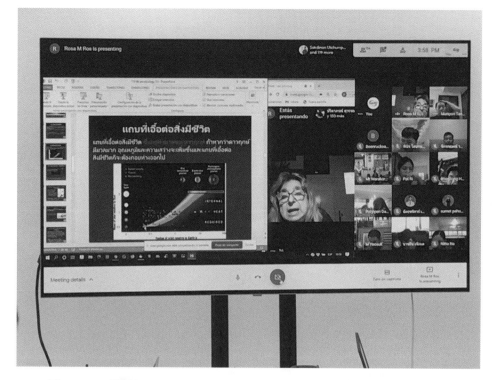

Figure 8. NARIT co-hosted with the IAU on the on-line NASE Workshop.

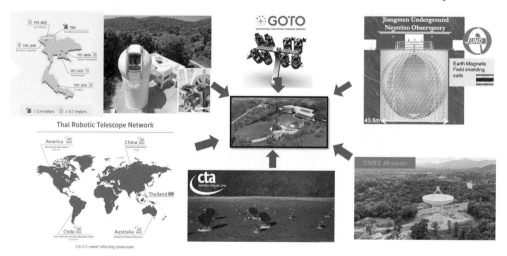

Figure 9. Data Intensive Astronomy for Education.

students, early career researchers through trainings, workshops and conferences on big data, data pipelines reduction and analysis, machine learning, artificial intelligence etc.

5. Conclusion

IAU Commission C1 is an important mechanism of the IAU under Division C to support the OAE, OAD and the OAO in nurturing and driving activities through astronomy education.

Figure 10. A High-Performance Computing (HPC) cluster at NARIT.

Several activities performed under Commission C1 working group such as activities under AstroEdu and NASE working groups etc. have been performed successfully and deserved to be continued. The OAE has been established since 2019 and subsequently has created National Astronomy Education Coordinators (NAECs) in many countries and continents. Therefore, an important mission of Commission C1 in the next few years (2021–2024) is also to support and promote several educational activities through the network of NEACs such as a promotion on Astronomy Education Research (AER), a project on preservation, digitization and access of publications of astronomy education, outreach and historic documents through the regional and inter-regional collaboration. Through the "International Training Centre in Astronomy (ITCA)" under auspices of UNESCO and the "IAU Southeast Asia Regional office of Astronomy for Development (IAU SEAROAD), the initiative of astronomy education activities through an international, young astronomy ambassador scheme on astronomy education and the collaborative activities on astronomy education for diversity, equity and inclusion are also planned.

The strategies of disseminating "astronomy literacy" for publics and building up career paths of human resources by providing a 21st century skills using "data intensive astronomy" can be a powerful driver for the future "astronomy education" in digital era. The astronomy activities regarding these 2 strategies should be concretely implemented and evaluated through the astronomy education research to show the impact of the strategic plans and their implementations. We plan to extend the activities on astronomy literacy and data intensive astronomy for education to promote human capacity buildings and human capitals in science and technology also through regional and inter-regional collaboration to support the fulfillment of the IAU Strategic Plan 2020-2030.

References

Zaki, T.2020 How your business can succeed in a digital transformation project, Information/ age, Editor's Choice, https://www.information-age.com/how-your-business-can-succeed-digital-transformation-project-123488683/

OECD, 2018–2019 PISA 2018 Worldwide Ranking-average score of mathematics, science and reading, FactsMaps, Editor's Choice, https://factsmaps.com/pisa-2018-worldwide-ranking-average-score-of-mathematics-science-reading/

https://en.wikipedia.org/wiki/Programme_for_International_Student_Assessment

Education and Heritage in the era of Big Data in Astronomy
Proceedings IAU Symposium No. 367, 2020
R. M. Ros, B. Garcia, S. R. Gullberg, J. Moldon & P. Rojo, eds.
doi:10.1017/S1743921321000739

NASE Workshop: Eclipses and Gravitation Lenses

Rosa M. Ros[1], Beatriz García[2]⑩, Ricardo Moreno[3]⑩, Noorali T. Jiwaji[4]⑩ and Mahdi Rokni[5]⑩

[1]NASE president, Polytechnical University of Catalonia, Barcelona, Spain,
email: rosamariaros27@gmail.com

[2]NASE vicepresident, ITeDA (CNEA, CNCE, UNSAM) & UTN, Mendoza, Argentina,

[3]NASE Secretary, Retamar School, Madrid, Spain,

[4]Open University of Tanzania, Dar es Salaam, Tanzania.

[5]ITAU, Bushehr, Iran,

Abstract. Astronomy is connected with the every day experiences of the people, since the observation of simple and repetitive phenomena, as the succesion of days and nights, untill events of high impact, as the total solar eclipses. In this sense, the Astronomy is a fascinating activity and can be used to inspire interest in sciences in general. In this contribution, we introduce the Network of Astronmy School Education as part of the IAU proposals connected with teaching training programs, and we highlight several examples on the specific topic of the eclipses: their importance and connection with the culture, that can capture students attention if we use the workshops as part of the classes.

Keywords. education of astronomy, eclipses, gravitational lens, NASE program

1. Introduction

The Network for Astronomy School Education, NASE, is a Working Group of the IAU-Commission 1. In the period 2010-2020 NASE has organised 205 training courses for secondary school teachers in the language of their countries. Currently NASE is present in 50 countries in Africa, America, Asia and Europe.

The program promotes Astronomy Education mainly at secondary and high school and in some cases at other levels, in order to:

(*a*) Encourage interest in science.
(*b*) Promote teaching innovation.
(*c*) Stimulate scientific vocations.

These three objectives are achieved by promoting innovative classroom activities, models and direct observations, trying to stimulate not only interest and reflection, but also the excitement and surprise in its execution, highlighting values and attitudes such as:
• Humility (it is necessary to recognize that in some cases we are wrong)
• Prudence (it is necessary to check and to verify)
• Rigor (it is not possible to be superficial)
• Constancy (it is not good give up)

In all the cases, the activities take into account the impact on the ecosystem, through special emphazis on climate change and light pollution.

In general, NASE presents all the main topics related with Astronomy and Astrophysics in the school, from the most common ones as eclipses and Moon phases, to those much more motivating such as exoplanets, neutron stars, black holes, pulsars and habitability zones among others.. The program attempts to make that students enjoy the emotion of understanding of the phenomena and the inter- and multidisciplinarych aracteristics of the Astronomy. Under this framework, this proposal not only gives information about science, but also intends to educate future generations for a new era of great discoveries.

Finally, as this symposium was coordinated with the total solar eclipse of December 14th, 2020, visible in Chile and Argentina, NASE prepared this workshop about some educational aspects on this topic and highlight its science and observational aspects.

2. Eclipses as gravitational lenses

Eclipses and transits are a special phenomena that offer humanity the opportunity to get knowledge from the Universe. In particular, the eclipse of May 29th, 1919, was the occasion showed that Einstein relativity theory was correct. Arthur Eddington measured the deviation of the position of stars whose rays passed close to the Sun. It confirmed Einstein's prediction about the geometric modification (curvature) of space, caused by the presence of the Sun (Ros, 2008)

The workshop shows a simple model of the visibility of a star almost aligned with the Sun during an eclipse. The Sun acts as a gravitational lens that curves the space around it and hence the light path from from the stars to the Earth. In 1936, Einstein published a short calculation showing that if two objects at different distances coincided exactly in the sky (are aligned), the image of the farthest would form a ring. He predicted that a foreground star could magnify the image of a background star. He was skeptical thatsuch a mirage could ever be seen and he dismissed the lineup as too unlikely to be of interest. In the last decades this topic became more and more important, because gravitational lensing enables high-precision mapping of dark matter distributions in galaxies and galaxy clusters, and, in this sense there are also demonstrations that can be used in the classroom (Huwe, 2015).

2.1. *Gravitational lens effects*

Light always follows the shortest possible path between two points. But if a mass is present, the space becomes curved and the shortest path between two points is a curve in the same way that when we are moving on the planet surface the minimum distance between two cities over the terrestrial sphere is an arc and not a straight line (Fig. 1) (Moreno, 2017).

Gravitational lenses produce four different effects:

Position changes and Multiplication: The deflector curves the light from the source and shifts the apparent position of the source (star or quasar). It also produces two or more images (Fig. 2). The gravitational lenses are not perfects lens and can produce multiple images and in some cases can produce the Einstein Cross.

Deformation and Magnification: When the source is an extended object (i.e. a galaxy), the result is a set of bright arcs. If the system is perfectly symmetrical, the rays converge and the result is a ring, knowing as the Einstein Ring (Fig. 3).

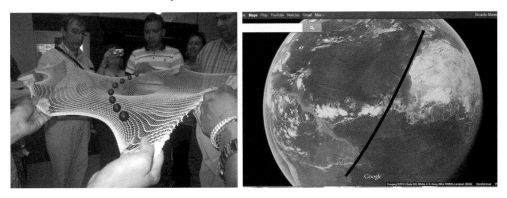

Figure 1. a)The path of a ball on a curved surface is not a straight line b) The path between two points on the globe is a curve.

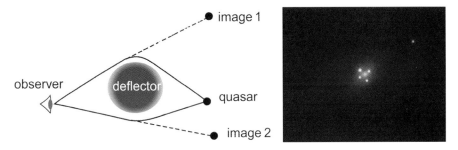

Figure 2. a) Bending of light from quasar by a deflector; b) Galaxy UZC J224030.2 032131, Credit Hubble Telescope NASA/ESA.

Figure 3. a) Bending of light from quasar by a deflector b) Galactic Cluster CL0024+1654 c) Galaxy LRG 3-757, Credit Hubble Telescope NASA/ESA.

2.2. *Simple model of gravitational lensing*

The proposed activity offers the production of a simple model of a gravitational lens using a glass goblet such a wine glass. The optical effects obtained with this piece of glass are similar to a gravitational lens. To develop this model there are two options:
1) the foot of a glass of goblet (Fig. 4), which acts as the deflector (star or galaxy or cumulus of galaxies in the Universe) and a grid paper, to see the effect of the lens; and
2) the glass of goblet with a traslucent liquid inside (which acts as deflector) and a light that crosses the glass (that will follow the path that indicates the deformation of space) (Fig. 5) (Moreno, 2017).

The foot of a glass goblet on a paper shows how the straight lines on the paper change to curved lines around the central point of the foot. Using both models, we can get easily arcs, Einstein rings and and with just a little difficulty the Einstein cross. After the demonstation, it is possible came back to the eclipse of 1919, or in general to all the eclipses. The changes in the astronomical coordinates of a very well know star that can

Figure 4. a) The foot of a goblet on a graph paper showing the b) Einstein rinng by means of the foot of a glass.

Figure 5. a) The glass of goblet with a liquid inside and the light of the mobile phone b) Images mutiplied where the central light appears in other places in a shape similar to a cross or a ring (the central light is actually not possible to observe when the deector is an opaque body).

be observed near the Sun is an evidence of the action of the Sun on space, the Sun is acting as a gravitational lens (Ros, 2008).

3. Conclusions

Eclipses are a fascinating field in Astronomy and a shocking phenomenon for society, wich can be used in education not only as a natural phenomena, but also as one of the oldest recorded events visible from the Earth without technology. The example of gravitational lens presented here is innovative not only as part of NASE program, but also as part of possible new approaches to science in the schools. In this sense, it is important to introduce in the school the topics that appear in the newspaper and media and try to pesent in some way the concepts in an adequate level. As was shown in this contribution, it is possible to teach and learn by doing, and it is necessary to start training new teachers for new students who in the near future will demand new teaching methods within the framework of knowledge that is expanding dramatically.

References

Huwe, P. & Field, S. 2015, Modern Gravitational Lens Cosmology for Introductory Physics and Astronomy Students, The Physics Teacher 53(5):266–270

Moreno, R., Deustua, S., Ros R.M. 2017, Expansion to the Universe, 14 steeps to the Universe. Astronomy course for teachers, (Rosa R.M. & García B. ed.), 2nd edition, NASE, UAI, 139, 141 *Antares*

Ros, R.M. 2008, Gravitational lens in the classroom, *Physics Education*, 43, 5, 506, 514.

Education and Heritage in the era of Big Data in Astronomy
Proceedings IAU Symposium No. 367, 2020
R. M. Ros, B. Garcia, S. R. Gullberg, J. Moldon & P. Rojo, eds.
doi:10.1017/S174392132100082X

A HOLOtta Fun:
Explaining Astronomy with 3D Holograms

Anne S.M. Buckner[iD]

School of Physics and Astronomy, University of Exeter,
Stocker Road, Exeter EX4 4QL, UK
email: a.buckner@exeter.ac.uk

Abstract. Traditional visual aids used in astronomy outreach (such as 2D telescope images) can fail to effectively convey complex and abstract ideas to lay audiences. With the advent of impressive CGI images widely available in film and other media, these aids may also not meet their expectations or visually engage people. To address this, we have been employing 3D holograms in lieu of 2D images for astronomy-based outreach activities both in-person (pre-pandemic), and virtually since the start of the pandemic. Here we demonstrate how the reader can make and incorporate holograms in their own virtual talks (no budget required) and present the feedback we've received so far.

Keywords. techniques: image processing, general: standards

1. Introduction

Astronomy can be challenging to effectively teach at the best of times, being an inherently complex and abstract research topic. Under normal circumstances a lecture with 2D telescope images is often the go-to for informal education settings, such as school workshops and public events. However, as visual technologies have made massive leaps and bounds in the past decade, so have people's expectations, and as a community we must keep up in physics if we want to continue inspiring the next generation. Prior to the pandemic we developed a successful 3D hologram workshop program which enabled researchers to provide fun, interactive talks to explain complex topics in an easy-to-understand format in informal education settings, all while capturing the imagination of participants.

Since the start of the pandemic, virtual talks and audiences have become the norm and so the need to employ technology to provide an innovative, powerful, highly effective and attention grabbing experience is now essential. In particular, studies suggest a positive correlation between childrens' participation in informal out-of-school learning activities with their understanding and general interest in the subject matter (Ofsted 2008, Bell *et al.* 2009, Stocklmayer, Rennie & Gilbert 2010, Dabney *et al.* 2012) but that to be successful, activities need to have the 'potential to amaze' (Walan & Gericke 2019). Therefore at a time when childrens' physics education is undergoing varying degrees of disruption, there is a strong argument for the creation of innovative and engaging virtual school outreach programmes. Our approach to this issue is the use of 3D holograms to explain astronomical topics in virtual workshops. Below we demonstrate how with a little bit of effort – and no budget – the reader can make and incorporate 3D holograms in their virtual outreach activities.

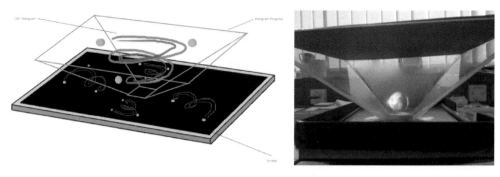

Figure 1. (Left:) Diagram of equipment used to generate holograms and (Right:) Photo of an Earth hologram.

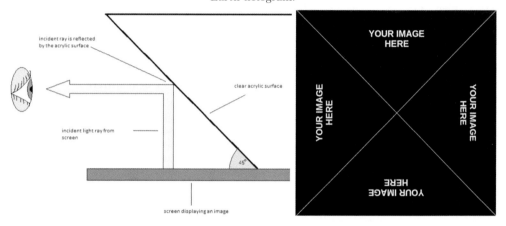

Figure 2. (Left:) Diagram showing the path of the light ray from the image on screen to the observer and (Right:) how the image is displayed on the screen. The white lines mark out the portions of the screen projected onto each side of the perspex pyramid.

2. What are Holograms?

Holograms are a light-based optical illusion in which 2D images gain the appearance of a physical 3D object to the observer. This is not a new concept, originating from the illusion known as 'Peppers Ghost' first conceived by scientist John Henry Pepper in 1862. Magicians and theatre productions have been doing this illusion for over 150 years. More recently, modern technological advances have enabled the enhancement of holograms making it possible to both create more complex objects and objects with a more realistic appearance, leading to use of holograms being adapted for other mainstream entertainment formats (e.g. music festivals). Still, they remain uncommon in education settings. As shown in Figures 1 and 2, modern holograms are generated by displaying images on a screen which are reflected off the surface a perspex pyramid (known as the 'hologram projector') placed centrally on top of the screen. Images can either be videos or pictures but must be carefully chosen (ideally they should have a high resolution and opaque colouring) to create sharp non-fuzzy holograms. Once an image has been selected it needs to be formatted such that it appears four times on the screen with a rotation of 0°, 90°, 180° and 270° (right panel Figure 2). Projecting the image onto all four side surfaces of the hologram projector enhances the illusion (when viewed) that the image is actually a physical 3D object inside the pyramid. We have published a free online tutorial† detailing how to properly format images.

† https://www.youtube.com/watch?v=Xv7YH-y7UzQ

Figure 3. Flowchart showing how holograms can be incorporated into virtual talks.

3. Using Holograms in Virtual Talks

Being star cluster researchers, we use holograms to show how they form and evolve – but holograms can be used as engaging visual aids for talks on any topic. The flowchart in Figure 3 details how we suggest they can be incorporated into virtual talks. In brief, the lecturer sends their hologram-formatted videos to each participant prior to the talk, along with instructions on how to make a hologram projector for a smartphone and/or tablet. Participants make their hologram projectors in advance of the talk, such that they can view the holograms on their own devices during the lecture when instructed to do so. Equipment needed by the participants (P) and the lecturer (L) are given below.

Phone or Tablet (P) To be the screen, as shown in the left panel of Figure 1.

Hologram projector template (P) The size of the projector scales with the size of the screen. Many templates are freely available online for screens of various sizes, and the participants can custom scale the pre-existing templates for a specific screen size. Our template of choice to send to participants is by Beals Science† as it contains two templates to fit typical phone and tablet screen sizes respectively, and includes a clear step-by-step guide on how to create the hologram projector.

Transparent acetate sheet or similar (P) To create the hologram projector. Alternatively any rigid clear plastic can be used e.g. a CD case, food packaging etc.

Scissors, tape and a pen (P) Tools required to create the hologram projector.

Video or picture (L) To convert into a hologram, either the lecturer's own or taken from a free database, for example NASA image archives, with accreditation.

Video editor (L) It is possible to convert videos/images to the hologram-format with many editors. Our free editor of choice is the VSDC Free Video Editor‡.

Hologram Gridlines Template (L) To guide the placement of images during the conversion process, as shown in the right panel of Figure 2. Our free template of choice is by Cyberlink§.

4. Student Feedback

Preliminary feedback from students has been positive. Initially we were concerned that generating the holograms on a phone/tablet would take away from the enjoyment of the workshop, as with our previous in-person workshop holograms were generated with 32" and 65" screens (i.e. the holograms were significantly larger). However this does not seem to be the case so far, with no negative comments regarding the holograms' smaller size, and prevalent expressions of excitement and awe towards their use.

References

Philip Bell, Bruce Lewenstein, Andrew W. Shouse, and Michael A. Feder 2009, *National Academies Press*, ISBN: 978-0-309-11955-9

Katherine P. Dabney, Robert H. Tai, John T. Almarode, Jaimie L. Miller-Friedmann, Gerhard Sonnert, Philip M. Sadler and Zahra Hazari 2012, *International Journal of Science Education, Part B*, 2,1, 63–79

The Office for Standards in Education, Children's Services and Skills (Ofsted), UK Goverment 2008, *Ofsted Report*, Ref: 070219

Susan M. Stocklmayer and Léonie J. Rennie and John K. Gilbert 2010, *Studies in Science Education*, 46,1, 1–44

Susanne Walan and Niklas Gericke 2019, *Research in Science & Technological Education*, 1–21, 2019

† https://preview.tinyurl.com/y3nkye43
‡ https://preview.tinyurl.com/l6vx6ox
§ https://preview.tinyurl.com/yxk7zkpa

Education and Heritage in the era of Big Data in Astronomy
Proceedings IAU Symposium No. 367, 2020
R. M. Ros, B. Garcia, S. R. Gullberg, J. Moldon & P. Rojo, eds.
doi:10.1017/S174392132100034X

Astronomy Data, Virtual Observatory and Education

Priya Hasan[iD] and S N Hasan

Maulana Azad National Urdu University, Hyderabad, India
email: priya.hasan@gmail.com

Abstract. We shall present with examples how analysis of astronomy data can be used for an educational purpose to train users in methods of data analysis, statistics, programming skills and research problems. Special reference will be made to our IAU-OAD project 'Astronomy from Archival Data' where we are in the process of building a repository of instructional videos and reading material for undergraduate and postgraduate students as well as interested participants. Virtual Observatory tools will also be discussed and applied. As this is an ongoing project, by the time of the conference we will have the projects and work done by students included in our presentation. The material produced can be freely used by the community.

Keywords. astronomical data bases, sociology of astronomy

1. Introduction

Virtual learning is a learning-teaching experience that is technology and communication enabled in a synchronous (web-conferencing) or asynchronous (self-paced) manner. Teaching is online and the teacher and learners are physically separated (in terms of place, time, or both). One of the biggest challenges to educators during the covid pandemic was virtual teaching which became the sole mode of communication. Virtual teaching has its advantages namely, convenience of access, wide global reach, inclusive content and collaborations across the globe. However, it has it's disadvantages too: essentially the lack of face-to-face interaction and the human element. The present talk describes our experiences with virtual teaching of astronomy with the help of astronomy data and Virtual Observatory tools.

In the early April 2020, we collaborated with Jana Vigyan Vedika (JVV) Telangana which is a member organisation of the All Indian People Science Network (AIPSN) and a member of the National Council for Science and Technology Communication (NCSTC) in India. The NCSTC is a scientific programme of the Government of India for the popularisation of science, dissemination of scientific knowledge and inculcation of scientific temper. JVV is an effective Science Forum in the state and has a very good reach in rural and urban Telangana.

We started off with a Six-Day Online Astronomy Course from the 18–23 May 2020 from 4:30–6:30 pm IST for School Teachers and interested students. We conducted a quiz at the end of the course and gave certificates to participants who scored above a threshold mark. We had 475 registrations and many were given certificates with JVV.

In May 2020, the International Astronomical Union-Office of Astronomy for Development (IAU-OAD) issued an Extraordinary Call for Proposals that could 'use astronomy in any form to help mitigate some of the impacts caused by COVID-19'. We proposed a project 'Astronomy from Archival Data' which involved Educational Projects for Under-Graduate and Post-Graduate Science Students (Possel 2020). The

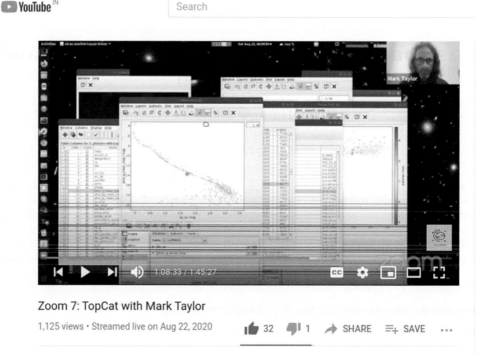

Figure 1. A Zoom Session in progress TOPCAT with Mark Taylor.

project trained students to use the high-quality astronomy data from various facilities. Participants were shown step-by-step techniques of accessing and analysing astronomy data from the internet, introduced to virtual observatory tools and programmimg techniques, and supported to formulate and develop projects. The training included special sessions on report writing, publishing and presenting in scientific journals†.

Almost 1000 students from 25 countries registered for the course, which was encouraging and which made it clear that students clearly require this kind of skill training. Figure 1 shows a mosaic of (1) Our website (2) Our YouTube channel (3) A zoom session in progress (4) One of the recorded sessions.

2. Modus Operandi

Although virtual teaching has been there for a while, this time it required more thought and planning. With the funds we has from OAD, we first set up a website design and content. To enable interaction, we used Social Media (Facebook), Telegram as well as regular email. At regular intervals we planned Homework, Quizzes and Polls so that we could get feedback from our participants.

The project consisted of three phases. The first was made up of live zoom sessions on Saturdays and Sundays from 10-11:30 am IST with lectures and interactive hands-on sessions with us and a variety of speakers from India and abroad on Virtual Observatory Tools, Data Archives and Science Cases. All the sessions were recorded and posted online. We now have over 60 video recordings online on YouTube. Reading material, quizzes, data sources, analysis techniques, etc. are available on our website (https://shristiastro.com/astronomy-from-archival-data/) as well as on YouTube. The

† More info at: https://shristiastro.com/astronomy-from-archival-data/

sessions ran from August to the end of November, 2020. The second phase was in December, when students could complete watching videos and select projects and interact with mentors. The third phase will be till the second half of February to give students time to work on projects of their choice. Students will present their results in the last week of February, 2021.

Apart from us, we had a wide variety of speakers like Mark Taylor, Luisa Rebull, Deborah Baines, Tim Hamilton, Ajit Kembhavi, Avinash Deshpande, Sushan Konar, Kaustubh Waghmare, Eeshan Hasan etc. to cover a wide range of junior and senior researchers. The talks ranged from Basics of Python, Astropy, TOPCAT (Taylor 2003), Aladin, ESASky, Machine Learning and Coding and how to use them. We also had talks on Science Cases for projects and some hand-held sessions on projects.

Figure 1 shows a Zoom Session in progress: TOPCAT with Mark Taylor.

3. Projects

Students were free to opt for one or more projects of their choice. Mentors have been having sessions with their groups to individually address issues and work on the projects. The list of projects offered are as follows:

- Exoplanets (Tim Hamilton/Priya Hasan)

This project involves analysis of Kepler light curves to derive parameters for Exoplanets including distance to the star, mass, radius and hence density. This and the next three projects were explained in a step-by-step procedure by Tim Hamilton and is available on our YouTube website. Enthusiastic participants are encouraged to try their skills with Transiting Exoplanet Survey Satellite (TESS) data.

- Globular Clusters Photometry (Tim Hamilton/Priya Hasan)

This is an exercise in doing Aperture Photometry on images of globular clusters from the SDSS data. Color-magnitude diagrams for the globular cluster are constructed.

- Globular Clusters Color-Magnitude Diagrams: Finding the ages and distances of Milky Way globular clusters (Tim Hamilton/Priya Hasan)

For a sample 11 globular clusters, participants had to plot the color-magnitude diagram showing the cluster data and the best-fitting isochrone together.

- Gravitational Lensing (Tim Hamilton/Priya Hasan)

For a given list, extracting close-in image of each gravitational lens system, showing the lens and the Einstein Ring is to be done. Students need to calculate the total mass, stellar mass, mass of dark matter and dark matter as a fraction of total mass.

- The Radio Pulsar-Magnetar Connection (Sushan Konar)

This project involved a comparison of the intrinsic properties of radio-pulsars and magnetars using available data.

- Star clusters with Gaia (TOPCAT) (Priya Hasan)

Extracting Gaia data and plotting the color-magnitude diagram using TOPCAT.

- Python and Gaia

Using python code with Gaia data (Priya Hasan)

- The spatial distribution of pulsars and the spiral structure of the galaxy (Avinash Deshpande)

- Galaxy Morphology (S N Hasan)

Studying galaxy morphology using decomposition of luminosity profiles using GALFIT (Peng 2010)

- Others

Participants were free to propose their own projects based on the material covered.

We hope that some of these students may even use this for applications in PhD programs/Graduate school as well as a start to serious research careers.

4. Lessons Learnt

This project is close to completion. The Internal Virtual Observatory Alliance (IVOA) supports this project. Participants will present their projects and their presentations will be added to the repository †.

In the course of our sessions we learned a few lessons which we would like to share.

• Programs like this are very useful to students since it shows them important steps in data analysis and techniques which are generally never taught as part of a course. To get the true flavour of a research career, it is very useful if students try their hand on some projects.

• Not all who register participate and hence there has to be constant monitoring of the participants. Interaction via social media, email, etc has to be maintained to keep in tune with issues faced by participants. This can be very time consuming as well as difficult.

• Planning of the talks has to be done carefully since most potential participants had their own online sessions on. We selected weekends since most students/participants do not have classes then. As time progressed, classes as well as examinations were on which made it difficult for students/participants to attend our classes.

• There are various issues like internet connectivity problems, power as well as overlapping sessions. Hence all our activities were recorded and made available on YouTube so that participants could attend at their convenience.

• A few of the participants were from non-English speaking countries, hence we enabled subtitles on our videos for their benefit.

• It is good to have a variety of speakers ranging from students, research scholars, post-doctoral fellows and junior and senior faculty since it can match the level of the participant and it makes the environment more inclusive.

• While sessions were on, gradually many students had regular classes and exams due to which the numbers dropped. Since the repository we created is available online, we hope that students can access the material at their convenience.

We have received approval for a new IAU-OAD project 'AstroSprint' which will be a follow-up of this project in the form of three workshops in the year 2021. The first will still be online and we hope the next two in July and November 2021 wgich will be offline where we can have interactions in-person. We think this was a very timely and purposeful initiative and it was a very good learning exercise. The support of the IAU-OAD is appreciated. We hope to continue with this work of developing a repository of instructive videos and study material that can be used for astronomy research.

References

Peng, C. Y., Ho, L. C., Impey, C. D. and Rix, H. 2010, *AJ*, 139, 6

Possel, M. 2020, *The Open Journal of Astrophysics*, 3, 1

Taylor M. B. 2005, *ASPC*, 347, 29

† As a team, we run Shristi Astronomy (https://shristiastro.com/). Details of our activities are on the website and the reader is encouraged to have a look. All the sessions have been recorded and are available on youtube https://www.youtube.com/c/shristiastronomy

Education and Heritage in the era of Big Data in Astronomy
Proceedings IAU Symposium No. 367, 2020
R. M. Ros, B. Garcia, S. R. Gullberg, J. Moldon & P. Rojo, eds.
doi:10.1017/S1743921321000697

Real data for astronomy class in the college and university

Ilhuiyolitzin Villicana Pedraza[1]⬤, Francisco Carreto Parra[2], and Julio Saucedo Morales[3]

[1]DACC, New Mexico State University,
Central Campus, Las Cruces, New Mexico, USA
email: ilhui7@nmsu.edu

[2]Dept. of Physics, New Mexico State University,
Main Campus, Las Cruces, New Mexico, USA

[3]Dept. of Astronomy, Sonora University,
Main Campus, Hermosillo, Sonora, Mexico

Abstract. Astronomy education is an efficient means of attracting more people to study science and not just astronomy. The diversity in the majors of the students allows us to expand the knowledge of astronomy to all fields. In this paper, we present our non-traditional method of using real data and observations for our College and University classes, it allows students to learn about the applications and how to use them to study the stars.

Keywords. education, real observations, exoplanets, asteroids

1. Introduction

The correct dissemination of the astronomical material for college students during their classes will help to engage the students in selecting STEM majors. In some cases, the students do not continue with STEM majors but they will be in love with subjects related to space sciences, in which case, they will continue to support astronomy and space sciences through key positions in their careers.

Typical laboratories for College and Universities use Stellarium for the classes. Our idea of use real observations for astronomy class in college and the early years of University born in 2017, when the regular laboratories seem to be boring and not attractive for the students. In this paper we show the details of some of our laboratories to be used by other teachers.

With the public observations and large data sets we decide use one of the catalogs of the Kepler Telescope published by Jason *et al.* (2014) in our course in 2017 and we presented the results in 2018 and were published in Cheung *et al.* (2020). The detailed laboratory is in section 2.

In 2018, we use astronomical observations focus on Asteroids. With the objective to find data to plot our light curve and interpret that, we use the Asteroid Light curve Photometry Database (http://alcdef.org/). The details of this lab are in section 3.

In 2019, we did observation and Analysis using the TMO (60cm) Observatory from NMSU. We observed asteroids during 3 days in remote mode from a classroom, we use different filters such as Blue (B), Visual (V), Red (R), Infrared (I), and H alpha. We obtained a light curve for one asteroid. Also, we use the SDSS for our classes in 2019 and 2020.

Table 1. Parameters of objects associated to exoplanets

KOI	Kepler ID	P (days)	Radius R*	Associated star
41.01	Kepler-100 c	12.815842 (0.000029)	2.25(0.11)	Kepler 100

During Fall 2020, the educative team of Vera Rubin Observatory (LSST), allowed to our students to collaborate with them as testers. So 24 students from the college of the New Mexico State University (DACC-NMSU), and 30 students from El Paso Community College (EPCC Texas), had opportunity to be the first to use a virtual application, as part of their class, and submit their comments to the creators. As instructors we had opportunity to evaluate the application and the impact in the knowledge acquired for the students. Furthermore, we saw how the students were excited when we mentioned that this work will be their contribution to a national astronomical and educative science project.

In section 4, we present a couple of innovative activities: The city in the space, and Searching one habitable planet.

2. Analysis of professional astronomical observations

2.1. *Exoplanets*

One exoplanet is a planet outside of our Solar System. Usually this kind of objects orbit other star, but there are exoplanets that are not related to a star, called rogue planets. In this laboratory we worked with exoplanets associated to one star.

The first confirmed detection was doing by Michel Mayor in 1995 with 51 Pegasi b, it is at 50 ly from Earth.

The Kepler spacecraft launched in March 2009 and spent a little over four years monitoring more than 150,000 stars in the Cygnus-Lyra region. The primary science objective of the Kepler mission was transit-driven exoplanet detection with an emphasis on terrestrial (Radius major than 2.5 Radius Earth) planets located within the habitable zones of Sun-like stars.

The objective of this Laboratory is learn how to analyze scientific information from literature, then create tables and graphics to compare the exoplanets with planets of our Solar System.

For this laboratory we use the published catalog by Jason *et al.* (2014). We can find information for the observation section and planet candidate sample. Also, information from the measuring Planet Parameters for this laboratory. The specific information to do this practice related to the planets is the identifier KOI, Kepler ID , P (period days), Rro (Radius planet), we obtain Teff (effective temperature) and R (Radius star). Also, Mj mass (in Jupiter Radius) and radius Rj (in Jupiter units), PC (planetary candidate), ro (stellar density) can also be used if you wanted a more complete study (pag.5,11,42, 43,57, 58). We create a table using the information from the page 42 of stellar parameters, the table should contain KOI (Kepler Object of Interest), Kepler ID, Temperature (Teff with errors), and R* (with errors). For example, Kepler 100 has associated two exoplanets in this work, Kepler 100-b and Kepler 100-c. Table 1 shows the parameters for Kepler 100-c, the Table 2 shows the parameters of the star associated to Kepler 100-c. Now the students can create a plot period vs radius (planet) using the information of the table 1 and other plot with Teff vs R* using the table 2, with the plots the students can interpret the plots. We selected information for the first twenty systems from the table of the paper for limit of time in class.

Previously we reported results of this lab in Cheung *et al.* (2020) but never the procedure with detail. The students compared the exoplanets systems with our Solar system. Some of the results of this comparison are for example: Kepler 65 is a star more massive

Table 2. Parameters of the star associated to exoplanet

KOI	Kepler ID	Teff (K)	R*rsun
41	Kepler-100	5825(75)	1.490(0.035)

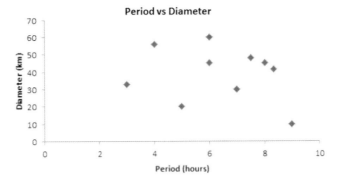

Figure 1. Relation between the Diameter and the Rotation Period for the asteroids laboratory.

than the Sun and has at three planets named (b, c and d). Kepler 65-c, with a radius of 2.57 ER, it is classified as Neptune size. Kepler 65-b with earth radius of 1.53 is a Super-Earth size which is similar to Kepler 65-d.

2.2. Asteroids

The objectives are to find a relation between the Diameter and the Rotation Period from the Photometric data. We use the Asteroid Light curve Photometry Database form the webpage http://alcdef.org/.

The students need to follow the instructions as follows: Click in the Search Database, it will go to other page for Asteroid search; In the search box write the name of the asteroid and click in submit. You need do that for all the asteroids of your list; At the bottom you will see a small table with the complete name of the asteroid, select the name of the asteroid and then display; Go to the bottom and find the summary table; Take note in one table of the diameter (km) and the period (hrs) of the asteroids from the bell asteroid zone. Fill other table for the NEAs (Near Earth Asteroids). Create a table in excel in order to plot the information, write the first column the value for Period and in the second column the value for Distance.The Period should be in X axes and Diameter in Y axes. Now the students can interpret the plot (See fig 1).

3. Innovation

3.1. The city in the space

The activity consists in write one essay in a multidisciplinary team. They need to explain what would be their contribution to an imaginary city a planet like Tatooine. The students need to assume the next considerations about the city: It is under construction, inside a dome with a diameter of 60 km, and with a supply of water from a region of craters. The settlement lacks pipes to transport water and the essential services. Also, it requires energy that will be obtained with solar panels. The area designated, already has some pre-built houses and one works as a school.

3.2. *Searching one habitable planet*

The Sun is dying and to get a ticket for a spaceship, students need to propose a close star with possibility of have an Earthlike planet. Students need to review concepts as habitable zones and type of stars for this essay. Students need to review extrasolar planet lists. Take in consideration distances and the interstellar conditions of the trip.This trip will last several generations, so students need to explain how their college skills will be necessary during this long trip.

After review all the project turned by the students, it call the attention that 75 percentage choose as the favorite candidate to escape of the death of the Sun and a possible future planet for humankind Proxima Centauri b. All of them use this option because it is an Earth like planet, close to a red star within the habitable zone. Concepts that they learned during the class. Few of them recognize that this kind of red dwarf present a high risk for the flares and the high levels of possible UV radiation. Some mention mechanisms of "evolution" to adapt a new environment.

Half of the works give a great summary of all the semester in their description. So this exercise help us to recognize how much our students learned during the semester. Also reading the references, they not only use the text book to do this work. The looked for books, magazines, webpages of different space agencies. Even science fiction was integrated in their works but they tried to be critical in what parts they can trust.

4. Conclusion

Our results show 3 educative objectives reached. The first one is an empirical mathematical interpretation: with the exoplanet systems activity, students learn and use Excel to create tables of data, plot a graphic, and practice interpretation of the information. This new skill is used with the next activity and empirically they have found relations between Diameter and Rotation period from Photometric data of asteroids. The second objective was to introduce the principles of research, and try to be critic: in order to look for a habitable planet, students red and look for different sources. They compared information, and surprisingly almost all reached the same conclusion, with some of them still mentioning some uncertainty. Finally, the third looks for the self-motivation to continue with their majors, they highlight their own skills, and we lead them to feel important in their future job, without import where they will be. In their essays they describe themselves as an important part of a crew.

References

Jason F. Rowel, Stephen T. Bryson, Geoffrey W. Marcy, Jack J. Lissaue, Daniel Jontof-Hutter, Fergal Mullally, Ronald L. Gilliland, Howard Issacson, Eric Ford, Steve B. Howell, William J. Borucki, Michael Haas, Daniel Huber, Jason H. Steffen, Susan E. Thompson, Elisa Quintana, Thomas Barclay, Martin Still, Jonathan Fortney, T. N. Gautier, Roger Hunter, Douglas A. Caldwell, David R. Ciardi, Edna Devore, William Cochran, Jon Jenkins, Eric Agol, Joshua A. Carter, and John Geary 2014, *ApJ*, 784, 45

Cheung, Sze-Leung, ...,Villicaña-Pedraza. I., Carreto-Parra, F.,..., & Shah, Priya 2020, *IAUGA*, 30, 510

Education and Heritage in the era of Big Data in Astronomy
Proceedings IAU Symposium No. 367, 2020
R. M. Ros, B. Garcia, S. R. Gullberg, J. Moldon & P. Rojo, eds.
doi:10.1017/S1743921321000600

IBSE-Type Astronomy Projects Using Real Data

Fraser Lewis[1,2] ⓘ

[1]Astrophysics Research Institute, Liverpool John Moores University, 146 Brownlow Hill,
Liverpool L3 5RF, UK.

[2]Faulkes Telescope Project, School of Physics and Astronomy, Cardiff University, The Parade,
Cardiff, CF24 3AA, Wales
emails: `fraser.lewis68@gmail.com`

Abstract. I present three examples of IBSE (Inquiry-Based Science Education) type activities for students and teachers using data and resources from the Faulkes Telescope Project and the National Schools' Observatory. Both projects have recently celebrated their 15th anniversary and both provide free access via the internet to 2-metre robotic telescopes to educational users throughout the World. Each activity contains supporting material and sample datasets in several aspects of astronomy as well as instructions on how to analyse data. These activities are designed to be 'teacher-free', extended projects for students.

They include the study of open clusters via astronomical images and population studies of exoplanets. I also present a Citizen Science project using data from Type Ia supernovae as discovered by the Gaia Alerts consortium. These data allow citizen scientists to develop their own Hubble Plot and begin to understand the link between Type Ia supernovae and the age of the Universe.

Keywords. Astronomy Education, IBSE (Inquiry Based Science Education), Big Data, Robotic Telescopes

---

1. Introduction

Based in South Wales, the Faulkes Telescope Project (FTP; http://www.faulkes-telescope.com/) provides free access to schools and other educators, via both queue-scheduled and real-time observations, to a global network of 2-metre, 1-metre and 0.4-metre telescopes. FTP was originally conceived by Dr Martin 'Dill' Faulkes as a way of promoting the teaching of STEM subjects (especially though the medium of astronomy) in schools in UK and Ireland.

Two 2-metre aperture telescopes of Richey-Chretien design were built by Telescope Technologies Ltd (TTL), a spin-off company of Liverpool John Moores University (LJMU). These two f/10 robotic telescopes are located at Haleakala on Maui, Hawai'i (FT North; FTN) and Siding Spring in New South Wales, Australia (FT South; FTS).

In 2006, FTN and FTS were bought by Las Cumbres Observatory (LCO) and since then, FTP has been an educational partner of LCO.

The LCO network now has a further nine 1-metre telescopes along with ten 0.4 metre telescopes. FTP provides free access (via the internet) to all of these telescopes via LCO's queue-scheduler and real-time exclusive access to two metre and 0.4 metre telescopes in both Hawai'i and Australia. This access is exclusively for educational users, predominantly in the UK and Ireland but also including educators and schools from other parts of the World.

FTP also provides resources (http://resources.faulkes-telescope.com/) on its website giving users information on what targets are suitable to observe, when they are visible, how to analyse the data and access the data archive. It currently has approximately 1200 registered users.

Complementary to the FTP is the work of the National Schools' Observatory (NSO; http://www.schoolsobservatory.org.uk/), which uses time on the 2-metre Liverpool Telescope (LT). The LT is based at the Instituto de Astrofísica de Canarias (IAC) at Observatorio del Roque de los Muchachos on La Palma in the Canary Islands, Spain. Also built by TTL, the LT features a broader range of instrumentation than FTN/FTS and is run by the Astrophysics Research Institute (ARI) at Liverpool John Moores University.

Established in 2004, the NSO provides schools in the UK and Ireland with access to the Liverpool Telescope through a guided observing system, using 10% of the LT's time. Its website contains astronomy related content, news and learning activities. With over 16,000 active users, including 13,000 school students, the NSO also allows non-UK/Ireland based schools and teachers to register affording free access to both the observing portal and resources.

2. The Activities

Since 2013, my work with the NSO as Operations Officer, was based around developing online resources which, to a large extent, would be teacher-free. As older school pupils develop independent learning skills, it was felt that we could provide activities which are based around aspects of astronomy that are both interesting and relevant to the curriculum. Astronomy is the perfect vehicle to explore science and STEM (especially Mathematics) and the existence of astronomical data collected by the Faulkes and Liverpool Telescopes presented a perfect opportunity to create useful, educational activities.

These activities are developed broadly in the Inquiry Based Science Education (IBSE; see https://allea.org/portfolio-item/aemase-inquiry-based-science-education-ibse/ for further details) format so that students are not presented with a series of numbered tasks to work through sequentially. Rather, they are presented with a set of webpages (in our activities, between 20 and 40 separate pages) which provide introductory and background material, as well as instructions on data analysis, sample datasets (and results) and an opportunity to reflect and share results. To give students some structure, navigation of the webpages is by a side bar menu. We firmly believe that this format provides users with an insight into the full scientific process from data collection, through analysis to presentation of and reflection on their results. We have recently introduced forums for students to share their work, accessible only via a login and moderated by NSO staff.

Depending on students' experience, we envisage that these resources will best serve students of 16 years and older, but it may be possible that younger students (i.e. 14 years +) will also enjoy and learn from participation in these activities, perhaps with additional teacher support.

We use real astronomical data, allowing students to explore the science of these objects as well as associated STEM (Science, Technology, Engineering, Mathematics) topics such as graph plotting and the measurement of uncertainties. These activities also aim to encourage the exploration of data archives. There is no requirement for specialist software - all our resources use software that is available free-of-charge and is generally usable across different platforms (e.g. MS Windows, Linux, macOS).

The first activity (http://www.schoolsobservatory.org.uk/discover/projects/clusters/main) was created in 2017. It allows students to learn about open clusters and HR diagrams as well as the technique of photometry. Including screenshots and screencasts, this activity teaches students how to analyse their data using the photomtry package,

Makali'i (https://makalii.mtk.nao.ac.jp/) and e.g. Excel and to upload their results in the form of a CMD with the aim of encouraging them to discuss their findings with other students. A template spreadsheet is provided along with a finder chart, to get students confident with a sample dataset.

Students can then choose between 28 datasets (Bessel-B and -V images from the FTP) of different open clusters but participants may wish to take their own observations of a suitable cluster of their choice with FTP or NSO. These datasets represent a subset of approximately 600 open clusters in our Galaxy that have not been intensively studied. There are even the opportunities here to go further and use either the FLOYDS (FTN/FTS) or SPRAT (LT) spectrographs to follow-up any object of particular interest, such as extremely bright stars or those which appear to be very red or blue. This gives a true flavour of IBSE in that initial studies can be seen as the launchpad to further, more in-depth research.

Our second activity focuses on population studies of currently known exoplanet systems (https://www.schoolsobservatory.org/discover/projects/exoplanets/main). It encourages students to explore the properties of exoplanets (such as radius, mass, orbital period) and investigate these to search for correlations between these properties. Students are encouraged to plot these data to search for correlations and grouping of data and to watch out for selection effects which may lead them to incorrect conclusions. By utilising the exoplanet.eu website, we are able to ensure that the list of exoplanets to explore remains up-to-date. As with our open clusters activity, extensive use is made of screenshots within our instructions.

Finally, I present a third activity which has been created as a crossover between IBSE and Citizen Science (https://www.schoolsobservatory.org/discover/projects/supernovae). This activity was funded by a grant from UK Research and Innovation (UKRI) as a pilot project using real data in Citizen Science. We recognise the huge success of the Zooniverse project (https://www.zooniverse.org/) and were keen to investigate whether users could participate more fully in the science, with both learning and enjoyment along the way. It uses data collected by FTP from Type Ia supernovae as discovered by Gaia Alerts (http://gsaweb.ast.cam.ac.uk/alerts/alertsindex). Users are instructed on how to perform browser-based photometry using an online browser-based photometry tool called JS9. From this photometry, users can add additional data-points to the Hubble Plot, enabling them to measure the expansion rate (the Hubble Parameter) and age of the Universe. Templates are provided in .xls format to allow users to plot their photometry and to create their Hubble Plot. Within the background material, students will see that their work replicates that which contributed to the Nobel Prize in Physics as recently as 2011 (https://www.nobelprize.org/uploads/2018/06/advanced-physicsprize2011.pdf).

3. Results

Since publishing these activities online (from 2017 onward), we have seen several hundred unique users working their way through the activities. We are keen to disseminate this work further via e.g. this symposium and the IAC's Astronomy Adventures in the Canary Islands (http://galileoteachers.org/astronomy-education-adventure-in-the-canary-islands-2020/), Global Hands-On Universe (GHOU; https://handsonuniverse.org/ghou2020/) and the ESA/Galileo Teacher Training Program (GTTP) (http://galileoteachers.org/esa-gttp-2020-a-journey-to-space-exploration-mission/) annual workshops and conferences.

One recent example of a student and teacher working with us is that of our follow-up imaging of Gaia Alerts targets. In 2018, the student (Jorgen Kolgjini, Eastbury Community School, London) assisted us with collecting data on the target, Gaia18aen, which was later found to be Gaia's first detection of a symbiotic star. We were delighted when Jorgen's contribution to this study (along with that of his teacher, Megan Greet) as an observer of the system using FT South, was recognised and they were both included as authors in a peer-reviewed paper in 2020 (Merc *et al.* 2020).

4. Conclusions

Each of these activities is available on the NSO website, each one using real data. This allows students to explore the science of these objects as well as the broader scientific process. It provides an insight into the idea of Big Data and incorporates STEM topics such as graph plotting, trend-lines and the measurement of uncertainties. The move toward online resources has sped up in light of the Covid-19 pandemic and we believe that resources such as those detailed here can contribute to increased understanding of astronomical techniques and how they sit within STEM.

It is envisaged that additional IBSE-type resources will be developed in coming years. From discussions with teachers, educators and students, we intend to base these resources on topics such as variable stars, black holes and spectroscopy. We always encourage interested parties to contact us with feedback and suggestions.

Reference

Merc et al., Gaia18aen: First symbiotic star discovered by Gaia, 2020, A&A, 644A, 49M

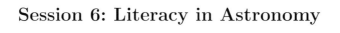
Session 6: Literacy in Astronomy

Education and Heritage in the era of Big Data in Astronomy
Proceedings IAU Symposium No. 367, 2020
R. M. Ros, B. Garcia, S. R. Gullberg, J. Moldon & P. Rojo, eds.
doi:

Introduction

Carolina Ödman begins this session with an invited talk for "Astronomy for Development - A story from South Africa." In her paper she describes the development of astronomy in Africa and its inspiration for young astronomers. She begins with its history and continues by outlining its development. Ultimately she says that the new astronomy heritage is the hands of the young and also includes Indigenous astronomy as a valued factor.

After the talk Claudio Pastrana asked:

By the end of this century, Africa will have almost half the world's population. I know of the efforts you make for education from your place. (You even help others who are on the same latitude).

Do you think that the rest of Africa will accompany the effort and bring the educational reality to the level of the reality of the expected increase in population? I'm asking to you for your very subjective, but very trusted insight.

Carolina answered:

It is a very difficult question you ask. I think we are not in control of what will happen, but I am optimistic! Wherever I look, there are fantastic young people taking on the challenges we have. And thank you for all the fantastic work you are doing and have been doing for so long!!

Rosa M. Ros, Beatriz García, Ricardo Moreno, Mahdi Rokni, and Noorali Jiwaji give us more insight into NASE with "NASE workshop: Eclipses with models and camera obscura." They explore the history of the camera obscura and both how it has been and how it can be used to better understand such as solar eclipses and Moon phases. This was correlated with the solar eclipse of December 14th, 2020. Lunar eclipses are discussed as well. The camera obscura is highlighted as a tool to fascinate people with such astronomy, and that eclipses play a significant role.

Walter Guevara asked following the talk:

In my work in solar radio astronomy, and I would like to make some experience of detecting the eclipse without seeing it, perhaps with the level of melatonin in the blood. I suggest putting a group of blind people with a watch, and recording their sensations. I have several students developing tools to record the average amount of light a week, and compare it with measurements during the eclipse, which will be partial from Peru.

Beatriz answered:

She has a project with Harvard University to transform light into sound. It was used last year so that blind people could follow the eclipse with the sound. They brought microphones and speakers. Other people, listeners, also wanted to continue with the sound, the eclipse. Regarding Walter's proposal, it is good to mention that since the temperature drop is great during the eclipse, it is not easy for blind people to detect other parameters, such as the melatonin level.

Rosa commented:

A few years ago of an activity to calculate the power of the Sun using the cheek as a detector. Blind people had a great sensitivity, greater than that of hearing people.

Manuel Núñez Díaz added:

In my center we has lasted all year, of drawing a solar analema, with stickers. We were occasions to explain concepts to the students, and to teach them that many times the result of an experiment or observation requires many months of collecting data.

David Gastelu continued:

In Uruguay, in my country there is a plan for the use of technology in the classrooms, and we are working with Arduino boards with light sensors, and the students are going to record changes in lighting during the eclipse.

Cristian Goez then added:

I attended the 2017 eclipse in the USA, and recorded the meteorological data with an SQM station. It was no possible to travel in 2019, and it will not be able to go now, and it was requests that someone collect the data and share it, so that it will be possible compare them.

Adita Quispe followed with:

From the Planetarium of the Geophysical Institute of Peru. It was a magnificent experience in observing the 2019 eclipse from Arequipa. Many people were exciting, and gathered at the stadium where they installed the instruments for observation. It is important to comment on the great capacity of the NASE Course Workshops to transmit knowledge in a practical and economical way. In particular, it is necessary to mention the NASE materials are adapted to the North or South hemisphere according each country

Next Franz Kerschbaum and Magdalena Brunner describe "Herschel and the invisible end of the rainbow." This article discusses the application of infrared radiation through diverse methods and media. A historical approach is taken and current method are discussed. Means of different and complimentary means of communication are offered.

Sara Ricciardi writes of "Engaging young people with STEAM : Destinazione Luna." In it Sara describes an initiative inspired by the 1969 NASA landing on the Moon. This was used to reach many children surrounding Bologna and Sara described its evolution. She describes evolution from STEM to STEAM and the importance of including Art. Anna continues that astrophysics is a history of light and related activity can inspire curiosity on students.

Following Sara's talk Claudio Pastrana asked:

How can I get the permissions to copy the model? I am thinking of doing it with polystyrene foam of the high density of about 5 cm of thickness, which would make possible the transport and would reduce the weight. Some written permission or some paperwork it's needed?

Sara responded:

No permission needed don't worry. It will be super nice if you can send me a picture. I can also send plans (I have a simple cad) and there are the files online if you what to 3dprint. Any question or comment send me a line sara.ricciardi@inaf.it

Durruty Jesús de Alba Martínez stated:

For the Constelacione Luna Project are considered stories and literature as Cyrano and Fontanelle?... We have one by Fr. Manuel Antonio de Rivas: Sizigías y cuadraturas lunares

and Sara Ricciardi responded:

We used contemporary picture book selected together with picture book expert for example Moon Man by Tomi Ungherer. Another we used a lot is Professor Astrocat and the frontier of Space. We got a bunch of book available and kids can choose. Every workshop was a bit different. The bibliography is very long (and we are adding). If you have other idea send me a line sara.ricciardi@inaf.it

The last article for this session is "Impact of language, culture and heritage on the way we learn and communicate Astronomy" by Basilio Solís-Castillo. In it the author describes the significance of language and culture in the way that people learn astronomy. The importance of learning astronomy in native language is emphasized for assuring inclusion, diversity, and equity. The article highlights the different perspective of learning astronomy from a Southern Hemisphere perspective. The article concludes with a discussion of the future of Chilean astronomy.

After the talk Walter Guevara Day stated :

The astronomical heritage of the Andean countries is wonderful, and there are still many places to discover.

Basilio replied:

There is so much to study and to learn from our ancestors. I think it is necessary to build our own identities... as countries and as cultures.

Education and Heritage in the era of Big Data in Astronomy
Proceedings IAU Symposium No. 367, 2020
R. M. Ros, B. Garcia, S. R. Gullberg, J. Moldon & P. Rojo, eds.
doi:10.1017/S1743921321000867

Astronomy for Development – A story from South Africa

Carolina J. Ödman [ID]

Inter-University Institute for Data Intensive Astronomy, University of the Western Cape,
Robert Sobukwe Road, Bellville 7535, Republic of South Africa
email: carolina@idia.ac.za

Abstract. In this article we describe the recent history of astronomy in South Africa from the perspective of development. We describe how all major astronomy initiatives have carried a component of development with them, be it capacity building or socio-economic development. We highlight some activities and conclude that South Africa's coherent and ambitious strategy has led to substantial changes in the astronomy research community in South Africa and that the young astronomers now starting their careers are taking possession of a bright future.

Keywords. Development, South Africa

1. Introduction

When we talk about astronomy for development, we combine two very different fields of human endeavour. One, broad ranging, with a strong focus on the well-being of people and of the planet, development. The other, a very specific area of pure science, rarely seen as applied, astronomy. For the purpose of this article, we adopt the United Nations definition of development embodied by the Sustainable Development Goals (SDGs, United Nations 2015). These are a set of 17 internationally agreed-upon goals summarised in table 1. The goals are decomposed into 169 targets and measured by 231 unique indicators (United Nations 2017–2020). The principles defining the SDGs can be summarised as follows: All human actions and activities must be inclusive and sustainable, all people and institutions must be resilient, and solidarity underpins the attainment of the goals, in particular solidarity with developing countries.

These principles are applied to human health, education, industry and work, and to the environment, with specific attention paid to water, land and food production. The last goal is one of global partnership, reinforcing the call for international solidarity. Some of the 169 targets are very clear, while some are more nebulous, describing an intention rather than a measurable number. This is to be expected, as many of the SDGs have to do with the layered and complex functioning of human societies and are therefore difficult to reduce to numbers. The indicators, however, are all numerical and translate the targets into measurable quantities. Notwithstanding the arguably reductionist approach, the SDGs with their targets and indicators form a global, tangible set of development goals, to which people and organisations can attach their work, and the achievement of which is expected to lead to the betterment of humanity and of planet Earth.

Table 1. The 17 Sustainable Development Goals.

Goal 1.	End poverty in all its forms everywhere
Goal 2.	End hunger, achieve food security and improved nutrition and promote sustainable agriculture
Goal 3.	Ensure healthy lives and promote well-being for all at all ages
Goal 4.	Ensure inclusive and equitable quality education and promote lifelong learning opportunities for all
Goal 5.	Achieve gender equality and empower all women and girls
Goal 6.	Ensure availability and sustainable management of water and sanitation for all
Goal 7	Ensure access to affordable, reliable, sustainable and modern energy for all
Goal 8.	Promote sustained, inclusive and sustainable economic growth, full and productive employment and decent work for all
Goal 9.	Build resilient infrastructure, promote inclusive and sustainable industrialization and foster innovation
Goal 10.	Reduce inequality within and among countries
Goal 11.	Make cities and human settlements inclusive, safe, resilient and sustainable
Goal 12.	Ensure sustainable consumption and production patterns
Goal 13.	Take urgent action to combat climate change and its impacts*
Goal 14.	Conserve and sustainably use the oceans, seas and marine resources for sustainable development
Goal 15.	Protect, restore and promote sustainable use of terrestrial ecosystems, sustainably manage forests, combat desertification, and halt and reverse land degradation and halt biodiversity loss
Goal 16.	Promote peaceful and inclusive societies for sustainable development, provide access to justice for all and build effective, accountable and inclusive institutions at all levels
Goal 17.	Strengthen the means of implementation and revitalize the Global Partnership for Sustainable Development

Notes:
* Acknowledging that the United Nations Framework Convention on Climate Change is the primary international, intergovernmental forum for negotiating the global response to climate change.

2. Science in Development

Certain areas of science have a direct footprint on development target. Examples include public health, energy, resource management, agricultural sciences, etc. Table 2 lists examples of applied sciences and their connection with the SDGs.

These fields of science are able to influence measures for development when they have a voice in policymaking. Evidence-based decision-making (EIDM) is increasingly favoured not just in health sciences but also in social sciences, economics and the humanities. When such informed policies are followed by effective implementation, it can be said that science had a direct impact on development, although some research seems to indicate that the gap between research and implementation in heath sciences for example, is around 17 years (Morris *et al.* 2011).

What then about fields of science that are not directly related to development? How do astronomy, quantum computing, pure mathematics or philosophy influence, let alone impact, on the development of nations and societies? An argument against those sciences is often that they represent a vain pursuit of knowledge for the sake of knowledge. Others believe that dreaming big is what gives rise to big ideas. That latter argument, when the budget for fundamental research is compared to that of social development and social services in government becomes quite compelling and has enabled countries other than high-income countries, to pursue some fundamental research.

What then about fields of science that are not directly related to development? How do astronomy, quantum computing, pure mathematics or philosophy influence, let alone impact, on the development of nations and societies? An argument against those sciences is often that they represent a vain pursuit of knowledge for the sake of knowledge. Others believe that dreaming big is what gives rise to big ideas. That latter argument, when the budget for fundamental research is compared to that of social development and social services in government becomes quite compelling and has enabled countries other than high-income countries, to pursue some fundamental research. One example of this is

Table 2. Connections between the sciences and the SDGs.

Science	SDGs	Example subfields with a direct link to the SDGs
Physics	6, 7, 12, 13, 15	Nanophysics, materials science, energy, fluid dynamics
Chemistry	6, 7, 12, 13, 15	Pharmacology, organic chemistry, Eeergy
Biology	2, 3, 6, 7, 13, 14, 15	Ecology, Biodiversity, Epidemiology
Mathematics	5, 8, 10	Statistics
Computer Science	4, 7, 9, 11	Machine learning, natural language processing, internet of things
Medicine	3, 6, 8, 15	Public Health, non-communicable diseases, epidemiology, mental health
Engineering	6, 7, 9, 11, 12	Civil engineering, electrical engineering, environmental engineering, robotics

Notes:
We acknowledge the multi- and inter-disciplinarity of the fields above. This classification is for illustration purposes.

found in the 1996 South African White Paper in Science and Technology (South African Government 1996), which states that:

> Scientific endeavour is not purely utilitarian in its objectives and has important associated cultural and social values. It is also important to maintain a basic competence in "flagship" sciences such as physics and astronomy for cultural reasons. Not to offer them would be to take a negative view of our future – the view that we are a second class nation, chained forever to the treadmill of feeding and clothing ourselves."

This perspective is often credited for South Africa's continued investment in modern astronomy, described in the section below. This white paper is now replaced with a new one (South African Government 2019), concerned more over the fourth industrial revolution, and concretising the benefits of investments in science, with a strong emphasis on capacity building, and astronomy has to demonstrate that it remains relevant in an evolving context.

When considering the contribution of astronomy to development, scientific education and the technical skills astronomers develop has been the most obvious element. As described in the IAU's 2010–2020 and 2020–2030 decadal strategies (IAU 2009, IAU 2019), education is at the centre of making astronomy contribute to a better world. And indeed, education leads to development in principle, but the studies demonstrating this link in the case of astronomy are short in supply. Another aspect of astronomy is inspiration. Indeed, it is considered a gateway science to other fields because it is so attractive. Astronomers produce beautiful images that capture the imagination and have set off many on the path of becoming astronomers. The experience of peering through a telescope can be life-changing, and as formulated in the 1994 White Paper, dreaming big should not be a privilege, but is essential to development of a nation.

3. Brief history of astronomy in South Africa

Modern astronomy in South Africa has colonial origins. The Royal Observatory in Cape Town was founded in 1820 by the British to map the Southern skies and to keep time. This was important for navigation and for the colonial trades, in goods and in people. The observatory became South African in 1972 when it merged with the Republic Observatory in Johannesburg and became known as the South African Astronomical Observatory (SAAO). It is now a research facility of the South African National Research Foundation (NRF). The Johannesburg Observatory had been set up in 1903 for meteorological purposes.

Radio astronomy did not begin in earnest in South Africa until later. In 1961, NASA set up a satellite monitoring dish at Hartebeesthoek, north of Johannesburg. NASA used the station, known as Deep Space Station 51, until 1975, when it was handed over to South African Council for Scientific and Industrial Research (CSIR) and became the Hartebeesthoek Radio Astronomy Observatory (HartRAO).

Fast forward to 2005, when the Southern African Large Telescope (SALT) was inaugurated (Physics Today 2006). The fruit of South African ingenuity and labour, it was opened as a symbol of Southern Africa carrying out its own science, with world-class instruments. SALT is the biggest single-dish optical telescope in the Southern Hemisphere and therefore provides a unique view towards the centre of our galaxy and its neighbours, the Magellanic clouds. It was constructed on a budget estimated to be 5 times smaller than a similar telescope elsewhere, mainly because it houses a spherical mirror. The mirror is segmented into 91 hexagonal elements, each about 1m in diameter. With a spherical mirror, all the elements are identical as opposed to unique for a parabolic mirror, and this is a major contributor to the reduced cost of the telescope. The telescope has some teething engineering issues in the beginning but is now contributing a steady flow of new research, and is part of international networks, such as the identification of optical and infrared counterparts to gravitational waves (B. P. Abbott *et al.* 2017).

The construction of SALT was accompanied by an innovative programme called the SALT Collateral Benefits Programme (SCBP). The goal of this programme was to ensure socioeconomic benefits locally, regionally, nationally and internationally from the existence of a world-class scientific facility. This was the first attempt at bridging the practice of a fundamental science with societal benefits directly. During the construction phase of the telescope, this ensured employment of local artisans and labourers, the upgrading of local infrastructure, and more. Once the telescope was up and running however, the question of how to develop the country through the presence of a telescope remained. SCBP has pursued and tested answers to this question since 2005. They have worked with teachers and schools with a national impact. They have unearthed the importance of cultural dimensions and indigenous knowledge and star lore to the world. They have pioneered the use of indigenous astronomy as an attractor to the modern science of astronomy. The SCBP has also constantly underlined the importance of science and mathematics for young people, an area where South Africa sadly underperforms (Reddy, V. *et al.* 2019).

To astronomers who are used to thinking that astronomical events are few and far between, the SCBP experience has demonstrated that when working with communities, any event is astronomical and that looking through a telescope at the moon remains inspiring and impactful. But the experience also shows that only through sustained activity can one truly impact communities. While the experience of SCBP is large, the evidence is still anecdotal. Academic research into the impact of SCBP was never part of its mandate, but social scientists and others are starting to show interest (see Walker, C. *et al.* 2019 and references therein) and ensuring local and regional socioeconomic development from astronomical facilities has become a flagship of the International Astronomical Union's Office of Astronomy for Development (OAD), which is hosted at the SAAO in Cape Town.

The next step in South Africa's history of astronomy is the participation in the Square Kilometre Array (SKA). The SKA, a radio telescope envisioned to have a square kilometre's worth of collecting area, was dreamt up and emerged from several proposed large scale radio telescopes in the 1990's. Ekers (Ekers, R. 2012) describes the history of the SKA. In the year 2000, a Memorandum of Understanding (MoU) between representatives of 11 countries in Europe, North America and Asia. This committed the parties to working towards the establishment of the SKA. In 2003, two years before the inauguration of SALT, and realising the quality of its sky, decided to enter the international competition to host the telescope. In 2006, the shortlisted countries were Chile, South Africa, China, Australia and New Zealand who were proposing to host the telescope jointly. After long years of proposals, technical documentation and site evaluations, the site selection was finalised in 2012, with the decision to build the SKA in both South Africa and in Australia. In 2018, on time, on budget and above specifications, the MeerKAT radio telescope was inaugurated. The MeerKAT is South Africa's 64-dish precursor radio telescope to the SKA. Since its inauguration, it has resulted in many discoveries already, including a paper in Nature (Heywood, I. *et al.* (2019)).

4. The development story of astronomy in South Africa

South Africa is a developing country. It is classed among low-to-middle income countries owing to its infrastructure, but is facing a number of development challenges, has high youth unemployment, a shrinking manufacturing sector and a large proportion of the population living in poverty. Nonetheless, the newly democratic government of South Africa from its transition from the Apartheid regime in 1994, had a vision and great ambitions, as quoted earlier from the White Paper on Science and Technology. But astronomy was not to be funded without a vision of development arising from its practice. And that's why, for example, the SCBP was originally set up. Similarly, for South Africa to engage in bidding for hosting the SKA, it had to show convincing arguments that this bid would lead to development, both to its own government and to its neighbours. These arguments were indeed convincing, and the African Union lent its support to the bid in 2010 (African Union (2010)).

In 2007, South Africa passed a piece of legislation, the Astronomy Geographic Advantage Act (AGAA, South African Government (2007)), which legally limited the creation of light pollution and radio frequency interference in a large area of the Karoo desert, where SALT is located, and where the SKA was proposed to be built. The AGAA was adopted in 2009 and defines core, central and coordinated astronomy advantage areas with regulations on the use of the electromagnetic spectrum, air traffic, land use, and more to shield astronomical facilities, built and planned, from any kind of interference. While the act has not made everyone happy in the affected areas, the seriousness of the legislation lent further credibility to the bid.

Such a law could be interpreted as a brake on development, but instead, it led to domestic technological innovation. A local cellular phone operator, wanting to provide service right to the edge of the allowed area designed new antennae that would emit in all directions but one, towards the protected area. In an interview†, the engineers explained:

> Mayhew-Ridgers and van Jaarsveld looked at a range of options, but settled on developing their own antenna technology to solve the problem. The solution was an antenna based on phased-array principles, providing omnidirectional coverage but also blocking the RF transmissions along a single direction (that would correspond with the bearing of the SKA core site). "The antenna has since been tested in the Karoo and performs extremely well. Trialling measurements have shown that the RF signal levels at the proposed SKA core site can be reduced significantly, while at the same time, much of the original GSM coverage can be retained," Vodacom said.

The South African commitment to the SKA also facilitated foreign direct investment, which is key to growth in developing countries (OECD (2002)). One example of that is cisco's investment to set up a Centre for Broadband Communications at Nelson Mandela University in the Eastern Cape province of South Africa. Cisco directly attributes the

† https://mybroadband.co.za/news/cellular/37130-quiet-cellular-antenna-technology-for-ska. html

decision to invest to the SKA efforts‡. This is not one of the few internationally famous South African universities, which is in itself worth noting.

South Africa's SKA programme has from the very beginning had a strong focus on human capacity development (HCD). The programme stems from a realisation that South Africa lacked the skills to benefit fully from the SKA should it be hosted in the country, both to build the facility and to do astronomy with it. SKA Africa (as it was known) embarked on a programme to fund study and professional development bursaries to build a critical mass of skilled young South Africans in all professions needed for the SKA: From vocational training in electronics to computer science to postdoctoral fellowships in astronomy. As of 2019, well over 1100 bursaries had been awarded. The SKA bursaries have also been higher than other government bursaries, ensuring good living conditions for the students (students on bursaries in South Africa live on shoestring budgets and are often struggling to make ends meet).

The bursary holders in astronomy and related fields are invited annually to a bursary holders' conference, which has become an inspirational key event on the international radio astronomy conference calendar. At that event, all bursary holders get to give talks about their research to an audience of their peers and international radio astronomers. This gives them critical practice in the skill of presenting their work as well as visibility and networking opportunities with the global well established radio astronomy research community and leadership. This has led to many South African radio astronomers having experiences abroad before coming back to South Africa to fill in research positions in academia and join the South African Radio Astronomy Observatory (SARAO). SARAO is the name of the NRF research facility under which all radio astronomy activities, SKA Africa and HartRAO have been consolidated. SARAO is tasked to manage the MeerKAT telescope and SKA-phase 1 currently under construction in the Karoo and the transition from construction to operations of the SKA is currently planned for 2028.

To teach this new cohort of radio astronomy students, the country also set up a generous SKA Research Chair programme aimed at attracting key international and national talent that can teach and supervise this new generation of astronomers. This has been very effective as South Africa is now firmly on the world map of excellence in radio astronomy, from having the humble beginnings of inheriting a NASA satellite monitoring dish in 1975.

The South African university system carries the legacy of segregated education that was in place under Apartheid. Many universities are what is called "Historically Disadvantaged Institutions" (HDIs). Much effort is put into trying to uplift those institutions. The University of the Western Cape for example, was founded in 1960 as a "coloured" university. It started an astronomy group just over a decade ago and in 2016, was ranked number one in Physical Sciences in Africa by Nature†. The Northern Cape Province of South Africa is both the largest in terms of land and the least populated one, due mostly to its unforgiving semi-desert climate and arid land. It is also home to the modern astronomical facilities. The Northern Cape is also known for its large mining industry, but until recently there was no university in the province. Inspired by the SKA's arrival as a big data machine, a new University, Sol Plaatje, was opened in Kimberley, the capital of the Northern Cape, with a specific focus on Data Science. It started off as a teaching university and is now building capacity in research as well.

As time went on, the need to develop the capacity in South Africa to handle the large amounts of data from the MeerKAT and eventually the SKA became urgent. Indeed, what would be the use of building a telescope if all the data just went abroad for ithers to

‡ https://www.cisco.com/c/en_za/about/press-releases-south-africa/2017/201712041.html
† https://www.uwc.ac.za/news-and-announcements/news/uwc-shines-brightest-of-all-in-physical-science-943

make discoveries with? This motivated the establishment of the Inter-University Institute for Data Intensive Astronomy (IDIA). A partnership of the University of Cape Town, the University of the Western Cape and the University of Pretoria, IDIA was set up to build within the South African university research community the capacity and expertise in data intensive research to enable global leadership on MeerKAT large survey science projects and large projects on other SKA pathfinder telescopes.

The Institute developed a cloud computing infrastructure installed at the University of Cape Town and allowing researchers in the partner universities and their collaborators to process MeerKAT data. To achieve that, fast and reliable data transport had to be guaranteed between the Centre for High Performance Computing, where SARAO stores and distributes the MeerKAT data from, and the IDIA infrastructure. To allow research to be done on the data, a large software development effort was also carried out, creating an imaging pipeline and visualization software for large radio astronomy data sets. Using web interfaces through Jupyter notebooks, researchers and graduate students are able to work with MeerKAT data in ways that were not possible before. The institute researches scientific software as well, developing immersive technologies for scientific data visualization using full digital dome projections and virtual reality for example.

In this data-intensive research initiative, the importance of development is not neglected. IDIA has an office for Development and Outreach that runs several capacity development programmes. In collaboration with SARAO, the OAD and Development in Africa using Radio Astronomy (DARA), a UK Newton Fund programme, IDIA has organised and hosted many big data research schools, where over a week, students from the Africa VLBI Network (AVN) countries – the original SKA Africa partners – come together and learn machine learning and artificial intelligence through diverse science projects. The institute also contributes industry skills training for the participants, helping them see themselves in private sector careers and in entrepreneurship. This includes role playing start-up pitches, industry-formatted workshops and CV labs, where students get advice on how to write their CVs for industry employment. While the skills acquired through astronomy are transferrable, they don't translate into other industries without an awareness of their applicability, so this is an important element of the training.

The institute also organises regular hackathons, shorter-formatted big data science training events, also including connections to the private sector data science context. In 2020 those events were held remotely, as all activities moved online because of the COVID-19 pandemic raging around the world.

These efforts of using astronomy to change the country is starting to bear fruit. The community of young professional astronomers and astronomers in training is reaching that critical mass where they start owning the space of astronomy research in South Africa. They come together boldly and confidently and do not let the rest of the world make them feel inferior as scientists. A great example of this is the "Astronomy in Colour" initiative, started by two graduate students and now beneficiary of a grant to run a series of talks given by trailblazers. Another initiative is the Supernova Foundation, started by a young South African astronomer that offers mentoring for female scientists by connecting them to more senior scientists across the globe.

What is also worth mentioning is that the new generation of scientists see outreach as a part of their work as scientists, not as an ad-hoc activity or an afterthought. They are very conscious that they are opening doors to younger generations to participate in science like their parents' generation could not. Children who see astronomers in South Africa today have role models that weren't there before, and those role models are as diverse as the population itself. This not only changes the face of astronomy but benefits the field enormously as well by generating interest and involvement of communities. So as we look into the universe with sharpened telescopic eyes, we also diversify the eyes that are peering into the distance.

5. Conclusion

The new astronomy heritage in the age of big data is, at least in South Africa, in the hands of young people. Decades of learning to use astronomy as an instrument for development through education, outreach, socio-economic development has proved that it is possible. We feel that it is important to mention role of indigenous knowledge in this conversation even if we have not covered in this article. Lessons learnt are that development is fundamentally interdisciplinary and the work benefits from the involvement of people in other disciplines, including social sciences and economics. But transforming the community and ensuring that the faces of those using astronomical facilities look like the people funding the telescopes is key, and it requires a, authentic effort. Finally, a concerted, coherent strategy to push scientific excellence has placed South Africa firmly on the world map of astronomy, and of science.

References

Abbott, B. P. et al., 2017 *The Astrophysical Journal Letters*, 848 L12.

African Union 2010, *Resolution* Assembly/AU/Dec.303(XV).

Ekers, R., 2012 *Proceedings of Science, Resolving the Sky – Radio Interferometry: Past, Present and Future* Manchester, UK, April 17–20, 2012.

Feder, T., 2006 *Physics Today* 59, 1, 35

Heywood, I., Camino, F., et al., 2019 *Nature*, 573, pages 235–237.

International Astronomical Union 2009, *Astronomy for Development: Building from the IYA2009.*

International Astronomical Union 2019, *IAU Strategic Plan 2020–2030.*

Morris, Z.S., Wooding, S., & Grant, J. 2011, *J R Soc Med.*, 104(12): 510–520.

Committee on International Investment and Multinational Enterprises 2002, *Foreign Direct Investment for Development: Maximising benefits, minimising costs.*

South African Department of Basic Education 2019, *TIMSS 2019: Highlights of South African Grade 9 Results in Mathematics and Science.*

South African Department of Arts, Culture, Science and Technology 1996, *White Paper on Science and Technology: Preparing for the 21st Century.*

South African Government 2007, *Astronomy Geographic Advantage Act 21 of 2007.*

South African Department of Science and Technology 2019, *White Paper on Science, Technology and Innovation: Science, technology and innovation enabling inclusive and sustainable South African development in a changing world.*

United Nations General Assembly 2015, *UN Resolution A/RES/70/1.*

United Nations General Assembly 2017, *UN Resolution,* A/RES/71/313 and *amendments,* E/CN.3/2018/2, E/CN.3/2019/2, E/CN.3/2020/2.

Walker, C., Chiningò, D., & Dubow, S. 2019, *Journal of Southern African Studies*, 45(4): 627–639.

Education and Heritage in the era of Big Data in Astronomy
Proceedings IAU Symposium No. 367, 2020
R. M. Ros, B. Garcia, S. R. Gullberg, J. Moldon & P. Rojo, eds.
doi:10.1017/S1743921321000727

NASE workshop: Eclipses with models and *camera obscura*

**Rosa M. Ros[1], Beatriz García[2]⬤, Ricardo Moreno[3]⬤,
Claudia Romagnoli[4]⬤ and Viviana Sebben[5]⬤**

[1]NASE president, Polytechnical University of Catalonia, Barcelona, Spain,
email: `rosamariaros27@gmail.com`

[2]NASE vicepresident, ITeDA (CNEA-CONICET-UNSAM) & UTN, Mendoza, Argentina,

[3]NASE Secretary, Retamar School, Madrid, Spain,

[4]Escuela de Posgrado. Facultad de Humanidades y Artes. UN de Rosario, Argentina.

[5]Escuela Normal Superior N° 34 "Nicolás Avellaneda", Rosario, Argentina.

Abstract. Ibn al-Haytham (known as Alhazen in occident), extensively studied the camera obscura phenomenon in the early 11th century. This instrument was used to obtain the projected image of a landscape on the screen and also was addopte by the scientists and famous painters along the centuries, to experiment with it until their final evolution as the modern photografic camera. The resource in the simple version of the "pinhole camera" can be used at the classroom to experience several phenomena, such us solar eclipses and Moon phases, and to each about optics and geometry. This contribution presents an application of this ingeniuos tool in the framework of solar eclipses, where the scale models are important to understand what really happens with the Sun-Earth-Moon system.

Keywords. education of astronomy, pinhole camera, Sun-Earth-Moon system, eclipses: production, eclipses: models in scale

---

1. Solar Eclipse

The eclipse of December 14, 2020 motivated that the IAU-Network for Astronomy School Education program (NASE) considered the possibility to organize a set of activites related to eclipses in order to give teachers and professor more tools to aply in the classroom that help to overcome the misconceptions on this topic.

Considering the gap that often can be find in the knowledge of the general public, it is possible to realize that not all the people accept that the solar eclipses occur when the Moon is in the new Moon phase, since it is necessary that the Moon be between the Earth, the Sun, with a very tight alignment. In that position, the Moon shows the Earth its dark part, while the illuminated part is logically towards the Sun.

To have an eclipse, the distance and sizes are important: the Moon is an small body and the Sun is a big one, but at the actual distance the apparent diameters are the same; but also there is a near perfect alignment when the Moon is new, and this occurs only a few times a year, because there is a angle of 5 degrees betwee the orbit plaes of the Moon and the Earth. On most occasions when there is a new Moon, it passes above or below the Sun, as seen from Earth.

This phenomena can be shown with a model in scale that can be easily built: the Earth will be represented by a small sphere of 4 cm. As the diameter of the Earth is about 12,700 km, at that scale, the Moon would be an sphere of 1 cm and the distance between

Figure 1. a) Simulating a solar eclipse using the model with the real Sun b) Detail of the solar eclipse on the Earth surface.

the Earth and the Moon would be 120 cm (the real one is in average 384,000 km). To contruct the final model, a stick of greater length is used as the base to mount the two spheres on pins, 120 cm apart. This model can be used outside on a sunny day (Fig. 1)to show and explain different concepts, including total an parcial eclipse because is possible to identify "umbra" and "penumbra" (Ros, R. M. (2017), Lanciano, N. (2011)).

2. Lunar Eclipse

Eclipses of the Moon can also be simulated by turning the model over. In this case the shadow cone of the Earth covers the Moon.This eclipses are visibles from all over the Earth where it is night, and only occurs when the Moon is in the full Moon phase.

The model is also used to see the ilumination of the Moon at day time, pointig the real Moon with the Moon of the model: the two spheres (the Moon of the model and the real Moon) with the same phase, since both are illuminated by the same Sun, and are seen from the same point of view. If we rotate the model, we can simulate the different phases of the Moon (Fig. 2).

Although it is more didactic to do the experiment with the real Sun, if there are clouds or the experiment is performed inside the classroom, the model can be used by illuminating it with a flashlight. In this case, is possibe to change the scale to 1/5 of the previous model. The Earth will be an sphere of 0.8 cm, the Moon will have a diameter of 0.2 cm, the distance between the Earth and the Moon would be only 24 cm and the flashlight of a mobile will be the Sun (Fig. 3). This model is specially interesting in online courses, but has not the same impact from the didactical point of view.

3. Camera Obscura, Dark Box or Pinhole Camera

The camera oscura can be a room or a box with an small hole through which the ligth enters and project an image in one screen. A simplest version of this tool, is a pinhole camera, a great device to observe the Sun, and also the solar eclipse. To construs a school version of this cara a large cardboard tube is recommended, but also a poster can be roll into a tube of about 10 cm in diameter. To fasten it, elastic bands are used. The length of the tube, determines the diameter of the image of the Sun (Ros, R. M.

Figure 2. a) Simulating a lunar eclipse with the real Sun b) Comparing the phase of the Moon with the real Moon.

Figure 3. Eclipse simulations: Lunar eclipse and solar eclipse.

Figure 4. Pinhole camera: a) measuring the diameter of the image; b) observing a parcial solar eclipse.

(2017)). One end of the tube must be cover with aluminum foil (hold it with an elastic band), and a small hole in the center with the wire of a paper clip will be the make. The other end of the tube must be cover with a semitransparent paper, which will serve as a screen. By directing the tube towards the Sun, with the end of the aluminum foil first, the projected image of the Sun will be seeing on the semi-transparent foil. By measuring the diameter of the image of the projected Sun, and the length of the tube, by simple geometry (similar triangles) the distance from the Sun can be calculated, knowing its diameter (or the diameter, knowing the distance).

Expanding this model, the outline of the Sun by projection can be observed in the "shadow" of a simple hole: on cardboard, with the fingers of the hand, kitchen objects (Fig. 5b), and even the holes between the leaves of the trees (Fig. 6a and 6b).

Figure 5. Eclipse images through the holes: a) fingers; b) kitchen objects.

Figure 6. Eclipse images through the leaves of trees; a) partial eclipse; b) anular eclipse.

It is interesting to note that the size of the hole influences the sharpness and intensity of the image, but not the size of the projected Sun, which depends only on the distance from the hole to the shadow.

As the eclipses are phenomena always impresive, the reccomendation for the teachers if there exits the possibility to observe one of this events, is to perare the activities previously to enjoy the moment with the students and all the communty.

4. Conclusions

Eclipses is a fascinating field in Astronomy. In the classroom it is a topic to teach Geometry, Mathematics, Physics and History, between other subjects. The concepts connected to the eclipses not only ae useful for a moment of the eclipses, but also to observe the environmet with new eyes: the spots of the Sun on the floor (always circular, as the full image of our star), the power of the Mathematics to estimate the diameter of the Sun or the distance between it and the Earth, the uses of the pinhole camera along different days at different times to show that the diameter of the Sun does not change at sunset, are only a few exampes of activities and project to develop as part of the new approach in education to estimulate the students and to present the Astronomy as an opportunity to a better access to science and technology, which are part of every day modern life.

References

Lanciano, N. 2011, Strumenti per i giardino del cielo, Edizioni junior, Spaggiari Eds., 165–168.

Ros R.M. 2017, Earth-Moon-Sun system: Phases and eclipses, 14 steeps to the Universe. IAU-NASE-Astronomy course for teachers, (Rosa R.M. & García B. ed.), *Antares*, 2nd edition, 76–84

Education and Heritage in the era of Big Data in Astronomy
Proceedings IAU Symposium No. 367, 2020
R. M. Ros, B. Garcia, S. R. Gullberg, J. Moldon & P. Rojo, eds.
doi:10.1017/S1743921321000351

Herschel and the invisible end of the rainbow

Franz Kerschbaum⑩ and Madalena Brunner

University of Vienna, Department of Astrophysics,
Türkenschanzstrasse 17, 1180 Vienna, Austria
email: `franz.kerschbaum@univie.ac.at`
web: `https://space.univie.ac.at/en/projects/rainbow/`

Abstract. The communication project "Herschel and the invisible end of the rainbow" features the year 1800 discovery and today's application of infrared radiation through diverse methods and different media in order to reach a wide audience. The discovery of the sun's infrared radiation by the Herschels is demonstrated in a creative way through the publication and performance of a theatre play and accompanying audio play. The documentation of the historical discovery, which changed both science and our daily life, is further supplemented by background information e.g. on the role of women in science in the late 18th and early 19th century. By this, the history of the discovery of infrared radiation becomes alive and easily comprehensible. Additionally, we carry out interactive experiments and demonstrations using a capable thermal infrared camera by which a mostly unknown and strange infrared world becomes visible for all generations. Our recent findings with the infrared space telescope Herschel are used to exemplify modern science use. With this colourful, diverse and interactive communication concept, which is easily extendable and adaptable, we already took part in several science festivals, workshops and training events.

Keywords. history and philosophy of astronomy, Sun: infrared, infrared: general

1. Introduction

While studying heat and colours, William Herschel discovered infrared radiation by chance in already 1800 (Herschel 1800 and other related papers in the same volume). It was the first *invisible radiation* that was not pure magic but was probed in a systematic way (Fig. 1). in 1801 already ultraviolet followed but other forms of invisible light only much later – radio waves in 1886 or X-rays in 1895. Today, infrared light is widely used for science and technology.

William Herschel was not working alone. Over most of his career his sister Caroline Herschel was a congenial partner. (Fara 2004, Hoskin 2005). Caroline started as assistant but over the years developed her own projects and published independent papers on e.g. comets, stellar clusters, nebulae and double stars. From 1787 on she got paid for her work by the crown, in 1828 she received the Gold Medal from the RAS, of which she became honorary member in 1835.

In order to complement our scientific work with ESA's Herschel Space telescope and our technical developments for its instrument PACS we initiated an FWF outreach programme (https://www.fwf.ac.at), which communicates the historical perspective of the original discovery of infrared radiation, provides educational hands-on experience with infrared radiation and spectroscopy, highlights the teamwork of the Herschel siblings, and the pioneering role of Caroline Herschel for women in science. The project uses a wide range of means to communicate – from art to experiment and combines several educational elements with outreach to the general public.

Figure 1. f.l.t.r. Herschels orginal experiment; our instruments; typical temperature readings.

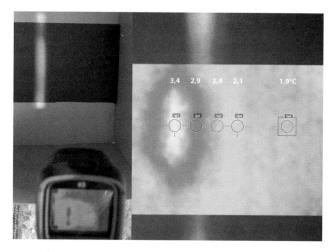

Figure 2. f.l.t.r. Liveview of *FLIR E8* thermal infrared camera of solar spectrum; superposition of visual and infrared spectrum incl. temperature readings.

2. Experimenting with visible and infrared radiation

Redoing the original Herschel experiments was the starting point of our activities. It began with research on cheap and accessible materials like thermometers and prisms and needed quite some finetuning to make repeatable solar measurements also in warmer environments (see Fig. 1). We were actually quite impressed how the Herschels made it happen when trying to replicate their experiment. Because of its relatively tricky setup with sometimes only marginal results it was not used further for our give-away experiments, the Aha!Boxes (see Fig. 5).

Today, we have other means to detect infrared radiation! With a thermal infrared camera like our *FLIR E8* (Datasheet: https://tinyurl.com/y3w4t3tx) the solar spectrum we used for our thermometer experiments can be simultaneously imaged in visible and infrared light. As shown in Fig. 2, the camera was used to measure the different temperatures of black paper illuminated by a winterly solar spectrum. An overlay of a visual image to the thermal one proves the spatial offset between the visual and infrared maximum of radiation.

Such a thermal infrared camera is also a perfect hands-on tool for the interested public. It directly shows how different the world looks like with *thermal eyes* as depicted in Fig. 3.

Figure 3. Typical FLIR E8 infrared images: f.l.t.r. Car; coffee mug; dancer in a park.

Figure 4. Demonstration of thermal infrared camera at public science fair.

People can experiment by themselves. Several applications in e.g. health or for thermal insulation purposes are easy to understand – and it is really fun!

Our thermal infrared camera was also the *star* at several science festivals and outreach events (Fig. 4) and reached very wide audiences and media coverage. The *Dance for Science* event combined art and science in a "moving" way (rightmost picture in Fig. 3). During the *Viennese Ball for Science* we streamed one of the ballrooms in infrared light and even had a *hottest dancer* contest. This entertaining part was accompanied by poster displays on related science projects.

Especially for kids hands-on experiments are crucial for impactful educational contributions. With our *Aha!Boxes* (https://ahaboxes.org/), which are small experiment boxes, we try to initiate experiments with light. As depicted in Fig. 5 one can do spectroscopy, spin a colour wheel, colour cartoons, and play Herschel theatre. The free give-away German and English boxes were produced together with the *Indian Manthan Educational Programme Society* (https://mepsindia.org/) in a social enterprise.

3. In the media and on stage

In parallel to the experimentally oriented activities, several online podcasts, interviews and videoclips round up our educational program on infrared radiation. Scientists speak about their research projects, historians highlight the societal context. In COVID-19 home-office and home-education times we also produced short feature video clips on infrared radiation and the science and technology behind.

From the beginning our aim was to offer as many as possible different and complementary means of communication. In doing so, one very special key element is our newly written 30min theatre play (Heger & Artacho 2018) where one can join the Herschel siblings during their experiments and learn about their work in an entertaining way. So

Figure 5. Aha!Boxes; solar spectrograph; colouring book; cut out "Herschels".

Figure 6. The authors playing Caroline and William Herschel, discovering infrared radiation.

far it was played by scientists (Fig. 6) at big science fairs and exhibitions. It is typically accompanied by topical talks on modern infrared science projects and hands-on activities with infrared cameras. All needed materials and scripts are available in German and English to everybody. Some first schools showed interest to stage the play themselves.

All elements of our project are documented and accessible via our webpages, both in German and English (https://space.univie.ac.at/en/projects/rainbow/). There one can learn about the historical discovery, redo the critical experiments, watch and listen to related podcasts and interviews, the theatre play and its scripts and get our *Aha!Boxes* and experiment yourself!

References

Fara, P. 2004 *Pandora's Breeches: Women, Science and Power in the Enlightenment* (Pimlico)

Heger, M. & Artacho, A. 2018 *Herschel and the invisible end of the rainbow, a play* (https://space.univie.ac.at/en/projects/rainbow/downloads/)

Herschel, W. 1800, *Philosophical Transactions of the Royal Society of London*, 90, 284

Hoskin, M. 2005 *Caroline Herschel: the unquiet heart. Endeavour* 29/1, 22

Education and Heritage in the era of Big Data in Astronomy
Proceedings IAU Symposium No. 367, 2020
R. M. Ros, B. Garcia, S. R. Gullberg, J. Moldon & P. Rojo, eds.
doi:10.1017/S1743921321001071

Engaging young people with STEAM: Destinazione Luna

Sara Ricciardi

INAF Italian National Institute for Astrophysics
OAS Bologna, Via Gobetti 101
email: sara.ricciardi@inaf.it

Abstract. On the occasion of celebration of moon landing (2019) we designed a set of educational activities for the youngest, based on the moon. We wanted to talk, play and engage young people reflecting and enjoying different points of view and demystifying the idea of science and scientists in a personal and meaningful journey. After this year of experimentation we also engaged with public schools co-designing and tailoring those activities despite the current sanitarian crisis. We will describe a physical artifact called 'Lunatario' and its 3d printable version together with cross-disciplinary educational activities and our tentative documentation. We will also describe how, with the help of a very diverse team, we embedded other media in our moon exploration working in particular with picture books and animation. We believe this is a great way to deeply engage young people with STEAM in a democratic way.

Keywords. STEM education, constructionism, Moon

1. Introduction

We believe INAF as a Public Institute could play a main role in developing digital and technological skills in education, as required by the Digital Education Action Plan [European Commission (2018)] of the European Commission also in a lifelong-learning perspective for teachers and educators and in general to cope with the needs of the so called "knowledge-based society", [Lisbon's strategy (2000)]. In the year of the anniversary of the Man on the Moon we feel the "moral obligation" to develop educational practices and outreach activities linked with such an amazing event in the history of humankind. Because of the universality of this event and the personal feeling often connected to the historical moment and to Space Exploration in general we had a great opportunity to engage with people that are not necessarily in our typical audience. We embraced this challenge and we chose to unfold this story in multiple ways, defining strategies and contents depending on the different audiences and occasions. To manage this challenge we partnered with the Cineteca di Bologna,a cultural institution and a world-renowned mecca for cinephiles, with Hamelin, a cultural association working with children and young adults through literature, comics, illustration and cinema, with Istituto Comprensivo 12, a public school complex (kindergarten to middle school) co-designing and testing the educational activities. We hosted a variety of activities: from very cozy workshops to giant events such as the final open air movie night with more than 5000 people in the main square of Bologna. For the rest of the paper we will focus on Destinazione Luna kids a portfolio of educational activities for kinder garden and primary schools.

Figure 1. The dedicated spaces for the 3 phases of Destinazione Luna KIDS as implemented at Cineteca di Bologna: the collective reading of picture books , the wooden Lunatario and the animation in Sala Cervi.

2. Destinazione Luna Kids (DLK)

Thanks to DLK, between January and April 2019, we managed to reach around 12 schools in the Bologna area and to meet over 300 girls and boys aged between 3 and 6. The twofold objective of our didactic experiment is to follow the children in a guided observation of the moon-object focusing on our perception and ability to infer the shape of the 3D object lighted by a beam of light and to stimulate their curiosity towards astrophysics; on the other hand we look at the moon through the lens of imagination, viewing – in the iconic *Sala Cervi* – short films dedicated to the Moon. Given the available time frame, the logistics and taking into account the cognitive development of the children at this stage we structured the workshop in a single two hours appointment and we provided follow-up materials for the teachers. We unrolled the party in three phases: a first moment dedicated to the collective reading of picture books, a second part devoted to the observation of the Moon and its phases with the wooden Lunatario and finally the experience of the movie theater (Fig. 1). We also developed other educational activities for the deployment of DLK in schools including a portable lunatario. All the resources are available on the INAF website devoted to innovation in education (PLAY website *https://play.inaf.it/destinazione-luna/* multiple languages)

2.1. *From STEM to STEAM to STREAM to STE*M: picture books and movie*

The classical definition of STEM (Science Technology Engineering and Math) used to group together those considered hard-science has been often replaced by the acronym STEAM. The A is for Art and basically represents the need to use STEAM disciplines as a true form of self expression and so self determination in your own learning because historically Art is connected with this idea. Also STEM disciplines are a way to self-express; even though scientific method is quite rigidly coded and the accuracy of its application grants the credibility itself of our research, there is an enormous space for creativity even if often this is not visible as much as in the artists' work. Another reason why not to segregate STEM disciplines from Art is the audience. Children are still not so affected by the useless atomization of knowledge in disciplines and they often learn in a "Leonardesque" fashion without worrying too much about disciplines. Finally also the research communities are recognizing disciplines as aggregate around questions, such as the twentieth-century sciences (cosmology, ecology, neuroscience) instead of sciences historically organized around funds, editors and University departments [Morin 1999]. Recently with the proposal of STREAM we add one more layer: the R is for reading and wRiting. Advocates of STREAM see literacy as an essential part of a well-rounded curriculum, as it requires critical thinking as well as creativity. This may be true but we connected STEAM and reading for other reasons; in our previous experiences we found that using a familiar object such as a picture book can help to spark the initial interest through a familiar and well known object and will help to keep a very short psychological distance between the children and the scientist or educator. The picture book is, in our opinion, the perfect object of transition from a familiar and protected world to the "realm of STEM" that for someone could be intimidating. What we want definitely to avoid is that the "WOW-effect" (intended as the feeling of wonder and excitement happening when people get to explore astrophysics) will become a sort of psychological obstacle; we believe that this "wonder" is something that is not always positive for everyone in the classroom not because those feelings are negative per se but because children could feel they are not smart enough to be involved in such wonderful (but maybe distant) ideas. Loads of evidences are collected showing e.g. the connection between self prejudice in STEM learning [Bian L., Leslie S.J., Cimpian A. 2017]. For all those reasons we believe the familiar picture book is one of the best trick to start a deep and passionate discussion about science that is not intimidating. We strongly believe in active, self-directed approaches in education and this has very often led us to value activities where the content cannot be 'transferred' or directly told or shown to the kids. Competences and beliefs cannot be taught but need a different medium (process) where they are achieved through the active construction of new and often unexpected meanings. This is always the case with modern picture books, where the meaning is not laid down in front of the reader but needs to be reconstructed in the composition of text and pictures, in the actual dynamics of the pages, where every new opening requests the active participation of the reader building a personal experience. Pictures do not only add to the text an iconography but are always necessary in the composition and are part of a whole. In the same way, in astrophysics, images are not only an explanatory addendum to the scientific content of research but are the place where the meaning is hidden and needs to be constructed through the test and the abstraction and the understanding of a scientific idea. Similar reflection guided the selection process of the short films taking care also principles of universal design e.g. we choose short film for its hermetical structure that makes it closer to poetry, we chose films without dialogues to facilitate kids' fruition. The full bibliography and filmography with guidelines are available online and on the booklet we publish.

2.2. *Lunatario design*

Astrophysics is largely a history of light even if recently gravitational waves are enriching our "view". Moreover light shapes our perception. We see things and we build models of the world around us observing the light scattering on objects. If we can't touch those objects, e.g. looking to an astronomical object as the Moon, this scattering is the only clue we have to construct our knowledge. So we decided to design an artifact that will explore this idea allowing us to visually explore the Moon and mimicking the actual observation from Planet Earth. As astrophysicists we always build artifacts [Papert 1980] such as telescopes or satellites to extract and share knowledge. Our attempt was to design an object that will allow us to tinker with the ideas of light, perception and our place in the (close) Universe. This artifact is basically a magic box, a platform for visual experiments: an octagonal wooden box (1.2 m of diameter) with 8 peeping holes to visually explore a 3D printed white moon fixed on the ceiling with a fishing string so in the dark it would appear suspended. With a light source (a torch positioned in one of the holes) the kids were able to experiment and "almost touch" moon phases. The facilitation of the Lunatario experience was playful,lightly guided and the kids build a common knowledge sharing insights between peers. The children were invited to explore the Lunatario;

 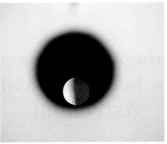 

Figure 2. A children looking at the Lunatario and his view. A little girl materializes her perceptions and ideas with paper and pencils.

freely playing and reflecting together they had the chance to independently guess what was inside the box, and how the appearance of the 3D printed moon changed according to the relative position between the light source and their position. When question or perplexity rose the educator encouraged the children to express their own idea, reformulating if needed and easing the share of the insights. When a first consensus was reached the educator changed the position of the Sun (torch). This gesture opens up more questions and thoughts and another phase of discussion. With enough time for exploration children build the idea of the importance of the relative position between the sun (torch) and the observer. We found very useful to provide some craft materials to "materialize concept" and thing a bit away from the box. For this purpose we provided black cards and white pencils so the children could choose to reproduce one or more "moon phases". This reflexive gesture was integrated as needed with new observations or "life drawing" with an eye on the Lunatario and an eye on the drawing. With the youngest this was very important because the concept of right/left and up/down dark/light may be not yet consolidated (Fig. 2).

We observed that this small and familiar moon sparkles new curiosity in children that immediately recall episodes of moon's observation and they express the desire to observe the real Moon again. The Lunatario is hopelessly exposed to the curiosity of the observers thanks to those small holes, the only way to reveal the unknown content. It's interesting to underline the analogy with so many optic toys and pre-cinematographic apparatus: from the chamber obscura to the zootrope, from the stereoscope to the Edison's kinetoscope. After all, the wonder, the "magic of vision", are the key elements in the cinema itself with reference to the relationship with the viewer. We also try other experimentation using Lunatario as a first playful step for a more complex workshop. As an example during the "Festival del Cinema ritrovato" Bologna June 2019 we animated the moon seen in the Lunatario using a stop motion app (Stopmotion Studio App *www.cateater.com*). The design of this exhibit and the associated activities are still evolving and they will be proposed and tested in other environments and targeting different audiences. We also prototyped a version of this artifact with Digital Fabrication (3D printing) for portability, to allow different solid figures in the Lunatario and with the hope to illustrate the pedagogical possibility 3D printing open up and how this technology could serve to construct knowledge and own it.

References

European Digital Education Action Plan, *COM/2018/022 final 2018*
European Council *Lisbon's strategy 2000.*
Morin E. 1999 Seuil *La Tête bien faite. Repenser la réforme, réformer la pensée*
Bian L., Leslie S.J., Cimpian A. 2017 *Science* Vol. 355, Issue 6323, pp. 389–391
Papert S.1980 Basic Books *Mindstorms: Children, Computers, and Powerful Ideas*

Education and Heritage in the era of Big Data in Astronomy
Proceedings IAU Symposium No. 367, 2020
R. M. Ros, B. Garcia, S. R. Gullberg, J. Moldon & P. Rojo, eds.
doi:10.1017/S1743921321000521

Impact of language, culture and heritage on the way we learn and communicate Astronomy

Basilio Solís-Castillo🆔

Escuela de Ingeniería, Universidad Central de Chile, Avda. Francisco de Aguirre 0405,
La Serena, Chile.
email: `basilio.solis@ucentral.cl`

Abstract. The language we speak, the culture in which we grew up and where we come from have a tremendous impact on the way we learn astronomy. Additionally, the historical predominance of Western culture has influenced the way our modern society sees the world, and of course, the sky. In this work, we will share author's experience working as science advisor in an outreach institution, where he explored different strategies to reach diverse communities and bring astronomy closer to broader audiences.

Even though the construction of world-renowned astronomical observatories in Chile has boosted the interest in astronomy on the community, many challenges have not yet been addressed. One of them is to raise awareness about the ancestral heritage of Chilean's first nations. Finally, we would like to highlight the importance of learning astronomy in our own language and therefore assure inclusion, diversity, and equity in our countries.

Keywords. Science education, Science communication, Astronomy outreach, Cultural astronomy

1. Introduction

Speaking a language different than English was, and sometimes still is, a disadvantage to learn astronomy. High impact science magazines, world-known science institutions and best-sellers books are predominantly produced and written in English, meanwhile other languages from comparable speaking population are left to play a secondary role. This is not only because English is the most spoken language with around 1.268 billion speakers around the world (according to *Ethnologue*∗) but also due to most authors believe that the language of science is and should continue to be English.

In the past, main interests of developing countries like Chile were focused only on economy, leaving education and science behind for decades. Additionally, that someone decided to study astronomy and become an astronomer was something considered to be very far from the Chilean standard and more often related to foreigners from Northern Hemisphere and first-world countries.

The present work will detail the author's experience as an astronomer and science communicator in Chile, how the current situation have changed during the last decades and what are the perspectives for the future of astronomy.

∗ https://www.ethnologue.com/guides/ethnologue200

2. Learning astronomy from the South of the world.

Looking through a standard astronomy textbook it is not a surprise to find that most references in the sky are related to Northern Hemisphere stars, constellations and visible to naked-eye objects like: Polaris, Vega, the Big Dipper, Cassiopeia and Andromeda galaxy. Their use to connect the reader to the night sky, is of course, very appropriate for the ∼90% of world population living in the Northern Hemisphere (according to *Worldometer†*), but the lack of references of Southern stars and constellations it is something that it was not considered.

Growing up with a night sky that is different from standard textbooks break the link between astronomy and the community. Even something normally considered to be very simple like finding Orion constellation can be a bit tricky for the untrained eye. One of the things we must have in mind is that the great hunter is seen up-side down from the Southern Hemisphere. Other Southern constellations such as Microscopium and Ara lack the easiness of recognition in the sky.

In the 60s, the construction of two observatories: Cerro Tololo Interamerican observatory‡ in 1967 (from National Optical Astronomy Observatory, hereafter NOAO§ and Association of Universities for Research in Astronomy, hereafter AURA¶) and Cerro La Silla observatory‖ in 1969 (from European Southern Observatory, hereafter ESO∗∗) changed things completely into another direction. Suddenly, astronomers wishing to work at the observatories became frequent visitors in the Northern part of the country. In the following decades, many other scientific institutions around the world put their eyes on Chile to build the next generation of telescopes. Soon, world's biggest observatory Very Large Telescope (VLT††) from ESO started to being built in Chilean soil together other important projects in consideration for the future. That was the country where the author was born.

Years later, working at the Planetarium from the University of Santiago‡‡, the author collaborated on a project called "The Universe to the south of the world: Our celestial heritage". In the context of the celebration of the bicentennial of the country's independence and with the contribution of Chilean historians, archaeologists and anthropologists, an exhibition was created to focus on what Chilean's first nations believe about the sky. Together with other initiatives along the country in the following years, Chilean community started to re-discover their celestial heritage. Words, ancient stories and traditions came to light to build again the link of the community with the sky. Nowadays, people commemorate the indigenous' new year every June for the winter solstice (in the Southern Hemisphere), the story about the Yakana (Solís-Castillo & Jaldín 2020), the celestial llama, is once again told to children in the North and the commonly known *Southern Cross* constellation is again identified as the "ostrich leg" in the Southern part of Chile. On the shoulders of our ancestors, astronomy have started to re-establish its bound with the community.

To teach about constellations in the Southern sky we must have in mind that ancient cultures did not recognize the same figures in the sky. Within the cultures that develops in the Andean region, the Inca culture was one of the most important in the Northern part of Chile before the European colonization. The Atacameños, current inhabitants of the Atacama Desert, inherited many of their beliefs and culture. In contrast with

---

† https://www.worldometers.info/geography/7-continents/
‡ http://www.ctio.noao.edu/noao/
§ https://www.noao.edu
¶ https://www.aura-astronomy.org
‖ https://www.eso.org/public/teles-instr/lasilla/
∗∗ https://www.eso.org/
†† https://www.eso.org/public/teles-instr/paranal-observatory/vlt/
‡‡ http://planetariochile.cl

Figure 1. Yakana, the celestial llama. Photograph taken by Alexis Jaldin.

the Western Hemisphere, these Andean cultures created constellations using the dark regions of the Milky Way, not stars. These are called *dark constellations* (Magaña 2006). Common Andean animals such as llamas, birds, foxes and snakes where represented on the sky filing the obscure parts of our galaxy. Figure 1 shows how the Yakana is depicted on the night sky, with Alpha and Beta Centauri as the eyes of the celestial llama. The appearance of the Yakana on the East warned the ancient Atacameños of the beginning of rainy season during the summer, and its disappearance on the West was considered the time when the celestial llama would drink water from the ocean and then bring it back in the summer. Its movement in the sky defined the calendar for these ancient cultures.

This celestial heritage is also present on the ground. In the Atacama desert, where the Atacama Large Millimeter/Submillimeter Array (ALMA†) is located in Northern Chile, the extreme conditions have preserved petroglyphs (Vilches 2005) and geoglyphs (Briones & Chacama 1987, Clarkson & Briones 2014) in several areas showing the interest on the sky of ancient cultures. Sun-like symbols, spirals, Saywas (stone altars) and straight lines on the desert's soil are commonly found to be aligned with celestial events like solstices and equinoxes (Moyano 2010). Interdisciplinary scientific teams now gather to study these traces of celestial connection of ancient cultures in Chile.

3. A brighter future for Chilean astronomy.

The exponential growth of astronomical infrastructure installed in Chile during the last decades has positioned the astronomy as a topic of interest for Chilean community and the government. Today is not so strange to find Chilean astronomers talking in TV shows, giving opinions about the future of astronomy in newspapers or science books as best-sellers in the main bookstores throughout the country. Along with public interest, astronomy research groups have been created in several universities and science institutes. On the other hand, cultural astronomy has become a relevant subject to study, gathering astronomers, archaeologists and anthropologists to study the different petroglyphs, geoglyphs and ancient ruins found along the country. These scientific teams are now recovering and re-constructing the ancient connection with the sky that was lost in the past.

Astronomy, as other sciences, is easier to learn when people can experience with their environment. Through known examples, parts of the sky are more easily identified and recognized by the community. In this context, recovering the constellations of Chilean's first nations help to bring the sky closer to people, because instead of a foreign mythological figure they are looking at something familiar and connected with their cultural heritage.

The future of astronomy in Chile seems brighter than ever with a new generation of telescopes planned to be operational in the next 10 years. Giant Magellan Telescope (GMT‡), Vera C. Rubin Observatory (former LSST§) and Southern part of the Cherenkov Telescope array (CTA¶) are international projects that will position our country as the world capital of astronomy. However, the greatest impact of all has been made on younger generations who have grown up knowing how special and unique our skies are, and how astronomy can help us understand where we come from and where we are heading. Our efforts, as scientists and science communicators, should be focused on bringing astronomy closer to communities whatever their culture, language and beliefs are. The making of a more inclusive astronomy is our greatest challenge for the future.

References

Briones, L. & Chacama, J. 1987, *Chungara, Revista de Antropología Chilena*, 18, 15–66

Clarkson, P. & Briones, L. 2014, *Diálogo andino*, 44, 41–55

Magaña, E. 2006, *Boletín del Museo Chileno de Arte Precolombino*, vol. 11, 2, 51–66

Moyano, R. 2010, *Chungara, Revista de Antropología Chilena*, vol. 42, 2, 419–432

Solís-Castillo, B. & Jaldín, A. 2020, *Fundación Chilena de Astronomía*, https://www.fuchas.cl/cielos-de-pueblos-andinos/

Vilches, F. 2005, *Boletín del Museo Chileno de Arte Precolombino*, vol. 10, 1, 9–34

† https://www.almaobservatory.org/
‡ https://www.gmto.org
§ https://www.lsst.org
¶ https://www.cta-observatory.org

Session 7: Big Data in Astronomy

Education and Heritage in the era of Big Data in Astronomy
Proceedings IAU Symposium No. 367, 2020
R. M. Ros, B. Garcia, S. R. Gullberg, J. Moldon & P. Rojo, eds.
doi:

Introduction

Introduction

This section for Session 7 on the fourth day of the symposium begins with the invited talk "The vigorous development of data driven astronomy education and public outreach (DAEPO)" by Shanshan Li, Chenzhou Cui, Cuilan Qiao, Dongwei Fan, Changhua Li, Yunfei Xu, Linying Mi, Hanxi Yang, Jun Han, Yihan Tao, Boliang He, and Sisi Yang. The authors stress the importance of education and public outreach (EPO) for astronomy and highlight data driven astronomy education and public outreach (DAEPO) as an important offshoot from the advances in Big Data and Internet technology. They describe the development of DAEPO, outline its many utilities, and stress its potential for the future.

After the presentation Anahí Caldú asked:

How do we address the large differences to technology access? I am thinking about Mexico, in which many regions have really limited access to internet or technology. Can we adapt big data projects to "simpler" versions?

Shanshan replied:

When we have this situation, we will bring data with us. Pre-download some data, install WWT on the laptop. But if the internet really limited, just download some picture and video will be great.

Juan Calos Terrazas asked:

For many students it is important to know the numerical-statistical methods with which the data can be processed. Are there seminars for data analysis?, for example with Python, R, C, etc. Thanks

Shanshan answered:

We just finish a Python seminars in China, It can be seen online but it's in Chinese. There are training last year also, you can check https://asaip.psu.edu/ this website.

"The role of Big Data in Astronomy Education" is a discussion of a primary theme of the symposium that is written by Areg Mickaelian and Gor Mikayelyan. They mention the range of astronomical observation over the entire electro-magnetic spectrum and that databases make such information available. They describe astrophysical Virtual Observatories (VOs) and an International Virtual Observatory Alliance (IVOA) to coordinate development of such technology. They point out that many astronomical education tools now use Big Data and that this greatly benefits the field.

"GALAXY CRUISE: Accessible Big Data of the Subaru Telescope for Citizen Astronomers" was written by Kumiko Usuda-Sato, Masayuki Tanaka, Michitaro Koike, Junko Shibata, Seiichiro Naito, and Hitoshi Yamaoka. They describe the GALAXY CRUISE project conducted by the National Astronomical Observatory of Japan (NAOJ) and that with it citizen astronomers classify and identify interacting galaxies on a computer screen. They point out some of the details and outline the project's value.

Following the talk Kentaro Yaji asked:

Why so many users of Galaxy Cruise in Russia and Ukraine? Are they the public people or young students?

Kumiko answered:

Articles were published on some online news sites. In addition, an influencial Russian YouTuber introduced GALAXY CRUISE.

Yusuke Tampo asked:

My question seems to related to every speaker, How the data calibration should have done for education and citizen science? Same as the data release for astronomers, or need more specified calibrations?

Baerbel Koribalski responded:

We use the same data, same calibration. What differs is the presentation of the data

and Kumiko answered: *We use the public released data, already calibrated by researchers who are in charge of the survey.*

The final paper in this section is "Creation of a MOOC and an Augmented Reality application on exoplanets for Exoplanets-A project" by Raphaël de Assis Peralta, Vincent Minier, and Pierre-Olivier Lagage. The authors describe the Exoplanets-A project and how results have been provided through the creation of a MOOC and an augmented reality application. They describe both and discuss the value for education.

Education and Heritage in the era of Big Data in Astronomy
Proceedings IAU Symposium No. 367, 2020
R. M. Ros, B. Garcia, S. R. Gullberg, J. Moldon & P. Rojo, eds.
doi:10.1017/S1743921321000594

The vigorous development of data driven astronomy education and public outreach (DAEPO)

Shanshan Li[1], **Chenzhou Cui[1]**, **Cuilan Qiao[2]**, **Dongwei Fan[1]**,
Changhua Li[1], **Yunfei Xu[1]**, **Linying Mi[1]**, **Hanxi Yang[1]**, **Jun Han[1]**,
Yihan Tao[1], **Boliang He[1]** and **Sisi Yang[1]**

[1]National Astronomical Observatories, CAS,
20A Datun Road, Chaoyang District, Beijing, China, 100101
emails: lishanshan@nao.cas.cn, ccz@nao.cas.cn

[2]Physics Department, Central China Normal University,
NO.152 Luoyu Road, Wuhan, Hubei, China, 430079

Abstract. Astronomy education and public outreach (EPO) is one of the important part of the future development of astronomy. During the past few years, as the rapid evolution of Internet and the continuous change of policy, the breeding environment of science EPO keep improving and the number of related projects show a booming trend. EPO is no longer just a matter of to teachers and science educators but also attracted the attention of professional astronomers. Among all activates of astronomy EPO, the data driven astronomy education and public outreach (abbreviated as DAEPO) is special and important. It benefits from the development of Big Data and Internet technology and is full of flexibility and diversity. We will present the history, definition, best practices and prospective development of DAEPO for better understanding this active field.

Keywords. Astronomical Education, DAEPO, Astronomical data, Education, STEAM, Citizen Science, WorldWide Telescope

---

1. Introduction

Astronomy is a natural science based on observation. From ancient time, people keep recording the observation results and forming original astronomical data. It seems natural to use these data in astronomical education and public outreach (EPO) activities. Educators have long realized that one picture shows a spectacular galaxy or a short video generated by numerical simulation which describes the evolution of the universe is better than thousands of words for student to understand. In the case of Internet, television, newspaper and other mass media, this effect is particularly obvious. For instance, people may not know what is M16 or the eagle nebula but have seen the picture of "pillars of creation"†, taken by the Hubble Space Telescope (HST). This is an example of astronomical data playing a role in EPO. It makes the public have an intuitive impression of celestial bodies like nebula.

These activities are not the major data driven astronomy education and public outreach (DAEPO) activities we want to discuss. Besides simple display as an image, astronomical data can play a more significant role in EPO activities. However, we still include them

---

† https://www.nasa.gov/image-feature/the-pillars-of-creation Feb. 23, 2018

in this paper because they impact the most extensive public and are the easiest accessible material for educators. This paper divides other common DAEPO activities into following categories: interactive data platform, scientific projects with amateur participate (mainly citizen science), interdisciplinarity study and activities. In each category, brief description, definition and best cases will be discussed.

2. Development of DAEPO

It is hard to clearly sort out the development of DAEPO, for this kind of astronomy EPO has great flexibility, diversity and contingency. It can be a small project like an interactive course based on astronomy data provided by professional astronomers or educators in local schools and museums. This kind of activities have strong randomness and the number of participants is relatively small. It also can be a multiuser online data analysis platform, with thousands of participants from all over the world.

For example, the national schools' observatory (NSO) from Liverpool John Moores University provides students around the world an opportunity to use the Liverpool Telescope (LT), the world's largest robotic telescope. It has an online interface and users can obtain different access permissions according to their roles. School teachers cooperate with NSO can carry out an interesting lesson and students can participate in the process of data acquisition. Although the threshold is high, NSO has over 16,000 active users.†

Besides the creativity and diversity of DAEPO activities, the rapid development of the communication means and the development of information technology make them even harder to collect and summarize. Even if the collection of these activities is completed right now, the information may be missing, outdated or inaccurate in the next second. The website CitizenScience.gov collectes crowdsourcing and citizen science activates across the USA. Some projects labeled "active" are actually suspended for various reasons. But through some representative events we can understand the general idea of DAEPO.

2.1. *The impact of Big Data on DAEPO*

The development of DAEPO is closely related to the growth of astronomical data volume and data processing requirements. Data records and analysis helps scientists understand many natural laws, like the Kepler's three laws of planetary motion. Compared with ancient time when people observing stars with naked eye, the inventions of astronomical telescope and photographic techniques make the observation more precise. After entering the 21st century, with the development of Big Data and Internet technology, astronomical data can be obtained, stored and transferred much faster than before and the amount of data increases every year.

Currently, astronomer use data from Sloan Digital Sky Survey (SDSS), Gaia mission, Large Sky Area Multi-Object Fiber Spectroscopy Telescope(LAMOST), Very Large Array(VLA) and many other observation facilities. SDSS, for example, Data Release 16, has 273TB in total‡. Gaia mission announced an EDR3 last year. The compressed CSV files is about 1.3TB released to all astronomer around the world (Gaia Collaboration *et al.* 2020). Chinese optical telescope LAMOST, has the highest spectral acquisition rate and the spectrum it has collected so far is more than the sum of other telescopes collected in the world. The Lamost team has already released over 10 million spectra (dr6.lamost.org). Soon, astronomy study will enter a new era of big data, not GB, TB, but PB, or even EB. The data sets acquired by telescopes will be too large to download and analyse using users' own facilities. Like the FAST, 500 meter Aperture Spherical radio Telescope in China began operation in science since 2020, will generate 20PB data

† https://www.schoolsobservatory.org/about, About the NSO project
‡ https://www.sdss.org/dr16/data_access/volume/, SDSS DR16 data volume, Dec. 2019

every year (Qian *et al.* 2019).The Legacy Survey of Space and Time (LSST) of Rubin Observatory plans to acquire 25TB every night (www.lsst.org).

With the data acquisition increase, astronomers do not have enough equipment and manpower to analyse and process all the data. They began to consider distribute some simple work and corresponding data to amateur astronomers, connected them through Internet and try to use their resources and wisdom to help scientific research. Among all these attempts, Internet-base public volunteer computing project SETI@home was one of the most famous. Launched in 1999, created by Berkeley SETI Research Center and hosted by the Space Sciences Laboratory, at the University of California, Berkeley, SETI@home aims to use the computing resources of volunteer to analysis observation data to detect intelligent life outside Earth (setiathome.berkeley.edu). Although until it stop sending out data, no sign of aliens was found, it is a significant attempt for astronomers to cooperate with amateurs. The enthusiasm and the power of amateur astronomers around the world impressed scientists. It also shows the potential of data related EPO activities.

The citizen science project Galaxy Zoo lunched in 2007 is one of the most important milestone of DAEPO. One reason for its design is also the needs for huge astronomical data processing (Masters & Galaxy Zoo Team 2020). Due to the reasonable design, public was enthusiastic with the project and some participants did make real discoveries.

2.2. *Virtual Observatory and DAEPO*

The development of DAEPO projects also benefit from the openness of astronomical data. Different from many other fields, most telescopes and astronomy projects will release their data to all the people around the world. Not only for astronomers, but also for scientists in other fields, engineers, teachers, students, etc. That means these massive data can be perfect material for EPO project.

Virtual Observatory(VO) must be mentioned when discuss the use of astronomical data. VO is a data-intensively online astronomical research and education environment, taking advantages of advanced information technologies to achieve seamless, global access to astronomical information (Cui *et al.* 2017). The resource linked by VO include observational data, computing resources, storage resources, software platforms and even observation equipments. In 2002, the International Virtual Observatory Alliance(IVOA) was formed and started to work on the development of VO standards and protocols. IVOA now has 21 members including VO projects from different countries and data centers, organizations (www.ivoa.net). With the efforts of all members, IVOA has already developed a set of standards and protocols, greatly improved the efficiency of data publishers and users. Major astronomical data centers and most observational projects in the world use IVOA protocol for data sharing and interoperability.

The concepts and protocols of IVOA not only help astronomers to better use data for scientific research, but also push the openness and sharing of astronomical data in more areas and scenarios. Data-intensively online astronomy EPO projects began to appear. Among them, the WorldWide Telescope(WWT) developed by Microsoft Research (Rosenfield *et al.* 2018), ESASky provided by European Space Agency(ESA) (Giordano *et al.* 2018) and the concept of data to dome proposed by International Planetarium Society (IPS: ips-planetarium.org) are relatively influential. Besides, with the continuous development of Big Data technology, such projects show good scalability and portability. For example, the interactive data visualization platform WWT can support normal PC, planetarium and virtual reality equipment. With the help of different media platforms and scenarios, astronomy knowledge will be transmitted more efficiently and accurately. The distance between the public and astronomical research are shortened.

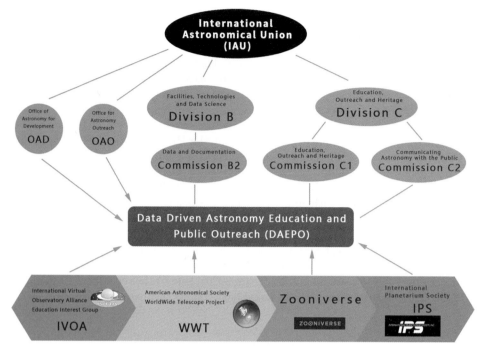

Figure 1. The organizational structure of IAU working group DAEPO.

With data related EPO projects become more and more active,people realized the importance of such projects to astronomy community and the development of astronomy EPO. In April 2017, Dr. Chenzhou Cui from National Astronomical Observatories, CAS (NAOC), PI of the Chinese Virtual Observatory, formally proposed the idea of DAEPO. He is the first one to use the phrase to summarize this kind of project. The International Astronomical Union (IAU) working group DAEPO established by him is hosted at the IAU Division B (Facilities, Technologies and Data Science) Commission B2 (Data and Documentation), and organized jointly with Commission C1 (Astronomy Education and Development), Commission C2 (Communicating Astronomy with the Public), Office of Astronomy for Development (OAD), Office for Astronomy Outreach (OAO) and several other non IAU communities, including IVOA Education Interest Group, American Astronomical Society Worldwide Telescope Advisory Board, Zooniverse Project (Fig. 1). It initially consisted of 16 members. All of them are professionals in related areas, including Dr. Beatriz Garcia (the chairman of IAU Commission C1), Dr. Pedro Russo (the chairman of IAU Commission C2), Dr. Sze-leung Cheung (the director of OAO), Dr. Kevin Govender (the director of OAD), Dr. Mark SubbaRao from IPS, Dr. Chris J. Lintott from Zooniverse Projct and Dr. Karen O'Flaherty from ESA. (daepo.china-vo.org)

According to the information on the official website of DAEPO working group, it has 3 major objectives. First, to act as a forum for people who interested in to discuss the value of the astronomical data in EPO, the advantages and benefits of data driven astronomy EPO, and the challenges it is facing. The working group also plan to provide guidelines, curriculum, data resources, tools, and e-infrastructure for DAEPO projects and activities. The third, it will provide best practices for reference. It is believed that with the promotion of this working group, the development of DAEPO will be more rapidly.

In order to elaborate DAEPO projects more clearly, some rough definition of DAEPO summarized here. In a broad sense, all activities, projects, groups, organizations, concepts or platforms use astronomical data to carry out education and public outreach can be called DAEPO. In a narrow sense, DAEPO projects are specially designed interactive activities to use astronomical data as core, with the purpose of spreading astronomical knowledge. It usually has 5 elements: organizer or designer, astronomical data, platform or tool, astronomical knowledge, target group or participants. In the following chapters, best cases of DAEPO will be introduced according to the classification mentioned early in this paper. In addition to project introduction, topics including the scope of the audience, the effect of knowledge dissemination, whether astronomical data is the core (indispensable), if the participants make any scientific discoveries will be discussed.

3. Application of data products in EPO

Let's go back to the "pillars of creation" we mentioned at the beginning. This kind of picture generated by astronomical data is very attractive and may be used in many EPO scenarios like physics classes, popular science articles, movies, museums, etc. Although strictly speaking these activates are not DAEPO, but they are the most widely used method. They cultivate the essential environment for all DAEPO projects. In most cases, this kind of activates is composed by data products from professional astronomer and communication activities. The content of data products is relatively simple, could be a picture, a short video or a piece of sound. These activities have the widest audience and flexible forms compared with other EPO activities. The audience of such activities usually passively accept information, and the dissemination efficiency of knowledge usually depends on the communication means used.

3.1. Data visualization

Vision is the most important feeling of human beings. At least 80 percent of the external information of a people is obtained by vision. Naturally, the results of data visualization provided by large astronomical telescopes become a major part of astronomy EPO activities. These astronomical data visualization results are mostly spectacular, bizarre, beautiful and beyond imagination. They embody the mystery of the universe and bring new stimulation and attraction to audience.

The famous Hubble Space Telescope of NASA launched in 1990 generates about 10TB of new data every year. As one of the most influential space-based optical telescope, it provides the world hundreds of pictures of planets, stars, galaxies, nebulae, asteroids, deep space objects, etc. These beautiful and magnficent pictures always spread quickly and attract public attention. After properly combining popular culture and using some imagination, these data products tend to be more influential. The image of Pluto released by NASA's New Horizons mission rapidly spread the whole world, because a heart-shaped area on it was noticed. Besides, the phrase "Pluto's heart" is mysterious and interesting enough to attract attention.

Another noteworthy instance is the first-ever black hole picture realeased by the Event Horizon Telescope (EHT) collaboration. Under normal circumstances, black holes are invisible because their gravity is so extrem, even light can not escape. But for many people with little background knowledge, it is too abstract to understand. This visualization generated by around 5PB data make people understand black hole more intuitively. The topic about this picture was viewed 860 million times on the Chinese microblog platform called Weibo. (Figure 2 UP)

#人类首张黑洞照片# 分享 申请主持人
阅读8.6亿 讨论41.7万

#国科大录取通知书带来宇宙声音# 分享 申请主持人
阅读1.1亿 讨论1.7万

Figure 2. (UP) Screenshot of Weibo topic of the first-ever black hole picture. (DOWN) Screenshot of Weibo topic of welcome package from University of Chinese Academy of Science.

3.2. *Data sonification*

Similar to data visualization, data sonification technology turns data to sound according to a certain internal logic. Some astronomers began to study it to help visually impaired people. And the meaningful products of the study can be also used in the field of EPO. For example, Jeff Cooke and his colleague from the Deeper, Wider, Faster (DWF) program build data sonification tools to help detect and study transient events in the Universe (Andreoni & Cooke 2019). CosMonic Projct developed by Ruben García-Benito uses data sonification to show astronomical theories. In 2020, the University of Chinese Academy of Science sent all freshmen an welcome package which including data sonification of 15 pulsar signals captured by FAST. The topic was very popular on Weibo and was viewed more than 110 million times in a short time. (Figure 2 DOWN)

3.3. *Means of communication*

Both data visualization and sonification products could be very simple and isolated. The audience with no research background in corresponding area can hardly interpret and learn anything from them. In this case, the astronomer, educator or media reporter who use these products need to provide external information to help people understand the knowledge behind the data. The final audience range and communication effect largely depends on these external information and the means of communication which closely related to the development of mass communication of human society. From the presses and publications, to science exhibitions, lectures, classes, summer camps, MOOCs, internet platforms, different communication way affect the efficiency of communication. But in general, the threshold for audience of these activities is very low. As long as the participants are interested in, they can be part of the process. And because audience usually passively accept information, the communication efficiency is not very optimistic.

Effected by the development of big data and Internet technology, people are immersed in all kinds of visual stimulation. Simple images of the universe in recent years were not able to attracted public attention as long as before. Stories and interesting details need to be connected with data products to create some social topics. In addition to pictures and sounds, video generated by data visualization may also be helpful.

4. Interactive data platforms

Interactive data platform usually is a data-intensive online platform with various of EPO functions and data built-in. Specially designed and developed by scientists and engineers, these platforms allow participants to interact with astronomical data flexibly and freely. Users can learn, review and apply scientific knowledge on platforms actively according to their own conditions. The emergence of these platforms are inseparable from the development of Big Data, Internet, cloud computing and information technology. It

Figure 3. WWT classroom of Lingyuan No.2 High School in China.

improves the efficiency of knowledge transfer but also raises the threshold of participation. To participate, people must be able to access specific hardware devices and have some basic computer skills.

4.1. *Data visualization platforms*

For those who have sustained insterest in astronomy knowledge and a certain self-learning ability, a platform centralizes massive astronomical data and related information can provide more possibilities. Participants can log in at anytime and anywhere to browse the data themselves or to use the platform in an EPO activity. For example, the astronomy software Stellarium first released in 2007 is an platform embedded visualization data and rendered realistic skies. It provides convenient way for users to get the star distribution at any time and any place. It is used by amateur astronomer all over the world and been translated to nearly 100 different languages (stellarium.org). Another example is ESASky developed by European Space Astronomy Centre (ESAC) Science Data Centre (sky.esa.int). It is a science-driven discovery portal allows user to simply access astronomical data through browser (Merín *et al.* 2017). Although the main purpose of ESASky is to visualize the metadata of ESAC archives and external partners, its explore mode provides an excellent opportunity for public to access real astronomical data. Currently, the platform has Spanish, English and Chinese virsions.

As the first designed astrophysical data exploration platform in the era of Big Data, WorldWide Telescope (WWT) is a tool included real observational data and simulation data for multi-purpose, like demonstration of scientific research results and interactive STEAM education (Rosenfield *et al.* 2018). It has a web portal and also a PC client which is able to cache data for offline use. With the world wide telescope concept orginated in 2001 (Szalay & Gray 2001), the platform with same name was launched by Microsoft Research in 2008 and open-sourced in 2015. The management of it was transferred to American Astronomical Society (AAS).

In China, WWT debuted in 2008. Since then,the Chinese Virtual Observatory team keeps introducing it to teachers and educators in China and published articles, books to help them use astronomical data to orgnize EPO activities. Because of the flexible forms and functions of WWT platform, it can adapt to different groups of audience with different types of EPO activities. But the threshold for organizers is high. Till the end of 2020, totally 11 WWT teacher training workshops, 2 online sessions were held in China and four national WWT tour contests have been organized. Over 2,000 science educators and students participated directly in these programs. 8 classrooms and 16 planetariums equipped with WWT were already put into use or under construction (Fig. 3). In November 2019, the download times of the China-VO version of WWT reached 40,000.

4.2. *Crowd Computing*

Originally designed and developed to support project SETI@home, Berkeley Open Infrastructure for Network Computing (BOINC) is an open-source middleware system for volunteer computing and grid computing. Now over 30 projects from diverse research areas are running on this platform, including astrophysics program like Asteroids@home, Einstein@home, Universe@Home, etc.† Benefit from the successful of SETI@home, volunteers from all over the world activily participate in these projects and help scientists with data analysis. Although this platform is successful in applying data processing, the participants are hard to learn anything from it because they only contribute their computer resources. Some of the projects have no detailed background, and the research related are too professional for general public to understood.

5. Scientific projects with public participation

Some DAEPO projects designed to let the public participate in the process of data acquire,classify and analyse. These projects usually proposed by astronomer in one particular field with a clear scientific goal. The organizer may set a requirement for participants and provide planty of background information. All people involed are consciously participating in the research process. According to the form of organization, there are two types of DAEPO projects fall into this category.

5.1. *School cooperation projects*

Astronomer from astronomical telescope or observatory can provide a certain observation time and data for EPO activities. They are usually familiar with telescopes' scientific objectives and are able to separate part of the research work to design some simple tasks for young students. In order to ensure the effect and for the convenience of management, some projects cooperate with interested and eligible teachers and schools. NSO mentioned before is one of the example. Although it is now open to users all over the world to apply observation time of Liverpool Telescope, cooperate teachers and students have higher authority.

Another example is the Pulsar Search Collaboratory (PSC) formed in 2007 to help scientists analyze more than 30TB data acquired by Green Bank Telescope.† Stared with local interested school and students, PSC has expanded nationwide (Williamson *et al.* 2019). Students have to attend an online training to participate and have actually dicovered new pulsars. Similar to it, PULSE@Parkes organized by Commonwealth Scientific and Industrial Research Organisation (CSIRO) Australia Telescope National Facility (ATNF) uses the observational time and data from Parkes radio telescope to study pulsars (Hollow *et al.* 2008).

5.2. *Citizen science*

The examples from above show there are some simple and repetitive tasks in astronomy research can be completed by students and the general public with certain training. Some astronomical discoveries in history have also proved this point. After some modification, projects like NSO and PSC can be easily extended to the public.

Galaxy Zoo is a DAEPO citizen science project originally designed for all the people around world. At the begining, it use the data set Data Release 7 (DR7) from Sloan Digital Sky Survey (SDSS) to let participants assist in the morphological classification of a large numbers of galaxies (Lintott *et al.* 2008). All galaxies are classified into one

---

† https://boinc.berkeley.edu/projects.php, Project list of BOINC
† http://pulsarsearchcollaboratory.com/home/about/, About PSC projects

of six categories – elliptical, clockwise spiral, anticlockwise spiral, edge-on , star/don't know, or merger†. Later, more data set was added in, including optical Hubble Space Telescope surveys, CANDELS survey, MaNGA, UKIDSS, etc. Within a decade, more than 1.4 million galaxies was classified (Masters & Galaxy Zoo Team 2020).

After Galaxy Zoo, Zooniverse launched dozens of similar citizen science projects. Many of them can be called DAEPO projects. For example, the one cooperated with NASA using Wide-field Infrared Survey Explorer (WISE) mission data to find Planet Nine. In most of these projects, participants only have to classify or tag the data. More than a million people around the world participated projects on this platform. Over 180 papers were published in space area including 67 from Galaxy Zoo series of projects‡. A bizarre object called Hanny's Object near the spiral galaxy IC2497 is one of the most famous discoveries of the project (Lintott *et al.* 2009).

Popular Supernova Project (PSP) designed and created by Xingming Observatory and China-VO pushed a little further for the data used in this project is from an amateur observatory. Launched in 2015, over 20,000 user registered. They can start to browse the image after they pass the basic online test. Totally 24 supernovae and extragalactic novae were discovered. The youngest discoverer only 10-year-old. (psp.china-vo.org)

5.3. *Amateur observation*

Some astronomy EPO projects call on amateur astronomers to submit their observation results to help form valuable astronomical research materials. Like International Meteor Organization (IMO) collects meteor observational records from people around the world. The International Occultation Timing Association (IOTA) teaches amateur observe and time occultations to help the discovery of new double stars and other astronomical research. There are also projects like JUNOCAM from NASA allow participants to upload telescopic image of Jupiter captured by themselves to help NASA plan future mission.

6. Interdisciplinary study and activities

With the increasing professionalism of DAEPO projects, the threshold for participation is rising and the number of participants will relatively reduce. But they can provide better experience and more meaningful achievement for participants. If keep increasing the requirement of participants, for example, they have to be a well trained programmer or engineer, an interdisciplinary study activities can be organized. In 2013, Galaxy Zoo Challenge held by Zooniverse and Kaggle have data scientists write algorithm to classify the morphologies of galaxies. Over 326 teams participated§.

In China, China-VO team held an astronomical data mining contest with Alibaba Cloud to classify selected spectrum data from LAMOST. All participants can use the resource provided by Alibaba Cloud and apply machine learning method to analysis data¶. Nearly 1,000 people participated. China-VO organized 2 AI competitions base on PSP data in 2019 with FUTURELAB and Kaggle. 483 teams from universities and colleges participated in these contests.

At present, due to the high resource requirements, such DAEPO contests are rare. However, this may be an important direction of DAEPO in the future. In these activities, the audience are professionals from other fields. They will not only learn experience and

---

† https://data.galaxyzoo.org/, Data of Galaxy Zoo
‡ https://www.zooniverse.org/about, About Zooniverse
§ https://www.kaggle.com/c/galaxy-zoo-the-galaxy-challenge, The information of Galaxy Zoo Challenge, Dec. 20, 2013
¶ https://tianchi.aliyun.com/competition/entrance/231646/introduction/, Tianchi astronomical data mining competition, Apr. 15, 2018

knowledge in astronomy, but also use astronomical data as research resources in their own fields. These DAEPO projects will help promote mutual exchanges and the progress in various scientific research areas. This is in line with the current development trend of astronomical research and the vision of IAU including OAD, OAO and the Office of Astronomy for Education (OAE). It will help further use of astronomy as a tool for development to benefit society. Currently, IVOA is working with OAD on utilization of astronomical data and related technologies.

7. Prospective development

Because of the diversity and flexibility of DAEPO, it is hard to predict the future development of it. But the trends can be discussed. With more and more people interested in DAEPO activates, professional, educator, students and public all begin to organize or participate in. For scientists, future research in some astronomy areas like time-domain astronomy may need the participation of amateur astronomers. Astronomy data scientists need to provide more available dataset and standards for DAEPO. For organizers, goals are to increase the scope of audience, reduce the threshold of participation and maintain scientific content. In order to achieve these goals, they need to work closely with scientists to carry out a delicate and appropriate design. For educator, suitable DAEPO projects can enrich teaching content and practice STEAM education. Combined with DEAPO projects and activities, some special tools and places like astronomical museum, interactive planetarium will be more common. In the future, The appear of more new media and technologies like virtual reality (VR), naked-eye 3D will bring more possibilities to DEAPO.

8. Acknowledgement

This paper is supported by National Natural Science Foundation of China (NSFC)(11803055), the Joint Research Fund in Astronomy (U1731125, U1731243, U1931132) under cooperative agreement between the NSFC and Chinese Academy of Sciences (CAS), the 13th Five-year Informatization Plan of CAS (No. XXH-13514, XXH13503-03-107). Thanks the teams of China National Astronomical Data Center (NADC), China-VO, and NAOC - Alibaba Cloud Astronomical Big Data Joint Research Center.

References

Gaia Collaboration, Brown, A. G. A., Vallenari, A., *et al.* 2020, arXiv:2012.01533

Qian, L., Pan, Z., Li, D., *et al.* 2019, Science China Physics, Mechanics, and Astronomy, 62, 959508. doi:10.1007/s11433-018-9354-y

Masters, K. L. & Galaxy Zoo Team 2020, Galactic Dynamics in the Era of Large Surveys, 353, 205. doi:10.1017/S1743921319008615

Cui, C., He, B., Yu, C., *et al.* 2017, arXiv:1701.05641

Rosenfield, P., Fay, J., Gilchrist, R. K., *et al.* 2018, apjs, 236, 22. doi:10.3847/1538-4365/aab776

Giordano, F., Racero, E., Norman, H., *et al.* 2018, Astronomy and Computing, 24, 97. doi:10.1016/j.ascom.2018.05.002

Andreoni, I. & Cooke, J. 2019, Southern Horizons in Time-Domain Astronomy, 339, 135. doi:10.1017/S1743921318002399

Merín, B., Giordano, F., Norman, H., *et al.* 2017, arXiv:1712.04114

Szalay, A. & Gray, J. 2001, Science, 293, 2037. doi:10.1126/science.293.5537.2037

Williamson, K., McLaughlin, M., Stewart, J., *et al.* 2019, The Physics Teacher, 57, 156. doi:10.1119/1.5092473

Hollow, R., Hobbs, G., Champion, D. J., *et al.* 2008, Preparing for the 2009 International Year of Astronomy: A Hands-On Symposium, 400, 190

Lintott, C. J., Schawinski, K., Slosar, A., *et al.* 2008, MNRAS, 389, 1179. doi:10.1111/j.1365-2966.2008.13689.x

Masters, K. L. & Galaxy Zoo Team 2020, Galactic Dynamics in the Era of Large Surveys, 353, 205. doi:10.1017/S1743921319008615

Lintott, C. J., Schawinski, K., Keel, W., *et al.* 2009, MNRAS, 399, 129. doi:10.1111/j.1365-2966.2009.15299.x

Education and Heritage in the era of Big Data in Astronomy
Proceedings IAU Symposium No. 367, 2020
R. M. Ros, B. Garcia, S. R. Gullberg, J. Moldon & P. Rojo, eds.
doi:10.1017/S1743921321000648

Construction of didactic devices that materialize the states of illumination of the ground and of the planet

Néstor Camino🆔

National Coordinator of NAEC Argentina – OAE IAU
Complejo Plaza del Cielo. CONICET-FHCS UNPSJB., Esquel, Patagonia, Argentina.
email: `nestor.camino.esquel@gmail.com`

Abstract. The original design of the IAUS367 included the development of a Workshop, thought to be a face-to-face instance, in order to interact with teachers and researchers, and to discuss the design, the theoretical foundations and the use of some didactic devices to observe and record the state of illumination and shadows casting variations on different objects, natural or artificial, through time, and its interpretation from a local, topocentric perspective, together with a global, planetary perspective. The pandemic forced us to convert this activity in a virtual one, which was very interesting anyway, with the participation of colleagues from different countries and with great theoretical wealth in the discussions. The Workshop lasted 2 hours, in Spanish language. We present here the initial proposal and a summary of what was discussed. (The complete video register can be accessed in the YouTube channel of IAUS367).

Keywords. Construction of didactic devices. State of illumination. Shadows. Local/Global vision. Didactics of Astronomy.

1. Key elements for the construction of didactic devices in Astronomy

The design, construction and systematic utilization of didactical devices to observe and register the diurnal and annual variations of the state of illumination of the ground, of natural and artificial objects of our environment, and of the shadows they cast, even local and of the planet as a whole, have some key elements, which are of great relevance for our proposal for the Didactics of Astronomy (Camino 2012, 2021; Lanciano 2019). **To experience and become aware of regularities** in the everyday environment (space and time), specially related with solar illumination (light and shadow). **To observe and register phenomena in a systematic way** in contact with the sky, and to learn how to construct questions, a task shared with others over time. **To ask ourselves critically** about what the purpose of recording regularities is (practical, cultural, etc.). **To learn how to materialize those regularities**, especially through markers in space and time, and to understand the existing ones, whatever material or social, generated by other cultures. **To understand that life in society**, even today, **is immersed and requires knowledge of the observable universe**, in particular the phenomena of the sky and their interaction with the place where we live.

2. Some examples of devices and the activities we have developed

Natural or artificial photometers. All objects can be used to compare the illumination states of their different parts. Each object could act, for this reason, as a "comparison photometer", which then allows us to utilize them as indicators of the position of the Sun in the local sky, and as devices to measure time. From mountains to buildings, their state of illumination brings us information of the real sky in real time. **Vertical gnomon**. A straight, vertical rod with a hole at its end, 1 m long (for the sake of simplicity in many mathematical calculations) is a simple but powerful device for the Didactics of Astronomy. Furthermore, we utilize threads of colour to materialize the rays of light that pass through the hole and to materialize the shadows of the gnomon on the ground. The diurnal and annual regularities in the length and orientation of the shadows of the gnomon make it possible to determine the spatial (meridian) and temporal (solar noon) symmetries at the observation site. The light and shadow structures generated by a gnomon during equinoxes and solstices form sections of three-dimensional cones in 3D space and conics on the ground, all of them beautifully materialized by the color threads. On the ground, the relationship between percentage of it covered by the threads that materialize the shadows is equivalent to the relation of the day-night length during that day (space and time linkage). We've developed many educational and research projects in the last decades utilizing these devices. In Patagonian Primary schools (Camino 1988–1989; Camino *et al.* 1998; Camino *et al.* 2020), many children worked weekly, during a two-years period, on the observation and recording of shadows of a gnomon, until they built a sundial. Other projects were developed in a collaborative way with students and colleagues of South America: during an equinox, the latitude of the observation site can be determined (Camino *et alii* 2009), as well as during equinoxes and solstices it is possible to determine the obliquity of the Ecliptic (Camino *et al.* 2014) and the determination of the Analemma (Camino *et al.* 2016) as well. These works highlight the awareness of what is common to all the observers on the Earth and what identifies each one depending on the place of observation and of the culture itself. **Smooth Sphere**. A sphere without any markings, being illuminating by the Sun, is a very important educational tool. Observing and marking on the sphere the position of the terminators (instantaneously maximum circle separating light and obscurity), day by day, as well as utilizing small gnomons casting little shadows, allow us to materialize on the surface of the sphere the observed regularities. During equinoxes and solstices, it is possible to construct the concepts: poles, meridians, parallels, polar circles, tropics, equator, and to estimate seasons, day/night relation, times of sunrise/sunset/noon, for each place in the world, in real time. Those markings on the planet gave rise to many social conventions, some still present nowadays, like the International Time Zone system. The Smooth Sphere brings the foundations of the most powerful tool for the Didactics of Astronomy, the Parallel Earth Globe (PEG). (Camino *et al.* 2020) **Parallel Earth Globe**. The use of a PEG makes it possible to build a dual vision (local and planetary) of the astronomical environment in which we live (Lanciano 2012). Decades of didactical work utilizing the PEG show it is a powerful device not just for the Didactics of Astronomy, but for a multicultural democratic worldview as well (access the official web site of "Globo Local: International movement for the liberation of the globe from its fixed support" project (Lanciano 2020). **The complete set of didactical devices for Didactics of Astronomy by naked eye**. The set of devices that we propose is made up of a gnomon, a smooth sphere and a PEG, located fixed and permanent in the place of observation, see Fig. 1, they should be used continuously through long periods of time, been scholar periods or extracurricular ones. To improve the teaching/learning process about the sky and of the linkage of humans and societies with the sky, it is of great relevance to recognize and to represent, previously, the local horizon by means of drawings, photographs and concrete models. Whenever it should be possible, the construction of sundials, in public spaces and in schools, is an educational and cultural element of great social relevance, and it could be considered as the materialization and conceptual synthesis of the didactical

Figure 1. Complete set of didactical devices (Smooth Sphere, equatorial sundial, gnomon, Parallel Earth Globe).

developments presented here (an equatorial sundial should be perfect, and a meridian line in-doors as well).

3. The possibilities offered by our conception of the Didactics of Astronomy

Our proposal for the Didactics of Astronomy bring some possibilities of great importance, not only for the relationship with the sky, cultural and personal, but for the process of knowledge construction in general. In first place, "knowing how to do, while living in contact with the sky", without sophisticated technology being a necessary condition, restrictive and away from real people; in second place, "knowing how to build knowledge", scientific or of any other type of knowledge, strengthening the bond of oneself with the universe and with others, improving the creativeness to formulate questions that lend us to further explorations. I like to think that through this conception of our relationship with the sky, and its educational and social relevance, we could talk and understand each other with the people of the cultures that we study nowadays through Cultural Astronomy. If all of today's technological society disappeared, who would still be able to build knowledge associated with the sky? I consider that nowadays astronomers should be at least as solvent as astronomers were centuries ago, with the ability to build knowledge by naked eye, and then dedicate ourselves to deepen what we've built utilizing very complex technology and modern theoretical constructs. We should have included in our "astronomical genetics" what the astronomers of the past did, and then as sons of our times do all the current scientific Astronomy.

4. Discussion during the Workshop

The participants highlighted their interest in working with the naked eye, in contact with the sky, together with other people, strengthening learning to observe, describe, conceptualize and explain, and critically generate questions. It is very important the link with schools, teachers and student's families to develop long-term projects on Didactics of Astronomy, e.g. in order to continue with the observations throughout the year, including vacations. Strengthen work with teachers, so that as people and as education professionals they know theoretically and experientially what they will later propose to students; the inclusion of Didactic of Astronomy in the initial training of teachers must also be strengthened.

The importance of the horizon and the local landscape is highlighted, and of the systematic observation and recording of the rising and setting positions of the Moon and the Sun, and of some stars, on that local horizon, as a basis for the work. The devices shown during the Workshop are built as the result of a long-term educational process, and they are built not as a recipe but as a concretion of what has been lived, and what is didactically worked on from what has been lived. Each object, natural or artificial, in its illumination states and in the shadows they cast, carry information about the astronomical environment, you just have to observe them, record them and build knowledge. Every shadow says something, every illuminated surface says something...

How to move from the recognition of the observable in place to the conceptualization offered by a Parallel Earth Globe, which already explicitly presents a conception of a spherical Earth, is a challenge not only to didactics, but also is highly relevant for works on Cultural Astronomy , since the practices of different indigenous peoples throughout history did not necessarily lead to the construction of a spherical model, with a global perspective, of the place where they live. Direct, experiential experience, over long periods of time, is essential to facilitate the passage of spatio-temporal conceptions and to work on the possible epistemological obstacles that could exist in each age group.

References

Camino, N. (1988–1989). Revista "El Gnomon Patagónico". Complejo Plaza del Cielo, Esquel, Argentina.

Camino, N. et al. (1998). "Construcción de las nociones de espacio y tiempo en segundo y tercer ciclos de la EGB. Aspectos conceptuales y didácticos de la determinación de la posición en el espacio y el tiempo mediante la construcción de un reloj de Sol", Parte I. Actas del SIEF IV. pp. 83–91. APFA, La Plata, Argentina.

Camino, N. et alii. (2009). "Observación conjunta del Equinoccio de marzo, Proyecto CTS 4 – Enseñanza de la Astronomía". Caderno N°31 (número especial), Sociedade Brasileira para o Progresso da Ciência.

Camino, N. (2012). "La Didáctica de la Astronomía como campo de investigación e innovación educativas". En Bretones, Paulo (compilador), Actas electrónicas del I Simpósio Nacional de Educação em Astronomia (SNEA I). Rio de Janeiro, Brasil.

Camino, N. et alii. (2014). "Determinación de la oblicuidad de la Eclíptica. Proyecto de observación conjunta entre Brasil y Argentina". Actas del Tercer Simposio Nacional de Educación en Astronomía, Curitiba, Brasil.

Camino, N. et alii. (2016). "Determinación observacional de la Analema. Proyecto de observación conjunta sudamericano". Actas del IV Simpósio Nacional de Educação em Astronomia. Goiânia, GO, Brasil.

Camino, N. et al. (2000). "Construcción de las nociones de espacio y tiempo en segundo y tercer ciclos de la EGB. Aspectos conceptuales y didácticos de la determinación de la posición en el espacio y el tiempo mediante la construcción de un reloj de Sol", Parte II. Actas del SIEF V. pp. 175–178. APFA, Santa Fe, Argentina.

Camino, N. (2021). "Diseño de actividades para una Didáctica de la Astronomía vivencialmente significativa". Góndola, Enseñanza y Aprendizaje de las Ciencias. 16 1.

Lanciano, N., (2019), Strumenti per i giardino del cielo, IV Ed., ed Asterios, Trieste.

Lanciano, N. (2020). "Globo Local" Project official web site: www.globolocal.net

Lanciano, N., Camino, N. (2012). "Le nuove visioni per il mondo nascono da nuove visioni del mondo/della Terra". En Falchetti, E., Utzeri, B. (curadoras), I linguaggi della sostenibilità. Nuove forme di dialogo nel museo scientifico. ANMS e-Books, Roma, Italia.

Education and Heritage in the era of Big Data in Astronomy
Proceedings IAU Symposium No. 367, 2020
R. M. Ros, B. Garcia, S. R. Gullberg, J. Moldon & P. Rojo, eds.
doi:10.1017/S1743921321000570

The role of Big Data in Astronomy Education

A. M. Mickaelian[iD] and G. A. Mikayelyan[iD]

NAS RA V. Ambartsumian Byurakan Astrophysical Observatory (BAO),
Byurakan 0213, Aragatzotn Province, Armenia
email: `aregmick@yahoo.com`

Abstract. We review Big Data in Astronomy and its role in Astronomy Education. At present all-sky and large-area astronomical surveys and their catalogued data span over the whole range of electromagnetic spectrum, from gamma-ray to radio, as well as most important surveys giving optical images, proper motions, variability and spectroscopic data. Most important astronomical databases and archives are presented as well. They are powerful sources for many-sided efficient research using the Virtual Observatory (VO) environment. It is shown that using and analysis of Big Data accumulated in astronomy lead to many new discoveries. Using these data gives a significant advantage for Astronomy Education due to its attractiveness and due to big interest of young generation to computer science and technologies. The Computer Science itself benefits from data coming from the Universe and a new interdisciplinary science Astroinformatics has been created to manage these data.

Keywords. Big Data, Astronomical Surveys, Catalogues, Databases, Archives, Multiwavelength Astronomy, Multi-messenger Astronomy, Astroinformatics, Virtual Observatories.

1. Introduction

In Astronomy we deal with vast number of objects, phenomena and hence, big numbers. Astronomy and its results also enlarge most of other sciences, as any research on the Earth is limited in sense of the physical conditions, variety of objects and phenomena, and amount of data. During the last few decades astronomy became fully multiwavelength (MW); all-sky and large-area surveys and their catalogued data over the whole range of the electromagnetic spectrum from γ rays to radio wavelengths enriched and continue to enrich our knowledge about the Universe and supported the development of physics, geology, chemistry, biology and many other sciences. Astronomy has entered the Big Data era and these data are accumulated in astronomical catalogues, databases and archives. Astrophysical Virtual Observatories (VOs) have been created to build a research environment and to apply special standards and software systems to carry out more efficient research using all available databases and archives. VOs use available databases and current observing material as a collection of interoperating data archives and software tools to form a research environment in which complex research programs can be conducted. Most of the modern databases give at present VO access to the stored information. This makes possible not only the open access but also a fast analysis and managing of these data.

Present astronomical databases and archives contain billions of objects, both galactic and extragalactic, and the vast amount of data on them allows new studies and discoveries. Astronomers deal with big numbers and it is exactly the case that the expression "astronomical numbers" means "big numbers". Surveys are the main source for discovery of astronomical objects and accumulation of observational data for further analysis, interpretation, and achieving scientific results. Nowadays they are characterized by the numbers coming from the space; larger the sky and (in case of spectroscopic surveys) spectral coverage, better the spatial (in case of spectroscopic surveys, also spectral) resolution and sensitivity (deeper the survey), larger the covered time domain, more data are obtained and stored. Therefore, we give the highest importance to **all-sky and large area surveys**, as well as deep fields, where huge amount of information is available.

2. Multiwavelength era in astronomy

During many centuries optical wavelengths were the only source of information from the sky. However, modern astronomical research is impossible without various multiwavelength (MW) data present in numerous catalogues, archives, and databases. MW studies significantly changed our views on cosmic bodies and phenomena, giving an overall understanding and possibility to combine and/or compare data coming from various wavelength ranges. MW astronomy appeared during the last few decades and recent MW surveys (including those obtained with space telescopes) led to catalogues containing billions of objects along the whole electromagnetic spectrum. When combining MW data, one can learn much more due to variety of information related to the same object or area, as well as the Universe as a whole.

In Mickaelian (2016a) we list most important recent surveys (those having homogeneous data for a large number of sources over large area) and resulted catalogues providing photometric data along the whole wavelength range, from γ-ray to radio. All-sky and/or large area surveys have been carried out in many wavelengths covering a very wide range, from 300 GeV energies (or 4×10^{-18} A) to 74 MHz frequencies (or 4 m), which means a wavelength/frequency/energy ratio of 10^{-18}. Given that H.E.S.S. Gamma-ray telescope may observe up to 100 TeV energies (or 10^{-20} A) and LOFAR is designed for up to 10 MHz frequencies (or 30 m), this ratio reaches 10^{-21}. MW approach is applied in astrophysical research. Based on our estimate, MW astronomy provides 96 photometric points, out of which 64 come from all-sky or large area surveys, which means that these data are available for most of the studied sources, depending on the sensitivity.

3. Big Data Era in Astronomy

During the recent 2 decades, a number of giant projects were accomplished in astronomy completely changing the numbers of available information and requiring new approach in research. Among the biggest projects in astronomy one should mention the digitization of POSS I and II (**DSS I and II**) and creation of very big catalogues, **SDSS** with its accurate optical images and spectroscopy, **WISE** with very accurate positional and NIR/MIR photometric data that revolutionized astronomy in this wavelength domain, and **Gaia** with its unprecedently accurate astrometric data. Out of upcoming projects we would like to mention **LSST** and **SKA**. At present the biggest astronomical catalogs are the following: SuperCOSMOS (All-sky survey based on DSS1/2; 1,900,000,000 objects), Gaia EDR3 (All-sky; 1,811,709,771 objects), USNO B1.0 (All-sky; 1,045,913,669 objects), GSC 2.3.2 (All-sky; 945,592,683 objects), SDSS DR16 (covered area: 14,555 deg^2; the photometric catalog has 932,891,133 objects), AllWISE (NIR/MIR, All-sky; 747,634,026 objects), and 2MASS (NIR, All-sky; 470,992,970 objects).

Astronomical surveys give so much information that huge catalogues, dedicated archives and databases are being built to store, maintain and use these Big Data (Mickaelian 2016b). It is estimated that there are some 400 billion stars in the Milky Way galaxy and some 125-500 billion galaxies in the Universe, so that we are very far to catalogue all these objects. Even after Gaia space mission we will have much more accurate astrometric and photometric data for the stars but not much more completeness of detections. LSST and SKA will provide significantly more numbers, but again, full coverage of our estimated numbers in the Milky Way (stars) and especially in the Universe (galaxies and QSOs) will not happen in the nearest future.

Optical, UV and NIR/MIR wavelength ranges give most of the information from the sky, however MW astronomy was born in the recent decades and makes huge steps toward the overall understanding of the Universe with its various manifestations from γ-ray to radio and in the nearest future most of the objects (e.g. in our Galaxy or all galaxies in the Local Universe) will have their counterparts in all wavelengths. At present, approximate numbers of catalogued sources at different wavelength ranges are the following: γ-ray – 10,000, X-ray – 1,500,000, UV – 100,000,000, optical – 2,400,000,000, NIR – 600,000,000, MIR – 600,000,000, FIR – 500,000, sub-mm/mm – 200,000, radio – 2,000,000. These numbers give a comparative understanding about the wavelength coverage of the observed Universe. Of course, many of these sources represent the same astronomical object, however cross-correlations are not made for all these sources and we can only estimate the total number of detected astronomical objects, which may be some 3 billion.

As various astronomical missions, surveys, catalogues, databases and archives give various types of information, the only way to compare their sizes is to give this information in bytes. Astronomers, together with nuclear physicists, reach the largest possible numbers and put new requirements for computer science. As an example, LSST every night will provide 30 TB of data, which is much larger than many archives created and complemented during many years. The increase in data providing is happening due to covered sky areas and data accuracy, i. e. both resolution and sensitivity, as well as due to many times coverage, i. e. creation of possibilities for time domain studies.

4. Virtual Observatories

Astrophysical Virtual Observatories (VOs) have been created in a number of countries using their available databases and current observing material as a collection of interoperating data archives and software tools to form a research environment in which complex research programs can be conducted. The science goals are to define key requirements for large, complex MW astronomy projects. Interoperability includes the development and prototyping of new standards for data content, data description and data discovery. VO technology is the study and prototyping of Grid technologies that allow distributed computation, manipulation and visualization of data. A number of national projects have been developed in different countries since 2000, and an **International Virtual Observatory Alliance** (**IVOA**; www.ivoa.net) was created in 2002 to unify these national projects and coordinate the development of VO ideology and technologies. At present it involves 19 national and 2 European projects.

On IVOA webpage, there is a section for Education, "he VO for Students and the Public" (Tutorials, etc.) at https://ivoa.net/astronomers/vo_for_public.html. The European project EuroVO has developed special versions of professional astronomical tools which are adapted for use by school students, including improved performance on older PCs, and offline use. They have also developed a set of tutorials showing example usage of these tools. The tools and the tutorials can be found at http://voforeducation.oats.inaf.it//eng_download.html. It is worth mentioning that Google Sky uses the familiar Google Maps interface to display sky survey data together with various "showcase" objects from astronomical projects. A much richer interface is available in Google Earth tool, which enables users to add their own data.

The Armenian Virtual Observatory (**ArVO**, www.aras.am/arvo.htm) was created based on the DFBS, Digitized Second Byurakan Survey (DSBS), and other digitization projects in Byurakan Astrophysical Observatory (BAO). ArVO project development includes the storage of the Armenian archives and telescope data, direct images and low-dispersion spectra cross-correlations, creation of a joint low-dispersion spectral database (DFBS / DSBS / HQS / HES / Case), a number of other science projects, etc. ArVO group at BAO was created in 2005 and it was authorized as an official project in IVOA also in 2005. An agreement on ArVO development between BAO and Institute for Informatics and Automation Problems (IIAP) was signed. The first science projects with DFBS/ArVO were the optical identifications of Spitzer Boötes sources in 2005. Joint projects were carried out between BAO and IIAP in 2007–2020. ArVO science projects are aimed at discoveries of new interesting objects searching definite types of low-dispersion spectra in the DFBS, by optical identifications of non-optical sources (X-ray, IR, radio) also using the DFBS and DSS/SDSS, by using cross-correlations of large catalogs and selection of objects by definite criteria, etc.

5. Summary

Modern astronomical data span the whole electromagnetic range (**multiwavelength astronomy**), as well as expand to neutrinos and gravitational waves and introduce **multi-messenger astronomy**. These are really Big Data with their volumes, variety of types and nature, velocity of generation and processing, and veracity of the quality (4 V-s). They provide the biggest data in all fields, as the Universe is much larger than the Earth and during the recent times astronomical technique became such that can observe, retrieve and analyze cosmic information in much better details than before.

Many of the modern **astronomical educational tools** use Big Data. This is because young people like computers, software and other computational tools, as well as because it is easier to make up online and automated tools for better distribution to the community. Using these data gives a significant advantage for Astronomy Education due to its attractiveness and due to big interest of young generation to computer science and technologies. The Computer Science itself benefits from Big Data coming from the Universe and a new interdisciplinary science Astroinformatics has been created to manage these data. And astronomical Virtual Observatories are the astronomical version of e-Science, which at present is being more and more developed.

References

Ahumada R., Allende Prieto C., Almeida A., et al. 2020, *ApJS 249, 3*

Gigoyan K. S., Mickaelian A. M., Kostandyan, G. R. 2019, *MNRAS 489, 2030. VizieR On-line Data Catalog III/266*

Lawrence A. 2007, *Astron. Geophys. 48, 3.27*

Markarian B. E., Lipovetsky V. A., Stepanian J. A., et al. 1989, *ComSAO 62, 5*

Massaro, E.; Mickaelian, A. M.; Nesci, R.; Weedman, D. (eds.) 2008, *The Digitized First Byurakan Survey, ARACNE Editrice, Rome, 78 p.*

Mickaelian A. M. 2016a, *Baltic Astron. 25, 75*

Mickaelian A. M. 2016b, *Astron. Rep. 60, 857*

Mickaelian A. M., Gigoyan K. S. 2006, *A&A 455, 765. VizieR On-line Data Catalog III/237A*

Mickaelian, A. M.; Nesci, R.; Rossi, C.; Weedman, D.; et al. 2007, *A&A, 464, 1177*

Véron-Cetty M.-P., Véron P. 2010, *A&A 518, 10*

Education and Heritage in the era of Big Data in Astronomy
Proceedings IAU Symposium No. 367, 2020
R. M. Ros, B. Garcia, S. R. Gullberg, J. Moldon & P. Rojo, eds.
doi:10.1017/S1743921321000764

GALAXY CRUISE: Accessible Big Data of the Subaru Telescope for Citizen Astronomers

Kumiko Usuda-Sato[ID]**, Masayuki Tanaka, Michitaro Koike, Junko Shibata, Seiichiro Naito and Hitoshi Yamaoka**

National Astronomical Observatory of Japan (NAOJ),
2-21-1 Osawa, Mitaka, Tokyo 181-8588, JAPAN
email: kumiko.usuda@nao.ac.jp

Abstract. The Universe is full of galaxies of various shapes; some galaxies have spiral arms and others don't. Why do galaxies show such diversity? How were galaxies formed and evolved? Galaxies are thought to grow by interacting and merging with other galaxies, and the galaxy mergers may be the key process creating the variety. GALAXY CRUISE is the first citizen science project conducted by National Astronomical Observatory of Japan (NAOJ) to unlock galaxies' secrets using the big observational data. We made the superior quality big data taken by the Subaru Telescope accessible to the public and invited them to participate in data classification. Here we report how we designed the website and its first-year progress.

Keywords. Citizen Science, Big Data, Online Resources

1. hscMap: Making the Latest Big Data Accessible to the Public

The Subaru Telescope is an optical-infrared telescope operated by NAOJ near the summit of Maunakea, on the Island of Hawai'i. It has a primary mirror diameter of 8.2 meters, making it one of the largest monolithic mirrors in the world. Hyper Suprime-Cam (HSC), the wide-field imaging camera mounted on the Subaru Telescope, has 870 million pixels and can cover nine times the area of the full moon in each exposure. The Hyper Suprime-Cam Subaru Strategic Program (HSC-SSP) is an unprecedented, extensive survey program started in 2014 to observe for 300 nights with HSC. Its first and second datasets were released to the public in February 2017 (Aihara *et al.* 2018) and May 2019 (Aihara *et al.* 2019), respectively. The second dataset includes 3.8 years of data which corresponds to 174 nights of observations. To make the big data accessible to the public, we released hscMap public version (http://hscmap.mtk.nao.ac.jp/hscMap4/), the user-friendly website to display the HSC-SSP data by modifying its original version for professional researchers (Usuda-Sato *et al.* (2018a) & Usuda-Sato *et al.* (2018b)). Fig 1 shows the initial screen of hscMap. When you start zooming into one of the green areas, a cosmic image captured with HSC appears. If you keep zooming deeper into the Universe, thousands of tiny points of light start to gush out, even from dark, starless areas. Each dot corresponds to one galaxy with hundreds of billions of stars. Using hscMap, anyone can easily explore the vast cosmic images taken by the Subaru Telescope.

2. GALAXY CRUISE: Galactic Journey with Citizen Astronomers

Among the countless galaxies in the vast cosmic images of HSC-SSP, many interacting galaxies are found affecting each other's shapes through their mutual gravitation.

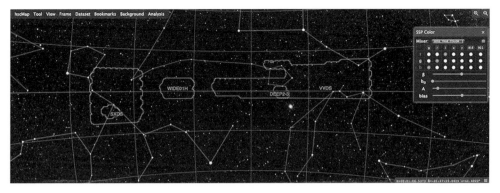

Figure 1. The initial screen of hscMap with the color mixer window. Anyone can easily explore the vast cosmic images taken by the Subaru Telescope inside the green areas using a mouse or a touch panel.

Figure 2. The GALAXY CRUISE website. Anyone can access the site through the internet using a PC or tablet to participate in the galaxy classification.

Studying the shapes of interacting galaxies and counting their number allows us to unlock the secrets of galaxy evolution and to understand the diversity of galaxies. However, it is very challenging for researchers to conduct such studies on their own because innumerable galaxies are found in the vast cosmic images.

GALAXY CRUISE (https://galaxycruise.mtk.nao.ac.jp/en/) is the first citizen science project conducted by NAOJ. Citizen Astronomers classify and identify interacting galaxies in the second dataset of HSC-SSP, which are displayed one after another on a PC or tablet screen. The Japanese site opened on November 1, 2019, and the English site (Fig 2) opened on February 19, 2020. Our project is likened to a cruise ship where many crew members sail together in the cosmic ocean. With the cruise map or the nautical chart developed from the observation map of HSC-SSP, we created the original world view of GALAXY CRUISE. There are four small and deep observation fields and six wide fields.

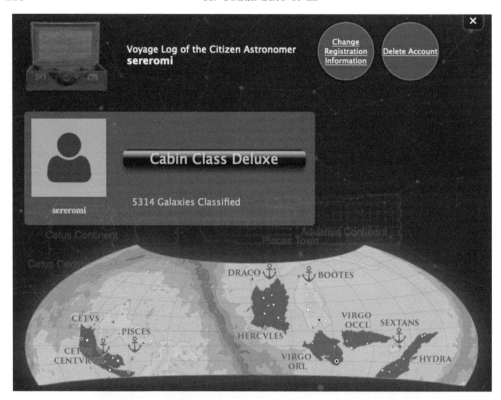

Figure 3. The voyage log of a Citizen Astronomer. The number of classified galaxies, the cabin class, and the cruise map can be seen. Every Citizen Astronomer starts with a fourth-class cabin and upgraded as the number of completed stages increases. On the cruise map, the anchor marks are added in the completed towns, and the completed continents are colored in brown.

In GALAXY CRUISE, we call small fields "towns", and wide fields "continents" and each town and continent corresponds to one stage. Citizen Astronomers classify galaxies while exploring the four towns and six continents to complete all ten stages.

The GALAXY CRUISE site has the following unique features to motivate many people to participate in this project and to maintain their interest.

(1) Thorough Training and Practice Menus
Before Citizen Astronomers register to start classification, they are required to complete the three training sessions to obtain a basic knowledge of galaxies. After login, the Practice Course is also available so that they can compare their classification results with those of the Captain (a galaxy researcher). These features enable non-professionals to classify galaxies confidently.

(2) Gamification Events
Every town or continent is divided into multiple areas. Citizen Astronomers can earn a souvenir (commemorative illustration) when they complete a certain number of areas. Also, when they complete a stage (town or continent), a departure stamp will be added to the passport. The voyage log (Fig 3), passport stamps, and souvenirs can be seen at the welcome page of each Citizen Astronomer.

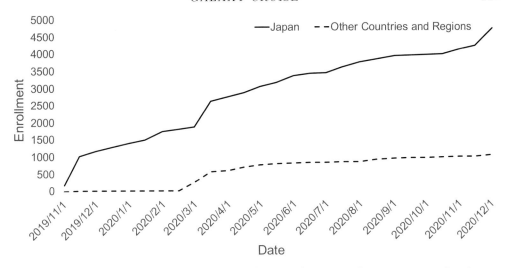

Figure 4. Change in enrollement in Japan (solid line) and in other countries and regions (dashed line).

(3) Exploration of the Vast Universe

As the hscMap engine is used for GALAXY CRUISE classification page, so that Citizen astronomers enjoy exploring the vast cosmic images captured by the Subaru Telescope.

To maintain active interaction with Citizen Astronomers, we upload a new topic as a NEWS article on each month's first day. In some months, we introduce received questions from Citizen Astronomers with the answers by the Captain. In other months, we report updates of the project. Sometimes we present unique shaped interacting galaxies reported by them on the official Twitter (@Galaxy_Cruise_e) and Instagram (galaxycruise_naoj) accounts. We also publish Citizen Astronomers' names who completed Stage 10 (the last stage) on the website recognition of their efforts.

3. The First Year of GALAXY CRUISE

Fig 4 shows the number of enrolled Citizen Astronomers. Within two weeks since the Japanese site opened on Movember 1, 2019, the enrollement reached 1000 thanks to many online news articles. When the English site opened on February 19, 2020, this event was covered by many online news sites and other websites in various languages such as English, Russian, French, and German. At the beginning of March 2020, most schools in Japan were temporarily closed due to the COVID-19 pandemic situation. GALAXY CRUISE was introduced as a recommended online science content on multiple sites. The number of people enrolled in Japan grew rapidly among the younger generation under 20. After April, the enrollment is gradually increasing.

As of December 1, 2020, a total of 5779 Citizen Astronomers from 80 countries and regions have registered. About 80% of them (4731 people) are from Japan. Outside of Japan, the countries with the highest enrollments are the Russian Federation (304 people), the United States (252 people), Ukraine (90 people), and Canada (54 people). The total classification results have exceeded 958,000. We will continue providing information and progress reports to Citizen Astronomers, and at the same time, will plan the next steps such as classfication of fainter galaxies as "Season 2", and combination Citizen Astronomy and machine learning to classify much more galaxies.

References

Aihara, H. et al. 2018, *PASJ* 70, S8

Aihara, H. et al. 2019, *PASJ* 71, 114

Usuda-Sato, K., Tsuzuki, H., & Yamaoka, H. 2018a, *Nature Astronomy* 2, 692

Usuda-Sato, K., Agata, H., Fujiwara, H., Horiuchi, T., Koike, M., Miyazaki, S., Naito, S., Tanaka, M., Yaji, K., & Yamaoka, H. 2018b, in: L. Canas et al. (eds.), *Proc. Communicating Astronomy with the Public Conference 2018 (NAOJ)*, p. 68

Education and Heritage in the era of Big Data in Astronomy
Proceedings IAU Symposium No. 367, 2020
R. M. Ros, B. Garcia, S. R. Gullberg, J. Moldon & P. Rojo, eds.
doi:10.1017/S1743921321001058

Creation of a MOOC and an Augmented Reality application on exoplanets for the Exoplanets-A project

R. de Assis Peralta[1]®, Vincent Minier[2] and Pierre-Olivier Lagage[1]

[1]CEA Paris-Saclay, DRF/IRFU/DAp/LDE3,
91191 Gif-sur-Yvette Cedex, France
email: `raphael.peralta@cea.fr`

[2]CEA Paris-Saclay, INSTN
91191 Gif-sur-Yvette Cedex, France

Abstract. The European Exoplanets-A project aims to provide a comprehensive view of the nature of exoplanet atmospheres, through an interdisciplinary approach.

Exoplanets-A includes a knowledge server where we provide the scientific results and educational resources gathered and developed during the project. In this proceedings, we present two such educational resources: a MOOC and an augmented reality application.

Keywords. Exoplanets-A, MOOC, Augmented reality application

---

1. Introduction

The last twenty years have witnessed an exceptionally fast development in the field of extra solar planets (exoplanets). While the detection of exoplanets is an ongoing process, the characterization of their atmospheres has just begun and is developing very rapidly.

The European Exoplanets-A project(Pye *et al.* 2020) aims to provide a comprehensive view of the nature of exoplanet atmospheres, through an interdisciplinary approach, which includes the integration of state of the art models of the star-planet interaction, atmospheric chemistry and dynamics, and planet formation.

The Exoplanets-A knowledge server (www.explore-exoplanets.eu) provides scientific and educational resources through two main pages: the Science page and the Learning page. The science page includes a knowledge base with direct access to all of the project's scientific products, such as novel methods, tools, and databases for characterizing exoplanet atmospheres. The Learning page was designed for the general public with educational resources based on the science products: videos, online courses, serious games, etc.

Within the Exoplanets-A context, we have developed a MOOC (Massive Open Online Course) on exoplanets as well as an augmented reality (hereafter "AR") application about the Solar System, exoplanets, and the James Webb Space Telescope.

2. MOOC description

Our MOOC is located within the Learning page of the knowledge server under the Online courses section (cf. www.explore-exoplanets.eu/course/mooc-exoplanets).

Through this MOOC, we invite users to acquaint themselves with the surprising universe of exoplanets. They will discover at their own pace their diversity, detection

Figure 1. View of the MOOC page in the Learning/Online courses section.

methods, and the next major space missions searching for and analyzing these new worlds (cf. Figure 1). The MOOC is divided into 9 themes (or modules):

- Module 1 is the general introduction to MOOC and the topics covered;
- Module 2 compares the planets of our Solar System with exoplanets, illustrating the amazing diversity of exoplanets;
- Module 3 provides notions of astronomy: overview of the different objects in the Universe and the distances between them;
- Module 4 addresses the stars, including ours (the Sun), their diversity, and their evolution over time;
- Module 5 presents the instrument with which we observe the Universe: the telescope. Understanding how it works and its limits is essential to grasp the difficulty of detecting exoplanets;
- Module 6 is dedicated to exoplanetary atmospheres: how do we observe them and what is the point of studying them?
- Module 7 presents the concept of habitability / climate of exoplanets and therefore the essential duo: star-planet. We also discuss exobiology.
- Module 8 focuses on the Trappist-1 stellar system. This system acts as a practical case compared to the previous modules.
- Module 9 opens onto future space missions and the complexity of interstellar travel. The in-depth part is dedicated to the future large space telescope: the James Webb.

Each module is divided into two or three difficulty levels (cf. Figure 2). The first one contains the main notions that people should retain, as well as a quiz allowing the learner to test him/herself. The other difficulty levels include more details; the learner can thus go deeper into the subjects that interest him/her the most.

Each level systematically contains external resources, with links to other educational content. We propose a wide range of educational content from both national and international research institutions (e.g., NASA, ESA, CNRS, CNES, CEA), in several different formats for a more entertaining approach. For instance, there are videos, scientific articles, serious games, conferences, citizen science, etc. Therefore, the originality of this MOOC is to bring together and organize pre-existing educational content.

Finally, this MOOC is open and free, without any subscription needed. Everyone can use it to learn about exoplanets or to get access to teaching and outreach resources.

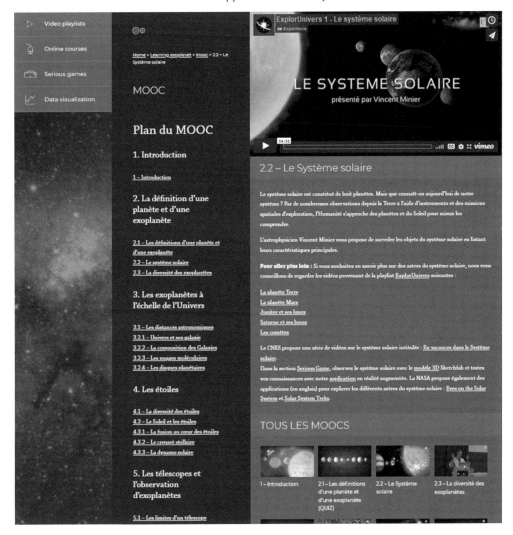

Figure 2. View of the MOOC outline.

To sum up, we propose a Massive (everyone can see it), Open (without registering or even leaving any contact), Online (on our knowledge server), Curated (resources has been curated and aggregated), Course (with learning outcomes and self-corrected quiz) : a MOOC[2].

In March 2021, an English and Spanish version will be also available.

3. AR application description

Within the Learning page, there is a section called "Serious games". A Serious Game is a tool using new technologies to teach in an entertaining way. Increasingly used, it is an excellent pedagogical tool.

With this intention, we developed an augmented reality application offering an interactive discovery of the diversity of exoplanets through 3D simulations (see www.explore-exoplanets.eu/game/augmented-reality-app/). The users can explore two extra-solar systems: 51 Pegasi, the first exoplanet system discovered, and Trappist-1, the famous 7-exoplanet system. There is also the possibility to explore the solar system

Figure 3. Use of the Augmented Reality application during the 2019 summer tour of the SpaceBus France association (Credits: SpaceBus France).

for comparison (cf. Figure 3). Finally, the user can discover a future Space mission: the James Webb Space Telescope. Three levels of discovery are proposed:

• Exploration: users can go through 3D models of the Solar System, two exoplanetary systems, and the JWST. Each model includes clickable points of interest, which open a detailed description.

• Self-assessment quiz: users are then invited to evaluate their knowledge by answering quiz questions.

• Evaluation of these answers: finally users can check their answers by revisiting the 3D models.

This application is for the moment only available in French and for Android platform. The application and its image marker are downloadable on the Exoplanets-A website. The printed image marker allows to use the app and "anchor" the scene in the room.

Our augmented reality application was broadly used by the French scientific outreach association SpaceBus France. This association proposes several astronomy-themed activities presented by professional astronomers for the general public. In 2019, during its month-trip throughout France, SpaceBus France proposed this AR application to more than 7000 people. Thus, it has already successfully been tested with public of different ages and backgrounds.

4. Acknowledgment

The research leading to these results has received funding from the European Union's Horizon 2020 Research and Innovation Programme, under Grant Agreement n° 776403.

Reference

Pye, J. P., Barrado, D., García, R. A., Güdel, M., Nichols, J., Joyce, S., Huélamo, N., Morales-Calderón, M., López, M., Solano, E., Lagage, P.-O., Johnstone, C. P., Brun, A. S., Strugarek, A., Ahuir, J., 2020, Exoplanets-A Consortium, *Origins: From the Protosun to the First Steps of Life*, 345, 202P

Education and Heritage in the era of Big Data in Astronomy
Proceedings IAU Symposium No. 367, 2020
R. M. Ros, B. Garcia, S. R. Gullberg, J. Moldon & P. Rojo, eds.
doi:10.1017/S1743921321000879

Open Astronomy and Big Data Science

Bärbel S. Koribalski⬤

CSIRO Astronomy and Space Science, Australia Telescope National Facility,
P.O. Box 76, Epping, NSW 1710, Australia
email: Baerbel.Koribalski@csiro.au

Abstract. Open Astronomy is an important and valuable goal, including the availability of refereed science papers and user-friendly public astronomy data archives. The latter allow and encourage interested researchers from around the world to visualise, analyse and possibly download data from many different science and frequency domains. With the enormous growth of data volumes and complexity, open archives are essential to explore ideas and make discoveries. Open source software is equally important for many reasons, including reproducibility and collaboration. I will present examples of open archive and software tools, including the CSIRO ASKAP Science Data Archive (CASDA), the Local Volume HI Survey (LVHIS), the 3D Source Finding Application (SoFiA) and the Busy Function (BF). Astronomy is international and includes or links to an incredibly wide range of sciences, computing, engineering, and education. Its open nature can serve as an example for world-wide interdisciplinary collaborations.

Keywords. Astronomical data bases, catalogs, surveys, ISM: galaxies, methods: data analysis

1. Introduction

For the purpose of education across the world, trustworthy and up-to-date public webpages, databases, software tools, etc., are essential. Ideally these are provided in many different languages and at a range of entry levels to ensure wide uptake. In our field of astronomy a large range of excellent sites exist and are regularly updated and expanded. Here I list a small selection of Open Astronomy sites that are indispensable in my field:

- ArXiv: *arxiv.org*
- Astrophysics Data System (ADS): *ui.adsabs.harvard.edu*
- NASA/IPAC Extragalactic Database (NED): *ned.ipac.caltech.edu*
- Lyon-Meudon Extragalactic Database (LEDA): *leda.univ-lyon1.fr*
- CDS Portal: *cdsportal.u-strasbg.fr*
- – SIMBAD Astronomical Database: *simbad.u-strasbg.fr*
- – VizieR: *vizier.u-strasbg.fr*
- – Aladin Lite: *aladin.u-strasbg.fr/AladinLite/*
- SkyView: *skyview.gsfc.nasa.gov*
- ESA Sky: *sky.esa.int*
- Astropy: *astropy.org*

Most astronomers would be familiar with many if not all of these sites, but often less well known are **Open Radio Astronomy** observatories, outreach & education resources, etc. In the following I briefly describe data portals from CSIRO's Australia Telescope National Facility (ATNF), which includes the famous 64-m Parkes Radio Telescope, the Australia Telescope Compact Array (ATCA) and the new Australian Square Kilometer Array Pathfinder (ASKAP), each regularly enhanced with innovative instrumentation.

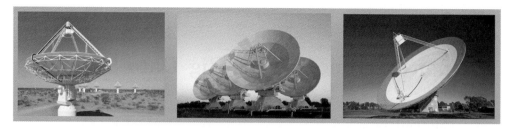

Figure 1. ATNF radio telescopes. — From left to right: ASKAP (36×12-m dishes equipped with wide-field Phased-Array Feeds (PAFs); 6-km diameter), ATCA (6×22-m dishes; 6-km diameter) and the 64-m Parkes Telescope as featured in the ATNF Daily Astronomy Picture (ADAP: *www.atnf.csiro.au/ATNF-DailyImage/* created by B. S. Koribalski).

Figure 2. Screenshots of the CASDA portal and the HIPASS Spectral Line Server.

2. ATNF Data Archives and Software

The Australia Telescope Online Archive (ATOA) contains unprocessed data from the ATCA (visibilities) and the Parkes Telescope (spectral line and continuum surveys) as well as unprocessed data and data products from the 22-m Mopra Telescope. ATCA observations are generally calibrated using the MIRIAD open software package: *www.atnf.csiro.au/computing/software/miriad/*. Observations with the Compact Array Broadband Backend (CABB; Wilson et al. 2011) on average add $\sim$25 TB yr^{-1} to the archive compared to ASKAP's expected data rate of 500 PB yr^{-1}.

Our fastest growing database is the CSIRO ASKAP Science Data Archive (CASDA), which is dedicated to holding the pipeline-reduced and validated output products (2D images, 3D image cubes, spectra, catalogs and in some cases calibrated visibilities) from ASKAP (see Figs. 1 & 2). Each ASKAP dish is equipped with Chequerboard PAFs which are typically used to form 36 beams on the sky providing an instantaneous field of view of around 30 square degrees at 1.4 GHz. CASDA contains excellent pilot survey data from WALLABY (Koribalski 2012, Koribalski et al. 2020) and EMU (Norris et al. 2011), including gas-rich galaxies, groups and clusters, distant black holes and giant radio galaxies. A recent addition is the Rapid ASKAP Continuum Survey (RACS; McConnell et al. 2020) conducted at 888 MHz, covering the entire sky south of declination $\delta = +51$ degr at 15 arcsec resolution and a median rms of 250 microJy beam^{-1}.

- CASDA: *data.csiro.au/dap/public/casda/casdaSearch.zul*
- ATOA: *atoa.atnf.csiro.au*
- HIPASS spectral data release: *www.atnf.csiro.au/research/multibeam/release*
- LVHIS Atlas and Database: *www.atnf.csiro.au/research/LVHIS*

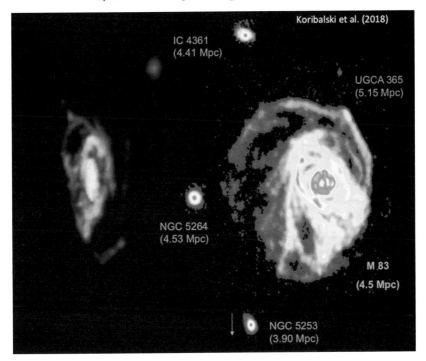

Figure 3. The galaxy M 83 and companions. — Left: 3D Visualisation of the stars and gas in the spiral galaxy M 83 using observational data. — Right: ATCA H I intensity distribution of M 83 and its neighboring dwarf galaxies from the LVHIS project (Koribalski et al. 2018).

3. The Local Volume HI Survey (LVHIS)

The LVHIS Atlas and Database (Koribalski et al. 2018) contain ATCA H I data cubes and H I moment maps for ∼80 nearby galaxies. These highlight the often very large, fast-rotating gas disks of spiral galaxies, as well as the wide range of dwarf galaxy morphologies. The 21-cm line of atomic hydrogen is an excellent tracer of past and on-going galaxy interactions and transformations (see Figs. 3 & 4). For ATCA-LVHIS we targeted a complete sample of southern gas-rich galaxies within 10 Mpc, selected from HIPASS (Koribalski et al. 2004). The raw (uncalibrated) data can be downloaded from the ATOA. Calibrated data and additional data products can be made available on request to the principal investigator. Figure 3 shows the large spiral galaxy M 83 and its nearest gas-rich neighbours. Figure 4 shows a collage of 20 LVHIS galaxies (not to scale), including M 83, ranging from low-mass dwarf companion galaxies to giant gas-rich spirals. Here we highlight the mean H I velocity fields obtained by combining data from typically three different ATCA configuration (∼30 hours on-source), with emphasis on the diffuse outer galaxy disks. Blue and red colours indicate the galaxies approaching and receding sides, respectively. While some of the displayed galaxies reveal reasonably symmetric velocity fields, most show peculiar features such as warps, bars, and tidal tails.

Similar quality H I images of nearby galaxies will be delivered by the ASKAP H I All Sky Survey (known as WALLABY). Early science and pilot survey data for WALLABY and other ASKAP science projects are already available in CASDA (e.g., For et al. 2019, Kleiner et al. 2019, Lee-Waddell et al. 2019, Koribalski et al. 2020), while the full H I 21-cm sky survey ($\delta < +30$ degr) with a resolution of ∼30 arcsec and 4 km s^{-1}) is expected to commence in 2021/2. We expect to detect at least 500 000 H I-rich galaxies out to redshift $z \lesssim 0.2$, including ∼5 000 well-resolved galaxies mostly known from HIPASS.

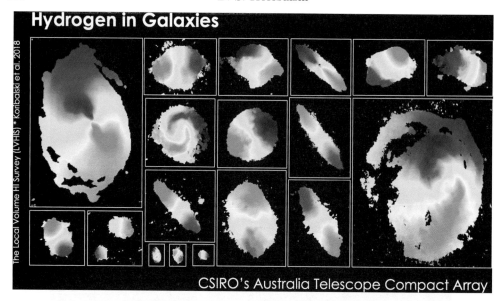

Figure 4. Collage of LVHIS galaxy velocity fields (Koribalski et al. 2018).

Our 3D Source Finding Application (SoFiA; Serra et al. 2015), which was developed for WALLABY (Koribalski et al. 2020), is an open-source software package with extensive user support. It is typically used for 3D source finding in spectral line data sets and includes a reliability estimate for every candidate. Among its outputs are source cubes, moment maps, integrated spectra and source properties. To enhance fitting of H I spectra as well as generation of realistic H I spectra for simulations we also developed the Busy Function (BF; Westmeier et al. 2014), which is available stand-alone (BusyFit) or as part of SoFiA (see *www.atnf.csiro.au/people/Tobias.Westmeier/tools.php*). These and other software packages as well as open multi-wavelength data archives are invaluable for research and education, here our ASKAP projects and future SKA projects.

References

For, B.-Q., et al. 2019, *MNRAS*, 489, 5723
Kleiner, D., et al. 2019, *MNRAS*, 488, 5352
Koribalski, B.S., et al. 2004, *AJ*, 128, 16
Koribalski, B.S. 2012, *PASA*, 29, 359
Koribalski, B.S., et al. 2018, *MNRAS*, 478, 1611
Koribalski, B.S., et al. 2020, *Ap&SS*, 365, 118 *(https://rdcu.be/b5Bfg)*
Lee-Waddell, K., et al. 2019, *MNRAS*, 487, 5248
McConnell, D., et al. 2020, *PASA*, 37, 48
Norris, R.P., et al. 2011, *PASA*, 28, 215
Serra, P., et al. 2015, *MNRAS*, 448, 1922
Westmeier, T., et al. 2014, *MNRAS*, 438, 1176
Wilson, W., et al. 2011, *MNRAS*, 416, 832

Session 8: Cultural Astronomy and Heritage

Education and Heritage in the era of Big Data in Astronomy
Proceedings IAU Symposium No. 367, 2020
R. M. Ros, B. Garcia, S. R. Gullberg, J. Moldon & P. Rojo, eds.
doi:

Introduction

Session 8 begins with an invited talk, "Cultural Astronomy: A scientific frame to understand academic astronomy as part of the Social World," by Alejandro López. In this paper Alejandro discusses cultural astronomy and how it not only studies the astronomy of other cultures, but also displays the cultural history of Western astronomy as well. He discusses the relationship of astronomy with other disciplines and how cultural astronomy links astronomy and the social world. He provides supporting arguments and stresses that cultural astronomy must play a key role in astronomical heritage and education.

Anahí Caldú asked Alejandro:

How do you draw a line with colonialism? Could bringing astronomy to communities be understood as a colonialist attitude?

Alejandro responded:

I understand that bringing academic astronomy to communities is not in itself colonialist. I think that if it is to do it without first speaking and knowing the communities and understanding their context, what they think of heaven, what they want and what they need. Avoiding colonialism also implies arriving with a humble attitude and sharing our knowledge and appreciating theirs, not going to enlighten the ignorant. But not in a condescending way but based on a real understanding of the provisional and situated character of all human knowledge, including what we think is excellent.

Of course, it also implies being aware of the context in which we would do this task, the institutions involved, what they are looking for, the conflicts that society faces, the role of our initiative in that scenario, etc.

Next, in another invited talk Antonio César González García writes about "Our Sky, the Sky of Our Ancestors." He describes how astronomy tends to be viewed from a perspective of modern science, but that other cultures have viewed the sky from different perspectives. Cesar discusses astronomy in culture and gives several related examples. He adds that astronomy in education can be used not only to introduce scientific concepts, but also engage local populations in a greater appreciation of their heritages.

"The astronomical heritage of pre-Hispanic societies in Venezuela: Total Eclipses of Sun reported in Petroglyphs" by Nelson Falcon and Alcides Ortega follows. They describe how relatively is known about the astronomy of pre-Hispanic societies in Venezuela that left no elaborate archaeological remains and had no written language. Petroglyphs, however, abound and the authors discuss examples of solar eclipses at two sites that could potentially correlate as inspirations for related petroglyphs.

Siramas Komonjinda, Orapin Riyaprao, Korakamon Sriboonrueang, and Cherdsak Saelee write of "Relative Orientation of Prasat Hin Phanom Rung Temple to Spica on the New Year's Day: The Chief Indicator for Intercalary Year of the Luni-Solar Calendar." In their paper they describe the ancient temple of Phanom Rung and how it may have been designed with calendrical purposes in mind. Orientations mentioned were those for Spica and the Moon.

"Cultural Astronomy for Inspiration" is described by Steven Gullberg. He describes how archaeoastronomy enthralls many and that this interdisciplinary field can be used to inspire, not only for astronomy but also for other disciplines as well. He uses the astronomy of the Inca Empire as an example and includes a number of field research photos that illustrate fascinating light and shadow effects. Steven also describes the Incas' use of dark 'constellations' in the Milky Way and depicts them in an illustration.

Gor Mikayelyan, Sona Farmanyan, and Areg Mickaelian next discuss "Armenian Astronomical Heritage and Big Data." They describe a rich astronomical heritage in Armenia and present efforts to preserve it, such as the Byurakan Astrophysical Observatory (BAO) Plate Archive. They continue their report with description of the Armenian Virtual Observatory (ArVO) database. They state that the electronic archive will be incorporated within the International Virtual Observatory Alliance (IVOA).

The section's last paper is "TIEMPEROS : Meteorological specialists from the pre-Hispanic indigenous cosmogony of Mexico, and the use of technology to promote astronomy and atmospheric sciences" by Cintia Durán. In it Cintia describes a personal investigation of request for rain and good weather rituals carried out in certain parts of central Mexico in consideration of the traditions and teachings of the local Tiempero. An education model came as a result and a prototype weather station was designed. Such stations then were built by several communities.

Education and Heritage in the era of Big Data in Astronomy
Proceedings IAU Symposium No. 367, 2020
R. M. Ros, B. Garcia, S. R. Gullberg, J. Moldon & P. Rojo, eds.
doi:10.1017/S1743921321001046

Cultural Astronomy: A scientific frame to understand academic astronomy as part of the Social World

Alejandro M. López[iD]

SIAC president, CONICET-UBA, Buenos Aires, Argentina
email: astroamlopez@hotmail.com

Abstract. In the past, Western academic astronomy has conceived in a very specific way its interests. However, in recent decades there has been a promising openness to the rest of the society, in the context of areas such as education, heritage and outreach. Despite this, there has not been an adequate scientific approach to do it, which would imply taking into account the social sciences and a truly interdisciplinary perspective. Here we want to develop the idea that this interdisciplinary approach already exists and it is called: Cultural Astronomy. Unfortunately, in the context of academic astronomy it has been only seen as a study of the "astronomies of others", intended as previous stages or failed attempts of Western academic astronomy. We will seek to show that Cultural Astronomy, as a critical reflection on the social character of the astronomical knowledge, is key to the success of these opening efforts.

Keywords. Cultural Astronomy, Interdiscipline, Education, Heritage, Development, Outreach

---

1. Astronomy and Development

"Astronomy for development" has been a priority for the IAU in recent years. The last two strategic plans, 210–2020 and 2020–2030 (IAU 2018; Miley 2009), are clear manifestations of this importance. We believe that this impulse is a huge step forward, since it puts academic astronomy in a position to assume a more active and leading role in the various challenges facing the contemporary world. In this sense, the IAU is generating an important expansion of its goals. Both strategic plans insist on the interdisciplinary nature of the effort necessary to carry out these ideas. One of the first things that we believe should be taken into account when considering an interdisciplinary work with the social sciences is the asymmetric relationships that exist between academic disciplines. These relationships have their roots in the complex history of the constitution of the "western" academic field and the knowledge policies that have articulated it. These asymmetries are today part of our academic common sense, and imply strong preconceptions about the importance and hierarchy of the different disciplines. An example of these can be found in the relationship diagram of astronomy with other disciplines that appear in the mentioned strategic plans (Fig. 1) (IAU 2018: 22, Miley 2009: cover page). There, it can be observed that while mathematics, chemistry, physics and biology are located under the title "Science and Research"; anthropology and history –in the first plan- are under the heading "Culture and Society". In fact, in the first plan (Miley 2009: cover page), they are placed next to "Perspective immensity of universe" and "Inspiration". But history and anthropology are sciences that carry out research. In any case society and culture could be proposed as their object of study, but in that case the other sciences should have been put under a label such as "living beings, matter, energy, patterns, etc." In fact,

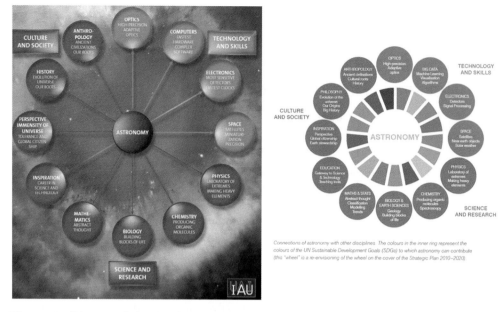

Figure 1. Diagram -"wheel"- of relationships of astronomy with other disciplines. On the left the version of the Strategic Plan 210–2020 (Miley 2009: cover page), on the right the version of the Strategic Plan 2020-2030 (IAU 2018: 22).

under anthropology, you can read: "Ancient civilizations. Our roots". But anthropology deals with human cultures and societies as a whole, not just those of the past. Moreover, its most characteristic methods, such as ethnographic fieldwork, are designed to study societies contemporary to the researcher. It is not about exploring only "our roots", it is about understanding ourselves and the others with whom we live.

This type of considerations is very common and can be seen in much of the terminology we use to classify the disciplines in dichotomous terms. In the table below, the left column contains terms that we use explicitly. The ones in the column on the right are their opposites and although, in general, they are not explicitly used; they are implicitly suggested when using the ones on the left.

Astronomy	Social sciences
Science	Non-science?
Pure	Impure?
Nature	Culture (Arbitrary?)
Hard	Soft?
Precise	Imprecise?
Objective	Subjective?
Truth	Opinion?
Developed	Non developed?
"We"	"They"?
Future	Past?

2. Cultural Astronomy

This contribution seeks to bring the perspective of cultural astronomy to the discussion of the links between astronomy and the social world. Cultural astronomy (Belmonte Avilés 2006; Iwaniszewski 1991; Ruggles and Saunders 1993) is an interdisciplinary area that studies the ways in which the different past and present cultures perceive the sky, the questions that are asked about it, the meaning they give to it, the practices that they develop related to it, the relationships that they build with that environment and what happens there; all as an integral part of its social processes of production, reproduction and transformation. Cultural astronomy is an interdiscipline dedicated to the study of social facts. This broad area of studies includes: Ethnoastronomy, Archaeoastronomy, History of Astronomy (some kind), Astronomical education (some kind), etc. We can say that Cultural astronomy is a perspective which is not searching for fragments of "our" astronomy in distant cultures. It is interested in the cultural context of astronomy (Aveni 1989; Zeilik 1983; Iwaniszewski 1989) but also studies astronomies as socio-cultural products (McCluskey 2005; Iwaniszewski 2009, 2011, López 2011a; Ruggles 2010, 2011). It is an area with ancient roots, whose first professional antecedent can be sought in the works of the British astronomer Norman Lockyer. Between 1980 and 1990 Cultural Astronomy achieved to establish solid methodological bases (Belmonte Avilés 2006; Iwaniszewski 1991; Ruggles and Saunders 1993). Archaeoastronomy, Ethnoastronomy and a socio-anthropological History of Astronomy are some of its branches. Cultural Astronomy have almost three big interacting "methodological traditions": the Green tradition (based on an statistical and general approaches), the Brown tradition (focus on specific cases and in the local and regional contexts); and the Blue tradition (interested in locally situated interpretations). Numerous professional astronomers have devoted themselves to this area of knowledge, including many members of the IAU. In fact, cultural astronomy has been present in the old Commission 41 of the IAU, as well as in the current C4 and C3 commissions of division C. The academics who dedicate themselves to cultural astronomy have gathered important experiences and developed an enormous conceptual baggage in reference to interdisciplinary work, especially with the social sciences. One of the issues that has become clear to astronomers working in cultural astronomy (Belmonte Avilés 2006: 46), is that interdisciplinary studies requires not only working in teams with specialists from all the disciplines involved. It is also necessary that each of the researchers learn their colleagues' language. For this reason, we believe that the perspective of cultural astronomy is essential when thinking about the links between professional academic astronomy and global society. Next, we will discuss some key ideas that cultural astronomy perspective can give to the academic astronomical community in order to rethink their role in the global society.

3. Knowledge as a social construction

We live in a socio-cultural-environmental reality. Our knowledge is a socio-cultural product (Bloor 1998), including astronomy. Social construction of knowledge does not mean neither simply "collective work", nor arbitrariness. Our world is a universe full of meanings, schemes of perception and metaphors organized into cosmovisions/worldviews and cosmologies, in which we are introduced by the other members of our society. Cosmovisions and cosmologies are poles of a continuum: Cosmovisions more linked to logics of practice, incorporated by daily activities (*habitus*), primary socialization, imitation, and day-to-day experience (Bourdieu 1997). This includes our ideas about the sky. Cosmology refers to more explicit and systematic elaborations. The experience of the senses limits the possibilities of any human cosmovision, but not to the point of generating a single compatible option, partly because the perception itself implies the prior imposition of socio-cultural patterns and presuppositions. The social character of the knowledge in question and the need to legitimize that knowledge and comply with accepted truth regimes imposes limits on the possible cosmovisions in a given society at a given time, but they are not absolute either and they do not unequivocally determine an only one possible cosmovision. A cosmovision not only implies a specific set of answers about how the world is, it also involves a set of questions, guiding objectives, truth criteria, etc. They were built historically and socially; any evaluation of their metaphysical and ontological assumptions would be from some equally historically and socially constructed system; they tend to be naturalized and seen as obvious and complete. Also, they are linked to social structures and essentially to power, this implies that in the context of the contemporary global order (or in any other order) the comparison between them is never a mere "epistemological" act, it is about a struggle to achieve legitimacy in a specific scenario. Knowledge systems are always imbricated in the general social field, with varying degrees of autonomy with respect to it and this implies that they are strongly crossed by power. The more important the sky is for a society, more their power mechanisms will be involved in managing the links with it. Every system of ideas and practices about the sky have a constitutively unfinished and changing character. A real dialogue between cultures, needs to assumes that they have hierarchical relationships with each other, linked to political, economic, ethnic, gender, age, religion, etc. inequalities. Because the construction of knowledge is crossed by power relations. As a brief sample of the possible cultural astronomy contributions, we can begin by analyzing the very idea of "development". In the social sciences the category of "development" is a complex and debated concept (Escobar 2005). In this sense there are two basic questions that we must ask ourselves about "development" in order to think about "astronomy for development". The first question is: development of what? The second question is: development for whom? In reference to the first of these questions we can point out three major areas recurrently mentioned in the debates on astronomy for development: education, heritage and economics.

4. Astronomy and Education

Regarding development in education, cultural astronomy has very important contributions to do. A true plural education in the contemporary world must be intercultural. That means, in the first place, that it must take into account the culture and ideas of those people who receive the education in question. Secondly, it supposes that it must understand that our world is formed by the permanent interaction of a great diversity of cultures, articulated by relationships crossed by power and inequality. Therefore, an astronomical education for development must dialogue and know the local systems of knowledge. It must prepare students to dialogue and build agreements in a diverse world. A true education should always be an intercultural education. For this, it is essential the contribution of researchers in cultural astronomy and in particular the methodological contribution of ethnoastronomy. Western Academic Astronomy assumes certain base metaphors, proposes a possible repertoire of emotions and attitudes towards the sky. These ideas are obvious for those who grew up socialized in them. For other human groups are profoundly opposed to their own way of experiencing and thinking about the sky. In this context an intercultural astronomy education implies a true reflexivity on behalf of the teacher. This implies denaturalizing the "western" sky. One of the ideas that we tend to take for granted is "natural laws." This idea is a legal metaphor for the conceptualization of the regularities of the sky, which is also considered the paradigm of "natural law". Those regularities are seen as the very foundation of reality. Therefore, alterations in these patterns generate concern. It is interesting to note how, on the one hand, the origin of this way of thinking about regularities comes from the realm of the social, but simultaneously the idea of a cosmic order is used to justify the social order. Another key point is the nature / culture divide. Division that is not shared by the views on the world of many human groups. It is a classification structure with fundamental consequences in the way of conceiving the world. Another important point is the way in which Western academia tells itself its own history and the history of the "development of universal thought." The way we usually tell ourselves about that past, especially in the institutions that train future scientists, is a linear story of progress. A story in which everything is a preparation for Western academic science. This linear history supposes the omission and deformation of a much more complex pattern of interconnections, diverse cultural projects, loans and flows of ideas, technology and people. All this shows us the importance of recognizing the culturally situated character of Western science. One of the obstacles to do this comes from the process of relative autonomization that grounded the modern scientific field in western Europe and USA. This process was linked to an imaginary of political and cultural autonomy of the scientific academy. This imaginary is part of the common sense of academics and block any attempt of thinking about the situated character of the Western academy. Natural sciences and specially astronomy have a leader role in this image of the academy. For this reason, Cultural Astronomy, making possible a new vision of academic astronomy, has an enormous potential to collaborate in a radical improvement of astronomical education, and scientific education in general. A more anthropological view of the history of Western Academic Astronomy, reveals its character as a socio-historical product. A deeper understanding of the astronomies of other cultures, which does not relegate diversity to a distant past, would be crucial to improve the teaching of astronomy in the world. Addressing the different ways of knowing the sky in greater depth, understanding its structure, allows us to appreciate the way in which the ideas and models with which humans seek to know the world are generated. Understanding the logics, metaphors, interests and observations in which the Western Academic Astronomy rest, can make it easier for educators to design strategies to approach the teaching of this astronomy in diverse cultural contexts. To do this we have some problems to solve. The first one is that a huge amount of material about the astronomy of different cultures used in the teaching and popularization of astronomy, do not have methodological rigor. Also, astronomies of other cultures are used in an "anecdotal" manner, as a kind of curious introduction to the strange things that were "thought" before the arrival of Western science. We have very few didactical materials about a socio-cultural perspective on Western astronomy for astronomy education. But we also have some key potential strengths: professionals in cultural astronomy and professional associations, as the Sociedad Interamericana de Astronomía en la Cultura (SIAC) – *http://siac.fcaglp.unlp.edu.ar/-*, Société Européenne pour l'Astronomie Dans la Culture (SEAC) – *http://www.archeoastronomy.org/-*, and the International Society for Archaeoastronomy and Astronomy in Culture (ISAAC) – *https://www. archaeoastronomy.org/-*. These institutions can give expert advice to astronomers involved in education projects.

5. Astronomical Heritage

As we have pointed out in a previous work (López 2016) regarding heritage, cultural astronomy plays a crucial role in the joint initiatives of the IAU and UNESCO on astronomical heritage. One of the things we have learned in this interaction is that today heritage has become a language for the expression of a great variety of conflicts, in a process similar to that which has occurred with environmental concerns (Leite Lopes 2019; López 2016). This occurs because local populations perceive that heritage and environment are issues that interest international organizations and national states. Therefore, it is important to have a broader view of what is at stake in each case where patrimonial

issues are discussed. This is especially relevant in the case of intangible heritage and
cultural landscapes, since the concept of heritage of international organizations tends to
privilege a static conservation, without changes. Today, heritage is an increasingly spread
concept. It has a great impact on many crucial areas: public politics, NGOs politics, public opinion, and aboriginal communities' strategies. The focus of the heritage concept is
the idea of "culture" as a value to protect. In particular, at the present we can see an
increasingly valorization of non-western achievements. But the concept of heritage has
strong links with a specific western juridical language and property conceptions. For this
reason, some of their key characteristics are: the demand of "authenticity"; the necessity
of a clear "definition" of the boundaries of every specific heritage; and the "preservation" of the integrity of the heritage. The use of the heritage concept has a tendency
to privilege the tangible aspects, the spectacularity, and the singularity of the proposed
heritage. This "western" bias has the consequence of and implicit hierarchization of the
different conceptions of humankind involved. The Western concepts have a very strong
tendency to prevail in the international definition of what is heritage and what is not.
Also we can see a strong tendency to use the concept of "culture" to refer to the diversity
of the human forms of life but hiding the power relations between the different societies,
making claims of political "neutrality". At present time, claims about world heritage
are, in many cases, claims about the ownership and rights, but in the case of aboriginal
communities the conflicts involved are also conflicts about different ways of think about:
the definitions of things, people and humankind; the idea of territory; the ownership; the
history, change, and identity. Heritage -as ecology- is now a new language or arena for the
display of the complex conflicts between societies, specially nation-states and the minorities within them. The ideas of "traditional" and "authentic" are conceptions frequently
applied to aboriginal populations. Usually they are grounded in the idea that that kind
of societies does not change (and if they change they lose their authenticity). They are
thought as societies that only enter history and change after the impact of the colonization processes (Sahlins 1988). This implies the conception of ethnic identity as linked to
some well-defined group of features like dances, clothes, specific ceremonies, or to well
define cosmological systems (Comaroff and Comaroff 2009). This does not fit very well
with the forms in which oral or predominantly oral societies function. An example of this
is the negative to understand the crucial role in present aboriginal communities in South
America of their own forms of Christianity, developed form the complex relationships
with western missionaries. Many western experts involved in world heritage initiatives
are looking for the "real aboriginal life" and do not pay attention to crucial cultural manifestations, with deep roots in the aboriginal cultures, because these manifestations are in
the contexts of aboriginal Christianities. In many cases this practices are not part of an
"acculturation" process, they are not a "mix". They are real cultural creations of these
groups, in the peculiar historical situation that they face. They are truly reinventions of
Christianism in terms of aboriginal logics, and are fundamental ways to legitimate – in the
context of the national society- important cultural forms, leadership mechanisms, social
organization, and conceptions about the relationships between human and not-human
beings (Altman 2015). Other very common idea is that aboriginal people lose their identity if they adopt western technology. But this is not necessary the case. For example, in
the Chaco region in South America, cell phones and computers make possible new versions of the oral culture of past centuries, reinforces old mechanisms of marriages making,
and expands the making of texts in aboriginal languages without the control of western
teachers or missionaries. In each case it is necessary to study these elements in context.
One of the great effects of patrimonialization is the practices of separation from the everyday that usually accompany it (Acuto and Flores 2019: 3). The "consecrational effect" of
this estrangement is part of the mechanisms of "monumentalization" and "enhancement"

that build the legitimacy of heritage as an approved form of memory. As mentioned by these authors, this is accompanied by the introduction of a regulated contact mediated by "experts" of what is thus separated. These are management practices of the "powerful" characteristic of the separation between "sacred" and "profane" and therefore we could consider heritage as a kind of materiality of a "secular form of the sacred" in the context of hegemonic modernity. The incorporation of some aboriginal cultural traditions into national states or international agendas implies in most cases the bureaucratization of these practices. This situation tends to reinforce the forced unification of practices that have a very wide spectrum of variation, according to the non-centralized nature of the societies involved. This often leads to the attempt to define clothing, movements, instruments, meanings, etc. An example of this from Argentinean Chaco region is the recently adopted ceremony for the "Moqoit new year" at the 15 of September, promoted by local government (Giménez Benítez *et al.* 2002; López 2011b). This new form of the spring ritual performances and calendrical conceptions of the Moqoit people have many effects. On the one side, promotes and gives public visibility in the non-aboriginal society to the Moqoit conceptions; and reinforce a dynamic of creation of networks at regional level between different communities. In the other side, implies the creation of a "domesticated" and simplified version of the Moqoit calendar. The actual form of this calendrical important period involves a complex group of signs (birds, flowers, stars, rains) which are not associated to one single day or to a fixed point in the western civil calendar. Also it's not exactly a Moqoit equivalent to "new year." In fact it is the visible manifestation of a new fertility cycle that began with another series of signs: The June solstice, the heliacal rise from the Pleiades and the first frosts. The Right to Free Prior and Informed Consent (FPIC) is a key principle for the relationships between aboriginal groups and national governments or international organizations. Many international regulations, as the Convention 169 of the International Labour Organization (1989), the UN Declaration on the Rights of Indigenous Peoples (2006), the American Declaration on the Rights of Indigenous Peoples, emphasize the rights of indigenous peoples to participate -prior to the decisions to be made- through their own decision-making structures in all actions that affect them, including patrimonialization processes. This must be free of pressures and manipulations -for example pressures using the promises of potential economical and touristic benefits-. In societies of low stratification, the processes of decisions usually involve the making of a consensus, and strong discrepancies between different communities and leaders. In many occasions the western agendas are not minded to tolerate these processes and their time scales.

6. Astronomy, economy development and coloniality

Finally, in reference to the economy, one crucial issue is that the academic astronomy community in general has supposed that the installation of large astronomical facilities drives local development. We must be especially attentive to the local impact of these great international astronomical facilities. These huge structures and the set of associated activities are driven by large international consortiums and involve huge amounts of money. We must be especially careful in its design and in the way in which its installation is agreed with the local populations. It must be a true dialogue, where the possibility of local inhabitants to refuse these facilities' install are real. The scientific community, when installing large facilities, should mark with its example the way in which other large undertakings should be build. But many recent cases (Casumbal-Salazar 2014; Herhold 2015; López 2018; Miller 2016; Mizutani 2016; Swanner 2013) show that we are following the ways of proceeding from the large extractive industries instead of setting agenda to them. As "knowledge industries" we should be a beacon for all other types of industries in terms of impact on the non-human ecosystem, but fundamentally in terms of impact on the local human society. And a crucial point in the evaluation of this impact is that local social actors, especially the most vulnerable and unprotected, must have an important voice and decision-making power. Science is proposed itself as dialogue and joint construction of knowledge. Therefore, we must be a school of dialogue and listening. In a world in which imposition and authoritarianism are often the easiest way out, the astronomical community has the opportunity to show that reason, dialogue, understanding and listening to the other are valid and efficient tools. Fighting the coloniality of knowledge implies recognizing that its character is inextricably linked to the human societies that produce this knowledge and to power relations. And with this in mind trying to make our own knowledge system less colonial. Also implies coexisting in the diversity of human worlds, seeing diversity as a wealth and not as a problem, but with a critical view of how diversity is articulated with inequality.

7. Final Remarks

The other big question is, as we said, for whom is development? The IAU documents show that there is quite a consensus that development should primarily favor local populations. And within that group to the especially vulnerable. This again implies that as a scientific community we must learn to listen to the demands and problems of these populations. As teachers know, all knowledge construction must start from previous knowledge, demands and interests. We need reflexivity to ask ourselves about our own systems of knowledge and practice about the sky in all its complexity, variation, historical and contingent character and as social productions. We need this to understand that "Western" academic astronomy is not a "trans cultural meta-system" of knowledge about the sky. As scientists we need a scientific approach to the Academic Astronomy relations with the world. That is why a profound dialogue with the social sciences is essential. As we see, a deeper interdisciplinary dialogue should be the next step to understand academic astronomy as part of the social world, and in the frame of the present IAU strategic Plan, in the construction of an astronomy for development. Cultural astronomy is the interdisciplinary academic space that has been working for years on the development of this type of methodological perspectives. For all of these we believe that it is of the first importance that the field of astronomy for development deepens its links with cultural astronomy. Cultural astronomy must play a key role in the articulation of the astronomical heritage initiatives. We need to be involved with people if we work with people. Cultural astronomy must play a key role in education initiatives, in thinking about academic astronomical education as a social-cultural situated enterprise. Also, to contribute to the world we must reflect on our practices and conflicts with the world. The symbolic struggle for the definition of the meaning of objects, places, times and practices, is not politically neutral. This struggle is marked by the force of colonial relations. If we want to have a more active role in the world and contribute to it, we cannot maintain a naive and uncritical attitude about the world's problems and our responsibility as scientists. The colonial logics are inscribed in our bodies and practices, we need to make a very strong epistemological vigilance to avoid the risk of reproducing colonial plunders in the name of science and culture. We can contribute to World development but the World can contribute to Academic Astronomy development too.

References

Acuto, F. A. & Flores, C. 2019, Patrimonio y pueblos originarios, patrimonio de los pueblos originarios: una introducción. In F. A. Acuto & C. Flores (eds.) *Patrimonio y pueblos originarios. Patrimonio de los pueblos originarios*. Buenos Aires: Ediciones Imago Mundi. 11–43.

Altman, A. 2015, Sky Travelers: Cosmos' Experiences Among Evangelical Indians From Argentinean Chaco. In F. Pimienta, N. Ribeiro, F. Silva, N. Campion, A. Joaquinito & L. Tirapicos (eds.) *Proceedings of 19° SEAC Meeting, "Stars and Stones: Voyages in Archaeoastronomy and Cultural Astronomy – A meeting of different worlds"*. London: Editorial British Archaeological Reports (BAR). 148–152.

Aveni, A. F. 1989, Introduction: whither archaeoastronomy? In A. F. Aveni (ed.) *World Archaeoastronomy* Cambridge: Cambridge University Press. 3–12.

Belmonte Avilés, J. A. 2006, La investigación arqueoastronómica. apuntes culturales, metodológicos y epistemológicos. In J. Lull (ed.) *Trabajos de Arqueoastronomía. Ejemplos de Africa, América, Europa y Oceanía*. Cambridge: Valencia: Agrupación Astronómica de La Safor. 41–79.

Bloor, D. 1998, *Conocimiento e imaginario social*. Barcelona: Editorial Gedisa.

Bourdieu, P. 1997, *Razones prácticas. Sobre la teoría de la acción*. Barcelona: Editorial Anagrama.

Casumbal-Salazar, I. J. A. 2014, *Multicultural Settler Colonialism and Indigenous Struggle in Hawai'i: The Politics of Astronomy on Mauna a Wākea*. Ph.D. Thesis. Hawai'i: University of Hawai'i at Mānoa.

Comaroff, J. L. and Comaroff, J. 2009, *Ethnicity, Inc.*. Chicago and London: The University of Chicago Press.

Escobar, A. 2005, El "postdesarrollo" como concepto y práctica social. In D. Mato (ed.) *Políticas de economía, ambiente y sociedad en tiempos de globalización*. Caracas: Facultad de Ciencias Económicas y Sociales, Universidad Central de Venezuela. 17–31.

Giménez Benítez, S., López, A. M. & Granada, A. 2002, Astronomía Aborigen del Chaco: Mocovíes I: La noción de nayic (camino) como eje estructurador. *Scripta Ethnológica* XXIII, 39–48.

Herhold, A. S. 2015, *Hawai'i's Thirty Meter Telescope: Construction of the World's Largest Telescope on a Sacred Temple*. M. A. Thesis. Oslo: University of Oslo.

International Astronomical Union 2018, *IAU Strategic Plan 2020–2030*. International Astronomical Union.

Iwaniszewski, S. 1989, Exploring some anthropological theoretical foundations for archaeoastronomy. In A. F. Aveni (ed.) *World Archaeoastronomy*. Cambridge: Cambridge University Press. 27–37.

Iwaniszewski, S. 1991, Astronomy as a Cultural System. *Interdisciplinarni izsledvaniya* 18, 282–288.

Iwaniszewski, S. 2009, Por una astronomía cultural renovada. *Complutum* 20 (2), 23–37.

Iwaniszewski, S. 2011, Cultural Impacts of Astronomy. In A. T. Tymieniecka & A. Grandpierre (eds.) *Astronomy and Civilization in the New Enlightenment. Passions of the Skies.Analecta Husserliana*. London-New York: Springer. 123–128.

Leite Lopes, J. S. 2019, Sobre processos de "ambientalização" dos conflitos e sobre dilemas da participação. *Horizontes Antropológicos* ano 12 (25 jan./jun), 31–64.

López, A. M. 2011a, Ethnoastronomy as an academic field: a framework for a South American program. In C. L. N. Ruggles (ed.) *Archaeoastronomy and Ethnoastronomy: Building Bridges between Cultures, IAU Symposium N° 278, Oxford IX International Symposium on Archaeoastronomy* Cambridge: Cambridge University Press. 38–49.

López, A. M. 2011b, New words for old skies: recent forms of cosmological discourse among aboriginal people of the Argentinean Chaco. In C. L. N. Ruggles (ed.) *Archaeoastronomy and Ethnoastronomy: Building Bridges between Cultures, IAU Symposium N° 278, Oxford IX International Symposium on Archaeoastronomy* Cambridge: Cambridge University Press. 74–83.

López, A. M. 2016, Astronomical Heritage and Aboriginal People: Conflicts and Possibilities. In P. Benvenuti (ed.) *Astronomy in Focus. As presented at the IAU XXIX General Assembly. Proceedings of the IAU. Vol. 11.* Cambridge: Cambridge University Press. 142–145.

López, A. M. 2018, Peoples Knocking On Heaven's Doors: Conflicts Between International Astronomical Projects and Local Communities. *Mediterranean Archaeology and Archaeometry* 18 (4), 439–446.

López, A. M. 2020, Cultural astronomy perspectives on "development". In T. Lago (ed.) *Astronomy in Focus. As presented at the IAU XXX General Assembly.IAU Symposium N°. XXX.* Cambridge: Cambridge University Press. 580–581.

Miley, G. 2009, *Astronomy for Development. Building from the IYA2009. Strategic Plan 2010–2020 with 2012 update on implementation.* International Astronomical Union.

Miller, S. 2016, Mauna Kea: Two Cultures and the 'Imiloa Astronomy Center. *Interdisciplinary Science Reviews,* 41(2–3), 222–245.

McCluskey, S. 2005, Different astronomies, different cultures and the question of cultural relativism. In J. W.Fountain and R. M. Sinclair (eds.) *Current Studies in Archaeoastronomy: Conversations across Time and Space.* Durham NC: Carolina Academic Press. 69–79.

Mizutani, Y. 2016, Indigenous People and Research Facilities for Space Science in the U.S.: Some Ideas for Outreach, Conflict Resolution, and Prevention. *Trans. JSASS Aerospace Tech. Japan* 14 (ists30), Po_5_1-Po_5_6.

Ruggles, C. L. N. 2010, Indigenous Astronomies and Progress in Modern Astronomy. In T. Lago (ed.) *Modern Astronomy. Accelerating the Rate of Astronomical Discovery (Special Session 5, XXVII IAU General Assembly). Vol. Proceedings of Science.* PoS(sps5)029. IAU. 1–18.

Ruggles, C. L. N. 2011, Pushing back the frontiers or still running around the same circles? 'Interpretative archaeoastronomy' thirty years on. In C. L. N. Ruggles (ed.) *Archaeoastronomy and Ethnoastronomy: Building Bridges between Cultures, IAU Symposium N° 278, Oxford IX International Symposium on Archaeoastronomy* Cambridge: Cambridge University Press. 1–18.

Ruggles, C. L. N. & Saunders, N. J. 1993, The study of cultural astronomy. In C. L. N. Ruggles & N. J. Saunders (eds.) *Astronomies and Cultures.* Niwot: University Press of Colorado. 1–31.

Sahlins, M. 1997, *Islas de Historia. La muerte del Capitán Cook. Metáfora, antropología e historia.* Barcelona: Gedisa.

Swanner, L. A. 2013, *Mountains of Controversy: Narrative and the Making of Contested Landscapes in Post-war American Astronomy.* Ph.D. Thesis. Harvard: Harvard University.

Zeilik, M. 1983, One approach to Archaeoastronomy: an Astronomer's View. *Archaeoastronomy the journal of astronomy in culture* 6 (1–4), 4–7.

Education and Heritage in the era of Big Data in Astronomy
Proceedings IAU Symposium No. 367, 2020
R. M. Ros, B. Garcia, S. R. Gullberg, J. Moldon & P. Rojo, eds.
doi:10.1017/S1743921321000740

Our Sky, the Sky of Our Ancestors

A. César González-García◉

Instituto de Ciencias del Patrimonio, Incipit-CSIC
Avda. de Vigo s/n, 15705, Santiago de Compostela, Spain
email: `a.cesar.gonzalez-garcia@incipit.csic.es`

Abstract. When we talk about Astronomy, we normally do not take into account that we are using a cultural specific way of understanding the sky. Astronomers, either professional, amateur or just lovers of the sky nowadays tend to approach the sky from the point of view of modern science. There, we approach the sky as something that needs to be explored, understood and explained.

However, this vision was not always like that, or even in other cultures is/was completely different. For centuries, the human being has comprehended the sky, its changes and constancies, as part of their world, as part of the environment, as part of their everyday life.

In this paper, I review a few of these different ways of approaching the sky in several cultures, from the Near East to Rome or the Andes and how we can use them today for education, outreach and heritage management.

Keywords. Cultural Astronomy, Archaeoastronomy, Past Astronomies, Heritage, Astro-tourism, Petra, Rome, Peru

---

1. Introduction

It is common ground in our days that the Western society lives away from the Sky. This seems true if we take into account important issues such as light pollution. We assume that past societies lived much more attached to the sky because they 'used' the sky to know when it was the time for harvesting or sowing. It was also used to know the route in long distance travels, either at sea or in land. And it was in several cases used as repository or as source of explanations, of metaphysics, for the everyday life.

I would argue that Western society is still very much attached to the sky. Perhaps in a different fashion than in other societies, and perhaps as another commodity to be consumed. It is today common that astronomical discoveries make it to the headlines, to be aware by the news of the next solar eclipse, the pass of a fireball, a singular planetary conjunction or a blue-moon. If we think about it more closely we might come into a paradox: the society that shatters the sky with sheer amounts of light at the same time asks astronomers for new and renewed knowledge about the Universe.

In a sense, Fig. 1 might be as a good example of such paradox. TA first look into this signpost in the Mediterranean island of Ibiza (Spain) might indicate that if we need such a signpost it is because we are indeed detached from the sky. Such reading does not take into account where this signpost is located or why it was placed there. The signpost is at the entrance of San Antonio, a population on the west shore of the island, world known for the sunset views. Watching a summer sunset at one of the famous bars near the seashore, while listening chill-out music composed for the occasion has become such a tourist attraction that this small population had to manage huge traffic jams every summer evening. To overcome the situation the town-hall decided to open a number of public parking lots, and installed these signposts to direct the drivers there.

Figure 1. Signpost in the Mediterranean island of Ibiza (Spain). Is this a clear sign of the lack of interest of our Western society on the Sky, or not?

In fact, then, the signpost could mean the contrary than we were arguing before: it appears as a necessary reaction by the local authorities to one of the ways our modern society interacts with the sky.

The study of how a society relates to the sky in general, taking into account the cultural context, is done by a discipline called Astronomy in Culture (González-García & Belmonte (2019), Iwaniszewski (2009), Ruggles (2015)). Particular ways to study such relationship could be gained by means of the material record, in particular that found at archaeological sites. Then we may talk of what has been called archaeostronomy(García Quintela & González-García (2009)). Another way is done through the ethnographic record, observation and participation on the actions of a given society. Then we talk of ethnoastronomy (López (2015)).

One example of the social interest of astronomy in culture and its usefulness for other disciplines, is the new visitors center at Stonehenge. This center presents the recent discoveries at Stonehenge and neighbouring sites, using the famous alignments towards summer solstice sunrise and winter solstice sunset as narrative vectors to reach the audience.

Such social demand often derives into popular conceptions (Krupp (2015)) and sometimes into trivialized and stereotypical images of the past (Iwaniszewski (2015)). One such could be the baseless claims at certain spots for alleged connections with the skies in the past. Penas de Rodas (Lugo, Spain; Vilas Estévez (2015)) is an example of that. This is a natural site near the ancient Roman town of Lucus Augusti (present day Lugo), were each summer solstice people crowds to witness summer solstice sunset between two large boulders, supposedly placed there by ancient people. The mixture of commoditization and consumption of the past with a lack of critical information or at least an answer from the academic world can be a fertile ground to fall into esoteric proposals as it was recently the case with the Maya end of the world in 2012. All these argues in favour of the interest to build the right discourse to be brought to our society to provide an answer to this demand.

One way IAU has been doing so in the past decade is the initiative 'Astronomical World Heritage', together with UNESCO (Ruggles & Cotte (2011), Ruggles (2017)). This initiative searches for human heritage, materials, knowledges and values, related to the sky at any epoch. In the past years a number of sites recognized as UNESCO World

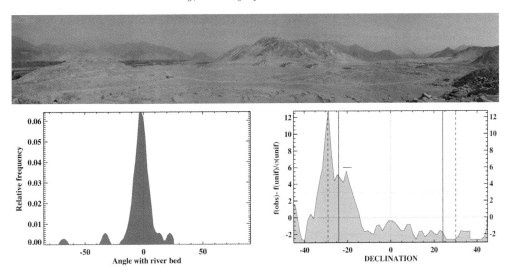

Figure 2. Top: panoramic view of the central part of Caral. The pyramidal buildings seem to bring the mountains to the urban space. At the same time, they are accommodated to the valley by following a line parallel to the river, as indicated by the angle of the buildings to the nearby river -bottom left-. The orientation of the buildings towards the eastern horizon seems to be the most important one, and concentrates at areas consistent with the southern rising of the moon -bottom right; vertical dotted lines indicate the equinoxes, vertical solid lines the solstices and vertical dashed lines the lunar extremes-. For details see text and González-García *et al.* 2021.

Heritage do include among the Universal Outstanding Values their relation to how the societies that built them understood the sky. Examples are not only modern observatories such as Jodrell Bank (UK), or ancient ones like Jantar Mantar (Jaipur, India) but also cultural landscapes like Risco Caído in Gran Canaria (Spain). Astronomy in Culture then provides new value to already existing cultural heritage, but also in some cases discovers new heritage to be valued and protected.

All these aspects make astronomy in culture a key vector to produce materials to be used by educators, outreach actors or heritage managers to their own goals. In the present paper we will review how this interest of our society in the Sky, in general, but in the past as well, can be used to the advantage of education, outreach and heritage management. In particular, we will see this from the results of three research programs I have been involved in together with my collaborators in the past decades.

2. Cultural Astronomy and Education, Caral

Allegedly the oldest city in the Americas (c. 3000 BC; Shady (2006)), Caral together with a number of pre-ceramic sites in the Supe and neighbouring valleys in the north central coast of Peru houses one of the earliest examples of astronomy and geometry in an ancient city (González-García *et al.* (2021)).

The inhabitants of the Supe Valley cultivated cotton, gourd and pumpkin for storing food making drinking vessels and waving fishing nets. The main protein supply was the fish from the near-by sea. They had networks of long distance trade, and carried out a heavy investment in building some of the earliest stone architecture (Burger & Rosenswig (2012); Haas & Creamer (2004). After a thousand years, the Supe society progressively diminished perhaps due to changing environmental conditions (Shady *et al.* (2001)).

According to the results of a campaign to measure a comprehensive set of buildings at the ten largest sites in the valley, the pyramidal buildings are mostly located either parallel or perpendicular to the local riverbed (see Fig. 2, bottom left). In this sense, the

Supe society accommodated to the terrain and the circumstances in a peculiar way that could parallel what has been advocated for at other areas in the world in the past (like ancient Egypt, see e.g., Belmonte (2012)). This is indeed a reflection of the understanding of the landscape by this society and possibly an example of geometry, understood in its Greek original meaning, as land measurement.

Besides, the ten sites are located within the valley in such a way that the main public buildings of these cities face the rising of the local winter full moon (González-García & Belmonte (2019); see Fig. 2, bottom right). To do so, they had to take into account the local landscape, as most buildings face a high mountainous landscape.

These winter full moons coincided with the period of sowing the cotton, and the end of one of the two fishing seasons in the near-by sea. Therefore, it appears as a propitious moment for performing social activities, like gathering for rituals at the central plazas.

The location and orientation of the pyramidal buildings do appear as a reflection of the understanding of the inhabitants of the Supe valley of their environment. Such included the local landscape, the terrain, but also the sky, through a proper orientation towards an important rising of the moon. This defined the right moments to carry out rituals, and all of these items provide a glimpse on the world-view of the Supe and Caral inhabitants.

The sites at the Supe valley, and particularly Caral, have had a long and difficult research history. Fist identified very early, they were finally investigated in their full right by prof. Ruth Shady. She had to face a number of threats and overcome the fierce resistance of part of the local population of present day Supe Valley. Despite being away of the arable lands, the archaeological prospections and the expropriation of lands for the archaeological works were often very difficult to explain.

The archaeological team engaged in an ambitious program of popular science and education, to bring their discoveries to the local population. They opened a number of small communal museums at the different communities. They celebrate several open doors days at the sites, with recreations of the ancient lifestyle, feasts and other events. Finally, Caral was recognized as a World Heritage Site in 2009 that has attracted an increasing number of tourists to the area, constituting a vector of development in the region. However, not all local people are thus engaged with the finding and the material remains.

In several instances, one of the problem is that the local population has moved to these areas in recent times due to the attractiveness of the indicated development. The uprooted population value their heritage as a tool to make a living, which is not a minor thing, but they do not consider it yet as their own heritage. It is perceived as the heritage of the archaeologists, despite the effort of the last.

One way the break this barrier is education, and the incorporation of the astronomical discourse as a guiding thread to provide a meaning to the material remains recovered by the archaeologists. The astronomical findings and their ritual interpretation provides a meaning, a value, the local people can get attached to from below as opposed to other discourses from the top. They can see the people of the past as their equals. People who had to face problems as they have to, and who recurred to rituals as they do today.

By using these results, and others, for education it would be easier to get the local population engaged and value their heritage. By using such approach from below new ideas on how to profit from the heritage they have could appear from the local population.

3. Cultural Astronomy, (astro-)Tourism and Outreach, Petra

Petra was the main city of the Nabataean realm for some centuries before and after the change of Era (Alpass (2013)). The Nabataeans were a nomadic people of alleged Arab ancestry, that by the first century BC had settled on the plateau east of the Jordan river. There, they controlled the end of the caravan routes that crossed the Arabian dessert and which brought incense, among other goods, to the Mediterranean shores. Besides, they developed sophisticated irrigation techniques for a dedicated agriculture in their harsh lands. They developed an original culture with a clear Hellenistic flavour that mixed with the Near Eastern and Arab elements. They were finally incorporated into the Roman Empire by Trajan in 106 AD. After that Petra still flourished for several centuries, until its final decay in the Middle Ages (Alpass (2013), Healy (2000)).

We have several ancient informations on the cultic calendar of the ancient Nabataeans and about their spiritual life, gods and goddesses. However, they come mostly for Graeco-Roman writers. They tell us that the main god was called Dushara, and possibly had a solar character among other attributes. Dushara was the son and at the same time consort of goddess Al-Uzza, aka Allat, who according to Panarion (Alpass (2013)) had a singular connection to Venus. There are also strong indications that there was a lunar god (or goddess) not yet fully identified, although there are a number of good candidates (Roche (1995)). All this information, pictures a religion where the heavenly bodies had a prominent role (Healy (2000)).

In fact, we know that the Nabataeans had a ritual calendar very similar to others found in the area (Belmonte *et al.* (2019)). It was a luni-solar calendar, where the moths followed the moon somehow, but had to introduce intercalary months now and then to keep the festivities in step with the sun and the seasons. The calendar possibly started at the month of Nissan, with the first crescent after the Spring Equinox. Other relevant moments were possibly the Winter Solstice and the Autumn equinox.

For the past decade, we have been investigating the astronomical aspects of the Nabataean life (see e.g. Belmonte *et al.* (2013); Belmonte & González-García (2017); Belmonte *et al.* (2020)). In particular, Petra is one of the focal points of our research so far. There we have discovered that the main public buildings related to ritual activities, such as temples, sanctuaries, tombs and cenotaphs, open such that they cluster in three main directions that are related to the equinoxes, and the two solstices.

In Petra, we argue that there was a main gathering and processions during winter solstice that may have involved several monuments, processional routes and moments along the whole city (see Fig. 3). It may have started at dusk on the full moon next to the winter solstice. At that time, the rising full moon could have been seen on top of the final end of the Gorge of the Siq from the magnificent façade of the Treasury (González-García & Belmonte (2019)). In fact, during major lunar standstill such full moon could illuminate the inner parts of the excavated chamber of these site. The procession could have started then moving to the central parts of the city through the colonnade street, with the next stop at the main temple of the city, the Winged Lions Temple, devoted to goddess Al-Uzza. Here, at dawn, and before sunrise, the people could witness the setting of the bright star Canopus. This is a prominent star in the southern skies used as a guiding beacon by the Bedouin to cross the dessert. The setting of these star heralded the moment to depart from the city center to climb an excavated processional route to one of the cities upper areas. Along the way, a small group, perhaps a chosen one, could witness an illumination effect only at winter solstice sunrise at the Lions Triclinum. Finally, the procession ended in front of the huge façade of the Monastery. This was possibly the cenotaph of one of the Nabataean Kings, but it could have housed as well one of the main worship sites to Dushara. There the setting sun would have illuminated only at this time the inner parts of the excavated chamber in a most peculiar way (Belmonte & González-García (2017)).

The entangled nature of these events, mixing ritual times, with astronomical events, and light and shadow displays do recall the spiritual minds of the inhabitants of this place in a way that might be a magnificent tool to engage a new and more sustainable type of tourism.

Figure 3. Some of the most prominent monuments in Petra could be connected to a processional route crossing the whole city. Such procession might have taken place at times close to winter solstice, possibly at the time of full moon. At dusk, the rising full moon would illuminate the façade and the inner parts of the Treasury (a-A). The procession then proceeded to the Winged Lionś temple at the city center. Before dawn, and heralding the new day, the setting of Canopus might have marked the start of the ascension through the excavated route towards the Monastery. A few hours after sunrise, a beam of light enters the Lions Triclinum illuminating the niche with an image of the deity. Finally, at the end of the day the procession would reach the Monastery to witness the double sunset event that illuminates the inner parts of this magnificent façade.

Petra, declared World Heritage Site in 1985, is nowadays literally flooded with hordes of tourists to capture selfies at the Treasury after a promenade by the Siq, in a perfect example of consumption of the past. Most tourists stop there before returning to their hotels. However, in this case, astronomy could be a new way to engage a different kind of tourism.

A trendy type of tourism these days is astro-tourism: the kind of tourism that searches the clearest skies to observe and perhaps photograph the stars. In a sense, we could advocate from astronomy in culture, a new kind of 'cultural astro-tourism', where the tourists may engage in the observation of the sites at particular moments like the winter solstice in Petra.

This type of tourism has already been active at many other sites worldwide (see e.g. ? for a review). To prevent manipulation, it would be desirable that this kind of tourism should naturally be more restricted, respectful to the different sites, it should engage the visitants more closely to the sites, and I would argue it could be more sustainable.

Indeed, the astronomical and cultural astronomical communities should engage in promoting such kind of tourism as alternatives to mass tourism in several areas. Mass tourism has one very damaging side effect for astronomy, as it is the building of huge tourist resorts next to the sites, with the possible direct impact on the night skies. A sustainable development of this tourism should imply an attention to the diversity, both gender and cultural, inclusion of non-western interpretations of the world and an outreach effort to attract the local communities. A high quality tourist, as a (cultural) astro-tourist, is likely to engage emotionally to the site and spend more time (and money) in the area per capita, than an average normal tourist. Finally, such tourist is likely more aware and respectful to the local environment than the average tourist.

4. Cultural Astronomy and Heritage Management, Cartago Nova

Heritage management related to cultural astronomy is experiencing a steady increase in several sites, especially thanks to the efforts of the promoters of the UNESCO-IAU Astronomical World Heritage initiative (Ruggles & Cotte (2011); Ruggles (2017)).

Heritage initiatives related to astronomy in culture are getting momentum. Stonehenge was already a World Heritage Site before the initiative took place, but the recent project to build a tunnel to divert the highway traffic at the area collided particularly with the protection of the site. Although the tunnel was planned to be built several kilometres away from the archaeological sites, one of its ends was designed to open directly in the solstitial alignment that has made Stonehenge famous worldwide. The cultural astronomical community, coordinated by Clive Ruggles, put a lot of effort to push the end to other area that had a lower visual impact on the site.

Another example of the fruitful interaction of the cultural astronomy community, the local astronomical amateur societies, and local academic and heritage management authorities comes from Cartagena (Spain).

In 2013 we were invited by the local authorities to develop a research program at the recently excavated areas of the ancient Carthaginian and Roman remains. Cartagena, a former industrial pole of the Spanish south-west, had suffered a profound industrial reform in the last decades that had left a deep fingerprint of unemployment and poverty. Local authorities tried to promote tourism, but the town had little tourist attractiveness at the time. However, Cartagena is one of the oldest towns in Iberia and with such a rich and long history, they designed a program of cultural development of the city. One of the key aspects of that development was the investigation of the archaeological remains.

In the last two decades, the archaeologists have recovered the remains of the Roman Theatre, the Roman Forum, the Punic city walls, and ancient temples among many other sites (see e.g. Noguera Celdrán & Madrid Balanza (2014)).

We were able to discover some very interesting links of the ancient Punic and Roman towns to the sky (González-García *et al.* (2015)). One is the orientation of the sanctuary of Athargatis at the ancient citadel. This sanctuary appeared on top of the Ars Hasdrubalis, the palatial complex of the Punic rulers of the cities. It faces one of the hills of the city, Sacred Mount, where the Greek chronicler Polybius says there was another sanctuary devoted to Chronos (Punic Baal-Hammon). We could verify that summer solstice sunrise, a moment particularly connected to the god Baal-Hammon (Escacena (2009)), was seen on the slopes of this hill illuminating the altar at the inner parts of the sanctuary.

The latter Roman city, built after the conquest of the town and its promotion as a Roman colony in the first century BC, followed an orthogonal grid. Such was the customary way of building Roman towns (Gross & Torelli (1994)), with a grid of orthogonal streets (Cardo and Decumanus), with often a central plaza (Forum) with public buildings.

The Romans followed elaborated rituals to establish the timeliness of the foundation, the sacredness of the city, and that it was built following the right directions. For the final step it was of paramount importance the observation of the heavens (see Rodríguez-Antón *et al.* (2019) and references therein). After nearly a decade investigating the orientation of Roman cities, we have discovered that, particularly for the cities built or reformed at the time of Augustus (Rodríguez-Antón *et al.* (2019); González-García *et al.* (2019)), the Decumanus had to follow very specific directions related to the image of the Emperor.

In particular, we have seen that most cities were built according to the equinoxes. This was one of the prescriptions of the Roman surveyors, but at the same time it was a moment of remembrance of the Emperor as he was allegedly born on September equinox (Espinosa-Espinosa & González-García (2017)). A significant number of towns also included orientation towards Winter Solstice sunrise, a moment of particular ritual prominence in the Roman calendar, and that also was connected to Augustus and his idea or restoration of traditional Roman order (González-García *et al.* (2019)).

Figure 4. Cartagena (Spain) has endeavoured a major program of recovery and restoration of several Punic and Roman remains in the last decade. Some of the highlights include astronomical events that are now used in Heritage management, outreach and tourism. The main image shows the summer solstice sunrise as seen from the Punic Atargatis sanctuary at the old citadel. This sunrise happens on the slope of present day Pico Sacro, where in antiquity there use to be a temple devoted to Baal-Hammon. Small inset: the Roman remains do include other relevant astronomical orientations, such as the Curia that would have been oriented towards the setting of Capricorn, a constellation liked to Augustus. The newly inaugurated museum incorporates a depiction of this at the main hall.

At Cartagena (Roman Carthago Nova) the new street layout done at the time of Augustus, and prominently the new Forum of the city, that included a temple possibly devoted to the Roman triad, a Curia and an Augusteum, faced Summer Solstice sunrise as seen on top of a hill that Polybius links with a mythical founder of the town. We argue that in this way the local authorities wanted to connect the new ruler and founder (Augustus) with the old one. But also, to the other side, the Curia, for instance would be facing the setting of the constellation of Capricorn (González-García *et al.* (2015)).

Capricorn appears in several material items, such as coins, gems and antefixes, associated to Augustus. It was the constellation that housed the Winter Solstice at the time and according to Augustus himself it was his 'sign' (Barton (1995)).

Interestingly, the local heritage management authorities have intensively used these results for the promotion of the newly found remains and its musealization.

On the one hand, they have promoted a number of activities with the local amateur astronomer association to disseminate among the local population the interesting rise and setting events at particular moment of the year. This has helped to engage the local population in the recovery and appreciation of the newly found heritage.

On the other hand, they are using these results to give new value to the discourses of the archaeological remains. To do so, for example, the new museum at the Forum, that recovers the volumes of the old Roman Curia hold a recreation of the constellation Capricorn with an explanation of the importance of this for the time (see Fig. 4).

5. Conclusion

Astronomy can be used in education, not only to introduce concepts of geometry, algebra, physics, chemistry or geology, to name but a few subjects. It can be, and in my view it should be, used as a way to develop a sustainable engagement of local populations to value pro-actively their heritage. Cultural Astronomy produces valuable scientific results that can be used by educators. Indeed, these must be developed further, in collaboration with these communities, so that they can be changed into materials useful for teachers and other actors in the education communities.

Cultural astro-tourism, as a particular type of astro-tourism, appears as a new way of sustainable high-quality tourism that should be promoted among the tourist destinations (see Iwaniszewski (2015)). In my opinion, it is the role of cultural astronomers to contact and engage with local authorities and stake-holders to disseminate their results and possibilities among them taking care of the attention to the diversity in a broad sense. The promotion of this type of tourism would be mutually beneficial. For the local stake-holders they will get a high quality, often engaged, tourism that may help develop the region sustainably. For cultural astronomer, this is a way to keep their areas of research protected for future investigations.

Engaging with local authorities to manage heritage helps conservation, but also provides value to that heritage that can be more easily and deeply understood by the public. Engaging the public in the conservation and preservation of heritage is one of the key issues of heritage management today. One of the few environmental factors that can be still today appreciated or at least modelled with rather high accuracy is the sky. A sunrise alignment today has sifted by a very small amount to what was usual several centuries ago, and can therefore be perceived with our own eyes, providing a touching tool to connect the public to that heritage we want to protect.

These lines of action open new areas where the expertise of our fellow colleagues could be of utmost importance. In this sense, it would be necessary that collegiate societies such as SEAC, ISAAC or SIAC were more active in these grounds following the example of the IAU at different levels (through Division C, the different Commissions and the several Working Groups). Our societies can play an active role by actively engaging in the identification and protection of Astronomical Heritage. Also, we can provide new and revised, bench-marked material for outreach and education. Some of these could be used as well for promoting sustainable tourism, like new programs and tools to visualize relevant astronomical events at particular sites employing augmented reality or virtual reality, that may be used at times when the event is not visible, and can be used by the disable, making the skies more accessible to everyone.

References

Alpass, P. 2012, *The Religious Life of Nabataea* (Leiden: Brill)

Barton, T. 1995, *Journal of Roman Studies* 85, 33

Belmonte, J.A. 2012, *Pirámides, Templos y Estrellas* (Barcelona: Crítica)

Belmonte J.A., González-García, A.C. & Polcaro, A. 2013, *Nexus Network Journal* 15, 497

Belmonte J.A. & González-García, A.C. 2017, *Culture and Cosmos* 21, 131

Belmonte J.A., González-García, A.C. & Rodríguez-Antón, A. 2019, in: G. Magli, A.C. González-García, J.A. Belmonte, & E. Antonello (eds.), *Archaeoastronomy in Roman World* (New York: Springer), p. 123

Belmonte J.A., González-García, A.C. Rodríguez-Antón, A. & Perera Betancor, M.A. 2017, *Nexus Network Journal* 22, 369

Burger, R. L. and Rosenswig, R.M. 2012, *Early New World Monumentality. University Press of Florida* (Gainsville: University of Florida Press)

Browning, I. 1989, *Petra* (London: Chatto and Windus)

Escacena, J. L. 2009, *Complutum* 20(2), 95

Espinosa-Espinosa, D. & González-García, A.C. 2017, *Numen* 64(5-6), 545

García Quintela, M.V. & González-García, A.C. 2009, *Complutum* 20(2), 39

González-García, A.C. et al. 2015, *Zephyrus* 75, 141

González-García, A.C. & Belmonte, J.A. 2019, *Sustainability* 11(8), 2240

González-García, A.C. & Rodríguez-Antón, A., Espinos-Espinos, D., García-Quintela, M.V. & Belmonte J.A., 2019, in: G. Magli, A.C. González-García, J.A. Belmonte, & E. Antonello (eds.), *Archaeoastronomy in Roman World* (New York: Springer), p. 85

González-García, A.C. & Belmonte, J.A. 2020, *Journal of Skyscape Archaeology* 5(2), 177

González-García, A.C. et al. 2021, *Latin American Antiquity* in press

Gros, P., & Torelli, M. 1994, *Storia dell'Urbanistica. Il Mondo Romano.* (Roma: Editori Laterza)

Haas, J. & Creamer, W. 2004, in: H. Silvermann (ed.) *Cultural Transformations in the Central Andean Late Archaic. In Andean Archaeology* (Oxford: Balckwell), p. 35

Healy, J.F. 2000, *The religion of the Nabataeans: a conspectus* (Boston: Brill)

Iwaniszweski, S. 2009, *Complutum* 20(2), 23

Iwaniszweski, S. 2015, in: C.L.N. Ruggles (ed.), *Handbook of Archaeoastronomy and Ehtnoastronomy* (New York: Springer), p. 287

Krupp, E.C. 2015, in: C.L.N. Ruggles (ed.), *Handbook of Archaeoastronomy and Ehtnoastronomy* (New York: Springer), p. 263

López, A.M. 2015, in: C.L.N. Ruggles (ed.), *Handbook of Archaeoastronomy and Ehtnoastronomy* (New York: Springer), p. 341

Laurence, R., Esmonde Cleary, S., & Sears, G. 2011, *The City in the Roman West c.250 BC-c.AD 250* (Cambridge: Cambridge University Press)

Noguera Celdrán, J.M. & Madrid Balanza, M.J., 2014 *Espacio, tiempo y forma* 7, 13

Roche, M.-J. 1995, *Transeuphratène* 10, 57

Rodríguez-Antón, A., González-García, A.C. & Belmonte, J.A. 2018, *Journal for the History of Astronomy* 49(3), 363

Ruggles, C.L.N 2015, in: C.L.N. Ruggles (ed.), *Handbook of Archaeoastronomy and Ehtnoastronomy* (New York: Springer), p. 353

Ruggles, C.L.N 2017, *Heritage Sites of Astronomy and Archaeoastronomy in the context of the UNESCO World Heritage Convention: Thematic Study no.2.* (Paris: ICOMOS)

Ruggles, C.L.N & Cotte, M. 2017, *Heritage Sites of Astronomy and Archaeoastronomy in the context of the UNESCO World Heritage Convention* (Paris: ICOMOS)

Shady, R. 2006, in: W.H. Osbell and H. Silvermann (eds.), *Andean archaeology III: north and south* (New York: Springer), p. 28

Shady, R., Haas, J. and Creamer, W. 2001, *Science* 292, 723

Vilas Estévez, B. 2015, in: F. Silva, K. Malville T. Lomsdalen, & F. Ventura (eds.), *The Materiality of the Sky* (Bath:Sofia Center Press), 23

Education and Heritage in the era of Big Data in Astronomy
Proceedings IAU Symposium No. 367, 2020
R. M. Ros, B. Garcia, S. R. Gullberg, J. Moldon & P. Rojo, eds.
doi:10.1017/S1743921321000685

The astronomical heritage of pre-Hispanic societies in Venezuela: Total Eclipses of Sun reported in Petroglyphs

Nelson Falcón[1] and Alcides Ortega[2]

[1]Dept. of Physics FACYT, University of Carabobo,
BOX 129 Avda Bolivar norte, Valencia, Venezuela
email: nelsonfalconv@gmail.com

[2]Direcciön de-Informätica, University the Carabobo,
Box 129 Avda Bolivar norte, Valencia, Venezuela
email: aortega@uc.edu.ve

Abstract. Astronomical observations have been documented in pre-Hispanic cultures. However, little is known about the astronomical heritage of pre-Hispanic societies in Venezuela, since there are no evidence of an ancestral material culture as other regions of Meso and South America, and these native Venezuelan groups lacked a written, alphabetic or ideographic language. There are innumerable petroglyph deposits in almost all regions of Venezuela, especially near Lake Valencia (Carabobo state) and those of the Andean foothills (Barinas state). By means of computerized simulation and archaeoastronomy techniques, the occurrence of a total solar eclipse is verified: in July 1, 577 at the Vigirima site; and another total solar eclipse at noon on May 16, 1398 at the Bum Bum site. It is concluded that the contemplation of the total eclipses of the Sun, must lead to the need for communication and recording, and lacking any type of writing, they drew the phenomenon, using the techniques and means at their disposal: engraving on rocks.

Keywords. archaeoastronomy: total solar eclipses, petroglyphs of Venezuela

---

1. Contextualization

In Venezuela there are no architectural remains similar in development and elaboration to the Aztec, Mayan or Inca cultures, and the native Venezuelan groups lacked a written, alphabetic or ideographic language. However there are many rock art stations in all the country. Close the lake of Valencia, (Carabobo state) in the central region is Virgirima archaeological complex (+10.302N −67.893W), the most important lithic site of Venezuela, with an area of approximately twelve hectares. Characterized by the profusion and variety of its rock formations, expressed through megalithic alignments, rocky piles used in the delimitation of spaces and above all by the existence a hundred rocks with petroglyphs in bas-relief. Also In the Venezuelan Andean foothills, of the Barinas state, there also are many sites with ancient rock art expression, all are petroglyphs in bas-relief, without pigmentation and isolates of the other archaeological evidence. The extensive area under study covers the northern or Andean foothills in three municipalities of the Barinas State, around of the Bum Bum town (+8.2674N −70.7339W). The petroglyphs in Bum Bum archaeological complex are not grouped together but rather scattered, mostly on streams and rivers, or in erratic blocks of the alluvial valleys in the plains east of the Andean mountain range. The Bum Bum petroglyphs present a greater profusion of details and a more elaborate abrasion technique; however stylistic similarities

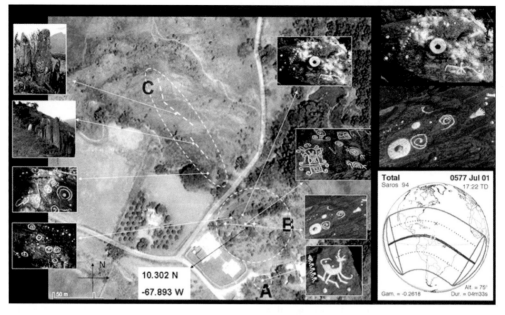

Figure 1. Virginia Archaeological station. Rock art representation of Total Solar Eclipse of the 577 Jul 01.

suggest that they were made by Arauquinoid groups that migrated to the western plains between the 10th and 12th centuries It is important to emphasize that the great majority of the petroglyphs are of a naturalist-realist type with a predominance of zoomorphic representations. The figurative features seem to indicate the importance of representing the surrounding natural world (fauna, natural phenomena, etc.) above that of the cultural world (anthropomorphic representations), and suggests a society with a close dependence on nature. Maybe in an egalitarian tribal organization, where the shaman (chief-priest) would have had an important role as a mediator between man-nature, and the woman as a symbol of the reproduction of the human force of work. The Venezuelan petroglyphs are engraved in bas-relief in erratic blocks that emerge naturally or in rocks arranged in their natural immersion slope. They were made by percussion and rubbing, and they lack pigmentation. They represent zoomorphic figures together with many geometric figures of coupled points, concentric circles, spirals, and few anthropomorphic representations. Its meaning, motivation and precise dating are unknown Albarrán (2012), beyond the free interpretation of its contents, without relation to the material culture of the ethnic groups that created it, mainly due to the absence of other evidence that allows its interpretation and dating. The petroglyphs do not show remains of any pigmentation; the false black or white color displayed (figs. 1, 2) is a photo manipulation to enhance the contrast. The grooves could have been made with hexagonal quartz crystal, abundant in the region, whose hardness is greater than metamorphic rocks of the substrate. The cuts, grooves and holes were made by successive drilling as evidenced by a half-finished rock found. Additionally there are several petroglyphs with representations of simple geometric figures (points, circles, semicircles and lines). One interpretation is their association with the sun, half moon and constellations, which could recall some possible and ancient eclipse of the Sun. Using the big date in NASA Eclipse web site Espenak (2006), we found which were the possible recorded total solar eclipses in the Venezuelan neo-Indian period: -100 BC until the Colon era.

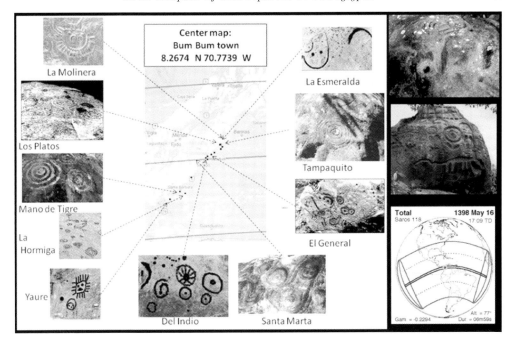

Figure 2. The extensive area of Andean foothills surrounding of the Bum Bum town. Rock art representation of conspicuous Total Solar Eclipse of the 1378 May 16.

2. Virgirima Petroglyphs

A calculation of the astronomical position allows a contrast of the relative position of the planets and brightest stars with the apparent position of the sun and the moon during the eclipse for the geographical coordinates of the Vigirima (Falcón 2013). In several petroglyphs a notable point is represented near the eclipsed sun.

Only the eclipse of the year 577 presents Venus very close to the solar disk. Also, the foothills of the coastal mountain range limit the topographic horizon and visibility of eclipses near dawn and dusk. Only the eclipse of the year 577 occurred at midday, and it just had Venus in position very near to the zenith. There are even petroglyphs with a hole through the rock, surrounded by a bas-relief and with the representation of a nocturnal bird (Fig. 1). The archaeological complex of Vigirima is within the zone of totality, almost on the maximum line The archaeological ceramic remains allow dating the occupation of the site between the 3rd and 8th centuries Falcón & *et al.* (2000). The ceramic chronology and the stylistic analysis of the petroglyphs coincide with the Eclipse of the 577 year.

3. Bum Bum Petroglyphs

In Bum Bum and their surrounding there are many petroglyphs with geometrical and abstracts symbol, maybe can be the representations about some Solar Total eclipse. The eclipse of the year 577 in Vigirima was not visible in that region, located more than 600 km away from the lake of Valencia. Using the big data of NASA web site Espenak (2006), there are five possibilities. The Andean mountain range to the northwest of the region limits the topographic horizon and the visibility of eclipses. These suggest the eclipse total of 1378 as the only possible. This eclipse was especially remarkable with almost 7 minutes of duration during the maximum, it occurred at midday with an altitude of 78° and with the 5 brightest objects in the sky after the sun and the moon. A phenomenon

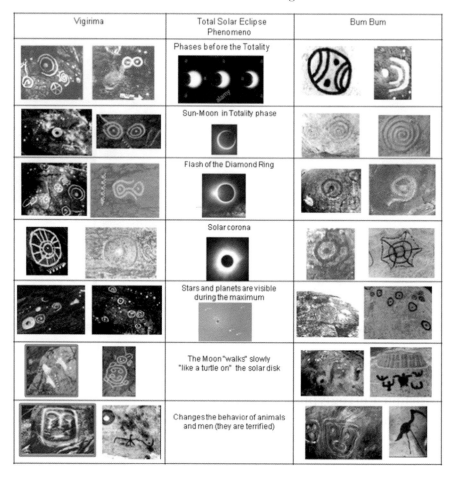

Figure 3. Comparison of the Vigirima and Bum Bum petroglyphs with the phenomena associated with a solar eclipse.

as unique and spectacular as that solar eclipse, happened at midday, had to have a mark on the collective memory and shamanic practices associated. The probable view of the sky in Bum Bum shows the eclipse at 1:17 p.m., and lasting a temperature drop of up to ten degrees in a that tropical region. Such an unusual event must have motivated its imperishable memory, with the only means available: engraving on rocks (Fig. 2). Notice that, as Vigirima site, There is even petroglyphs with a hole through the rock, surrounded by a bas-relief (solar corona) and footprints, which may signify the slow transit in the sky.

In fig. 3 the interpretation of some petroglyphs found in relation to the phenomena associated with a Total Solar eclipse is synthesized

4. Conclusions

The high frequency of representations of nature at a figurative level, in comparison with the anthropomorphic representations, indicate the importance of representing the surrounding natural world and suggests a society with a close dependence of the individual on nature. The stylistic similarity in the Vigirima and BumBum petroglyphs shows that they were made by ethnic groups of Arauquinoid origin that moved from the basin of Lake Valencia to the Andean piedmont between the X-IV centuries in accordance with the ancestral settlement of Venezuela. We have seen how big data in astronomy also allows the reconstruction of ancient ethnology, understanding the archaeological significance and cultural heritage of a particular region. The contemplation of the total solar eclipse, must lead to the need of communication and recording, and due to the lack of written language, in the societies of the Venezuelan neoindio, these aboriginal groups employed the techniques and means at their disposal: the draw engraved on rocks (petroglyphs). Also the using of big data in astronomy also allows the reconstruction of ancient ethnology, understanding the archaeological significance and cultural heritage of a particular region.

References

Albarrán, Y. 2000, *Boletín Antropológico*, 30, 84, 150.

Falcón, N., León, O., Delgado, Y. 2002, *Mañongo*, 15, VII, 299.

Falcón, N. 2013, *Virtual Archaeological Review*, 4, 8, 155.

Espenak, F. & Meeus, J. 2006, *Five Millennium Canon of Solar Eclipses: -1999 to +3000*, NASA TP-2006-21414.

Education and Heritage in the era of Big Data in Astronomy
Proceedings IAU Symposium No. 367, 2020
R. M. Ros, B. Garcia, S. R. Gullberg, J. Moldon & P. Rojo, eds.
doi:10.1017/S1743921321001034

Relative Orientation of Prasat Hin Phanom Rung Temple to Spica on New Year's Day: The Chief Indicator for the Intercalary Year of the Luni-Solar Calendar

Siramas Komonjinda[1][iD]**, Orapin Riyaprao[2],**
Korakamon Sriboonrueang[2], and Cherdsak Saelee[1]

[1]Department of Physics and Materials Science, Faculty of Science,
Chiang Mai University, Chiang Mai 50200, Thailand
email: `siramas.k@cmu.ac.th`, `cherdsak.s@cmu.ac.th`

[2]National Astronomical Research Institute of Thailand, Maerim, Chiang Mai 50180, Thailand
email: `orapin@narit.or.th`, `korakamon@narit.or.th`

Abstract. Prasat Hin Phanom Rung, located in Buriram Province of Thailand, is an ancient temple that had been built between the 10th and 13th century. The temple, which is off east-west orientation by 5.5° towards north, has unveiled the astonishing phenomena exhibiting both astronomical and architectural intellect of the ancient builders. The phenomena involve perfect quarterly-alignments of the sun through all the fifteen doorways of the temple. The phenomenal orientation of this ancient architecture has been elucidated by several scholars— including historians, archaeologists, and astronomers—that it might be related to solar or lunar events only. However, our studies have otherwise found a clue to this mystery that it may be based on how the ancient intelligence used stars in the zodiacal constellations to regulate agricultural calendars. In this study, we find that Phanom Rung was oriented with respect to Spica such that on the day Spica set on the west-side doorway at dawn, the sun was entering *Mesha Rashi* (Aries). This day has a direct connection to a New Year's Day of Saka calendar (Śaka Era), presently called *Thaloeng Sok* Day. Furthermore, we have found the relationship between Spica and the full moon of *Caitra* from which the intercalary month-year (*Adhikamas*) was detected.

Keywords. Phanom Rung, Temple Orientation, Spica, Intercalary Month-year, *Adhikamas*

1. Introduction

Prasat Hin Phanom Rung, a spectacular Khmer ancient sanctuary situated on extinct volcano in Buriram, is one of the most significant ancient monuments of Thailand. This temple had been built between the early 10th and 13th century. The oldest structures remaining in compound, which can be dated back to the 10th century during the King Rajendravarman (*r.* 944 – 968), are the two brick sanctuaries: one facing the east and another facing the south. The main sanctuary was built and dated back to the 11th – 12th century by Narendraditya. The last structures built are the two libraries and the pavilion during the King Jayavarman VII (*r.* 1181 – 1218) (Fine Arts Department 2005).

The Phanom Rung temple, which is off east-west orientation by 5.5° towards north, has unveiled the phenomena involving perfect quarterly-alignments of the Sun through all the fifteen doorways of the temple: during April 3rd – 5th and September 7th – 9th for sunrises; and during March 5th – 7th and October 5th – 7th for sunsets. The significance of this phenomenal sunrise and sunset events has been previously studied by many scholars (Chunpongtong 2010; Komonjinda 2011; Mollerup 2007). Most of these studies explain such phenomena in relation to the Sun and/or the Moon. However, our studies of over a decade have otherwise proposed another answer to this mystery orientation that it may be based on how the ancient intelligence used stars in the zodiacal constellations to regulate agricultural calendars.

According to Yano (1986), the first step in the planning and construction of ancient Hindu temple is to determine cardinal directions by either observing fixed stars or observing the shadow of a vertical column or gnomon. Moreover, in ancient times, the risings or settings of stars, along with the Sun and the Moon, were used to determine the date of farming activities. Those stars including Spica, Regulus, Aldebaran, Antares, and Pleiades were marked by ancient Chinese (Chu 1947), Greek (Penrose 1893, 1901), Indian (Rajani & Kumar 2019; Rao 1992), Javanese (Magli 2020) and Khmer (Magli 2017). Therefore, we hypothesize that the orientation of the Prasat Hin Phanom Rung may be used as an astronomical instrument to observe the bright marked stars.

Ancient Khmer adopted the Saka Calendar (Śaka Era) in which a bright star had been used to mark the day the Sun entering *Mesha Rashi* (Aries) as the New Year's Day (NYD) this day is currently known as the *Thaloeng Sok* (TLS) Day of present *Culasakaraj* (CS) (Saelee *et al.* 2018). Based on Hipparchus First point of Aries, a marked star was Spica or Alpha Virginis (α Vir) (Saha & Lahiri 1992); therefore, in this study, we examine relative orientation of Phanom Rung to Spica by taken the precession of equinox in to account to reveal the dates and azimuths of Spica rising and setting at the temple as detailed in Section 2. Furthermore, we have investigated the relationship between Spica and the full moon of *Caitra* from which the intercalary month-year (*Adhikamas*) were detected as detailed in Section 3.

2. Relative Orientation of Prasat Hin Phanom Rung to Spica

In this section, we explore the relation between Spica and the Saka's NYD from the Hipparchus time (140 B.C.) to the present day (A.D. 2020) and propose that the off east-west orientation by 5.5° towards north is related to the position of Spica. When the First Point of Aries was introduced by Hipparchus, the ecliptic longitude of Spica was located almost 180° opposite to this point in 140 B.C., but it changes its position roughly 30° (the precessional cycle in a period of 26,000 years or 1° every 72 years) in A.D. 2020.

From the observer point of view, the star will rise or set earlier each day because the sidereal day is shorter than the mean solar day by about 4 minutes. The timings when Spica is rising and setting at Phanom Rung (14.53189° N, 102. 94028° E; GMT+7) are investigated for the year 2020. There are four dates known as: (1) "heliacal rising"—the first day (18 Oct) at which Spica first becomes visible in the east at dawn; (2) "acronychal rising"—the last day (12 Apr) when Spica (after a period when it was visible at night) rises in the evening after sunset; (3) "cosmic setting"—the first day (15 Apr) at which Spica after becomes visible in the west at dawn; and (4) "heliacal setting"—the last day (15 Oct) when Spica (after a period when it was visible) sets after sunset. Among these four dates where both Spica and the Sun are seen on the horizon, only two dates are found related to the orientation of the temple, *i.e.*, the cosmic setting (Spica set south of west at dawn in April) and heliacal setting (Spica set in the evening in October). However, the cosmic setting is considered coinciding with the Saka NYD and the phenomenal sunrise.

Using the Stellarium 0.20.3, we obtain the dates and azimuths of Spica setting in the west while the Sun rising in the east at a horizon. From Fig. 1 (*left*), the orientation of

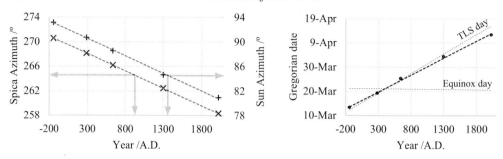

Figure 1. Left: The azimuth of Spica (×) when setting compared to the azimuth of the Sun (+) during rising. Right: The date when Spica sets as the Sun rises.

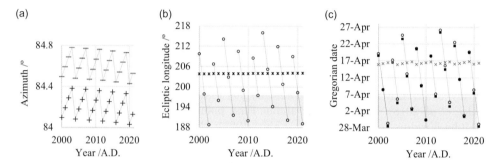

Figure 2. *a*) The azimuth of the Sun when rising on the 3rd (−) and the 4th (+) of April that the Sun can be seen through the 15 doors. The full moon of *Caitra* (○) in relation to *b*) Spica (∗) and *c*) the TLS day (×). The full moon date on the Thai calendar is symbolized in solid square. The years in shaded areas are *Adhikamas*.

Phanom Rung (az. 84.5° and 264.5°) coincides with the azimuth of Spica around A.D. 930 which is matched with the first structure planning period (*c.* A.D. 900 − 1000) and the azimuth of the Sun about A.D. 1360 in which all the structures were already constructed. Fig. 1 (*right*) also shows that the shift on such dates, eventually due to the precession of equinox, is the same as the calendric TLS day of CS, but slightly different as a longer length of the sidereal year was used. Therefore, we postulate that the orientation of the temple is likely to aim at Spica around A.D. 930 and cosmic setting of Spica may have been used to mark the NYD of Saka Calendar.

3. Indicator for Intercalary Year of the Luni-Solar Calendar

The Prasat east-west line is oriented off the east towards the north by 5.5° or equivalently at the azimuth = 84.5°. We find that the dates in which the sun rises with azimuth closing to 84.5° are the 3rd and the 4th of April (Fig. 2*a*). The Sun's azimuth of these two dates in each year differ by approximately 23.7′ or 0.4° (decreasing or shifting towards north), whereas the azimuth of the Sun of the same date each year is shifted back approximately 5.7′ except in the solar leap year.

We find that in 2000 − 2021, the full moon of *Caitra* exhibits the similar pattern of its longitude and date as shown in Fig. 2*b* − 2*c*. The pattern is repeated every 19 years (metonic cycle). In each year, the full moon of *Caitra* shifts backward 11° because ordinarily the lunar year is shorter than solar year by 11 days. As a result, in one solar year, the full moon date in *Caitra* shifts about 11 days. The *Adhikamas* years, which were assigned on the official Thai lunar calendar year 2002, 2004, 2007, 2010, 2012, 2015,

2018 and 2021, can be noticed by both moon's longitude and date on the full moon of *Caitra* as highlighted with the shaded area.

Fig. 2*b* shows the relationship between the actual full moon of *Caitra* and Spica from which the *Adhikamas* year were detected. Based on the intercalary month-year of the present CS calendar, if full moon's longitude of the year Y-1 is between $197° - 208°$ (the Spica's longitude of $204°$ is in this range), the year Y will be an *Adhikamas* year. It is to be noted that Saka and CS calendar based their TLS day calculation on the original *Surya Siddhanta*.

Fig. 2*c* shows the relationship of the calendar full moon date of *Caitra* and the TLS day. If the calendar full moon of *Caitra* year Y-1 occurs (April 7th – 16th) before the TLS day, the year Y will be an *Adhikamas* year. In other words, the full moon of *Caitra* occurring on/before the TLS day within a maximum of 10 days can be used to indicate that the next year is an *Adhikamas* year. It is important to note that the calendar full moon in the year 2011 is on April 18th, which is after the TLS day, and thus the year 2012 will not be an *Adhikamas* year. This result is conflicted with the official Thai lunar calendar and the detailed argument is already given in Saelee *et al.* (2018).

4. Concluding Remarks

We can conclude that Prasat Hin Phanom Rung may have been built as a religious monument and also as an astronomical instrument as follows:

(1) The orientation of the Phanom Rung is believed to be used to observe Spica setting on the west-side doorway at dawn as the mark for the sun entering *Mesha Rashi* (Aries), and thus a direct connection to a New Year's Day of Saka calendar (Śaka Era). This day is presently called the *Thaloeng Sok* Day.

(2) The orientation that is shifted away from east towards north by $5.5°$ could come from the observation of the chief Spica setting opposite to the rising sun (cosmic setting) at the azimuth $264.5°$ around A.D. 930, which coincided with the time the first temple layout was planned.

(3) The case study during the year $2000 - 2021$ reveals that if the full moon of *Chitra* occurs within $10 - 11$ days before the New Year's Day, then the following year will be an *Adhikamas* year. It is possible that the ancient Khmer may used the full moon of *Chitra* in relation to the cosmic setting of Spica as an indicator for adding an intercalary month to an *Adhikamas* year.

Acknowledgements. This archaeoastronomical research is an extension to the research conducted by the late Assoc. Prof. Samai Yodinthara, Asst. Prof. Mullika Thavornathivas, and Assoc. Prof. Sanan Supasai. We are grateful for Assoc. Prof. Boonrucksar Soonthornthum, Worapol Maison, Voranai Pongsachalakorn, Dr. Direct Injan and The Fine Art Department of Thailand. This work was supported by the National Astronomical Research Institute of Thailand and the Physics and Astronomy Research Group, Chiang Mai University.

References

Chu, C. 1947, *Popular Astronomy*, 55, 62–77.
Chunpongtong, L. 2010, *J. of Royal Society of Thailand*, 35(4), 718–729 [in Thai]
Fine Arts Department. 2005, *Prasat Phnom Rung* (Burirum: Vinai Printing)
Komonjinda, S. 2011, *Proceedings of the International Astronomical Union*, 7(S278), 325–330.
Magli, G. 2017, *Studies in Digital Heritage*, https://doi.org/10.14434/sdh.v1i1.22846
Magli, G 2020, in *Archaeoastronomy* (Cham: Springer), pp. 245–259.
Mollerup, A. 2007, *Mueang Boran Journal*, 33(2), 31–44.
Penrose, F. C. 1893, *Nature*, 48(1227), 42–43.
Penrose, F. C. 1901, *Nature*, 63(1638), 492–493.

Rajani, M. B., & Kumar, V. 2019, *J. of the Society of Architectural Historians*, 78(4), 392–408.

Rao, N. K. 1992, *Bulletin of the Astronomical Society of India*, 20, 243–254.

Saelee, C., Tawonatiwas, M., & Yodintra, S. 2018, *Chiang Mai J. of Sci.*, 45(6), 2491–2508.

Saha, M., & Lahiri, N. C. 1992, *History of the Calendar in Different Countries Through the Ages* (New Delhi: Council of Scientific & Industrial Research), pp. 199–200.

Yano, M. 1986, *Indo-Iranian Journal*, 29(1), 17–29.

Education and Heritage in the era of Big Data in Astronomy
Proceedings IAU Symposium No. 367, 2020
R. M. Ros, B. Garcia, S. R. Gullberg, J. Moldon & P. Rojo, eds.
doi:10.1017/S1743921321000612

Cultural Astronomy for Inspiration

Steven R. Gullberg🅾

Director for Archaeoastronomy and Astronomy in Culture, College of Professional and
Continuing Studies, University of Oklahoma, USA.
email: `srgullberg@ou.edu`

Abstract. Cultural astronomy is the study of the astronomy of ancient cultures and is sometimes called the anthropology of astronomy. The many ways that astronomy was used by ancient cultures are fascinating and this can be used to inspire interest in all astronomy, as well as astronomy in culture. Archaeoastronomy is interdisciplinary and among its practitioners are not only astronomers and astrophysicists, but also anthropologists, archaeologists, and Indigenous scholars. Much can be learned about ancient cultures though examination of how and why they used astronomy. This paper will highlight several examples that can capture public attention.

Keywords. archaeoastronomy, cultural astronomy, Inca astronomy, Machu Picchu

1. Introduction

Cultural astronomy examines the astronomy used in ancient cultures, including orientations found at sites and structures of ancient peoples. Many people find cultural astronomy to be fascinating, and this can be advantageous as an opener to inspire further interest in other areas of astronomy. Several examples follow from the astronomy of the Incas. .

2. Astronomy of the Inca Empire

Astronomy was at the center of the Incas' religion and agriculture. The Incas were the children of the Sun and believed their emperor to be the Sun's direct descendant (Bauer, B. 1995). Solar worship was the official religion of their empire. The first emperor of the conquest, Pachacuti, imposed it across the realm, maintaining that he was the son of the Sun and his wife the daughter of the Moon. The ruling Inca was the central figure in solar worship, supporting the assertion that he was the descendant of the Sun (Zuidema, R. T. 1964). The Incas learned the cycles of solstices and equinoxes and used this knowledge as a key component of their annual crop management activities, as well as for determining dates for religious celebrations.

The Huaca (Shrine) of Kenko Grande

Kenko Grande is a large carved limestone outcropping north of Cusco. On its upper surface are two gnomons carved and situated to create a specific light and shadow effect. At the June solstice sunrise, light passes through a carefully designed fissure aligned to illuminate first one of the gnomons and then the other, with both casting shadows that create an image. The result is known as "the awakening of the puma" (Gullberg, S. R. 2020) Fig. 1. Pumas were one of the three most sacred creatures in the Inca cosmos, the other two being the condor and the serpen.

The Huaca of Lacco

Lacco is an even larger limestone outcropping to the northeast of Kenko Grande. Its prominent astronomical features are three caves with astronomical orientations

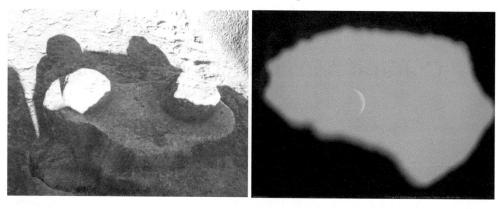

Figure 1. a) Awakening of the Puma b) Crescent Moon (Author's photos)

Figure 2. a) June Solstice Sunrise b) Illuminated altar (Author's photos)

(Gullberg, S. R. 2020). Lacco's southwest cave contains a small altar and a light-tube aligned for illumination. Fig. 1b looks outward through the light-tube at the crescent Moon.

Lacco's northeastern cave opening is aligned for June solstice sunrise. The sunrise position on the horizon daily draws nearer from the right until at the time of the solstice it stops, centered on the opening of the cave (Fig. 2a). Light from the Sun illuminates an altar and reflects into much of the rest of the cave (Gullberg, S. R. 2020).

Lacco's southeast cave has carved steps that descend into the outer chamber with a serpent carved into the entrance wall. The inner chamber includes an altar and a vertical light-tube. The cave is called the Temple of the Moon but was found as well to illuminate the altar at the time of the zenith Sun (Gullberg, S. R. 2020) (Fig. 2b).

Huaca 44
Huaca 44 is a small huaca located between Kenko Grande and Lacco. It includes two large carved circles that could be used to indicate the direction for each of the six primary solar horizon events. Tangential lines drawn traced between the larger and smaller circles indicate the directions for viewing June Solstice Sunrise, December Solstice Sunset, December Solstice Sunrise, and June Solstice Sunset. A line traced across the two indicates the directions for an Equinox Sunrise and an Equinox Sunset (Gullberg, S. R. 2020) (Fig. 3a).

The Palace of Q'espiwanka
There may have been as many as 16 towers on the Cusco horizon erected for sunrise calendrical purposes, but they were all destroyed during the Spanish extirpation of in-

Figure 3. a) Huaca 44 b) Pillars (Author's photos)

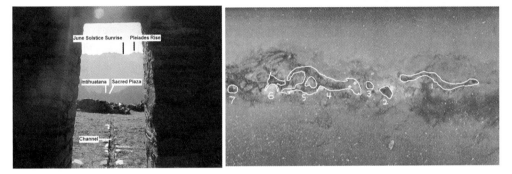

Figure 4. a) Llactapata b) Dark 'constellations' (from Author's photo and painting by Jessica Gullberg)

digenous religion. Near the north of the Sacred Valley, however, two such pillars survive on the Cerro Saywa ridge above the present village of Urubamba. The pillars mark the rise of the June solstice Sun when viewed from a sacred granite boulder on Huayna Capac's palace grounds called Q'espiwanka. As the sunrise moves on the horizon from right to left it rises first over the right pillar giving warning and then continues to the left pillar by the solstice (Gullberg, S. R. 2020) (Fig. 3b). This extant example helps to support that similar pillars once surrounded Cusco.

The Royal Estate of Machu Picchu and Nearby Llactapata

The greater significance of Machu Picchu and its surrounding area began to emerge with research conducted at Llactapata in 2003 (Malville, J. M. 2006). The Llactapata Ridge lies five kilometers from Machu Picchu across a deep gorge below. Well over one hundred structures are engulfed there within the cloud forest. The Llactapata Sun Temple, however, is now kept clear and exhibits solar orientation. The Sun Temple, the River Intihuatana at the base of the gorge, and the Sacred Plaza of Machu Picchu all lie on an axis of the June solstice sunrise/December solstice sunset.

A ceremonial channel built from the Sun Temple's central door points across the River Intihuatana to the Sacred Plaza (Gullberg, S. R. 2020) (Fig. 4a). Beyond, on the horizon, is where the June solstice Sun rises, and as well the nearby heliacal rise of the Pleiades. The observed brilliance or dullness of the Pleiades at this time is used to this day to forecast crops and planting with regard to an impending El-Niño drougt (Orlove, B. S. 2000).

Incan Use of the Milky Way

The Incas recognized "dark constellations," or the shapes of beings formed by dark clouds in the visible band of a section of the galaxy prominent in the Southern Hemisphere. They saw great cosmological characters meant to guide them in their daily lives. Most are animals that figure prominently in Andean cosmology and myth (Urton, G. 1981), (Gullberg, S. R. 2020). Machacuay, the serpent, leads the procession of dark constellations in the Milky Way. The serpent travels head before tail across the sky (Urton, G. 1981). Hanp'atu, the toad, follows closely behind Machacuay. Toads were thought of as bad omens created by the devil. Hanp'atu is a much smaller and is po- sitioned to the left of the snake (Urton, G. 1981). Tinamou are indigenous birds of ancient lineage. Yutu, the Tinamou, is what Western astronomy knows as The Coalsack and follows Hanpatu in the Milky Way (Urton, G. 1981).Yacana, the llama, figures prominently in Inca tradition and was thought to animate llamas on the Earth (Gullberg, S. R. 2020). Yacana dominates the section of the Milky Way used by the Incas for dark constellations and is situated between Centaurus and Scorpius. The prominent stars Alpha and Beta Centauri are thought to be its eyes (Urton, G. 1981). Below Yacana is a smaller dark constellation called Uñallamacha, a baby llama suckling its mother (Urton, G. 1981). After Yacana and Uñallamacha is another small constellation called Atoq, the fox. The Sun enters Atoq during the December solstice. Urton (1981) explained that the Milky Way and Atoq catch up and rise with the December solstice Sun in the southeast during the same period of time that terrestrial baby foxes typically are born, around 15–23 December. The final dark constellation is a second Yutu. This additional Tinamou completes the celestial procession (Urton, G. 1981, Gullberg, S. R. 2020), (Fig. 4b).

3. Conclusions

Cultural astronomy is a fascinating field and it adds great insight to what is known about early cultures, most of which used astronomy. The University of Oklahoma offers well-developed online programs with graduate and undergraduate degree opportunities, certificate programs, and individual courses. The Inca examples presented here are just a few of a vast many that can be used to capture the attention of non-astronomers and further inspire them. Archaeoastronomy can be a great opener with the public. It intrigues them and can inspire greater curiosity about much of astronomy and astrophysics in general.

References

Bauer, B. and Dearborn, D. 1995, *Astronomy and Empire in the Ancient Andes: The Cultural Origins of Inca Sky Watching*, University of Texas Press.

Gullberg, S. R. 2020, *Astronomy of the Inca Empire: Use and Significance of the Sun and the Night Sky*, Springer Nature

Malville, J. M., Thomson, H., & Ziegler, G. 2006, The Sun Temple of Llactapata and the Ceremonial Neighborhood of Machu Picchu, in *Viewing the Sky Through Past and Present Cultures*, eds. T. Bostwick & B. Bates, City of Phoenix Parks, Recreation, and Library, 327–339.

Orlove, B. S., Chiang, J. C., & Cane, M. A. 2000, orecasting Andean Rainfall and Crop Yield from the Influence of El Nino on Pleiades Visibility, *Nature, 403*, 68–71.

Urton, G. 1981, *At the Crossroads of Earth and Sky: An Andean Cosmology.* Austin, University of Texas Press.

Zuidema, R. T. 1964, *The Ceque System of Cusco: The Social Organization of the Capital of the Inca,* Leiden, E.J. Brill.

Education and Heritage in the era of Big Data in Astronomy
Proceedings IAU Symposium No. 367, 2020
R. M. Ros, B. Garcia, S. R. Gullberg, J. Moldon & P. Rojo, eds.
doi:10.1017/S174392132100048X

Armenian Astronomical Heritage and Big Data

Gor A. Mikayelyan⬚, Sona. V. Farmanyan and Areg M. Mickaelian⬚

NAS RA Byurakan Astrophysical Observatory (BAO),
Byurakan 0213, Aragatzotn Province, Armenia
email: `gormick@mail.ru`

Abstract. Astronomy in Armenia was popular since ancient times and Armenia is rich in its astronomical heritage, such as ancient and medieval Armenian calendars, records of astronomical events by ancient Armenians, the astronomical heritage of the Armenian medieval great thinker Anania Shirakatsi, etc. Armenian astronomical archives have accumulated vast number of photographic plates, films and other careers of observational data. The Digitized Markarian Survey or the First Byurakan Survey, is the most important low-dispersion spectroscopic database. It is one of the rare science items included in UNESCO "Memory of the World" Documentary Heritage list. The Byurakan Astrophysical Observatory (BAO) Plate Archive Project (2015–2021) will result in digitization and storage of some 37,000 astronomical plates and films and in creation of an Electronic Database for further research projects. Based on these data and archives and development of their interoperability, the Armenian Virtual Observatory was created and joined the International Virtual Observatory Alliance.

Keywords. Archaeoastronomy, FBS, DFBS, Plate Archive

1. Introduction

Armenia is one of the cradles of ancient science, and astronomical knowledge was developed in ancient Armenia as well. Contrary to its small territory and relatively small population, Armenia was and is rather active in astronomy. Astronomy in Armenia was popular since ancient times: there are signs of astronomical observations coming from a few thousand years ago. Among the astronomical activities that have left their traces in the territory of Armenia are: the rock art, ruins of ancient observatories, the ancient Armenian calendar, astronomical terms and names used in Armenian language since II millennia B.C., sky maps from Middle Ages, and most important, one of the largest modern observatories in the region, the Byurakan Astrophysical Observatory (BAO) with its 2.6m and 1m Schmidt telescopes.

The Byurakan Astrophysical Observatory (BAO) Plate Archive consists of 37,500 photographic plates and films, obtained with the 2.6 m telescope, the 1 m and 0.5 m Schmidt telescopes, and other smaller telescopes during 1947–1991. Its most important part, the famous Markarian Survey (or the First Byurakan Survey, FBS) 2000 plates were digitized in 2002–2007 and the Digitized FBS (DFBS) was created. New science projects have been conducted based on this low-dispersion spectroscopic survey data. In 2015, we started a project to digitize the whole BAO Plate Archive and its scientific products. It is aimed at digitization, extraction and analysis of archival data and building an electronic database and interactive sky map. The Armenian Virtual Observatory (ArVO) database will contain all new data. ArVO will provide all standards and tools for efficient usage of the scientific output and its integration in international databases.

2. Armenian Archaeoastronomy

Studies of the Armenian rock art present in the territory of modern Armenia show that the Armenians were interested in heavenly bodies and phenomena. The Earth, the Sun, the Moon, planets, comets, Milky Way, stars, constellations are reflected in these pictures drawn on rocks in mountains around Lake Sevan and elsewhere in Armenia.

According to investigations by Badalian, Tumanian and Broutian, the Armenian calendar was one of the most ancient in the world. Armenians used Lunar, then Lunisolar calendar, and since mid the 1st millennium B.C. they changed to Solar calendar, which contained 365 days (12 months by 30 days and an additional month of 5 days). The new year began in Navasard (corresponding to August 11), when the grape harvest was underway and the constellation Orion became visible in the night sky. Together with the months, all days of any month also had proper names. The year 2492 B.C. was adopted as the beginning. The Armenian Great Calendar was introduced in VI century, and the difference with the Julian one was re-calculated (Mickaelian 2014a, Farmanyan & Mickaelian 2015, Mickaelian & Farmanyan 2016). It is remarkable that the Mkhitarians from Venice are the oldest regular publishers of the Armenian and world calendars (since 1775).

3. Armenian Astronomy in the Middle Ages

One of the most remarkable scientists in the Middle Ages was Anania Shirakatsi (612–685), who had rather progressive astronomical ideas for those times. He has left a few books and writings that survived up to nowadays. Anania Shirakatsi knew about the spherical shape of the Earth. He accepted also that the Milky Way consisted of numerous faint stars, could correctly interpret Lunar and Solar eclipses, and had a number of other progressive astronomical knowledge for that time. Anania Shirakatsi's works serve as the main source for establishing the ancient Armenian astronomical terminology, including the names of constellations and stars (Harutyunian & Mickaelian 2014, Mickaelian & Mikayelyan 2014, Farmanyan & Mickaelian 2017).

According to Prof. Pskovskiy, the 1054 Supernova was first seen and recorded in Armenia in May 1054 (and only later in summer in China). Interestingly, its remnant, the famous Crab nebula has been studied in detail in the Byurakan Astrophysical Observatory and was one of its famous objects of investigation.

Ghukas Vanandetsi and Mkhitar Sebastatsi lived and worked in Europe in 17th-18th centuries and are known for their detailed charts of the heavens. Lukas Vanandetsi made astronomical instruments, published the first sky chart with Armenian names of constellations in Amsterdam at the beginning of 18th century. Mkhitar Sebastatsi was the person who founded the Armenian Catholic Church community in St. Lazar island near Venice, a touristic site for many visitors (Mickaelian 2014a, Mickaelian & Farmanyan 2016).

4. Modern Armenian Astronomy

The modern astronomy in Armenia begins with the foundation of the Byurakan Astrophysical Observatory (BAO). It is one of the most important astronomical centers in Eastern Europe and Middle East region, both by its scientific instruments and achievements. The Observatory was founded in 1946 on the initiative of Viktor Ambartsumian. The main scientific instruments at BAO are: 2.6m telescope, 1m and 0.5m Schmidt telescopes (Mickaelian 2016).

Being one of the largest telescopes in European, Asian and African region, the Byurakan 2.6m telescope allows to make detailed spectral, photometric and other investigations of interesting faint objects. The 2.6m telescope was installed in 1975 and is in operation since 1976. It is a classical Cassegrain system telescope. During 1976–1991, the primary observations have been carried out on the morphological study of Markarian galaxies, investigation of star clusters, groups and clusters of galaxies. The observed 5000 slit spectra on this telescope are of stellar objects of the First Byurakan Survey, T Tau and flare stars, objects of the Second Byurakan Survey. During the observations of 1996–2007, new interesting results have been obtained. Faint objects in a 1° field are being observed to find high-redshift primordial galaxies. Hundreds of IRAS and SBS galaxies, as well as many non-stable stars and young stellar objects have been studied spectroscopically. A systematic search for emission-line objects is being carried out with the 2.6m telescope, too.

The BAO 1m Schmidt telescope is one of the largest Schmidt-type telescopes in the world and one of the most efficient astronomical telescopes in general. The telescope was installed in 1960 in the main territory of the Byurakan Observatory. One of the 1m Schmidt telescope's advantages was the presence of its three objective prisms (1.5°, 3°, and 4°), which made possible wide-field spectroscopic observations with various dispersions. The objective prisms can rotate in the position angle that allows obtaining spectra of any orientation.

The First Byurakan Survey (FBS) is the most famous work done with this telescope. More than 2000 photographic plates were obtained. 1500 objects were selected, which are known at present as Markarian galaxies (Mickaelian 2014b). The survey involved the largest ever astronomical study of the nearby universe and is considered one of the most important achievements of the 20th century astrophysics. In 2011 The First Byurakan Survey has been included in UNESCO's "Memory of the World" International Register (Mickaelian *et al.* 2019a).

5. The Digitized First Byurakan Survey

The Digitized First Byurakan Survey (Mickaelian *et al.* 2007a, Massaro *et al.* 2008) is the digitized version of the First Byurakan Survey. It is the largest spectroscopic database in the world, providing low-dispersion spectra for 20 million objects. DFBS is a joint project of the Byurakan Astrophysical Observatory (BAO), Cornell University (USA) and Universita di Roma "La Sapienza" (Italy). The DFBS has been created in 2002–2003 as a result of digitization and reduction of 1874 FBS plates.

High-accuracy astrometric solution has been made for each plate. At present all plates have astrometric solution. The typical rms accuracy is 1 arcsec. Dedicated software allows quick access to any field by given position and extraction of the needed spectra, their calibration, classification and study (Nesci *et al.* 2007, Mickaelian *et al.* 2007b). The DFBS is free for the astronomical community. DFBS is the largest Armenian astronomical database and one of the largest in the world.

6. Byurakan Astrophysical Observatory Plate Archive

Byurakan Astrophysical Observatory (BAO) Plate Archive is one of the largest astronomical archives and is considered to be BAO main observational treasure. Taking into account decades hard work of Armenian astronomers and the work of BAO telescopes, as well as the results of their activities, we can say that BAO Plate Archive is one of our national scientific values. Due to Viktor Ambartsumian's brilliant ideas and the mentioned observational work, the Armenian Government has recognized BAO as National value. Today BAO archive holds about 37,000 astronomical plates, films or other carriers of observational data (Mickaelian 2014c, Mickaelian *et al.* 2019b).

The digitization project is aimed at compilation, accounting, digitization of BAO observational archive photographic plates and films, as well as their incorporation in databases with modern standards and methods, providing access for all observational material and development of new scientific programs based on this material (Mickaelian *et al.* 2016, Mickaelian *et al.* 2017). The electronic archive will be a part of the Armenian Virtual Observatory (ArVO) and hence, will be incorporated in the International Virtual Observatory Alliance (IVOA).

References

Farmanyan, S. V.; Mickaelian, A. M. 2015, *in IAU General Assembly, Meeting #29, id.2256634*

Farmanyan, S. V.; Mickaelian, A. M. 2017, *ComBAO, Volume 64, Issue 1*, p. 52

Harutyunian, H. A.; Mickaelian, A. M. 2014, *in Astronomical Heritage in the National Culture, Proceedings of Archaeoastronomical Meeting, Armenia. Yerevan, "Gitutyun" Publishing House of the NAS RA*, p. 87

Massaro, E.; Mickaelian, A. M.; Nesci, R.; Weedman, D. (eds.) 2008, *ARACNE Editrice, Rome, Italy*, 78 pages

Mickaelian, A. M.; Nesci, R.; Rossi, C.; Weedman, D.; et al. 2007a, *A&A, 464*, 1177

Mickaelian, A. M.; Sarkissian, A.; Dubernet, M.-L.; Le Sidaner, P.; et al. 2007b, *in JENAM-2007, "Our non-stable Universe", held 20-25 August 2007 in Yerevan, Armenia. Abstract book*, p. 85

Mickaelian, A. M. 2014a, *in Astronomical Heritage in the National Culture, Proceedings of Archaeoastronomical Meeting, Armenia. Yerevan, "Gitutyun" Publishing House of the NAS RA*, p. 44

Mickaelian, A. M. 2014b, *in Multiwavelength AGN Surveys and Studies, Proceedings of the IAU S304*, p. 1

Mickaelian, A. M. 2014c, *in Astroplate 2014, Proceedings of a conference held in March, 2014 in Prague, Czech Republic*, p. 109

Mickaelian, A. M.; Mikayelyan, G. A. 2014, *in Astronomical Heritage in the National Culture, Proceedings of Archaeoastronomical Meeting, Armenia. Yerevan, "Gitutyun" Publishing House of the NAS RA*, p. 160

Mickaelian, A. M. (ed.) 2016, *Byurakan Astrophysical Observatory. Yerevan, NAS RA "Gitutyun" Publishing House*, 64 pages

Mickaelian, A. M.; Farmanyan, S. V. 2016, *Mediterranean Archaeology and Archaeometry (MAA), Vol. 16, No 4*, p. 385

Mickaelian, A. M.; Abrahamyan, H. V.; Andreasyan, H. R.; Azatyan, N. M.; et al. 2016, *Astronomy in Focus, Proceedings of the IAU XXIX GA, Volume 29A*, p. 130

Mickaelian, A. M.; Gigoyan, K. S.; Gyulzadyan, M. V.; Paronyan, G. M.; et al. 2017, *ComBAO, Volume 64, Issue 1*, p. 102

Mickaelian, A. M.; Sargsyan, L. A.; Mikayelyan, G. A.; Erastova, L. K.; et al. 2019a, *in Astronomical Heritage of the Middle East. ASP Conference Series, Vol. 520*, p. 105

Mickaelian, A. M.; Erastova, L. K.; Gyulzadyan, M. V.; Ohanian, G. B.; et al. 2019b, *in Astronomical Heritage of the Middle East. ASP Conference Series, Vol. 520*, p. 117

Nesci, R.; Cirimele, G.; Mickaelian, A. M.; Sargsyan, L. A. 2007, *in JENAM-2007, "Our non-stable Universe", held 20-25 August 2007 in Yerevan, Armenia. Abstract book*, p. 94

Education and Heritage in the era of Big Data in Astronomy
Proceedings IAU Symposium No. 367, 2020
R. M. Ros, B. Garcia, S. R. Gullberg, J. Moldon & P. Rojo, eds.
doi:10.1017/S1743921321000715

TIEMPEROS: Meteorological specialists from the pre-hispanic indigenous cosmogony of Mexico, and the use of technology to promote astronomy and atmospheric sciences

Cintia Durán[iD]

Tlaloque, Medellín 264, Mexico City, Mexico
email: carrillo.du@gmail.com

Abstract. The cult of the mountains, the wind and the request for "good rain" constitute today,the fusion of pre-Hispanic religious beliefs and meteorological knowledge in the agricultural development of central Mexico. Understanding this cult of the earth, from an indigenous perspective, led by certain specialists who have extensive knowledge of the landscape and meteorology, called Tiemperos, is a fundamental and necessary feature for the development of atmospheric sciences and the inclusion of rural villages in environmental research, carried out in certain areas of Mexico. Understanding the world in which these specialists are inserted is complex if one does not have a joint vision of the ethnographic data and the social relevance that the Tiemperos have on the communities. During 2018 I carried out an investigation on the request of rain and "goodweather" rituals that are carried out year after year in certain areas of central Mexico. From that initiative we developed an educational model and a prototype weather station that could be designed, built and adapted to the needs of each community, considering the traditions and teachings of the local Tiempero. Making use of microcontrollers, basic electronics, and a traditional indigenous technique, each station was built and designed with the people of the community where it would be installed, with the idea of involving and enriching scientific meteorological knowledge, which could be useful for each community. The project, still in development, included meteorological stations designed by the author and built by the communities, a series of educational exercises for children involved in the project and the proposal of a "goodweather" ritual using the data collected by the meteorological stations, with the intention of using technology and science-based information with traditional indigenous practices giving way to new forms of research and inclusion of science in remote communities in Mexico.

---

1. We call it Earth, they call it Tlalli

What is Earth exactly? A planet, a rock spinning in space, a home. In Nahuatl† The place where there is fire, and therefore there is heat, is called "**Chantequitl**" and at the same time the word assigned for "home" is "**Chantli**". The Earth, that chaotic and mutant place that protects us, is seen in many cultures as a living, divine entity that feels and decides for us. From planetary sciences specifically geology, rocksbear witness to events that happened millions of years ago, the study of "geological time" is essential to know our home and ourselves. Human culture has observed the earth and its changes since we began to become aware, what were rocks, plants, clouds, rain, cold, wind, fire like in those days? What is different in our way of approaching it, to observation? At present,

† Nahuatl is a macro-language spoken in Mexico, it has existed since at least the 5th century and is used in at least 16 states of the country.

the development of science opens new paths for us, many of them on the foundations and excavations of women and men before us who, from curiosity and work, have managed to reach essential conclusions. Technological advancement gives us an insurmountable advantage over our ancestors. Could it be that curiosity is not enough?

This same curiosity led us in 2018 to begin an investigation on the rituals of requesting rain and "good weather" that are carried out year after year in certain areas of central Mexico. From that initiative we developed an educational model and a prototype of a meteorological station that could be designed, built and adapted to the needs of each community, considering the traditions and teachings of the local *Tiempero*. Making use of microcontrollers, basic electronics and a traditional indigenous technique to work the clay. Each station was built and designed with the people of the community where it would be installed, with the idea of involving and enriching scientific meteorological knowledge, which could be useful for each community.

The project, still under development, included meteorological stations designed by us and built by the communities, a series of educational exercises for children involved in the project and the proposal of a ritual of "good weather" using the data collected by the stations, with the intention of using technology and science-based on traditional indigenous practices, opening the way to new forms of research and inclusion of science in communities far from Mexico's science research centers.

The development of this project led us to several important conclusions in our own way of approaching citizen science, hand in hand with a community rooted in historical, religious and indigenous traditions. During the exercise in the city of Guadalajara, we worked with the community of the municipality of Tonalá, in the state of Jalisco, Mexico. Tonalá is a small town that is part of the metropolitan area of Guadalajara, (the second largest city in Mexico), and where year after year a historical-religious tradition takes place where a large part of the people participates: The dance of the Tastoanes.

In the approach of our project, the inclusion of typical traditions of the places was not contemplated, at that moment we failed to consider the fundamental contributions that knowing their traditions would bring to our project, however, the work with the community took our research to know and include their own traditions, under which astronomical observation and scientific experimentation were indisputably reinforced.

At that point of the investigation, two interesting things happened: On the one hand, we learned about the tradition of the dance of the *Tastoanes* and the importance it represents for the inhabitants of the city. When we first arrived in Tonalá, we came across that the only true way to establish a relationship with the community was by participating in its traditions.

2. Technology as a substitute for divinity? Or the inclusion of social traditions to incorporate society into science.

This idea, as obvious as it may seem, was the trigger for several significant points in the development of our research, under three important points:

- **Working in collaboration with the local artist.** Angel Santos craftsman of the town, helped us design a new meteorological station, built using the traditional technique of burnished clay and with whom you can share and learn about the questions that the inhabitants of the town ask themselves when living on a hill of great historical tradition.
- **The astronomy group with the village children.** "Sky Observation Club" where we were able to carry out a knowledge exchange with children from 4 to 10 years old, participants of the traditional dance of the *Tastoanes*.
- **The community work of the citizenship**. Involve the neighbors in the construction and obtaining of atmospheric data that the meteorological stations were releasing.

These three new axes formed a new basis for our (still growing) social-artistic-science experiments and gave me new questions:

- Have we failed, assuming that the public is not interested in heaven?
- Is our approach to the popularization of astronomy and meteorology too western?
- How can we involve the public in modern astronomical research?

It will probably be difficult to get out of the security niche that the academy provides, it is easier to assume that society is unwilling but even within those invisible dividing lines, there are cracks and holes where we can cross, and establish new communication bridges where information, experimentation, curiosity but also multidiscipline, can manage to traverse and flow freely.

Session 9: Astronomy and Inclusion

Education and Heritage in the era of Big Data in Astronomy
Proceedings IAU Symposium No. 367, 2020
R. M. Ros, B. Garcia, S. R. Gullberg, J. Moldon & P. Rojo, eds.
doi:

Introduction

Session 9 on Day 5 begins with "Women in science: the need for a global cultural change" by Karen Hallberg. Karen begins by mentioning the relatively low participation by women in STEM and give as an example the number of female Nobel laureates versus the number of males. She then discusses examples of some strong women in science, but concludes that there still is much to be done to inspire more girls to pursue science.

Next, from Jarita Holbrook, is another invited talk, "ASTROMOVES: Astrophysics, Diversity, Mobility." Here Jarita describes the ASTROMOVES project that she has created to gather data regarding women needing to work abroad longer or have had more positions abroad to advance their careers to parallel those of male colleagues. The project is intended to study astrophysicist career decision making and changes of positions in terms of relocations and changes in job titles while staying in one place. She discusses methodology and gives several examples. She concludes by discussing plans for the future.

Following the talk Anahí Caldú asked:

With Covid-19 I would expect that the situation is special for the people that are looking for jobs this year (I know administrative processes have slowed down). It would be interesting to compare these results to post-pandemics ones. What do you think?

Jarita answered:

Yes, it would be a good future comparison. Interestingly, those that I have interviewd that are mentoring postdocs/grad students at that transition are talking about what is going on due to the Pandemic.

Anaely Pacheco Blanco asked:

Are you using social media to look for people to interview? or there is a webpage were there is an elegibility list (as I understood is focused in EU)?

Jarita replied :

I am not using social media to find interviewees because the users are skewed to those that are younger, not 8 years past PhD. Second it would give me too many people to interview – given that I'm slow to get people scheduled, etc. ALSO anyone can volunteer to be interviewed if they make the criteria.

Silvia Casu, Gian Deiana, and Emilio Molinari describe "The inmate sky – Astronomy in a juvenile detention institute," which is a discussion of the "Open up, Sky!" project for astronomy education at a juvenile detention institution in Sardinia, Italy. They give an overview of the project, outline its implementation, and close with a discussion of results.

José Gómez asked Silvia:

What was the main problem when working with young people in prison?

Silvia replied:

The main difficulty is related to the extreme heterogeneity of the target group. Of 12 youg involved, 6 were from abroad (Africa and East Europe). They have different ages (from 16 to 23 years old), different school, familiar and religious background. Some of them were in jail for severe crimes, other for drug dealing or theft. Another problem is their possible attitudes. Some of them have oppositive attitudes, and they do not want to show interest in something which could be related to "school". You need some time to build an emotional relationship, to show them that you do not judge them or do not confond them with their crimes. Another problem is gender related: I am a woman, they are men, and they feel they havo to show me they are "strong" and "real men". We spent some time to discuss abouth gender roles misconception.

The section continues with "Accessible Astronomy Activities for the Blind and Visually Impaired in Puerto Rico" by Gloria Isidro Villamizar, Carmen Pantoja Pantoja, and Mayra Lebrón Santos. They describe the development in Puerto Rico of astronomy activities accessible for the blind that can include three-dimensional tactile material to help the users conceptualize a topic. They highlight many activities that were employed at the workshops that have been conducted.

"Inclusive Eclipse: a sensorial experiencing along Chile" was written by Paulina Troncoso Iribarren, Carlos Santander, Javiera Diaz, Henry López del Pino, Erika Labbé, Ignacio Schacht, Carlos Morales Marín, and Angie Barr. It continues the discussion of the use of LightSound in Chile during the eclipse of 2019. They describe the device and how it functions

This section concludes with "Women in Astronomy and in Sciences: cracking the code with a third culture" by Anna Curir. Anna also talks about the under-representation of women in STEM. She cites a UNESCO report, "Cracking the code: Girls' and women's education in STEM" Anna relates that girls are often brought up to believe that STEM topics are more masculine with women better suited for the humanities. She talks about two cultures, scientific and humanistic, and adds that a third culture of astronomy could help due to the interdisciplinary study of its ancient history.

Education and Heritage in the era of Big Data in Astronomy
Proceedings IAU Symposium No. 367, 2020
R. M. Ros, B. Garcia, S. R. Gullberg, J. Moldon & P. Rojo, eds.
doi:10.1017/S1743921321000788

Women in science: the need for a global cultural change

Karen Hallberg⓪

Centro Atómico Bariloche and Instituto Balseiro (CNEA, CONICET, UNCuyo),
Bustillo 9500 (8400) Bariloche, Argentina
email: `karen@cab.cnea.gov.ar`

Abstract. This paper summarizes the talk given at this conference in which the cultural aspect of the low participation of women in science, mainly in STEM (Science, Technology, Engineering and Mathematics) areas, is emphazised. A few personal recollections will be presented and some some striking numbers to illustrate the current situation will be given. In addition, some thought provoking ideas on what is known as "neurosexism" are explicited and a tribute is made to three women that overcame the challenges posed to them in different times in history (including current times) and helped paved the way to the new generation. However, there is still a long way to go. The inclusion of women and of other relegated sectors of society in scientific and technological activities is an important pending issue which will be achieved when our society as a whole reaches the necessary cultural maturity.

Keywords. women in science, inclusion, cultural change

1. Introduction

To illustrate the situation, some personal experiences are mentioned. Several years ago, while I was a postdoc at the Max-Planck Institute for complex systems in Dresden, Germany, I was about to submit a paper on my recent work on a new numerical method to calculate quantum electronic properties of novel materials, which I had written on my own. So I passed it on to a (male) colleague and he said he thought it was ok but strongly suggested me to sign with my initials and not with my full name, "so that you have a greater chance that the paper gets published because it's strange that a scientific paper is signed by a woman alone", he said. I thought this was somewhat strange and, as I have a strong character and like challenges, I decided to write my full name. After some struggle with the referees the paper was accepted [Hallberg (1995)].

Some time ago I had been invited to lecture at a conference on my work on quantum interference mechanisms in nanoscopic systems. After the talk, my host came up to me saying he had liked the talk very much and that he was positively surprised because, he said, "the whole talk was very rational, clear and logical, you seem to think like a man" (no comments).

In the beginning of this year (my last trip in 2020 due to the pandemic) I was invited to a workshop on Entanglement in Strongly Correlated Systems in Benasque, Spain. I wasn't surprised to count only 5 women among 55 participants (see Fig 1), a situation which has ben like this since I began my professional career in this topic.

This could be also the story of many young girls who follow STEM careers. I invite you to look around, to think about the meetings you've participated in and to reflect on the proportion of women in those forums, not only on the numbers, but also on their role in leadership and in decision making. Women are not only a minority, but they are far

Figure 1. Typical low participation of women in physics conferences (yellow circles), for example at the Workshop on Entanglement in Strongly Correlated Systems, Benasque, Spain Feb. 2020 (photo from the Centro de Ciencias de Benasque Pedro Pascual web site http://benasque.org/2020scs/cgi-bin/pictures.pl)

from being protagonists in decision and policy making. Few exceptions confirm the rule and men continue to dominate the field.

Worldwide in science, according to UNESCO [UNESCO (2019)], less than thirty percent of scientists are women. And this includes all sciences, natural, biomedical, exact and social sciences. If we consider natural, technological, engineering, mathematical, and computational sciences alone, the numbers are even lower since the proportion of women in biomedical and social sciences greatly outweigh the other fields.

Look, for example, at the Nobel laureates in science. To date it has been awarded to: 12 women in Physiology and Medicine (out of 210 men), 4 women in Physics (out of 212 men), 2 women in Economics (out of 82 men) 7 women in Chemistry (out of 179 men). In total it has been given to 25 women in sciences out of 683 men! See Fig. 2

And in mathematics, the Fields Medal was awarded to only one woman (Maryam Mirzakhani) among a total of 60 and the Abel Prize to one woman (Karen Uhlenbeck in 2019) among 20 recipients. This is really striking and shows that, as a society, we are evidently facing a problem which we must face as a whole.

2. Discussion

Do we need to change this situation? In that case, why? some would ask. Maybe women have different preferences, maybe we think differently, maybe we have a different brain? According to several reliable publications on this topic, there is mounting evidence that the brain is not gendered, "not more than the heart, the kidneys or the liver"[Elliot (2019)]. It is a well installed myth that women's brains are wired for empathy and intuition, whereas male brains are supposed to be optimized for reason and action. We don't even know what consciousness is, how we reason, how we store information, and we are looking for differences between men and women's brains? This has been dubbed neurosexism [Rippon (2019), Elliot (2019)] by Gina Rippon in her book *Gendered Brain: the new neuroscience that shatters the myth of the female brain*, where she states that only a gendered society will produce a gendered mind.

Others argue that we have different perspectives and that it is important to have more women scientists to increase diversity. Of course diversity and the inclusion of different perspectives is good for science, however there is more diversity between men stemming from different cultures than between men and women en general. So this argument is a valid one when discussing the cultural reach and inclusion of modern science, but not for

Figure 2. Nobel Prizes given to women in the scientific fields as compared to those given to men (updated to Dec. 2020).

gender issues. There is no reason whatsoever there shouldn't be an equal participation of women and men in science and technology, in all fields. The fact that there is such an enormous gap is a strong warning that we still have serious cultural, social, economic and political biases. And the burden should not be put on women's shoulders but on society in general, on our culture, on our political structures.

Of course, there have been women that have challenged the barriers and have made history in science. I'd like to pay tribute to some of those women who have paved the way for other women to follow in astrophysics and astronomy.

3. Some conspicuous examples

<u>*Caroline Lucretia Heschel*</u> (Hanover, 1750-1848) was a pioneer of her time: an educated woman who would catalogue stars and nebula and discover comets including the periodic comet35P/HerschelRigollet, which bears her name. She was the first woman astronomer to earn a salary, 50 pounds sterling a year (equivalent to 6,400 pounds sterling in 2020) as an assistant to her brother, the astronomer William Herschel (who discovered Uranus in 1781 which he mistook for a comet), with whom she worked throughout her career. She was the first woman in England to hold a government position, to publish scientific findings in the Philosophical Transactions of the Royal Society. She was also the first woman to be named an Honorary Member of the Royal Astronomical Society (1835, with Mary Somerville) and to be awarded a Gold Medal of the Royal Astronomical Society (1828). She was also named an honorary member of the Royal Irish Academy(1838). On the occasion of her 96th birthday in 1846, the King of Prussia presented her with a Gold Medal for Science.

In 1802, the Royal Society published Caroline's catalogue in its Philosophical Transactions of the Royal Society A under William's name. With her brother, she discovered over 2,400 astronomical objects over twenty years. The asteroid 281 Lucretia (discovered in 1888) was named after Caroline's second given name, and the crater C. Herschel on the Moon is named after her. The telescopes they polished and made were the best in those times [Herschel (0000)].

Jocelyn Bell Burnell (Dame, N. Ireland, 1943) She discovered the first radio pulsars in 1967 as a postgraduate student. She found a signal which was pulsing with great regularity, at a rate of about one pulse every one and a third seconds. Temporarily dubbed "Little Green Man 1" (LGM-1) the source (now known as PSR B1919+21) was identified after several years as a rapidly rotating neutron star [Bell (1968)]. The discovery was recognised by the award of the 1974 Nobel Prize in Physics. However, despite being the first person to discover the pulsars she was not one of the recipients of the prize. In 2018, she was awarded the Special Breakthrough Prize in Fundamental Physics for her discovery of radio pulsars. She decided to give the whole of the 2.3million prize money to help female, minority, and refugee students seeking to become physics researchers (the Bell Burnell Graduate Scholarship Fund).

Andrea Ghez (USA 1965) American astronomer and professor at the Department of Physics and Astronomy, University of California, Los Angeles. Her research focuses on thecenter of the Milky Way galaxy. In 2020, she became the fourth woman to be awarded the Nobel Prize in Physics, (shared with Genzel and Penrose) for their discovery of a supermassive compact object, now generally recognized to be a black hole, in the Milky Way's galactic center with a mass of Sgr A* is 4.1 ± 0.6 million solar masses [Ghez *et al.* (2008)].

Using the world's largest telescopes, Genzel and Ghez developed methods to see through the huge clouds of interstellar gas and dust to the centre of the Milky Way. Stretching the limits of technology, they refined new techniques to compensate for distortions caused by the Earths atmosphere, building unique instruments. These images of the Galactic Center are at very high spatial resolution and have made it possible to follow the orbits of stars around the black hole Sagittarius A* (Sgr A*). The partial orbits of many stars orbiting the black hole at the Galactic Center have been observed.

4. Conclusions

There still is an enormous cultural barrier that keeps many girls from science (and also girls and boys, stemming from less advantaged or discriminated sectors of society). We are witnessing an important moment in history concerning women's rights and empowerment, in all fields. However, although the situation has improved somewhat compared to around half a century ago thanks to many proactive actions, there still is a very long way to go. The path must be paved to include all that talent we are dismissing and to allow anyone to choose a scientific career. We definitely need a big cultural change. It is, mainly, a question of justice, non discrimination and equal opportunities.

References

Hallberg, Karen A. 1995 *Phys. Rev. B* 52, R9827(R)

UNESCO 2019 http://uis.unesco.org/sites/default/files/documents/fs55-women-in-science-2019-en.pdf

Elliot, Lise 2019, *Nature*, 566, 453, doi: https://doi.org/10.1038/d41586-019-00677-x

Rippon, Gina 2019. *Gendered Brain: the new neuroscience that shatters the myth of the female brain*, London: The Bodley Head Ltd. ISBN 9781847924759.

Herschel, Mrs. John (ed.). Memoir and Correspondence of Caroline Herschel. London: John Murray, Albemarle Street. http://digital.library.upenn.edu/women/herschel/memoir/memoir.html

Hewish, A., Bell, S., Pilkington, J. et al. "Observation of a Rapidly Pulsating Radio Source" 1968 *Nature* 217, 709–713 https://doi.org/10.1038/217709a0

Ghez, A. M.; Salim, S.; Weinberg, N. N.; Lu, J. R.; Do, T.; Dunn, J. K.; Matthews, K.; Morris, M.; Yelda, S.; Becklin, E. E.; Kremenek, T.; Milosavljevic, M.; Naiman, J. 2008. "Measuring Distance and Properties of the Milky Way's Central Supermassive Black Hole with Stellar Orbits". *The Astrophysical Journal* 689 (2): 1044–1062. arXiv:0808.2870. Bibcode:2008ApJ...689.1044G. doi:10.1086/592738. S2CID 18335611.

Education and Heritage in the era of Big Data in Astronomy
Proceedings IAU Symposium No. 367, 2020
R. M. Ros, B. Garcia, S. R. Gullberg, J. Moldon & P. Rojo, eds.
doi:10.1017/S1743921321000909

ASTROMOVES: Astrophysics, Diversity, Mobility

Jarita C. Holbrook🆔

Science, Technology & Innovation Studies, University of Edinburgh, United Kingdom
emails: Jc.holbrook@ed.ac.uk

Abstract. The US astronomy/astrophysics community comes together to create a decadal report that summarizes grant funding priorities, observatory & instrumental priorities as well as community accomplishments and community goals such as increasing the number of women and the number of people from underrepresented groups. In the 2010 US National Academies Decadal Survey of Astronomy (National Research Council 2010), it was suggested that having to move so frequently which is a career necessity may be unattractive to people wanting to start a family, especially impacting women. Whether in Europe or elsewhere, as postdocs, astrophysicists will relocate every two to three years, until they secure a permanent position or leave research altogether. Astrophysicists do perceive working abroad as important and positive for their careers (Parenti, S. 2002); however, it was found that the men at equal rank had not had to spend as much time abroad to further their careers (Fohlmeister, J. 2012). By implication, women need to work abroad longer or have more positions abroad to achieve the same rank as men. Astrophysicists living in the United Kingdom prefer to work in their country of origin, but many did not do so because of worse working conditions or difficultly finding a job for their spouse (Fohlmeister, J. 2014). In sum, mobility and moving is necessary for a career in astrophysics, and even more necessary for women, but astrophysicists prefer not to move as frequently as needed to maintain a research career. To gather more data on these issues and to broaden the discourse beyond male/female to include the gender diverse as well as to include other forms of diversity, the ASTROMOVES project was created which is funded through a Marie Curie Individual Fellowship. Though slowed down by COVID-19, several interviews have been conducted and some preliminary results will be presented.

Keywords. Cultural Astronomy, History And Philosophy Of Astronomy, Sociology Of Astronomy, Social Studies Of Science, Qualitative Research.

---

1. Introduction

The ASTROMOVES project is to study the career decision making of astrophysicists and those in connected fields, the changes of positions in terms of relocations and changes in job title while remaining in the same place, and if and how these are related to intersectional identities. 'Intersectional identities' attempts to capture the many axes of difference embodied by individual astrophysicists and is in reference to 'intersectionality' defined as:

> *The interconnected nature of social categorizations such as race, class, and gender, regarded as creating overlapping and interdependent systems of discrimination or disadvantage (Oxford English Dictionary n.d.)*

When considering intersectional identities there is space to consider nationality, mother tongue and disabilities/abilities along with race, class and gender as mentioned in the definition. Astrophysicists have long been tackling how to become more diverse in terms of including women and members of underrepresented groups (Cowley, A. 1974). Most of the social science studies of astrophysicists have focused on women; to summarize the findings of a few of these studies women and men both need to relocate multiple times for postdocs but the outcome, securing a permanent position, is different for women and men. Every ten years, there is an USA based exercise where the larger astrophysics community comes together to assess the current state of the discipline and to lay out the priorities for the coming decade; the resulting report is referred to as the Decadal Survey. In the 2010 Decadal Survey (National Research Council 2010), it was suggested that having to move so frequently which is a career necessity may be unattractive to people wanting to start a family, especially impacting the careers of women. This suggestion is supported in part by the American Physical Society's (Ivie, R. 2015) study which showed that women in physics & astronomy associate having children with negatively impacting their career and slowing their rate of promotion. Also supported in part by Fohlmeister, J. 2012 study which found that "Those who became mothers feel very restricted in mobility and point out that it is harder to combine job and family." In that same study, the second most cited reason for moving was to be near family. These findings point to conflicts between the need to relocate often for career advancement and the needs of women to have stability and external support to maintain a family and career. Men astrophysicists face the same family issues but they rate the importance of these issues to their careers as much lower than the women astrophysicists. Another issue is where astrophysicists relocate to; in the European context; solar astrophysicists perceived working abroad, that is working outside of their home country, as important and positive for their careers (Parenti, S. 2002).

However, Fohlmeister, J. (2014) found that the men at equal rank had not had to spend as much time abroad to further their careers. By implication, women need to work abroad longer or have more positions abroad to achieve the same rank as men who have worked abroad. Astrophysicists living in the United Kingdom prefer to work in their country of origin, but many did not do so because of worse working conditions or difficultly finding a job for their spouse (Fohlmeister, J. 2014). Summarizing these research results for studies of women in astrophysics and their careers, moving is necessary for a career in astrophysics, and working abroad is even more necessary for women, but there is some evidence that astrophysicists prefer not to move as frequently as needed to maintain a research career. There are a few studies of intersectionality and astrophysicists such as Ko, L.T. 2013, which found that women of color in some cases prioritized life work balance in general, including family life, over remaining in their chosen field and revealed the positive role that activism plays, as well as importance, in women of color's lives even when doing activism is not academically rewarded and not considered as part of advancement. This and other studies considering intersectional identities in astrophysics are not focused specifically on career moves.

ASTROMOVES' focus on mobility and career decision making and its use of qualitative interviews with astrophysicists provides a deep dive towards surfacing previously unidentified factors as well as greater comprehension of known factors important for navigating astrophysics careers. One of the goals of ASTROMOVES is to leverage the research findings to make recommendations to the astrophysics community for retaining and promoting their diverse members. Though in its beginning stage, the project has preliminary results that reveal decision making factors previously not recorded, identified a new gender related outcome and has expanded previous studies by including greater gender diversity.

2. Population

ASTROMOVES is a study of astrophysicists and scientists in adjacent fields such as space scientists, planetary scientists, etc. Each person had to be at least two career moves past their doctorate. Given that the project was funded with European monies, each person had to have lived, worked or studied in Europe at some point in their career. Targeted populations include people that are ethnically and nationally underrepresented in astrophysics, gender diverse people, people with physical disabilities, and people with invisible disabilities or different abilities such as dyslexia or an autoimmune disorder. Ethically, everyone who chooses to participate in ASTROMOVES volunteers and they have a say in how their data is used as well as how it appears in print.

The ASTROMOVES initial population of seven scientists was divided into four males and three females, but their self-reported gender identities include two bisexuals, one homosexual and one asexual, thus over half are embodying gender diversity. Gender diverse astrophysicists were identified using the Astronomy & Astrophysics Outlist (see https://astrooutlist.github.io (Mao, Y.-Y. 2013)]. Among the seven scientists are people of African descent, Indian descent, Middle Eastern descent, and European descent. Snowball sampling was used, where people were asked to recommend other people for ASTROMOVES interviews, and so on.

3. Data Collection and Analysis

The primary method of data collection is interviews of between 30 minutes to 1.5 hours depending upon how many career moves the person being interviewed has made. These qualitative interviews cover the details of each career move and their decision making, as well as including places they did not apply to and offers they did not accept. Each interview was video recorded or voice recorded and transcribed. As a social political statement reminding astrophysicists of their debt to indigenous communities for allowing telescopes to exist on their sacred mountaintops, each person is given a Hawaiian pseudonym which is used across ASTROMOVES publications.

Publicly available documents are used to contextualize the interviews. To get the sequence of career moves, publicly available CVs are primarily used. When a CV is not available, the career sequence can be approximated by noting the affiliations given for their scientific articles found on the Astrophysical Data System archive (https://ui.adsabs.harvard.edu/), which is publicly available, also. The Astrophysics Rumour Mill (http://www.astrobetter.com/wiki/Rumor+Mill+Faculty-Staff+2019-2020) is another public source of information about job offers that is fairly complete back to 2000. These documents are used as prompts for the interviews.

Sensitive information related to intersectional identities was recorded or not depending upon the desires of the interviewee. Such information was written down and sent after the interview; interviewees had the option to write down any additional information that they wanted to keep private but that they thought was important for the analysis. In addition to personal information about their intersectional identities, the astrophysicists were hesitant to name departments with negative reputations or bad reputations, such as those that treated women postgraduate students badly by not supporting them or graduating them; this information was written as private information.

Qualitative analysis software was used to analyse the contents of the interviews. To anonymize information on specific institutions, their Shanghai designations for physics, which includes astrophysics, were used (Shanghai Ranking's Global Ranking of Academic Subjects 2020 – Physics — Shanghai Ranking - 2020, n.d.). Repeated themes, topics of discussion, and experiences were compared across the set of interviews. Given the small sample size of seven interviews all results presented in the following section are preliminary.

4. Results and Discussion

"You know, it was a different environment than I was used to. It was perhaps a little more (regional stereotype on being more formal). [USA Top 50 location] is very laid back. There it's very, it's very congenial. There's a lot of interaction between faculty and students and postdocs, and it's feels like more of a community. [USA Top 10 location] is a little more hierarchical. And it was, you know, there was the top people and then there was, you know, others. And, yeah, I didn't, I felt like I didn't, there were some people... You know, individually, I liked many of the people there, but the overall atmosphere I felt was more sort of competitive and not so friendly. And, so I, I wasn't that sad to leave. You know. And I had visited the [2nd USA Top 50 location] and I knew that... some of the people there. So, I knew I would have... like a nice community there".

The above excpet from one of the interviews is a true transcription of what was said without correcting for filler words nor repetitions. When the interviewee was asked about attaching their name to this quote, the answer was only if it were grammatically corrected, thus it is presented anonymously. As for the content, many of the astrophysicists spoke about the academic environments that they preferred or didn't. Only one person stated that they enjoyed and thrived in competitive environments, most indicated that they preferred less competitive environments similar to the quote above.

"The end of my most recent research fellowship, then three months of unemployment and it was pretty harrowing: I'VE NEVER BEEN UNEMPLOYED BEFORE!"

All three of the females interviewed have been unemployed at some point since earning their doctorate. Two of the three were recently unemployed connected to COVID-19. During their period of unemployment, two had to live on their own personal savings since they were not eligible for public unemployment funds; the third was eligible for and used some public unemployment assistance funds. In contrast, none of the four males interviewed had been unemployed since their doctorate. The fact that 100% of the females had experienced unemployment indicates greater job insecurity; in fact, one person felt strongly that if they had been male, then the situation which lead to unemployment would not have happened.

If female astrophysicists do have higher job insecurity, then their supervisors should consider securing additional funds to continue to employ them until they obtain a new position. Two of the male astrophysicists mentioned having positions made with them in mind, but neither was on the verge of unemployment at the time, rather the positions were made to attract them to a new institution.

Kamea: "I've actually got to [the] interview stage for a couple of jobs and, and they were just unceremoniously called off...months later they called me back and said, 'We'd like to resume the interview process with you.'"

Maka'ala: "I'm not happy about how this has affected my, you know, my life and my ability to be with people that I care about in terms of my productivity, it has had no effect or even a positive effect, because all the things like travel and hanging out with friends or, you know, just whatever that I would do, I don't have any more. So, I have fewer distractions."

The females' experiences of unemployment cannot be disentangled from the COVID-19 pandemic. To disentangle this, future interviews would have to be done with more astrophysicists that were on the job market during 2020 as well as interviews with other astrophysicists that experienced unemployment prior to 2020. Checking the Astrophysics job page in the fourth quarter of 2020 (Anonymous, 2020), of the 11 jobs that mention COVID-19, 2 temporarily suspended their searches, 7 cancelled the job and 1 offer was rescinded. The quote by Kamea shows how Kamea was part of the group of astrophysicists that were effected by the astronomical job market's fluctuations due to COVID-19.

Maka'ala is in a permanent position, so COVID-19 did not challenge their job security; rather, Maka'ala speaks of the isolation brought about by the pandemic. Most of the astrophysicists that lived alone were unhappy about their isolation to the point that one, Hema, self-described as an extrovert, regularly did meet people socially even thought it was against the local government regulations.

Hema: "It's not so much about the number of persons, a day that I need to speak [to]. It's more about the time of day... I spend the entire day running around, because I have a job. I've work to do. I have training in the evening after work, which requires a lot of my energy - physical energy. But if I come home at 7 pm. And I don't have a plan of seeing someone that's what makes me feel really lonely... I can talk to people at work all day. But if I'm home alone in the evening with nothing to do. That's the hard part for me. That's the vulnerable moment for me, it's sort of being home alone in the evening with no one to talk to..."

Hema and another astrophysicist, Haoa, were both given permission to work in their place of employment during COVID-19. Hema requested permission; whereas Haoa didn't have a choice because of Haoa's leadership role in an observatory related construction project.

The combination of teaching online and doing research was having a negative impact on Maka'ala:

"there's just a lot to keep up with and I can't, you know, if it's... if it's... if I'm working on a paper I can always say, 'No, I'm just going to stop at this point and take a break,' but I can't... I feel like I've lost my agency in terms of setting my schedule. Even if I, even if I might still be working long hours in the summer when I was doing research versus now when I'm teaching so... that's been a little bit mentally tiring."

It should be noted that a few articles have been published on how COVID-19 is exacerbating the differences between the careers of women and men in astrophysics, such as Inno, L. (2020) and Venkatesan et al. (2020). Both articles point out the drop in publication submissions from women during the pandemic will have the long term impact of women being less competitive, because currently astrophysics competitiveness is attached to publication rates. They advocate that the pandemic impact be considered in future hiring decisions.

Contemplating unemployment and the females interviewed, financial considerations have to be considered among the factors important for career decision making in astrophysics. Postdocs and other fellowships come with a fixed salary, which means everything is equal across the sexes. However, in most cases for permanent positions, salaries are negotiable as are what are called 'start-up packages', the onus is on the new faculty member to ask for what they need to be successful. In their study of women scientists that had taken a career break, which included astronomers, Mavriplis, et al. (2010) found that women were the least confident about negotiating for their salary and start-up package. This initial negotiation, which women are not confident about, may be the start of building resentment and discontent if and when they discover that others have negotiated for and gotten more both before and after they were hired. Finances were discussed during the interviews and most said that they did not make career decisions based on salary or cost of living. However, some did note when they had taken a pay cut. One of the women did provide information about negotiating for salary and start-up, which is still being analysed. None of the females had purchased homes; whereas three of the four males had purchased homes in the place where they thought they would remain, even though they later may have moved. If this same information is considered in terms of marital status, two of the males that purchased homes were married; however, these two were the only married people among the astrophysicists interviewed.

5. Conclusions

ASTROMOVES, though still in its beginning phase, has produced interesting results that have to be tempered by the small number of interviews completed. Nonetheless, the most troubling findings are related to sex where the female astrophysicists had all experienced unemployment post PhD. Generally, the astrophysicists did not consider salary, cost of living or start-up package in their career decision making. COVID-19 has changed the lives of astrophysicists; many of those interviewed that were living alone spoke of their unhappiness connected to loneliness, and two were impacted due to the fluctuating job market due to COVID-19.

As ASTROMOVES continues, the interview pool will be increased to 50 scientists; an effort is being made to include those with physical disabilities and invisible disabilities. Already, there is enough gender diversity to have three gender categories for future analysis: cis-men, cis- women and other gender; which will be the first time such categories are possible in discussions of the mobility and career decision-making of astrophysicists.

6. Acknowledgements

This research has made use of NASA's Astrophysics Data System Bibliographic Services. This project is funded by the European Union's Horizon 2020 research and innovation programme under the Marie Skłodowska-Curie individual fellowship program H2020-MSCA-IF-2019 grant # 892944.

References

Cowley, A., Humphreys, R., Lynds, B., & Rubin, V. 1974, Report to the council of the AAS from the working group on the status of women in astronomy-1973. *Bulletin of the American Astronomical Society*, 6, 412–423.

Fohlmeister, J., & Helling, C. 2012, Career situation of female astronomers in Germany. *Astronomische Nachrichten*, 333(3), 280–286.

Fohlmeister, J., & Helling, C. 2012, Careers in astronomy in Germany and the UK. *Astronomy & Geophysics*, 55(2), 2.31-2.37.

Inno, L., Rotundi, A., & Piccialli, A. 2020, COVID-19 lockdown effects on gender inequality. *Nature Astronomy*,, 4(12), 1114–1114.

Ivie, R., & White, S. 2015, Is There a Land of Equality for Physicists? Results from the Global Survey of Physicists. *La Physique Au Canada*, 71(2), 69–73.

Ko, L. T., Kachchaf, R. R., Ong, M., & Hodari, A. K. 2013, Narratives of the double bind: Intersectionality in life stories of women of color in physics, astrophysics and astronomy. *AIP Conference Proceedings*, 1513(1), 222–225.

Mao, Y.-Y., & Blaes, O. (n.d.). 2013, *Astronomy and Astrophysics Outlist*, Retrieved 19 January 2021

Mavriplis, C., Heller, R., Beil, C., Dam, K., Yassinskaya, N., Shaw, M., & Sorensen, C. 2010, Mind the Gap: Women in STEM Career Breaks. . *Journal of Technology Management & Innovation*, 5(1), 140–151.

292   J. C. Holbrook

National Research Council 2010, *New Worlds, New Horizons in Astronomy and Astrophysics.*, National Academies Press.

Oxford English Dictionary n.d., *'intersectionality, n.'*.

Parenti, S. 2002, The European solar physics community: Outcome from a questionnaire. *Solar Variability: From Core to Outer Frontiers,* 506, 985–990.

Shanghai Ranking's Global Ranking of Academic Subjects 2020–Physics — Shanghai Ranking–2020 n.d., Retrieved 19 January 2021.

Venkatesan, A., Bertschinger, E., Norman, D., Tuttle, S., & Krafton, K. 2020, The Fallout from COVID-19 on Astronomy's Most Vulnerable Groups. *Women In Astronomy.*

Education and Heritage in the era of Big Data in Astronomy
Proceedings IAU Symposium No. 367, 2020
R. M. Ros, B. Garcia, S. R. Gullberg, J. Moldon & P. Rojo, eds.
doi:10.1017/S1743921321000636

The inmate sky Astronomy in a juvenile detention institute

Silvia Casu⍟, Gian Luigi Deiana and Emilio Molinari

INAF-Osservatorio Astronomico di Cagliari
Italy
email: `silvia.casu@inaf.it`

Abstract. We here present the general outline of the project "Open up, Sky!", a pilot project of astronomy education in a Juvenile detention institution in Sardinia, Italy. The project is still in progress, and we report here the first preliminary results.

Keywords. inclusion, astronomy literacy, prison, astronomy education

1. Astronomy inside detention institutes

Talking about astronomy inside a jail could be perceived as a bit paradoxical. For most people the word astronomy evokes first of all images of open dark sky, perfect infinite space and beauty and probably the idea of an ultimate freedom. It could be not easy to talk about astronomy to people who live a distressing situation, locked in a confined space, deprived – for a short or long period – of their own freedom. So does it make sense to do an astronomy educational project for inmates?

Let us take a step back. Is it important to do STEM education in jail? The word "education" comes from the latin word e-duco, which means *lead out*. What are we supposed to lead out with the education? The answer is quite easy: compentences, soft skills, principles, values, rules of behaviour, valid for daily life, not only within the detention institution, but in the external world where the inmates will come back, once the sentence is over. In a broader sense, educating in prison means educating to freedom: to live in a profitable way for oneself and for others, to take responsibility for their own choices, and so, definitely, to grow. Educating through astronomy, in particular, means using the sky as something that brings back to elementary, primary experience for all. It often represent a place with which you have an affective relationship (let us think, for example, to our emotivional relationship with the moon), and leads us to reflect on transversal themes. And this is perfectly in agreement with the meaning of the IAU tag *#astronomyforabetterworld*.

Unfortunately, at least in Italy, STEM education inside jails and juvenile detention institutes seems not very common. From a recent survey Report Antigone (2017), it comes out that only a very small percentage of educational activities are dedicated to science: among these, most activities regard to basic computer science and robotics, little physics, poor astronomy. Among these very few experiences, the Italian National Institute for Astrophysics (INAF) for about ten years proposed some pilot outreach projects, such as the astronomy session of the "Free to read" project in Bologna Pratello juvenile detention institute (2011), or the astronomy discussion with Samantha Cristoforetti inside the Beccaria jail in Milan in the framenwork of the "Close encounters" project (2015), where the parallelism between the life of reclusion in jail and in space station has been profitably used. More recently, we inaugurated the Italian tour of the Inspiring Stars exhibition (https://sites.google.com/oao.iau.org/inspiringstars) at the Rebibbia Jail in Roma and signed a formal agreement with the Tuscany Regional Penitentiary Administration for a series of astronomy seminars. All these projects had a great success and aroused a lot of interest and participation by detaines. Using these encouraging premises, we hence decided to propose a more structured educational project to the Sardinia Center of Juvenile Justice.

2. Project Overview and Implementation

After a lot of preliminary meetings, in February 2020 we signed a formal agreement between our Insitute and the Sardinia Center for Juvenile Justice for the implementation of the project *Open Up, Sky!*, an astronomy education project, whose general lines and specific methodology have been designed in agreement between INAF staff and Quartucciu Juvenile Detention Institute educators. In particular, we adopted EBL (Enquiry-based Learning) and PBL (Project-based learning) approaches, following constructivist and constructionism principles, favoring a learner centered and collaborative approach. We also selected a list of possible contents, ranging from light properties to universe phenomenology to space science to astrobiology. Attendance of meetings has been fixed once a week, in the afternoon, outside the common school activities, on a voluntary basis. We also decided not to fix the activities program order, preferring to follow the particular interests of the young detained.

The project was supposed to start in March 2020, but has been stopped due to the pandemia. We discussed about the possibility to start anyway with on-line meetings, but, since one of the key points of the project is the establishment of a direct relationship between inmates and tutors, we agreed that the physical presence of the astronomers was essential. So we started on August 2020. At the present moment, 18 meetings have been held in presence, involving 12 young and 3 internal educators in total.

During the meetings we use different educational tools:

• **hands-on activities**. We follow the educational constructivist and constructionism principles of "learning by doing": instead of simply listening to an astronomy lecture, we foster the engagement with the subject matter to solve a problem or create something. We hence use hands-on activities to introduce astronomy objects and phenomena.

• **story-telling**. We introduce almost all meetings with several narrative elements: tales of scientist biographies (Galileo, Newton, Edison, Maxwell, Curie, Bell Burnell,...), stories of particular discovery achievement, history of space conquest, etc, to tickle the imagination and promote the creation of a emotional link with inmates lives and experiences;

• **tinkering** The tinkering approach is characterized by a playful, experimental, iterative style of engagement, in which makers are continually reassessing their goals, exploring new paths, and imagining new possibility (Resnick & Rosenbaum 2013). We propose several tinkering activities (scribbling machines, circuit boards, chain reactions) to introduce research methods and processes;

• **coding bases** In order to promote computational thinking formation or empowerment, we introduce the basic principles of coding with unplugged activities and the offline version of Scratch.

We also succeeded in organising a star gazing event, inside the jail court and with all the obvious limitations due to artificial security lights. During the event, inmates and prison guards took the opportunity to familiarize with a small telescope and to observe Saturn and Jupiter. Another event is scheduled to observe the Moon. Unfortunately, for security and privacy problems, we do not have pictures of the various activities done till now. We just obtained to take some pictures, checked and approved by the institution

Figure 1. Left: a moment of the stargazing event. For all the inmates it was the first time they approach to a telescope, and for most of them the first time they really watch the sky. And they were completely excited by this. **Right**: The scribbling machines activity. They spent almost two hours totally committed to the activity: they had a lot of fun and created wonderful and very creative objects!

responsibles, only in particular moments. In Fig. 1 we show the approved ones, relative to the stargazing event and the scribbling machine tinkering activity.

3. Preliminary results

As previously said, the project is still on going. Anyway, we can summarize here some of the faced difficulties, the initial learned lessons and some outcomes.

General and specific difficulties.
Participants The involved inmates represent a very heterogeneous target: they come from different nations (Senegal, Bosnia, Romania, Egipt, Tunisia, Italy) and have hence different social, school level and religious background. Moreover, they are of different ages (the youngest being 16 years old, the oldest 23) and present different committed crimes and distressing past experiences. Again, some young went out from the juvenile detention institution during the project, because of transfer procedures or changes in sentence application, causing some educational discontinuities.
Physical limitations Inside a jail there is a general lack of real autonomy, and, in addition, we cannot use freely any kind of materials (scissors, cutter, ...) and tools (no connection). Moreover, most of them are not allowed to leave the institute even for a couple of hours, so we could not let them visit our observatory/planetarium. The practical organization of the star gazing event itself took a long time and a number of authorizations.
Attitude Even if the participation to the project is voluntary, some of the young experienced (or pretende to show) an initial mistrust and/or indifference to the project. Of course, they could be limitated by the fear of judgment and by the lack on knowledge of different life styles. So it took time to establish an emotional educational relationship with them. Again, their psychological state is very unstable, and even the most motivated sometimes failed to join the activity. At the same time, it is not easy to be involved in their lives and stories without risking being emotionally overwhelmed.
To face all these problems, we benefit of the presence of almost one educator of the Institute during all the project meetings and of the professional advice of a psychotherapist during all the project phases.

Figure 2. The Open Up, Sky! project logo, designed by Laura Barbalini (INAF-OA Brera).

Lessons learned. As a general vision, we promote the use of Astrophysics and Space Sciences to encourage and support the self-determination and self-expression of the individual, regardless of gender, social status and culture of origin. And, of course, of committed crime. So a project inside a juvenile detention institute was perfectly matching this thought, and, even if it is very complicated enter inside prisons, it is very important – we would say, fundamental – to talk about astronomy to detained people. Even a hour-long meeting could be useful for prison inmates but, when possible, try to develop a longer project based on active participation.

Again, it is important to be organized but flexible. You should have in mind very structured and well designed plans, but it is more important to be able to follow the young interests and to change plans at the very last moment.

Moreover, it is extremely important to be very patient: even if they seem to not listen, do not panic! Often they are only pretending, because they do not know how to "play" with you. You should remember that most of them did not have a proper childhood and adolescence, but they always have inside a child spirit to talk with. Indeed, we obtained the most important engagement with "playing" activities such as tinkering, unplugged coding, rocket constructions.

Unfortunately, we could not yet perform a proper evaluation of the project and its outcomes. But some initial indications seem very positive and encouraging. First of all, the level of attention of the involved young is increasing meeting by meeting. Their participation is more active, and now we can have real discussions about some scientifical issues. Moreover, one of the young started a school program with very good grades.

In conclusion, we started this project with the idea to open some new horizons to young people confined in a closed world. By providing new points of view and new perspectives for involvement and participation in culture and knowledge formation processes, our initial aim was to to help people to find and maintain an inner way to have complete control over their own decisions and their own lives. We still do not know if we succeded in this, but it is worth it because, as Paul Freire said, *"education is freedom"*. And the truth is that we learned more that we tried to teach.

4. Ackowledgments

We warmly thank all the Quartucciu Juvenile Detention Institute staff for their constant collaboration and patient availability. We also thank Laura Barbalini (INAF-OA Brera) for the project logo design (Fig. 2)

References

Associazione Antigone 2017, *XIV rapporto sulle condizioni di detenzione*, http://www.antigone. it/quattordicesimo-rapporto-sulle-condizioni-di-detenzione/istruzione-e-formazione/

M. Resnick & E. Rosenbaum, 2013, *Designing for Tinkerability, in: Design, Make, Play: Growing the Next Generation of STEM Innovators*, Taylor & Francis

Education and Heritage in the era of Big Data in Astronomy
Proceedings IAU Symposium No. 367, 2020
R. M. Ros, B. Garcia, S. R. Gullberg, J. Moldon & P. Rojo, eds.
doi:10.1017/S1743921321000429

Accessible Astronomy Activities for the Blind and Visually Impaired in Puerto Rico

Gloria M. Isidro Villamizar[1] ⓘ**, Carmen A. Pantoja Pantoja**[2] **and Mayra E. Lebrón Santos**[3]

[1]Caribbean University, Bayamón Campus
Department of Liberal Arts
167 Ave. Km. 21.2 Bayamón, PR, USA, 00960-0493
email: gisidro@caribbean.edu

[2]University of Puerto Rico-Río Piedras
Natural Sciences Faculty, Department of Physics
17 Ave. Universidad Ste. 1701 San Juan, PR, USA, 00925-2537
email: carmen.pantoja1@upr.edu

[3]University of Puerto Rico-Río Piedras
General Sciences Faculty, Department of Physical Sciences
14 Ave. Universidad Ste. 1401 San Juan, PR, USA, 00925-253
email: mayra.lebron3@upr.edu

Abstract. We present the design and development of accessible activities in astronomy for blind persons in Puerto Rico. We design for a diverse audience that sees from different perspectives, but with the same purpose: to know and discover the Universe. We adapt tactile materials to develop themes that require visual images. We design and develop three-dimensional tactile material to offer blind people the opportunity to get the conceptual idea of the specific topic under consideration. Listening and designing bearing in mind the voice of blind people with their different life experiences is essential. Through years of experience (2006 – 2020) we have learned to use new strategies in the design and development of tactile materials. We recognize that what we have achieved to date has been possible through the exchange of efforts, collaboration, and volunteering. In recent months, we have been publishing videos with each of the tactile materials, with the purpose of contributing to the literacy of astronomy worldwide.

Keywords. Accessibility, astronomy, blind people, person with disabilities, tactile materials, popularization of astronomy.

---

1. Introduction

Discovering the night sky has been of interest to all mankind, since ancient times. Even without vision, it is still of great interest to imagine and discover celestial spectacles and to have knowledge of our Universe. Raising awareness of the general public and integrating blind people and people with low vision into astronomy outreach activities to offer "everyone" the opportunity to know and discover the Universe is our purpose. Since our beginnings, we have listened to the voice of blind people (collaborators) for whom our activities are directed.

2. Tactile materials

We adapt tactile materials to cover topics that require visual images. In some cases, it has been necessary to design and develop three-dimensional tactile material to offer

Figure 1. Workshop in Israel.

blind people the opportunity to get the conceptual idea of the specific topic under consideration. For other cases we have used available commercial materials to aid in the description of concepts like relief drawing paper, Braille books, Braille paper and stylus, among others. We have used available educational materials like the ALMA 12-Meter Antenna Paper Models from the National Astronomical Observatory of Japan (NAOJ 2021). We constructed 66 paper models of the ALMA antennas and organized inclusive events for their use. We designed a model of the Arecibo radio telescope that can be assembled, with the purpose of making blind people, together with their family or friends, to participate in the visit to the Arecibo Observatory radio telescope. The model integrates mathematical concepts in its structure, and is accompanied by three "Visitors Guide" booklets, printed in large print, in Braille, in Spanish and English (Bartus *et al.* 2007). We are saddened by the news of the recent collapse of the radio telescope.

3. Workshop experience

The area of accessibility in science and mathematics benefits greatly of the dialogues and exchanges in experience between the interested parties. We have participated and organized workshops for teachers, students, astronomers, and popularizers of science. We organized a workshop (Alonso *et al.* 2008), with the topic of the phases of the Moon. This workshop included blind participants. It is important to encourage events in which the blind and the sighted can participate to promote awareness in the public towards disabilities. We developed an inclusive workshop to prepare a group of interdisciplinary college level students to work as volunteers during the planned events for the International Year of Astronomy (IYA2009). Our group of students included two totally blind students and they all learned about tactile materials and used them during our outreach events in 2009 (Isidro 2009). Isidro was invited to participate of the IYA2009 Israel, and offered a workkshop for future teachers in Special Education in Afula (Fig. 1).

4. Collaboration with other experts

One of the challenges in the development of accessible materials is the large spectrum in which a disability can be manifested. We have collaborated with different groups of experts in the design and evaluation of models. We have collaborated in the tactile Moon project of the Astronomical Observatory of the University of Valencia (Ortiz Gil, A.,

Figure 2. Mars with Art

et al. 2012). We evaluated the book "The Sky at your Fingertips" together with blind undergraduate students at UPR Río Piedras. We had on loan the tactile semi-spheres and presented the Planetary Show "El Cielo en Tus Manos" at the University of Puerto Rico.

5. Astronomy through Art

We have employed a technique to create astronomy tactile models that are sculpture to focus and simplify the astronomy concepts we wish to present. We designed the model of the tactile Sun, a sculpture piece containing eruptions, flares, and a sunspot in relief. This strategy was made possible by participating (Gloria Isidro) as a volunteer in the "Manos Que Miran" sculpture workshop under the direction of the blind sculptor Luis Felipe Passalacqua at the Puerto Rico Art Museum. (Isidro and Pantoja *et al.* 2014). Using this technique we developed a tactile model of a Spiral Galaxy and presented it during the activity "Stars for All", as part of the activities of the International Year of Light 2015. We created "Mars with Art" a sculpture that contains artistically expressed selected features of the Martian surface (Fig. 2).

6. Activities for associations of blind people

We develop Astronomy activities with older adults who are new to reading and writing Braille. The different textures on tactile materials and Braille books are an excellent motivational strategy with Braille writing. We have collaborated developing accessible activities with the Puerto Rican Association of the Blind, Inc., in Río Piedras, and with the "Luz de Amor" Association of the Blind, Inc., in Bayamón, Puerto Rico (Fig. 3).

7. Accessible Astronomy Activities & COVID19

During this time of global pandemic we have shifted from displaying tactile materials to describing already developed materials. To "touch" is to "see", and to "hear" is to "imagine". We are making video recordings with each of the tactile materials already developed and sharing our experiences, so that they may be useful in disseminating astronomy for the blind. We continue to develop 3D materials as future projects.

Figure 3. "Asociación Puertorriqueña de Ciegos"

8. Conclusions

We design for a diverse audience that sees from different perspectives, but with the same purpose: to know and discover the Universe. Listening to the voice of blind people with their different life experiences is essential when adapting or designing tactile materials. Through years of experience (2006 – present) we have learned to use new strategies in the development of tactile materials. We recognize that what we have achieved to date has been possible through the exchange of efforts, collaboration and volunteering. Our biggest obstacle is the time to design, develop, evaluate new strategies, publish them, give them continuity and follow-up. We have been successful in creating tactile models that are sculptures for which we have carefully selected features to highlight, simplify and focus astronomy themes. We recommend to join forces to create and develop new accessible activities and materials.

References

Alonso, J., Pantoja, C. A., Isidro, G. M. & Bartus, P. (2008). *Touching the moon and the stars* Astronomy for the Visually Impaired. ASP Conference Series, 389.

Bartus, P., Isidro, G. M., La Rosa, C., & Pantoja, C. (2007). *Arecibo Observatory for all* Astronomy Education Review, Vol, 6 Issue 1, p. 1-14.

IsidroVillamizar,G.(2009,Aug18). *Liderando la Accesibilidad AIA2009PR* [Video]. https://youtu.be/CStzUGJ7T1Q

Isidro, G. M. and Pantoja, C. A. (2014). *Tactile Sun: Bringing an Invisible Universe to the Visually Impaired* CAP Journal 15, pp. 5-7.

National Astronomical Observatory of Japan(2021) . *ALMA 12-Meter Antenna Paper Models* https://www.nao.ac.jp/en/gallery/weekly/2015/20150811-alma.html

Ortiz Gil, A., Ballesteros Rosselló, F., Moya, M. J., Lanzara, M. (2012). *Tocar la Luna* Época No. 160. Universidad de Valencia, España.

Education and Heritage in the era of Big Data in Astronomy
Proceedings IAU Symposium No. 367, 2020
R. M. Ros, B. Garcia, S. R. Gullberg, J. Moldon & P. Rojo, eds.
doi:10.1017/S1743921321000946

Inclusive Eclipse: a sensorial experiencing along Chile

P. Troncoso-Iribarren[1] , C. Santander[2] , J. Díaz[3], H. López[4],
E. Labbé[4], I. Schacht[5] , C. Piña[1], C.A.L. Morales Marín[5,6],
H. Drass[3,7], A. Barr Domínguez[8]

[1]Escuela de Ingeniería, Universidad Central de Chile, Avda. Francisco de Aguirre 0405,
La Serena, Chile. email: paulina.troncoso@ucentral.cl

[2]Departamento de Física, Facultad de Ciencias Físicas y Matemáticas, Universidad de Chile,
Avenida Beauchef 850, Santiago, Chile.

[3]Instituto de Astrofísica, Pontificia Universidad Católica de Chile, Avda. Vicuña Mackenna
4860, 782-0436 Macul, Santiago, Chile.

[4]Núcleo de Astronomía, Facultad de Ingeniería y Ciencias, Universidad Diego Portales,
Av. Ejército 441, Santiago, Chile.

[5]Coperativa de Trabajo Geodésica, Tirso de Molina 05264, Temuco, Chile

[6]Agencia Regional de Desarrollo Productivo de La Araucanía, Arturo Prat 0221,
Temuco, Chile

[7]Centre for Astrophysics, University of Southern Queensland, West St, QLD 4350, Australia.

[8]Centro de Investigación Multidisciplinario de la Araucanía, Facultad de Ingeniería,
Universidad Autónoma de Chile, Avenida Alemania 01090, Temuco, Chile.

Abstract. An Eclipse is an astronomical event that convenes a large audience. Few days before it, most of the community is aware of the event and the press is activated fully on it. The alignment recovers our most intrinsic human aspects, the curiosity, and enthusiasm for a natural phenomenon. This work is focused to enjoy and perceive it in three different ways: visually, listening, and in an artistic expression.

We focused on the construction of more than one hundred LightSound devices, which the main purpose is to record the light intensity and transform it into different tones. Besides, we created an artistic representation of the Eclipse motivated by the ancestral culture of the people residing in the totality zone. This music adds a sensorial joy to the eclipse event.

Keywords. eclipses, instrumentation: miscellaneous, instrumentation: photometers

---

1. Introduction

Bieryla et al. (2018) designed the electronic device, hereafter LightSound, which transforms the light intensity into different tones. This device permits to record and listen the differences of light intensity of certain light source. Pointing this device to the Sun, it emits an high-pitched sound, while during the Sun eclipse, and its gradual occultation, the sound changes from a high to a low-pitched sound. This device was used in the USA, African, and Argentinian/Chilean Eclipse in 2017, 2018, and 2019, respectively. In Chile, we received from the LightSound IAU100 Project† twelve devices, which we distribute

† http://astrolab.fas.harvard.edu/LightSound-IAU100.htm

actividades astronomia chile
discapacidad dispositivo
divulgacion eclipse escuchar
guardaria inclusiva lightsound
personas
usarlo uso visual

Figure 1. *Left:* LightSound device for the Eclipse 2020 in Chile. The case was specifically design to adjust all electronic supplies and printed in 3D. In front, the light detector is placed. To the left side, the switch on/off bottom is located, the blue switch change between the tone and musical mode. The logo of our association "Astronomía Inclusiva" is on front of the image. *Right:* Thirty most used Spanish words in our application form to request a LightSound.

along the country. The enthusiasm of the Chilean people, requesting the device to different channels and persons dedicated to astronomical outreach, motivated this team to assemble the devices in Chile. For the Eclipse 2019, we were able to build five devices, so we recorded the Eclipse sound in 17 points along Chile. With these data, plus the one collected in Argentina, Allyson Bieryla and Soley Hyman built the Eclipse Sound Map‡. An extensive description of the original design and its use can be find in Bieryla et al. (2020); Soley et al. (2019). This development motivated our Chilean team to develop, build and communicate the concept to the Chilean community. We were awarded two project from the ESO-Comite Mixto and GEMINI funds to build one hundred LightSound (Fig. 1) and another one hundred Orchestar. The Orchestar device converts the light color into different musical tones, as it is explained in details in Bieryla et al. (2018).

2. Overview

In its original version, the LightSound transforms the light intensity into different tones, to listen this sound visit the following website https://youtu.be/_BM6IjDj-AM. We explore the possibility of considering different instruments for marking the intensity levels, as it can be listened here https://www.youtube.com/watch?v=DtWXAyS-4s0.

Transferring the concept to the community and device distribution: The LightSound and Orchestar devices can be used to explain various science topics using sounds. Yet, since the concept is unknown, people relate this device to a specific scientific experiment. Our main idea was to use the device during the Eclipse to reach a large part of the community, introduce the concept and continue with other experiments during 2021. Our original plans were to distribute the electronic supplies and build the devices in hands-on or co-work session with the interested people, specially in centers dedicated to astronomical Outreach and Universities. Due to the pandemic, we had to adapt this plan, distributing the assembled devices to the people, and rely in our developers team. The devices were distribute along Chile, from Calama to Punta Arenas, and in various Outreach centers such as Museo Interactivo Mirador, Las Campanas Observatory, Universidad Católica, Universidad de Chile, Universidad de Concepción and Universidad Austral.

‡ https://youtu.be/RraNpZkSxNY

Description on how to implement MIDI sounds.. The LightSound works as an musical instrument. It plays the sounds based on the MIDI system, therefore it depends on the note, instrument and channel. The MIDI chip allows 15 channels, so there can be 15 instruments playing at the same time. The original version of the LightSound uses one channel which reproduce tones depend on the light intensity: more light intensity means higher pitched sounds, less light intensity means lower pitched sounds. In this version, presented in this contribution, we added a new feature: more light intensity means more instruments. To do this we employ MIDI songs. However, the LightSound reproduce the song in real-time using the MIDI commands, therefore it can not use the MIDI song itself, it is necessary to extract the instruments in the song before use it. We wrote a script to do this, which is available in the following website https://github.com/cjsantander/SongToLightSound. The output are the instruments of the song, where each file has the MIDI commands that LightSound can read. In this way, the LightSound read the instrument one by one and decide the volume of each instrument depending on the light intensity recorded in the sensor. We have to modify the original version of the arduino code to implement this new feature. The new code to play tones or multiple instruments is available in this website https://github.com/HenryDelMal/LightSound_ESP32.

Music creation and its connection with the Eclipse light curve. We have chosen two musicians dedicated to the etno-music, these persons have been reproducing the sound of antique instruments that has not survive until nowadays. They re-created the sound of these instruments with the historical information recompiled. The music created for this purpose is available in this website https://portaldisc.com/disco.php?id=26771. and use the following instruments, Antares (from Nazca culture), Rukos pinkullos (from the Cañari people of Ecuador), drum of agricultural use, and ceramic whistles.

Live transmission. We broadcast the sound of LightSound from the totality zone, in the locality of Lanco, Los Ríos Region. We organized this event dubbed "Eclipse Inclusivo" in collaboration with Centro de Estudios Científicos (CECs), and Centro Cultural de Promoción Cinematográfica de Valdivia (CPCV). Throughout the transmission, with interviews with authorities and representatives of the communities of people with disabilities, the device was playing in the background, capturing the variations of light in the environment, and every so often the transmissions were stopped to listen to it. This was replicated by 39 fanpages around the world, reaching 10,000 live reproductions, it is available in this website https://www.facebook.com/ficvaldivia/videos/384101929485674, a version with only image and sound was also broadcasted and it is available in https://www.youtube.com/watch?v=qCg7M33CkiM, for those who wanted to enjoy it.

3. Implications

This project was fully motivated for the enthusiasm of the Chilean people to get to know the concept of how the light can be transferred into sound. Besides the principal purpose of listening the Eclipse, most of the people found a second application of the device related to their daily life context. For example, some visually impaired persons said that they will use it to listen the sunrise/sunset or to know how dark/clear are their clothes. Most of the teachers commented on how to implement this device in their classroom, it is a experiment to mark the difference and feasible to do it online. The persons working on the Outreach centers mentioned their wish to include it in their regular activities related to Astronomy and other Sciences. The officer of the national governmental service dedicated to people with some declare impairment, SENADIS, mentioned how positive it is for their labour to include this type of device in their collections and offer it permanently to the community. Hereafter, we report some of the comments and use that persons declare to give to this device with their own words.

- Cecilia: I will use it to know when is sunrise and sunset, and share it with the members of the association for Blind people of the Araucanía.
- Alfredo: I will it use in two FONDART projects related to experiencing inclusion through theater. His records of the Eclipse in the Araucanía are published on the website https://www.facebook.com/alfredo.s.santibanez/videos/10223220581358485.
- Arturo: I will use it to detect lights switch on/off at home, to know if the clothes are dark/pale or the day is sunny/cloudy. His record of the Eclipse in Santiago are published on the website https://www.facebook.com/astroinclusiva/videos/177093007481688.
- Fundación Pequeñas Grandes Estrellas: We will incorporate the LS in our 2021 activities! aiming to make Astronomy closer to visually impaired kids.
- LCO outreach officer: We use it to broadcast the Eclipse in our youtube channel, https://www.youtube.com/watch?v=c84K0lceGVY&feature=emb_logo. Post Eclipse, we will work in collaboration with Dedoscopio.
- Marlene: I will keep it and try to listen the Moon and stars.
- SENADIS officer: We will keep it in our governmental records, as one of the project that impact positively to the Inclusion. The device will be available to individuals who request to use it.
- Maritza: I will incorporate this device to the set of activities that we have already create in the network of teachers "Nodo Andino de Astronomía"
- María: I have plenty of plans to use it post-eclipse. Once the face-to-face activities are back, I will add to the AstroBVI activities such as: inclusive station in the mobile planetarium of Science Faculty of PUCV and vulnerable schools in rural areas of Valparaíso.
- Lorenzo: I will use it in a school of the Elqui Valley, in which we teach to old people and some are visually impaired.
- René: I will use in the Optics lab for the Engineering students.
- Rodrigo: During the Eclipse, we will transmit live from the Radio & TV of Villarrica. Later we will incorporate it to the Dedoscopio activities.
- Juan: Enjoy the sunsets, and explain the eclipses in activities organized by my association "Astronomía en las calles".
- Carla: I will perform some experiments for the eclipse and in outreach activities related to the Sun, mechanics of the sound waves and electromagnetism. I will transfer this concept to Peruvian associations dedicated to teach science to visually impaired.
- Jennifer: I will use it in the activities that I organize as an outreach officer of CATA and Universidad de Chile.
- Sonia: AUI/NRAO will use it in our activities focused on inclusion and gender equity.
- Museo Interactivo Mirador: we will incorporate it to our regular activities of the museum. In particular as a complement for the "Plaza Solar" of MIM.

Acknowledgements

This is project is executed by part of the members of Astronomía Inclusiva group, financed by ESO-Comite Mixto 2019 and GEMINI 32190018, sponsored by Universidad Central and Universidad Autónoma de Chile. Wanda Díaz Merced is our endless source of inspiration, energy and improvements. We also thanks the companies Imprimitech and Altronics which helps us to lower the device costs from 70 USD to 25 USD.

References

Bieryla, A.; Diaz-Merced, W.; Davis, D.; Hart, R. *AAS*, 2312,1606
Bieryla, A.; Diaz-Merced, W.; Davis, D 2018, *AAS*, 2322, 0903
Bieryla, A.; Diaz-Merced, W.; Davis, D.; et al. 2020, *CAP*, Issue 28, 38.
Hyman, Soley O.; Bieryla, Allyson; Davis, Daniel *AAS*, 23325511, 2019

Education and Heritage in the era of Big Data in Astronomy
Proceedings IAU Symposium No. 367, 2020
R. M. Ros, B. Garcia, S. R. Gullberg, J. Moldon & P. Rojo, eds.
doi:10.1017/S1743921321000405

Women in Astronomy and in Science: cracking the code with a third culture

Anna Curir⬤

INAF – Osservatorio Astrofisico di Torino, I -10025 Pino Torinese (Torino), Italy
email: `anna.curir@inaf.it`

Abstract. Girls' under-representation in science, technology, engineering and mathematics (STEM) education is deep rooted and puts a detrimental brake on progress towards sustainable development. Both education and gender equality are an integral part of the 2030 Agenda for Sustainable Development, adopted by the United Nations General Assembly in 2015, as distinct Sustainable Development Goals (SDGs). The UNESCO Report: *Cracking the code: Girls' and women's education in STEM* provides a global snapshot of this underrepresentation and the factors behind it. The fight against stereotypes to 'crack the code', or to decipher the factors that hinder or facilitate girls' and women's participation in STEM (and particularly in Astronomy) education must take into account the persisting dichotomy between the so called two cultures.

Keywords. history and philosophy of astronomy, sociology of astronomy, miscellaneous

1. Introduction

C. P. Snow lamented in his book *The Two Cultures and the Scientific Revolution* the gap between scientists and literary intellectuals (Snow 1969).

The two cultures, the scientific and the humanistic, do not talk each other. Between the two languages there exists a wall that only a limited number of people with a deep knowledge of both are able to overcome. One could say that a bridge needs to be built between the two cultures, namely a common background which favours reciprocal understanding. The gap between Humanities and Sciences has been deeply investigated by the intellectuals of the so-called *third culture*, who want to promote a new alliance between the scientific and humanistic culture (Brokman 1995). Iinside such a third culture the two communities of intellectuals could have a fruitful dialog. Each community would take advantage of the point of view of the other. For example philosophers could benefit from the rigour and nature-centred point of view of scientists, and scientists could take advantage of the critical attitude of philosophers.

A third culture would also unify arts and sciences, because the artists' and scientists' creative processes are very similar. The scientist, the poet, the painter, the philosopher create their works using similar inner processes which mainly consist in making a bricolage of elements already known. These elements are combined to give a totally new representation of nature, of man or of life. Therefore both artists and scientists are involved in intuiting change of perception of the world and materialising it for others to experience, see and eventually change. In such a creation, the scientist, the poet, the painter, the philosopher are deeply emotionally involved. A third culture would integrate sensitivity and Sciences, by introducing the awareness of the emotions involved in the scientific discovery (Curir 2014). According to Théodule Ribot *an idea is nothing more* than an idea, a simple fact of knowledge, it does not produces nothing, it can't do any-thing; it only acts if it is felt, if there is a emotional state to accompany it, if it awakens trends, motor elements (Ribot 1896).

A third culture would be more multidisciplinary and less sectorial, a scientist of the third culture would be similar to a 'natural scientist' of the 18th century (Curir 2018).

An uneven gender distribution amongst the two cultures can be observed: amongst academics in the fields of hard Science, men are clearly predominating. In some fields of the Humanities, there is a predominance of women. The support to a third culture would be a support to equity for women in Science.

2. UNESCO: cracking the code – girls' and women's education in STEM

The UNESCO report *Cracking the Code – girls' and women's education in STEM* tell us that gender differences in STEM education at the expense of girls are already visible in early childhood care and become more visible at higher levels of education. Girls appear to lose interest in STEM subjects with age, and lower levels of participation are already seen in advanced studies at secondary level. By higher education, women represent only 35% of all students enrolled in STEM-related fields of study.

Neuroscience reports differences in brain structure and functions between men and women (Ruigrok et al. 2014) but few differences have been found between boys' and girls' brains relevant to learning (Eliot 2013). Studies on the neural basis of learning have not found that boys and girls master calculation or other skills differently and no dif-ference in brain composition can explain gender differences in mathematics achievement (Spearman & Watt 2013).

Girls are often brought up to believe that STEM are 'masculine' topics and that female ability in this field is innately inferior to that of males, whereas they do not feel so inferior in humanities studies. This can undermine girls' confidence and willingness to engage in these subjects. Supportive learning environments can increase girls' self-confidence and self-efficacy in STEM. The fight against stereotypes to 'crack the code', or to decipher the factors that hinder or facilitate girls'y and women's participation in STEM (and particularly in Astronomy) must also take into account the persisting dichotomy between the two cultures and the uneven distribution of women between them. Teaching Astronomy as a third culture discipline will surely have a central role in determining girls' interest in this subject and in providing equal opportunities to access.

3. Astronomy as a third culture

Most ancient cultures saw pictures in the stars of the night sky. The earliest known efforts to catalogue the stars, with cuneiform texts and artifacts, date back roughly 6000 years. These remnants found in the valley of the Euphrates River, tell us that our ancestors observing the heavens saw the lion, the bull, and the scorpion in the stars. By the 5th century B C most of the constellations had come to be associated with myths, and the Catasterismi of Eratosthenes completed the mythologization of the stars. At this stage, the fusion between Astronomy and mythology was so complete that no distinction could be made between them. Therefore in ancient Astronomy the narration, the myths, and the observations and measures were interlaced: Astronomy is the first third culture.

Modern Astronomy and Cosmology are important examples of sciences which were progressively abandoned as sciences based on human centrality in favor of a perspective where man is projected in a stream of time. Indeed, Astronomy and Cosmology put us in a continuous contact with a reality which is deeply 'historical and evolutionary'. An atomical physicist can ignore the history of the reality of which he is studying the model, but a cosmologist must always take into account the history and the evolution of the phenomena he is dealing with. Therefore the astronomer and the cosmologist are forced to think in terms of a Natural History which links together Cosmology, Biology and Historical Sciences. Also because of such a structural opening, modern Astronomy and Cosmology have the connotation of a third culture.

4. Metaphors in Science

Evelyn Fox Keller stressed the ubiquity of metaphors in scientific discourse, necessary for generating knowledge about a world not yet known. The scientist seeks understanding of the not yet intelligible by comparison with that which is already familiar. Nevertheless, what we confront is something new, not contained in the world we already know, and so the analogical relation is not quite sufficient. Indeed, the essence of a metaphor is precisely the juxtaposition of similarity and difference. And the instability it generates by virtue of its insistence on both similarity and difference is essential to the logic of scientific discovery (Fox Keller 2020). Fox Keller put emphasis on Bohr's acceptation of both the competing metaphors about light: light as particles and light as waves. The new quantistic paradigm was created by overcoming this duality and accepting both the metaphors. Through their dynamic interaction, Keller points out, metaphors define the realm of the possible in science (Fox Keller 1995).

But the objects of Astronomy are many metaphors at a time: a black hole is an exact solution of the Einstein's equations, but also an active galactic nucleus, an X-ray emitter, a radio source... Moreover, the astronomical research involves many disciplines: Thermodynamics, Fluid Dynamics, Nuclear Physics, Chemistry, Special and General Relativity, to tell only some of them, but also, as we mentioned, Historical Sciences, Biology, Archeology. Astronomy is intrinsically a third culture: because its primordial roots are interlaced inside narration and measure, mith ad observation. And because its modern version is integrating different disciplines and metaphors of interpretation and these interpretations are embedded in time stream dimension.

5. Some hypothesis about women's intelligence

Our *gendered world*, Gina Rippon says, shapes everything, and generates *gendered brains*. With the discovery of the brain plasticity, we know now that the brain is much more a function of experiences. If you learn a skill your brain will change, and it will carry on changing. Once we acknowledge that our brains are mouldable, then the power of gender stereotypes becomes evident. If we could follow the brain evolution of a girl or a boy, we could see that right from the moment of birth these brains may be set on different roads (Rippon 2020).

Although neurobiological explanations about male and female thinking and learning styles are controversial, it has been argued in some Cognitive Psychology studies that whereas men tend to look for abstract and theoretical arguments, dissociating them from any distracting information, women are more apt to see and make connections between ideas and the larger context (Halpern & Lamay 2000; Kimura 1999; Kimura 2004).

It is often argued that women are more oriented toward and better at assimilating diverse forms of information whereas men prefer to isolate explanations and excel in tasks requiring more local processing (Wyer *et al.* (2001)).

If women intelligence is more naturally multidisciplinary, the perception of Astronomy as a multidisclipinary third culture should encourage women to challenge themselves in such a discipline. Moreover, overcoming the gap between Humanities and Sciences to approach a third culture will provide a fertile ground for women's sensibility and for a holistic approach to nature.

This approach cannot be costless. We know that for a successful scientific career the best performance is actually the one focused on a specific field. Becoming very specialized is now more profitable than being involved in different fields and in the connections between them. We must hope for the future a deeper awareness of the importance of a third culture and of a multidisciplinary approach to Science: the development of a richer view of the world.

Rothen and Pfirman in their exploratory paper wanted to focus on intersection between women and interdisciplinary Science. They conclude that gender gaps are complex issues and a justifiable concern for reasons of social equity in Science. Thus, making better use of the talent of female scientists has been a prominent policy objective for at least a quarter of a century. As the data may suggest, it does seem that interdisciplinarity could serve as a strong entry point into scientific studies for women (Rhoten & Pfirman 2007).

There are several examples of women that were interested in Science and Humanities: I mention here Sofia Kovaleskaja, outstanding mathematical scholar and literary writer, Patricia Churchland, who worked on the interface between Neuroscience and Philosophy, and Evelyn Fox Keller who did research in Physics and Biology but also in the History and the Language of Science.

6. Conclusions

It could be that women are well positioned to make major advances in interdisciplinary research; they may like to integrate across fields and approaches. Using interdisciplinarity to attract women, as well as other underrepresented minority groups into Science, could lead to a stable pathway through scientific and academic careers. This point could be more enhanced in the field of Astronomy, due to its intrinsic nature of third culture Science, and interdisciplinary Science.

References

Brockman, J. 1995, *The Third Culture; beyond the Scientific Revolution* (New York: Simon and Schuster)

Curir, A. 2014, *Les Processus Psychologiques de la Découverte Scientifique* (Paris: L'Harmattan)

Curir, A. 2018, *L'emergence de la troisiéme culture et la mutation létale* (Paris: L'Harmattan)

Eliot, L. 2013, *Sex Roles: A journal of Researc*, 69, No. 7-8, 1-19

Fox Keller, E. 1995, *Refiguring Life: Metaphors of Twentieth-century Biology* (New York: Columbia University Press)

Fox Keller, E. 2020, *Interdisciplinary Science Reviews*, 45, 3, 249

Halpern, D. F. & LaMay M. L. 2000, *Educational Psychology Review*, 12, 229

Kimura, D. 1999, *Sex and Cognition* (Cambridge, MA: MIT Press)

Kimura, D. 2004, *Sexuality,Evolution and Gender*, 6, 1.

Rhoten, D.& Pfirman, S. 2007, *Science Direct*, 36, 56

Ribot, T. 1896, *La psychologie des sentiments* (Paris: Alcan)

Rippon, G. 2020, *The Gendered Brain* (London: Vintage Publishing)

Ruigrok, A. N. V., Salimi-Khorshidi, G., Lai, M. C., Baron-Cohen, S., Lombardo, M. V., Tait, R. J. and Suckling D. 2014, *Neuroscience & Biobehavioral Reviews*, 39, 34–50

Snow, C., P. 1969, *The Two Cultures and the Scientific Revolution* (Cambridge: Cambridge University Press)

Spearman, J. and Watt, H. M. G. 2013, *Learning Environment Research*, 16, No. 217, 217–238.

Wyer, M. et al. (eds) 2001, *Women, Science, and Technology. A Reader in Feminist Science Studies* (New York/London: Routledge)

Session 10: Informal Education in Astronomy

Education and Heritage in the era of Big Data in Astronomy
Proceedings IAU Symposium No. 367, 2020
R. M. Ros, B. Garcia, S. R. Gullberg, J. Moldon & P. Rojo, eds.
doi:

Introduction

Introduction

Session 10, the last of the symposium, begins with the invited talk "Embracing a culture of lifelong learning - in universities & all spheres of life" by Edith Hammer where she discusses the important role of life-long learning. She describes this as being key to addressing the many challenges that humanity faces. Edith describes that learning to learn is a fundamental skill to be mastered in order to master continued learning throughout life. She states that lifelong learning is for all and it is imperative to create learning opportunities that can reach everyone.

Edith was asked by Walter Guevara Day:

What do you think should be done so that this lifelong learning is extended to all ages and especially to the very young and the elderly, who are increasingly displaced in some countries.

and she responded:

There is a broad range of actions to be taken to make LLL a reality, including the integration of the concept of LLL in education and development policies and strategies, strengthening educational institutions with a LLL perspective, establishing flexible educational pathways and mechanisms for the recognition of prior learning, among many others. In a recent report on the futures of LLL, the UNESCO Institute for Lifelong Learning has proposed 10 key messages and a set of actions points to "Embrace a culture of lifelong learning". These 10 points are: Recognize the holistic character of lifelong learning; Promote transdisciplinary research and intersectoral collaboration for lifelong learning; Place vulnerable groups at the core of the lifelong learning agenda; Establish lifelong learning as a common good; Ensure greater and equitable access to learning technology; Transform schools and universities into lifelong learning institutions; Recognize and promote the collective dimension of learning; Encourage and support local lifelong learning initiatives, including learning cities; Reengineer and revitalize workplace learning; and Recognize lifelong learning as a human right. The report is available here: https://unesdoc.unesco.org/ark:/48223/pf0000374112?posInSet=1 & queryId=6f6d2b66-c054-4f6a-8563-293f7697c717

William Waller asked:

In the absence of a local university, what would you recommend to engender LLL in a community?

Edith answered:

In urban areas, usually there is a good educational infrastructure, including universities and other educational institutions. In rural areas, this may not be the case. Many different stakeholders can contribute to provide lifelong learning opportunities to people of all ages. For example, community centers and community libraries have helped to engage local people in learning processes in countries around the world. Some good examples can be found in UIL's publication on "Communities in Action: LLL for sustainable development", available here: https://unesdoc.unesco.org/ark:/48223/pf0000234185

Also, many universities and platforms such as Coursera offer online courses (MOOCs), which are available to anyone with an internet connection. This can also be a chance for individual learners in structurally less developed areas to engage in higher education.

"Turn on the Night! Science and Education on Dark Skies Issues" was written by Constance Walker, Richard Green, Pedro Sanhueza, and Margarita Metaxa. In it they talk about the IAU's Dark and Quiet Skies working groups and some of the latest dark skies protection issues. They also discuss dark skies education and heritage.

Andrea Sosa writes of "Let's turn off the lights and turn on the night: to the rescue of starlight in an age of artificial lighting." Andrea relates that an estimated one-third of the population of the world has never seen the Milky Way due to light pollution and the fundamental loos experienced as a result. She outlines reasons to work to minimize light pollution and discusses some global initiatives to do so.

Next is "The search for extraterrestrial intelligences and the Fermi Paradox" by Nikos Prantzos. Here Nick talks about the search for extraterrestrial intelligence and he includes discussions of the Drake Equation and the Fermi Paradox. He outlines current thought and states that the plurality of worlds is more controversial today than ever before and includes that the detection of an inhabited planet would be one of the major events in human history.

Harufumi Tamazawa discusses "Sunspot observation by the cooperation of amateur astronomers and researchers in Japan in early 20th century as early citizen science program." He emphasizes the merit of cooperation between astronomers and amateur observers and the increasing popularity of doing so. Harufumi gives examples that underscore the value of working together in such projects.

"Solar Eclipses in India's Cultural and Political History" is described by Ramesh Kapoor. He discusses early records of eclipses in India and how they can help to fix timelines in history. He gives several examples dating form as early as 322 CE. The paper show eclipses to be interwoven with life an culture in India.

The final paper, "Estrelleros: Astronomy in hospitals," was written by Gloria Delgado-Inglada, Diego López-Cámara, Alejandro Farah, Jorge Fuentes-Fernández, Orlando García, Carolina Keiman, Tita Pacheco & Jaime Ruíz-Díaz-Soto. The authors describe an initiative designed for children who must endure long hospital stays. The effort is not only entertaining, but even more so is designed to increase their scientific knowledge and promote the pursuit of scientific vocations. Examples from Mexico are given and the authors discuss future expansion of the project.

drom Milagros Vera asked:

How do you finance the project?

and the response was:

We have obtained financing mainly from the Mexican Consejo Nacional de Ciencia y Tecnología through two different programmes in 2019 and 2020. One to promote scientific vocations and the other to carry out activities of social appropriation of scientific knowledge. Also our institute support us in many different ways. For example, bringing all the material and volunteers to the hospital.

Walter Guevara Day asked:

Now, With the pandemic problem, how are they doing?

with the response:

For obvious reasons, we have not been able to visit hospitals this year. We have concentrated on generating new material, creating a manual for the activities and preparing everything to be ready for the moment when we can visit hospitals again. In 2021 we plan to carry out virtual visits. We still do not know exactly how to do it and which activities and hospitals are the most adequate. It is a big challenge but we are ready to face it.

Susana Deustua asked:

Have you been in contact with others who carry out these types of activities? Donald Lubowich has many years working with patients in NY.

and was answered with:

Not yet but in this conference I have met some interesting people such as Donald Lubowich. I plan to talk with them to improve and expand our project.

Education and Heritage in the era of Big Data in Astronomy
Proceedings IAU Symposium No. 367, 2020
R. M. Ros, B. Garcia, S. R. Gullberg, J. Moldon & P. Rojo, eds.
doi:10.1017/S1743921321001010

Embracing a culture of lifelong learning – in universities & all spheres of life

Edith Hammer

Programme Specialist, UNESCO Institute for Lifelong Learning (UIL)
email: `e.hammer@unesco.org`

Abstract. This presentation will provide an introduction to the concept of lifelong learning, exploring its relevance and potential for future human development. Lifelong learning – comprising learning in formal education, in non-formal contexts and informal ways – plays an increasingly important role within society and also within the higher education sector. Universities have a social responsibility towards society, conducting research that benefits society, making research results widely and openly available, communicating research to the wider community, and providing learning opportuntities for people of all ages and social backgrounds.

Increasingly, the global community acknowledges that lifelong learning – available to all, at every stage and in every sphere of people's lives – is key to addressing the multiple challenges faced by humanity. Lifelong learning fosters people's capacity to deal with change and to build the future they want. This is profoundly important given the disruption and uncertainly resulting from the familiar threats and opportunities of demographic change, the climate crisis, the rapid advance of technology and, more recently, the COVID-19 pandemic. In such uncertain times, new ways of learning provision have to be found.

Learner autonomy is the foundation of this lifelong learning culture. Learning to learn has become a basic competence, as has managing one's own learning journey and creating one's own learning biography. Understanding all levels of learning as learner-centric presents a crucial shift to fundamentally thinking and planning education with demand in mind. This way, learners are active agents rather than passive recipients of prescribed knowledge. They co-design and use any learning process and its outcomes actively to realize their potential as fully as possible.

Learning is understood as a collective process, taking place among peer groups, within communities and across generations. Education emphasizes becoming global citizens who care about each other, other communities and the planet. Consequently, lifelong learning is for all, and learning opportunities can be created for and with the most excluded learners. There is a global learning ecosystem, built collectively to inspire and empower learners with a plethora of opportunities. The learning ecosystem integrates diverse learning modalities fluidly, including all digital-based and real-life experiential learning as well as blends of formal, non-formal and informal learning. The learning opportunities allow for planned or spontaneous, individual of collective learning. There is easy access to face-to-face and online learning opportunities as local infrastructure, global connectivity and sponsored devices are available for all. New pedagogical and andragogical principles have been developed, including innovative blended learning concepts that integrate digital and face-to-face elements while considering learners' specific needs.

The presentation will link these future-oriented ideas of lifelong learning with the higher education context, the open science movement and explore ways in which research-based knowledge can be provided to learners in different contexts.

Keywords. Education, Lifelong learning, UNESCO

1. Defining lifelong learning

The concept of lifelong learning implies that a learning process is no longer constrained within a prescribed life span or the walls of a formal institution, but rather extends throughout life and occurs in various settings. There are five essential elements to the UNESCO definition of LLL (UIL, forthcoming):

All age groups. Lifelong learning is a process that begins at birth and occurs across the whole lifespan for people of all ages and origins with learning opportunities and activities, depending on their needs and professions.

All Levels of education. Lifelong learning links all levels and types of education to build adaptable pathways between them. This includes early childhood care and education (ECCE), primary and secondary school education, higher education, adult and non-formal education, as well as technical and vocational education and training (TVET).

All learning modalities. Lifelong learning recognizes all modalities: formal (institution-alized, leading to recognized qualifications), non-formal (institutionalized, alternative or complementary to formal education, usually not leading to recognized qualifications) and informal (not institutionalized, on a self-directed, family-directed, community or socially directed basis).

All learning spheres and spaces. Schools are just one part of a wide learning universe, a space which also includes families, communities, workplaces, libraries, museums, and other online and distance learning platforms. Hence, building bridges between formal and diverse non- formal education to create new opportunities for learners.

A variety of purposes. Lifelong learning is people – oriented and human – rights based, providing people with opportunities to develop their potential throughout life, regardless of the starting points, addressing a wide range of learning needs and contributing to the development of an inclusive society and an advanced economy.

2. Relevance of lifelong learning within the higher education sector

During the past decades, the mandates of universities and higher education institutions have changed, addressing new groups of learners and responding to their social respon-sibility. Given the current demographic and socio- economic transformations, the role of higher education in addressing societal challenges and projecting positive impact have increased. Universities now have a distinct role in (re-)educating highly qualified people to meet the changing demands on labour markets, thus requiring quality education with a focus on the needs of diversified groups of people.

Universities, on top of their traditional missions of teaching and conducting research, are expected to fulfil roles that reflect economic, social and cultural contributions to their local environment. The so-called 'engaged university' plays a local development role by offering lifelong learning and other services that contribute actively to shaping or reshaping the social, cultural and economic situations of local communities. That can be done through outreaching to a wider scope of non-traditional students, such as working professionals, older people, people of low socio- economic backgrounds, migrants, ethnic minorities, people with disabilities, people from remote areas, among others.

3. Transforming universities within a framework of lifelong learning

Transformations of universities to become lifelong learning institutions occur at insti-tutional level and encompass modifications in policy frameworks and support. In the last 15 years, several international and regional frameworks and recommendations have been drafted, emphasizing an important role of universities in promotion of lifelong learning. They furthermore, outline major areas of transformations.

In 2008, the European Universities Association (EUA) issued the European Universities' Charter on lifelong learning, which called on universities to make 10 clear commitments to lifelong learning. The EUA Charter also recommended concerted action from governments in providing the appropriate legal and financial frameworks to promote lifelong learning widely (EUA 2008).

A year later, the 2009 UNESCO World Conference on Higher Education stressed the role of higher education in lifelong learning and stated "the knowledge society needs diversity in higher education systems, with a range of institutions having a variety of mandates and addressing different types of learners. This includes promoting research for the development and use of new technologies and ensuring the provision of technical and vocational training, entrepreneurship education and programmes for lifelong learning" (UNESCO 2009).

In 2015, the United Nations' 2030 Agenda for Sustainable Development, specifically Sustainable Development Goal 4 (SDG 4), called on the world to "ensure inclusive and equitable quality education and promote lifelong learning opportunities for all". It is further specified to "ensure equal access for all women and men to affordable and quality technical, vocational and tertiary education, including university" by 2030 (Target 4.4).

4. Embracing a culture of lifelong learning

During May and June 2020, the UNESCO Institute for Lifelong Learning (UIL) invited 12 experts from different fields (including demography, economics, education, philosophy, public health, neuroscience and sociology) to take part in a three-weeks online consultation to reflect on how lifelong learning can contribute to building a desirable future by 2050 and to propose concrete measures. This collaborative work resulted in a report entitled "Embracing a culture of lifelong learning" (UIL 2020). The report is a future-focused vision of education, which requests a major shift towards a culture of lifelong learning by 2050. The challenges that humanity faces, resulted from various changes in climate, technology and demographics, and changes caused by COVID-10, call for societies to take action. Thus, education has to be learner-centric with a demand-led approach, allowing for learners of all ages and backgrounds to co-design and use a learning process to achieve their full potential. At the same time, learning has to be a collective process that acknowledges the value of peer and integrational learning.

The report includes 10 key messages to realize a culture of lifelong learning (Fig. 1). These messages strive to rethink the purposes of education and the organization of learning. They also contain tips for actionable directions such as translating visionary ideas into policy, research agendas and initiatives. All key messages presented in the following illustration are linked with each other. Together they help to realize a new vision for lifelong learning by 2050. Also, they are relevant to transforming the field of higher education:

(a) **Recognize the holistic character of lifelong learning**
   To provide a holistic perspective on learning, two dimensions can be stressed: First, there is a need to recognize the anytime principle of learning and its materialization through learning pathways. Second, the anywhere dimension refers to the vision of an ecosystem of learning with decentralized and diversified learning provisions, as the boundaries between formal and non- formal and informal learning have been blurred and non-formal learning may take place in a context of a formal education institution. Moreover, technological advancement brought a larger selection of means and modalities to facilitate learning pathways for every group and every person, allowing for combined education and training, formal, non-formal and informal

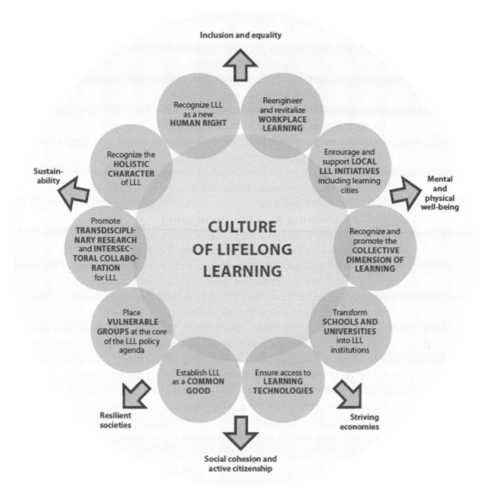

Figure 1. Key messages for fostering a culture of lifelong learning (UIL (2020)).

learning, face-to-face and distance education, and directed and self-directed learning. Within the context of higher education, information and guidance services for students and potential students are key to support flexible learning pathways into and throughout higher education, as well as to ensure possible transitions into other fields of education or the employment sector.

(b) **Promote transdisciplinary research and intersectional collaboration for lifelong learning**

Lifelong learning has been widely recognized as a powerful tool for developing more sustainable societies, economies and living environments. To harness its potential to create a sustainable future for the next generations, fostering transdisciplinary approaches and intersectional collaboration will be a key. Transdisciplinary and intersectoral collaboration should include joint research, as well as practical implementation of innovative initiatives, such as development of blended pedagogies to bridge conversational and digital learning. From the perspective of LLL, one of the main challenges for education system is to offer flexible learning pathways, within and outside formal education, allowing individuals to accumulate different learning experiences throughout life.

(c) **Place vulnerable groups at the core of the lifelong learning agenda**
Lifelong learning includes a diversity of learning modalities, so that people that were previously excluded should be able to join. Inclusive education encourages an active role and the participation of learners, their families and other communities. As lifelong learning is for all, vulnerable groups such as migrants, refugees, older people, youth and adults in risk, and people with disabilities must have equal opportunities to learn. Recognizing the value of interdisciplinarity, inclusive education aims to strengthen the links between schools and society to enable families and their communities to participate in and contribute to the educational process. In order to foster an inclusive and just society, vulnerable, disadvantaged and marginalized groups need to be placed at the core, and that is achieved with policies and instruments, and mainstreaming the focus in the entire legal, policy, delivery and funding framework.

(d) **Establish lifelong learning as a common good**
To ensure the availability and accessibility of learning opportunities for all, the trends towards market regulation of education provision should be gradually replaced by a commons approach through which voluntary social collaboration in open networks is used to generate social-environmental value. Establishing education resources and related tools, including IT solutions, as common goods allows institutions to manage them as commons, sustainably and equitably, in terms of participation, access and value. To widen access, the concept of a commons (such as open access, open source, open educational resources and co-operative online platforms) should be applied to lifelong learning initiatives.

(e) **Ensure greater and equitable access to learning technology**
In a rapidly developing world digital skills are paramount in achieving key policy goals, including employability, meeting labor market needs in a forth industrial revolution, strengthening social inclusion and contributing to a vibrant democratic life. Digital technology can offer innovative ways to engage, support and assess learners of all ages, providing them with opportunities for more adaptive, personalized and responsive learning for everyone, even the ones prior excluded. Lifelong learning plays an important role in contextualizing the use of technology and highlights its application in different parts of life and learning. The distinction between formal, non- formal and informal learning has been a key element in making sense of what is being learned through digital technology.

(f) **Transform schools and universities into lifelong learning institutions**
The lifelong learning perspective recasts the role of educational institutions and highlights the need to transform schools and universities into lifelong learning institutions. This process requires a revision of curricula, more flexibility in terms of study programmes and learning formats, including the introduction of stackable credits as an alternative to full degree programmes. Also, access to higher education should be widened through multiple learning pathways, depending on learners' abilities and needs. By shifting their attention from 'what to learn' to 'why and how to learn', higher education institutions can better conceptualize and link students' prior knowledge and experiences. The transformation of pedagogies is another essential element of this transformation. This implies that teaching and learning are guided by the principle of epistemic pluralism, meaning that teachers cultivate a critical and explorative attitude among students and support the development of different perspectives on particular subjects. Last but not least, universities should cater to the needs of the wider community by making their facilities and resources available for adult learning and education. They should reach out to and involve local communities in teaching, learning and research activities.

(g) **Recognize and promote the collective dimension of learning**

Learning as a collective endeavor is deeply rooted in all cultures and is evident in concepts such as learning neighborhoods, learning circles, and leaning communities and families, all of which acknowledged as social dimension of learning. The collective dimension puts renewed emphasis on face-to-face learning, particularly in public spaces, and also acknowledges the potential of new technologies, enabling digitally linked learning collectives with similar interests. Furthermore, it emphasizes that education is guided by the idea of educating to care, not to compete. Such understanding promotes social cohesion and is a crucial aspect in building learning cities or working towards the creation of a learning planet.

(h) **Encourage and support local lifelong learning initiatives, including learning cities**

Cities are favorable settings for promoting lifelong learning for all with their population density, available infrastructure and capacities. The concept of 'learning cities' is a people-centered and learning focused approach, which stresses a collaborative, action-oriented framework for working on the diverse challenges that cities increasingly face. Some cities managed to foster a lifelong learning culture by connecting education and training institutions and cultural institutions and engaging a wide range of partners. At community level, learning opportunities often include spontaneous forms relevant to the specific needs of communities, such as initiatives to promote environmental sustainability, to support women's health or to forge new types solidarity. These community-driven initiatives are key to achieving the potential of lifelong learning and often encourage integrational learning.

(i) **Reengineer and revitalize workplace learning**

Workplaces across sectors, including for the self-employed and those working in the informal economy, are potentially important learning environments, even more so if enterprises become learning organizations. Workplace learning is a crucial driver for lifelong learning and becomes increasingly important, considering the ongoing transformation of the nature of work and the changes taking place in the labor market. For workplace learning to reflect a culture of lifelong learning, there is a need to individualize learning opportunities in response to each worker's needs and to follow progressive models, focusing on developing workers' autonomy at work, as well as flexible, short courses and various incentives, including financial rewards and options for recognition, validation and accreditation.

(j) **Recognize lifelong learning as a human right**

The right to education is no longer limited to accessing the school system, but rather serves to guarantee continuity of learning throughout life, including relevant guidance and digitally portable assessment of all learning outcomes. Lifelong learning is strongly connected to the idea of learning freedom with an ecological dimension. On the grounds of it being a human right, lifelong learning could also serve as an indicator of social justice. As such, lifelong learning would not be defined only as an individual right, but as a social right universally accessible to all citizens. Lifelong learning is guided by three imperatives: access to learning always (across countries and languages); resilience (an educational commons that can withstand different crisis); and transparency (learning resources and facilities, including software and technology, must be open and part of the public domain).

5. Final remarks

Making lifelong learning a reality requires more than innovative policies, more funding or better technology. It demands a radical change, a cultural transformation involving all stakeholders, governments, individuals and employers, as well as urban communities, notably learning cities. Higher education institutions are crucial stakeholders in this transformation process by including people of different background and of all ages in learning processes. They play a vital role in addressing societal challenges and serving their communities.

Transforming higher education institutions within a framework of lifelong learning will require some effort. One of the first steps is to revise curricula and develop new concepts for teaching and learning. That also includes an increased use of technologies and digitalization of education processes to meet the requirements of the modern world. Making the internet and AI central to the lifelong learning agenda and applying a commons approach to education involves making technology gradually open and accessible to all. Flexible learning pathways have to be introduced in order to respect the diversity of individual learning biographies and more opportunities should be available for people to continue their education at any point of life. Lifelong learning is a collective endeavor, it is about learning from others and with others. To ensure a high relevance of learning opportunities, learners (in particular vulnerable groups) should be involved in co-creating knowledge and learning resources. The promotion of open access resources for teaching and learning are also crucial for transforming higher education.

Openness shall be a guiding principle of higher education: Within the Open Science movement, openness refers to several factors: an important issue is the accessibility of knowledge, as well as teaching and learning resources. This is related to the need of more diverse open access practices in the higher education sector and beyond, to make the results of (often publicly funded) research openly accessible (e.g. through national and/or institutional open access strategies). Open science also refers to the principles of openness to different types of knowledge, including from traditionally excluded group (indigenous groups, scientists from the Global South, etc.) as well as diversifying publishing languages. Another important aspect is the promotion of community-based participatory action research and citizen-science approaches in higher education.

This brief summary report of the presentation on "Embracing a culture of lifelong learning" aimed to illustrate that reimagining the future of education from a lifelong learning perspective can help us think our way to a future that is more cohesive, sustainable, inclusive and generally brighter.

References

European University Association 2008, *European University Association. European Universities' Charter on Lifelong Learning.*

UIL 2020, *Embracing a culture of lifelong learning. Contribution to the futures of education initiative.*

UIL 2019, *Forthcoming. Making lifelong learning a reality: A handbook*

UNESCO 2009, *World Conference on Higher Education: the New Dynamics of Higher Education and Research for Societal Change and Development; communiqué.*

Education and Heritage in the era of Big Data in Astronomy
Proceedings IAU Symposium No. 367, 2020
R. M. Ros, B. Garcia, S. R. Gullberg, J. Moldon & P. Rojo, eds.
doi:10.1017/S1743921321001125

Turn on the Night! Science and Education on Dark Skies Issues

Constance E. Walker[1], Richard Green[2], Pedro Sanhueza[3] and Margarita Metaxa[4]

[1]NSF's NOIRLab, USA
email: connie.walker@noirlab.edu

[2]Steward Observatory, Dept. of Astronomy, University of Arizona, USA

[3]Oficina de Proteccion de la Calidad del Cielo del Norte de Chile

[4]National Observatory of Athens and Arsakeio School of Athens, Greece

Abstract. The "Turn on the Night" associated event had presentations on the latest dark skies protection issues considered by the IAU's Dark and Quiet Skies working groups. Presentations were also made on dark skies education programs and cultural/scientific heritage.

Keywords. dark skies protection, astronomy, dark skies oases, bio-environment, education, cultural and scientific heritage, nightscapes

1. Introduction and Science Update on Dark Skies Issues

The "Turn on the Night" associated event was divided evenly between science updates on the latest dark skies protection issues and education programs important to preserving the cultural heritage of starry night skies. For the science updates, topics included protection of ground-based observatories, dark skies oases and the bio-environment, as well as the impact of satellite constellations on astronomy. Material for the latter topic was drawn from research conducted by the Dark and Quiet Skies (D&QS) working groups sponsored by the IAU for the United Nations. This research, along with recommendations for mitigation solutions, can be found in the full report here: https://www.iau.org/static/publications/dqskies-book-29-12-20.pdf, and as a summary for the United Nations here: https://www.iau.org/static/publications/uncopuos-stsc-crp-8jan2021.pdf. For the dark skies education programs and cultural/scientific heritage section, the whys and hows were explored utilizing the international citizen-science program, Globe at Night, as a means to bring awareness of issues on and solutions to light pollution. The authors are members of Commission B7 on the Protection of Existing and Potential Observatory Sites.

1.1. Protection of Ground-Based Observatories

Ground-based observatories are critical to astronomy and are complementary to space astronomy missions. They can be built at a substantially larger scale and lower cost than those launched into orbit, and remain the engines of cutting-edge discovery. Rapidly growing artificial sky glow puts world observatories under threat. In the past decade, the globally averaged rate of increase of sky glow was 2% per year, roughly double the rate of world population growth. Even the darkest remote observatory sites are impacted by human activity. In particular, solid-state LED lighting threaten astronomy through its higher blue content and low cost, which fuels demand for more light.

The IAU has defined the upper limit of artificial light contribution for a professional site adequate for true dark-sky observing to be 10% at an elevation of 45 deg in any azimuthal direction. The newest professional observatories are significantly below that limit. The goal of the model regulatory framework proposed for the UN is to slow, stop, and reverse the rate of increasing artificial skyglow specific to each major professional observatory in no more than a decade. The approach to achieving this goal is to design to match the illumination level to need, to limit unnecessary spectral content, to use precise optics to minimize spill light, and to employ active control to reduce light levels when usage is low. Professional observing sites should have a protected near zone of typically 30 km radius, within which both lighting levels and color rendition are sharply limited. For those urban areas within 300 km of observatories impacted by sky glow, the tightest limits on the range of recommended best practices and standards are applied, along with full shielding and curfews.

Provisions for the recommended regulatory framework are: 1. Exclusive use of luminaires with no light above horizontal; 2. Limiting lamps' spectral content in the blue region; 3. Limiting the brightness to the minimum required; 4. Implementing curfews and light level controls; 5. Making sure the light falls on the intended surfaces; 6. Designing lighting to minimize light propagating toward observatories; and 7. Lumen caps in the context of a regional lighting plan.

1.2. *The Impact of Satellite Constellations on Astronomy*

As the number of satellites continues to grow toward over 107,000 in the next decade, astronomy is facing a tipping point situation where increasing interference will jeopardize astronomical science. While regulatory and technical mitigations are possible, no combination can fully avoid impacts on astronomy. Astronomers are submitting a report to the United Nations on the need to create an international approach to equitably managing light and radio emissions from space and preventing undesired impacts. Without immediate action, all of humanity will lose a clear view of the Universe and its secrets. Without the implementation of the proposed recommendations to the UN, these satellites will be detectable by even the smallest optical or infrared telescopes, depending on the hour of night and the season. Moreover, up to several hundred satellites may be visible even to the unaided eye, particularly low on the horizon and in twilight hours.

Initial studies of the unmitigated effects of satellites show a variety of impacts on astronomy from minor to severe, depending on the nature of the telescope and satellite system. (Walker *et al.* 2020; Hainaut & Williams 2020; Ragazzoni 2020; McDowell 2020; Tyson *et al.* 2020). Observations with telescopes that view large portions of the sky will be severely impacted without substantial mitigations. While telescopes with narrow fields of view are less impacted, observations with long exposure times and particularly in the hours close to twilight and low on the horizon are still significantly affected. Wide-field astrophotography will be affected, too. The impacts are not limited to ground-based observatories; space-based telescopes in LEO will suffer as well, and in those cases, mitigations are more challenging to implement.

The recommendations toward mitigation solutions for the impact of satellite constellations on astronomy can be found in the full report and summary report for the UN mentioned in Section 1.

1.3. *Protection of Dark Skies Oases*

A dark sky oasis (also often referred to as a "dark sky place") is a location where the darkness of the night sky is protected by an outdoor lighting policy or other legal bodies. In the DSO, the cultural, scientific, astronomical, touristic values and the natural landscape are preserved.

The main bodies giving accreditation for dark sky oases are the International Dark-sky Association (IDA), based in Tucson, U.S. and the Starlight Foundation, based in Tenerife, Canary Islands. Worldwide, there are 223 dark sky places in 27 different countries, constituting a total land area of just over 20 million hectares. The International Union for the Conservation of Nature operates the Dark Skies Advisory Group (IUCN-DSAG) which aims are, among others, to preserve the ecological integrity of natural environments, ensure the full enjoyment of a wilderness experience and appreciate the integrity, character and beauty of rural landscapes.

The D&QS Oases Working Group proposes the following main recommendations: the levels of sky brightness considered to be appropriate for different dark sky places are the ones defined by the IUCN DSAG and, in addition, assume the default condition of no artificial light for all protected dark sky oases. Blue light should be avoided or minimized. Lighting should be strictly controlled and switched on only when it is needed. All exterior lights should only distribute light below the horizontal, and the upward light output ratio (ULOR) should be no minimized. LED lights should have a central management system (CMS) to reduce or extinguish light output in off-peak hours. No development in or near highly ecologically sensitive sites should be permitted. Monitoring of nighttime conditions in/near dark sky oases is encouraged. Active management of nighttime darkness as a natural source is encouraged through recognized conservation best practices. Finally, restoration plans should be implemented when sky brightness thresholds are routinely exceeded.

1.4. *Protection of the Bio-Environment*

The vast majority of life on Earth needs darkness at night to thrive. Many species of fauna and flora show strong sensitivity to daily light and dark cycles. The growing body of scientific research shows that ALAN causes significant negative effects the health of humans and flora and fauna, including changes in habitat use, migration, reproduction, predator-prey relationship, ecosystem functions and services (food sources, scavengers, pollination, water clarification), effects in the immune response, and fatalities at significant enough levels to pose extinction threats so some species. Put simply, ALAN is a risk factor for biodiversity.

The DNA in human cells include multiple "clocks" that work on circadian timescales and ultimately regulate many of our most important functions, including hormone secretion, sleep, digestion, and metabolism. Short-term effects on sleep and cognition are no longer in dispute. The effects of artificial light at night on humans include glare, retinal damage provoked from short wavelength blue light and melatonin suppression. The evidence being collected shows more prevalence of cancer (breast and prostate, maybe others), obesity (altered leptin and ghrelin), diabetes (glucose metabolism) and mood disorders (depression, bipolar).

Recommendations: Promote the adoption of environmentally friendly lighting regulations for regions, countries, municipalities and communities. The strategy should be summarized in the four pilar of the efficient lighting scheme order to prevent environmental harm: "The right light, at the right place, at the right amount, for the right duration". Specifically, reduce radiant flux (and irradiance), improve control of directionality of the radian-luminous flux and minimize melanopsin-activating blue content within the radiant/luminous flux.

2. Dark Skies Education and Heritage

2.1. *Globe at Night Citizen Science*

Everyone can undertake reasonable light pollution mitigation strategies, one of which is to participate in the Globe at Night program. Hosted by the NSF's NOIRLab, Globe at Night invites citizen-scientists worldwide to measure and submit (with their smartphone) their night sky brightness observations in an effort to raises awareness of the impact of light pollution. Other dark skies education programs and resources were mentioned during the associate event and URLs provided. The presentation also discussed what the issues are, what the solutions are and how to participate in Globe at Night to make a difference.

The international Globe at Night citizen-science campaign has run for 15 years. In that time over 660,000 people have been involved from 180 countries. Our team has given hundreds of workshops and developed educational programs, curricula and kits, hundreds of which were given at no cost to people in at least 50 countries (e.g., during IAU100). Globe at Night was part of a cornerstone in the International Year of Astronomy and the International Year of Light. Every year Globe at Night is involved with Global Astronomy Month as well as many other events worldwide. The program has made significant differences as a result in changing mindsets and encouraging action.

For more information, please visit http://www.globeatnight.org.

2.2. *Night Sky: A Scientific and Cultural Heritage*

Through the night sky, humanity managed to develop astronomy. The history of astronomy stand as a tribute to the complexity and diversity of ways in which people rationalised the cosmos and framed their actions in accordance with that understanding. This close and perpetual interaction between astronomical knowledge and its role within human culture underscores the importance of protecting dark skies.

The night sky is part of our cultural heritage, handed down from earliest times through to the present day. The recording and transmission of the astronomical activities is manifested in legends, folk tales, sacred landscapes, etc (https://www3.astronomicalheritage.net/index.php/tangible-fixed-heritage-category). These are the products of scientific activities in their cultural context.

The scientific and technological dimension of a starry night on the other hand is an essential part of the legacy of the sky. Dark skies are an essential condition to maintain windows to knowledge of the universe (Marin 2009).

Thus beyond the importance of the scientific and cultural legacy related to astronomy and starlight, there is a landscape dimension and the conservation of nature in relation with the beauty and the quality of the night sky, the related human activities and the nocturnal ecosystem. Thus the nightscape is of great importance and an international action in favor of intelligent outdoor lighting is urgently needed.

3. Summary

It takes a community which understands what is at stake to accomplish successfully a goal that benefits all of society. We are grateful to the organizers of IAU Symposium 367 for allowing us the opportunity to provide information that would lead toward that goal. For an update on future information, please stay in touch with IAU Commission B7 on the Protection of Observatory Sites or contact Connie Walker at connie.walker@noirlab.edu.

References
Hainaut, O. R., & Williams, A. P. 2020, *A&A*, 636, A121
Marin, C. 2009, *Proceedings of the IAU*, 5(S260), 449

McDowell, J. C. 2020, *ApJL*, 892, L36
Ragazzoni, R. 2020, *PASP*, 132, 114502
Tyson, J. A., *et al.* 2020, *AJ*, 160, 226
Walker, C., Hall, J., *et al.* 2020, *Bulletin of the AAS*, 52(2)

Education and Heritage in the era of Big Data in Astronomy
Proceedings IAU Symposium No. 367, 2020
R. M. Ros, B. Garcia, S. R. Gullberg, J. Moldon & P. Rojo, eds.
doi:10.1017/S1743921321000910

Let's turn off the lights and turn on the night: to the rescue of starlight in an age of artificial lighting

Andrea Sosa[iD]

Centro Universitario Regional del Este, Universidad de la República,
Ruta nacional 9 y ruta 15, 27000, Rocha, Uruguay
email: asosa@cure.edu.uy

Abstract. Our ancestors contemplated an inspiring night sky of science, philosophy, art ... today, it is estimated that one third of the world's population have never seen the Milky Way. The progressive degradation of the quality of the night sky due to an inappropriate use of the artificial light at night, as well to other sources of sky pollution, must be considered as the fundamental loss of a scientific, cultural and environmental heritage of humanity.

In this public talk we summarized the most relevant aspects of light pollution, the reasons for promoting good lighting to protect dark skies, and some of the initiatives at a global level that are being developed to preserve the darkness of the night sky.

Keywords. astronomical heritage, dark skies, artificial light at night, light pollution

1. Introduction

1.1. *What is meant by a dark sky?*

A dark night sky is the one that shines only naturally due to the weak emission of molecules from the Earth's atmosphere (known as 'airglow') and the brightness of faint natural sources of light such as the zodiacal light, *gegenschein* (the sunlight scattered by interplanetary dust particles concentrated around the ecliptic), the Milky Way, and diffuse celestial objects. Of course, the natural sky darkness must be evaluated in the absence of moonlight and twilight (so when the Sun is about 18 degrees or more below the horizon, so their refracted radiation by the atmosphere cannot longer reach our visual).

To asses the brightness of the sky the Bortle scale is commonly used, which subjectively characterizes the quality of the night sky in a given location, by giving an arbitrary class number to a given site and set of observing conditions (Bortle 2001). A sky brightness nomogram by H. Spoelstra † relates the Bortle scale to the visibility of the Milky way, the limiting visual magnitude, and the number of stars visible above the horizon. For instance, class 1 corresponds to a natural, unpolluted dark sky, with a Milky Way rich in details, an approximate visible stellar magnitude of 7, and between 5.000 – 6.000 visible stars at the hemisphere. A class 9 corresponds to the most polluted skies. For classes 8-9 the Milky Way is not visible, the approximate visible stellar magnitude is between 0 and 4, and the number of visible stars at the hemisphere is between 5 and 300.

As stated in the report by the Dark Sky Oases Working Group of the *Dark and Quiet Skies for Science and Society* Conference (see Section 2.2), a quantitative way of expressing sky brightness is to use physical units for surface brightness, given that the night sky can be considered as a hemispherical surface with the observer at the center.

† http://www.darkskiesawareness.org/nomogram.php

A popular unit of night sky brightness is magnitudes per square arc second (mag/sq arc sec). A star of visual magnitude 21 is a million times fainter than a naked eye star of magnitude 6. Although stars are essentially point sources of light, we can imagine that light being spread over a tiny square in the sky, whose side in angular measure is 1 arc second. If the whole sky had the same surface brightness as this tiny area, then the sky brightness would be 21 mag/sq arc sec. The darkest possible skies are about 22 mag/sq arc secs, and this figure represents a typical value of the natural airglow which is always present. A Bortle class 1 corresponds to a sky brightness of about 21-22 mag/sq arc secs. A sky which is 20 times brighter than this natural airglow background would be about 18.5 mag/sq arc secs and this would be a typical value in many urban environments. If the sky is 100 times brighter than the natural airglow background, then the brightness would be 16.7 mag arc secs, a value found in the central areas of the world's large cities with Bortle class 8.

1.2. *What is light pollution?*

Light pollution is caused by the excessive or inappropriate use of artificial light at night. As stated in the report by the Dark Sky Oases Working Group of the *Dark and Quiet Skies for Science and Society* 2020 Conference (see Section 2.2), Light pollution is "the sum of all adverse effects of artificial light at night, consisting of spill light emitted by a lighting installation which falls outside the boundaries of the property for which the lighting installation is designed and because of quantitative or directional attributes, gives rise to annoyance, discomfort, distraction, or a reduction in ability to see essential information". As a result of the artificial light scattered by the air molecules or aerosol particles in the Earth's atmosphere, the brightness of the night sky is increased causing the stars to be less visible as a result of reduced contrast.

Although concern about light pollution arose in the astronomical field, since the beginning of this century studies have been developed that show how light pollution also harms our health, wildlife and ecosystems. These problems, added to the waste of energy resources, show the importance of properly regulating artificial lighting at all levels.

1.3. *The darkness of the night sky as a scientific and cultural heritage in danger*

A dark sky is essential for astronomical observations carried out from ground-based facilities. Without dark skies, astronomers are unable to receive the faint signals of light from distant objects in outer space. Dark skies are also an important part of the cultural and natural heritage of all civilizations. Many astronomical observatories are built in remote locations in an effort to escape the light of cities and towns. Even so, these observatories are threatened by light pollution (Cheung 2018). Today, it is estimated that a huge fraction of the world's population – including millions of children – have never seen the Milky Way. The progressive degradation of the quality of the darkness of night sky must be considered as the fundamental loss of a scientific, cultural and environmental heritage of humanity. The dark night sky must be regarded as an endangered natural resource.

2. Overview

2.1. *Good reasons to fight against light pollution*

In addition to the brightening of the night sky which harms observational astronomy, the light pollution of the night sky has adverse effects on human health, the bio-environment and biodiversity, as documented by the International Dark-Sky Association (see Section 2.2) and the World Health Organization. For a comprehensive review of publications on this topic see the report of the Bio-Environment Working Group of the *Dark and Quiet Skies for Science and Society* 2020 Conference. Light pollution can be regarded as one of the most widespread forms of environmental pollution. It also causes electricity waste (so energy waste), as it is evidenced by the *New World Atlas of Artificial Night Sky Brightness* (Falchi *et al.* 2016).

Circadian rhythms of roughly 24 hours regulate many of the most important functions of the human body, such as hormonal secretion, digestion, sleep, and metabolism. Many species of fauna and flora show strong sensitivity to the daily light-dark cycles imposed by the earth's rotation, with a vast majority of animals (including crucial pollinators, many mammals and migratory birds) being nocturnal. It should come as no surprise that the uncontrolled sweep of the night darkness due to the increasing light pollution alters the physical and psychological health of humans and animals. For instance, the exposure to light at night decreases the production of melatonin, which as an agent of the immune system helps to suppress many hormonal cancers.

Despite their better energy performance and more modern technology, light-emitting diodes (LEDs) do not help decrease light pollution (Cheung 2018). The low cost white LEDs have a strong blue component in their spectrum, which scatters easily into the atmosphere. For this reason, traditional outdoor Sodium lamps are better to reduce light pollution. These warmer colored lights also have the advantage of a spectrum with relatively narrow bands that can be filtered during astronomical observations.

Over the centuries, astronomy has evolved, becoming one of the sciences that most uses and promotes the development of space exploration and new technologies. Paradoxically, some of these technological developments are turning against it. The sky is also polluted by artificial radio signals, including the increasing contribution of electronic devices such as cell phones, sensors, etc. The aircraft routes also constrain the astronomical sites. But recently a bigger threat has emerged: the mega constellations of satellites. These are large conglomerates of telecommunications satellites in low-altitude orbits that would be starting operations at the end of 2020, and whose brightness and frequency will have a huge negative impact on scientific astronomical images as well to the night sky landscapes. Mitigation techniques are currently under study, such as reducing the brightness of satellite surfaces, or the development of algorithms to remove satellite trails from CCD images.

2.2. *Global Initiatives to preserve the darkness of the night sky*

Two leading international organizations dedicated to protecting the night sky and promoting astronomy-related activities are the *International Dark-Sky Association* (IDA)†; a non-profit organization founded in 1988 based in Arizona, and the *Starlight Foundation*‡; a non-profit organization created in 2009 by the Canary Islands Astrophysics Institute, supported by the UNESCO, the IAU and the UNWTO, to promote the *Starlight Initiative*§. The International Astronomical Union (IAU) has Commissions and Working Groups on Dark and Quiet Sky Protection under Division C.

La *Ley del Cielo* (Law of the Sky; Law 31/1988) of Canary Islands was a pioneering law in defense of the night sky. The United Nations Office for Outer Space Affairs and Spain, jointly with the IAU, organized on October 2020 the *Dark and Quiet Skies for Science and Society* Conference¶. The event will result in a document that describes what measures Governments and private enterprises can adopt to mitigate the negative

† https://www.darksky.org/
‡ https://www.fundacionstarlight.org/
§ https://www.starlight2007.net/
¶ https://www.unoosa.org/oosa/en/ourwork/psa/schedule/2020/2020_dark_skies.html

impact of technological implementations on astronomy (e.g. urban lighting, radio broadcasting and satellite constellations' deployment) without diminishing the effectiveness of the services they offer to citizens. The final outcome document, intended to become a reference to further analysis of the situation, will be presented to the intergovernmental Committee on the Peaceful Uses of the Outer Space (COPUOS) for consideration.

The International Dark-Sky Association proposes five principles for responsible outdoor lighting: 1. All light should have a clear purpose, and before installing or replacing a light, determine if light is needed, and its impact on wildlife and the environment; 2. Light should be directed only to where needed, and use shielding and careful aiming; 3. Light should be no brighter than necessary; 4. Light should be used only when it is useful, and use controls such as timers or motion detectors if needed; and 5. Use warmer color lights where possible, and limit the amount of shorter wavelength (blue-violet) light to the least amount needed.

Excessive lightning, or poorly designed outdoor lighting, can cause glare and therefore unsafety. We must bear in mind that good outdoor lighting is not about lighting *less* but about lighting *better*.

Education and outreach are essential to create public awareness of the importance of promoting good lighting and preserving the darkness of the night sky. In this sense, some of the most successful and recognized citizen science projects, with the endorsement of the IAU, are: the *International Year of Astronomy Dark Skies Awareness*†, *Globe at Night*‡, the *International Year of Light Quality Lighting Teaching Kit*§, and the *IAU100 Dark Skies for All*¶.

2.3. *Dark sky oases and astro-tourism*

Dark sky oases, also known as dark sky places, are areas where the night sky has some form of legislative protection from the effects of the artificial light. Such protected areas are certified by a internationally recognized accreditation organization, mainly the International Dark-Sky Association and the Starlight Foundation. Dark sky oases can be used for different purposes such astronomical research, astro-tourism, heritage values, wilderness areas for public education and outreach, etc. The Aoraki Mackenzie International Dark Sky Reserve in New Zealand, with IDA accreditation, became in 2012 the first recognized dark sky oasis in the southern hemisphere, and the largest in the world. La Palma Starlight Reserve in the Canary Islands was the first to be certified by the Starlight Foundation in 2012.

Astro-tourism can be regarded as an innovative form of sustainable tourism based on a natural, free and infinitely renewable resource, provided that the darkness of the night sky is preserved. Contributes to revalue the cultural, historical and environmental heritage, constitutes an engine of social-economic development for rural areas and creates quality employment, besides promoting astronomy and science in general.

References

Bortle, J. 2001, *Sky and Telescope*, 101, 126
Cheung, S. (Ed.) 2018, *IAU Journal of Light Pollution*. International Astronomical Union Office for Astronomy Outreach and National Astronomical Observatory of Japan. April 2018.
Falchi, F., Cinzano, P., Duriscoe, D., Kyba, C., Elvidge, C., Baugh, K., Portnov, B., Rybnikova, N., & Furgoni, R. 2016, The New World Atlas of Artificial Night Sky Brightness, *Science Advances*, 2, 1–25

† http://www.darkskiesawareness.org/
‡ www.globeatnight.org
§ www.noao.edu/education/qltkit.php
¶ https://darkskies4all.org/

Education and Heritage in the era of Big Data in Astronomy
Proceedings IAU Symposium No. 367, 2020
R. M. Ros, B. Garcia, S. R. Gullberg, J. Moldon & P. Rojo, eds.
doi:10.1017/S174392132100106X

The search for extraterrestrial intelligences and the Fermi Paradox

Nikos Prantzos🅾

Institut d'Astrophysique de Paris

Abstract. Of all the questions that Man asks himself about the Universe, the one concerning the possibility of the existence of an extra-terrestrial life form, and even more of an extra-terrestrial civilization, is probably the most fascinating. We cannot answer it today, but we can imagine the implications for the human species and its future in the Universe.

1. The debate on the "Plurality of Worlds": chance or necessity?

Advances in 18th century astronomy and the understanding that stars are suns like ours reinforced the idea that countless Inhabited Earths exist in the Universe. However, at the beginning of the 20th century, the debate on the Plurality of worlds was enriched by arguments inspired by biology. Alfred Russel Wallace, the co-founder with Charles Darwin of the theory of evolution, was the first to use this kind of argument against the notion of another intelligent life form in the Universe. In the 1905 edition of his book *Man's Place in Nature*, Wallace noted that Man is the result of a series of unique and unpredictable events in the long chain of evolution. The likelihood of this same series of events happening elsewhere, even in Earth-like environments, is remote. This argument also applies to intelligent life.

Adopted by many biologists, Wallace's argument introduced into the debate on the Plurality of worlds the "sense of history": a series of individually unimportant events, the effects of which are amplified over time to the point that the end result becomes completely unpredictable. Indeed, the traditional presentation of Darwinian evolution emphasizes the progressive complexification of matter, as if it were an inevitable process. The passage from bacteria to multicellular organisms, from fish to reptiles, and from mammals to humans is considered a one-way street. Along this path, natural selection rewards those who adapt best to their environment with the survival of their lineage. However, as the American biologist Stephen Jay Gould points out, this conception of evolution can be totally wrong. Natural selection is not the only factor determining the evolution of species, and it does not always take small steps. Catastrophic phenomena have wiped out species that seemed well equipped to survive by natural selection.

The most famous example is undoubtedly that of the dinosaurs: after a reign of 130 million years, these "terrible lizards" disappeared 65 million years ago, due to the collision of the Earth with a large asteroid. The survivors of such disasters in Earth's history did not always display greater complexity than those who disappeared, and their comparative advantage was not, a priori, obvious. From this point of view, mammals owe their survival only to their good fortune and not to any "superiority" over dinosaurs. Is then the emergence of man and of intelligence, during these last millions of years, a matter of pure chance? These considerations have extremely important implications for the existence of other intelligent life forms in the Universe.

2. The number of extra-terrestrial civilizations is... N

The scientific study of the subject "Extra-terrestrial Intelligences" has a short history; it dates back only about sixty years. In an article published in 1959 by the British journal *Nature*, American physicists Giuseppe Cocconi and Philip Morrison suggested that the microwaves (radio waves of high frequencies) are the best means for interstellar communication. Micro waves penetrate not only the Earth's atmosphere but also the clouds of gas and dust pervading the Galaxy. In contrast, visible photons, our traditional "window" to the Universe, are absorbed by these clouds; thus, optical telescopes see much less far into the Milky Way's disk than radio-telescopes. Moreover, microwaves have another advantage: they carry little energy, which means that sending a message by this type of wave is preferable from an energy point of view.

These considerations inaugurated the modern era of the debate on the Plurality of worlds by opening the perspective of a scientific study of the problem. It was from this period that the acronym ETI (Extra-Terrestrial Intelligence) was born. The first to put these ideas into practice was American astronomer Frank Drake, who started in 1960 the first programme of radio-signal research with the radiotelescope of Green Bank, USA. In 1961 Drake organized the first conference on radio communication with extra-terrestrials at Green Bank. Preparing the agenda of the conference, he tried to assess the likelihood of success of that research, by attempting an estimate of the number of technological civilizations today present in our Galaxy. His formula, the famous "Drake Equation", has generated a phenomenal amount of work and analysis over the past half century.

Drake's equation describes a steady state, where the number of communicating civilizations remains approximately constant (those that disappear are replaced by an equal number of new ones). This number N is given by:

$$N = R_\star f_{\text{PLA}} n_{\text{e}} f_{\text{LIFE}} f_{\text{INT}} f_{\text{TEC}} L \qquad (1)$$

where $R_\star$ is the star formation rate in the Milky Way (number of new stars formed per year), f_{PLA} gives the fraction of stars with planets, n_{e} is the number of telluric planets in the continuously habitable zones of these stars (at distances where the heat of the star allows temperatures favoring the presence of liquid water) for billions of years (so that complex life has time to appear), f_{LIFE} is the fraction of habitable planets on which life actually appeared, f_{INT} provides the fraction of these planets where evolution has produced intelligent beings, f_{TEC} is the fraction of planets whose beings are capable of communicating by radio signals, and L represents the average lifespan of these technological civilizations.

It is clear that Drake's formula has no predictive power: only the first three factors are more or less known today, which is a definite progress compared to the XXth century, where only the first one was known. The value of $R_\star$ is estimate at around 10 stars per year and has not changed much over the last billion years. For f_{PLA} and n_{e}, the latest statistics on the number of terrestrial planets detected around nearby stars, through the observations of the NASA's space telescope *Kepler* in the last decade suggests that such planets occur around 10% of solar type stars, which make up one tenth of the total number of stars. Thus, the product of these three factors in an astrophysical term $R_{\text{ASTRO}} = R_\star f_{\text{PLA}} n_{\text{e}}$ roughly gives 0.1 telluric planets formed each year in the Galaxy.

By grouping the following three into a biotechnological factor $f_{\text{BIOTEC}} = f_{\text{LIFE}} f_{\text{INT}} f_{\text{TEC}}$, the Drake equation is written: $N = R_{\text{ASTRO}} f_{\text{BIOTEC}} L$ (see **?**). The advantage of this writing is to allow the visualization of the impact of factors f_{BIOTEC} and L, after having fixed the value of R_{ASTRO} by observations. By assigning f_{BIOTEC} its maximum value (= 1, being the product of three fractions), Drake's formula is simplified to $N = 0.1L$: the *maximum* number of civilizations in the Galaxy is one tenth of their lifespan, expressed in years. But it is expected to be much lower, because f_{BIOTEC} can only be (much) smaller than 1.

Contrary to what is often claimed, the value $N < 1$ (N less than 1) is allowed, but it has only a statistical significance. It means that the emergence of a civilization is a rare event in the Galaxy, with two successive civilizations being separated in time by a period longer than their lifespan. Thus, "alone in space" does not necessarily mean "alone in time" nor "to be the first", nor "to be unique". Thousands of technological civilizations appeared, perhaps, in the Galaxy, lived a long time – thousands or millions of years – and even colonized their neighborhood, being "alone" in the Galaxy during that time; and they disappeared, unable to communicate with others, unaware that others have preceded them and unknown to their successors. Our own civilisation may just be such a "lonely heart" in the Milky Way.

So far, ETI research has yielded two results, one likely definite, the other perhaps provisional. The probes sent to explore our Solar System did not signal any form of life in our close vicinity; however, it is not yet ruled out that microscopic life forms have appeared in the past on Mars or that they exist today in the icy oceans of certain satellites of giant planets, such as Europe or Enceladus. In addition, listening to the sky in radio frequencies did not result in any detection of an extra-terrestrial signal; taking into account the difficulty of the task (where to look? in what frequency? how long? with what sensitivity? etc), this result is not surprising. However, even if we manage to listen to the hundred billion stars of our Galaxy over ten billion radio channels for several centuries, what conclusion could we draw from the absence of an artificial signal? Quite simply, that none of these hypothetical civilizations is currently broadcasting in our direction, which does not really settle the debate on the existence of ETI.

3. The Fermi Paradox

There is another fact of observation, the importance of which is difficult to measure: the absence of the slightest trace of an ETI on our planet or in the Solar System. The late 1940s saw the first wave of reports of flying saucers and other unidentified flying objects (UFOs), especially in the United States. During a visit to the Los Alamos military laboratory in 1950, the Italian physicist Enrico Fermi –Nobel Prize winner in Physics– engaged in a discussion on this subject with his colleagues. Discussion shifted to the more general topic of extra-terrestrial civilizations and interstellar travel. "But where are they?" Fermi suddenly asked his interlocutors, meaning that if "they" exist, "they" should have visited us several times already in the past. According to Fermi, the absence of traces of such a visit did not necessarily imply the non-existence of extra-terrestrials; it could result either from the impossibility of interstellar travel, or from the too short lifespan of a technological civilization, probably self-destroyed after the discovery of the secrets of the atom (the period of "the equilibrium of terror" between the United States and the Soviet Union had just begun at the time).

This discussion remained virtually unknown for a long time. In 1975, astronomers Michael Hart and David Viewing rediscovered independently Fermi's arguments. Hart radically concluded that the absence of extra-terrestrials on Earth meant that we are the only technological civilization in the Galaxy and therefore the search for radio signals would only be a waste of time and money. It was after this provocative article that the subject was baptized "the Fermi paradox". Hart's pessimistic conclusions opened a period of passionate debate around ETI, particularly in the United States, a controversy that continues to this day.

4. Cosmic loneliness?

Any paradox rests on the invalidity of one (at least) of its underlying hypotheses, but it is impossible to present here all the arguments of the supporters and opponents of ETI on the Fermi paradox. The arguments most often discussed do not concern the "physical" aspect of the problem (feasibility of interstellar travel, construction of self-reproducing robots, etc.) but its "sociological" aspects. Some think that extra-terrestrials are not interested in space travel or expansion in the Galaxy: their civilization would quickly turn to spiritual values (contemplation, meditation, etc.), or it would have adopted the "zero growth" dear to certain common e environmentalists, which would have prevented them from space colonization. Others, like Fermi, fear that the longevity of a technological civilization is too short for any significant colonization of its vicinity.

Another class of sociological arguments, generally known as the "zoo –or quarantine– cosmic hypothesis" was formulated in the early twentieth century by the father of astronautics Konstantin Tsiolkovsky, and independently rediscovered in 1984 by American astronomer John Ball: the aliens would have arrived in our solar system, in the recent or distant past, but would limit themselves to observing us from afar for various reasons: they would consider us too "primitive" and would wait for our "maturation" to include us in their "galactic community" (according to Tsiolkovsky), or they would not want to interfere with our development (principle often invoked in the famous television series Star Trek under the name of "prime directive"). According to a variant of this argument, extra-terrestrials would have even contributed to the development of intelligence in our ancestors (the most famous version being undoubtedly that proposed by the science fiction writer Arthur Clarke and his famous "black monolith", appearing in Stanley Kubryk's movie *2001, A Space Odyssey*).

There is a common weak point in all sociological arguments. It is hard to accept that they apply to *all* extraterrestrial civilizations, without any exception. If hypothetical civilizations are numerous, at least one should have escaped annihilation, mastered space travel, and embarked on a galactic colonization program. The behavior of animal species on Earth shows us that they always go through a phase of expansion, favored by natural selection, because it maximizes their chances of survival. Moreover, at least one of these civilizations should have transgressed the "taboo" of avoiding all contact with our own. If none of them did, we would be "atypical", because we would be the only ones wanting to communicate with other civilizations. But it may be that the typical lifespan of civilizations, even as great as several million years, is too short to allow them to explore a large enough fraction of the Milky Way and find us.

The Plurality of Worlds is more controversial today than ever. The arguments on both sides ("We are unlikely to be alone in this vast Universe" and "Where are they?") are of statistical nature. Therefore, their value is extremely low, since one cannot make statistics on the basis of a single known case, life on Earth. The detection of an inhabited planet – and even more of an extraterrestrial civilization – would constitute one of the major events in the history of the human species. The non-detection of ETI signals, even after several centuries of research, would not prove the non-existence of extra-terrestrial civilizations. It should, however, prepare us for a life of cosmic solitude ...

Reference

Prantzos, N. 2020, *MNRAS*, 493, 3464–3472

Education and Heritage in the era of Big Data in Astronomy
Proceedings IAU Symposium No. 367, 2020
R. M. Ros, B. Garcia, S. R. Gullberg, J. Moldon & P. Rojo, eds.
doi:10.1017/S1743921321000776

Sunspot observation by the cooperation of amateur astronomers and researchers in Japan in early 20th century as early citizen science program

Harufumi Tamazawa[1,2] 🆔

[1]Fucalty of Fine Arts, Kyoto City University of Arts,
13-6 Kutsukake-cho, Oe, Nishikyo-ku, Kyoto 610-1197 Japan

[2]Graduate School of Letters, Kyoto University,
Yoshida Honmachi, Sakyo-ku, Kyoto 606-8501, Japan
email: `tamazawa@kwasan.kyoto-u.ac.jp`

Abstract. The development of astronomy has been developed by the cooperation of amateur astronomers and researchers. Sunspot observation is a good example of Extreme citizen science in early days. Issei YAMAMOTO (1889–1959), organized "Oriental Astronomical Association (OAA)," Yamamoto's materials (now in Kwasan observatory) include solar observation data sent from many observers in Japan. From the viewpoint of today's Citizen Science, collaborative observation of sunspot between researchers of solar physics and amateur astronomers in Japan has clearly a context of social mission rather than mere academic interest. From the viewpoint of science communication, we can see that Yamamoto's call includes a social mission to promote astronomy in Japan, and that amateurs responded to Yamamoto's call by participating in the observation network. It can be said that this collaboration have not only "cultural" aspect but also "civic" or "practical" aspect.

Keywords. history and philosophy of astronomy, sunspots

1. Introduction

Citizen science, collaborative research between professional researchers and general public, is becoming more and more popular. Along with environmental science, history and biology, astronomy is one of the most popular fields of citizen science. Various researchers have pointed out the hierarchy of citizen science. Haklay (2013) categorized citizen science according to the depth of participation: crowdsourcing, distrusted intelligence, participatory science, and "extreme citizen science". At the stage of extreme citizen science, participants of citizen science project do not only cooperate in collecting data, but also think together with researchers, identify problems, analyze data, and write papers. The collaboration between professional researchers and amateur observers in astronomical observation is one of the best examples of extreme citizen science. The discovery of supernovae, novae, and comets by amateur observers and detailed analysis by professional researchers has led to several papers with amateur astronomers as co-authors (for example, Abdo *et al.* 2010; Kawabata et al. 2010). It is important to examine the participants of Citizen Science to understand why they participate in the project in order to make future projects more useful. Ikkatai *et al.* (2020) investigated people's willingness to participate in citizen science, focusing on their level of interest in science and technology in Japan, and pointed out that "Contribution," "stimulation of intellectual curiosity," and "latest knowledge" were identified as motivating factors. Shen (1975) pointed out three categories of scientific literacy (practical literacy : knowledge for solving practical problems, Civic literacy: Understanding of scientific terms, concepts, and methods, and Cultural literacy: the ability to use scientific information as intellectual entertainment). It is useful to analyze the three categories of scientific literacy from the participants' behavior. The participants in professional-amateur joint observations in astronomy nowadays are clearly participating with cultural literacy. It is useful to look at the history of the establishment of the system to understand why it became this way. In this paper, I will examine how amateur astronomy was created in Japan, and consider the prospects for citizen science, from the viewpoint of scientific literacy.

2. Issei Yamamoto and Amateur Astronomers in Japan

Issei Yamamoto was the first director of Kwasan Observatory of Kyoto University. Kwasan Observatory was open to the public from its inception in 1929, a pioneering move at that time. Staffs of Kwasan Observatory, researchers of astronomy, has been actively involved in science communication activities from the early stages of its opening, for example, by holding observing sessions for the public. In addition to public outreach, Yamamoto cultivated and interacted with amateur astronomers in Kwasan Observatory. These are Kwasan Observatory is said "sacred place for amateur astronomers in Japan." In 1920, before the opening of Kwasan Observatory, Yamamoto organized "Oriented Asia Astronomical Society (OAA)", an organization of amateur astronomers in Japan, and worked to foster amateur astronomers. In 1940, after retiring from Kyoto University, Yamamoto established a private observatory, which is now called Yamamoto Observatory. There are many records of his interactions with amateur astronomers at Yamamoto Observatory, which are now in the possession of Kwasan Observatory. Among them are sunspot records by amateur astronomers from all over Japan. Amateur astronomers sent their observation records to Yamamoto through the OAA, and Yamamoto and his colleagues compiled the records.

3. Sunspot Observation by Amateur in Japan in early 20th

Reports on sunspot observations in the Yamamoto Observatory were archived, and the oldest data is in 1935. Reports of sunspot observation in 1935 were submitted in a format prepared in advance (Fig.1). Looking at the records collected in 1943, we can find not only formatted papers, but also scraps of notebooks, postcards, and letterheads. In addition to the observation rules, there are also personal letters from the observers to Yamamoto. Looking at these records, we can see that the observers were trying to make observations even under the WWII.

In the first issue of "The Heaven" (In Japanese, "Tenkai"), a member magazine of OAA, it was stated that in order to carry out continuous observations, not only the United States and Europe but also Asia needed to participate in the observations. A cooperative system is essential for continuous observations, and it must be organized (Yamamoto & Furukawa 1920). As for sunspot observations, it takes time to obtain sunspot data collected in Belgium at that time, so it is desirable to observe sunspots by oneself if it is possible. Yamamoto's comments (Yamamoto 1926) imply that the training of amateur astronomers was necessary for astronomical observation in Japan, and that there was an practical and civic aspect. This is in contrast to today's activities, which have a strong cultural aspect.

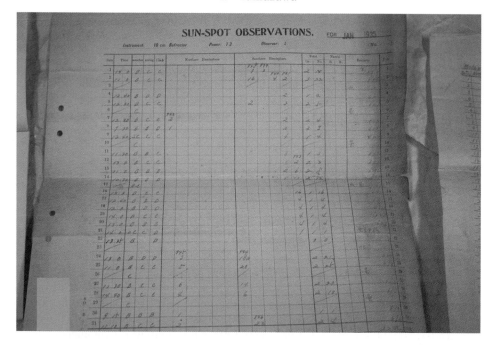

Figure 1. An example of sunspot observation record written in 1935.

4. Archiving of historical observation records

Unraveling the history of pro-amateur collaboration in astronomy is a perfect example of extreme citizen science, therefore by using them we can think about the future of science communication and citizen science. The International Astronomical Union (IAU) has pointed out the importance of archiving historical observation records (IAU 2018). Historical observation records are not only useful for astronomy itself, but also for understanding the history of astronomy, the social theory of science and technology, and the history of science communication.

This work was supported by JSPS KAKENHI Grant Numbers JP18H01254 (PI: H. Isobe) and JP20K20338 (PI: K. Yamori).

References

Abdo, A. A., et al., 2010, *Science*, 329, 817.
Haklay, M. 2013 Citizen science and volunteered geographic information: Overview and typology of participation, in Sui, D., Elwood, S., and Goodchild, M. (eds.), *Crowdsourcing geographic knowledge* (Springer), pp. 105–122.
International Astronomical Union(IAU) 2018, IAU 30th GA RESOLUTION B3
Ikkatai, Y., Ono, Y., Udaka, H., & Enoto, T 2020, *Japanese Journal of Science Communication*, 27, 57–70
Kawabata, K. S., Maeda, K., Nomoto, K., et al. 2010, *Nature*, 465, 326.
Shen, B. S. *Science literacy and the public understanding of science.* In Communication of scientific information (pp. 44–52). Karger Publishers. 1975
Yamamoto, I, & Furukawa, R. 1920, *The Heavens*, 1,1,1-2
Yamamoto, I. 1926, *The Heavens*, 6, 63, 1–22

Education and Heritage in the era of Big Data in Astronomy
Proceedings IAU Symposium No. 367, 2020
R. M. Ros, B. Garcia, S. R. Gullberg, J. Moldon & P. Rojo, eds.
doi:10.1017/S1743921321000569

Solar eclipses in India's cultural and political history

R. C. Kapoor🄳

Indian Institute of Astrophysics, Koramangala, Bengaluru 560034, India
email: rckapoor@outlook.com, rck@iiap.res.in

Abstract. This contribution presents earliest records in the Indian stone inscriptions and literature that specifically mention the eclipse as total or annular and the eclipses that timed with wars. Solar cult temples can be found all over India. In a few, the assigned dates coincide or are close to the dates of solar eclipses of large magnitude in the area.

Keywords. Solar eclipses, Indian Stone inscription records, Solar Cult Temples, War Eclipses

1. Introduction

Eclipses have been an integral part of India's cultural and political history and occasions for the Hindus to engage in ablution and charity with hope for a better after−life. The first right words on eclipses came from Aryabhatta (476−550 CE) in his monumental work Aryabhatiya (499 CE). He showed how the eclipses are a play of shadows of the Moon and the Earth and gave an algorithm for their prediction and calculation (Shukla 1976 : Ch IV). Only rarely did the classical astronomy works carry actual observations of eclipses. As an exception, the Siddhantadipika by Parameśvara (ca. 1380−1460 CE) carries observations of solar and lunar eclipses that he himself observed through 1393−1432 CE (Hari 2003). The first reference to annular eclipse in an Indian text is found in the Brhat Samhita (BS; 505 CE), an encyclopaedic work by Varahamihira (485−587 CE) where he mentions of a madhye tamah − obscuration in the middle (BS: V:51; Bhat 2010). Such a statement is possible only if an annular eclipse had been actually seen. There were two such that happened near Ujjain where he lived, in 496 CE and 550 CE. No astronomer ever predicted an eclipse as annular or total in the region. That is why there is hardly a record where the eclipse is specifically referred to as being annular or a total. The Tantrasangraha (1501 CE; Ramasubramanian & Sriram 2010) of Nilakantha Somayaji mentioned that when the angular diameter of the Sun is greater than that of the Moon, the eclipse is annular. The eclipse calculation acknowledges this disparity but stops short of eliciting how it may arise.

2. The earliest Indian records

The eclipses help fix the timelines in history arrived at by other means. There are innumerable stone and metal inscriptions recording genealogy, religious activity, heroic acts, gifting of land made on the cardinal days or on eclipses to increase religious merit, etc. from the middle of the first millennium (Shylaja & Ganesha 2016). Except for the dates, there is no eclipse phenomenology to be found in the records.

Presently, the earliest record of a solar eclipse belongs to the Nagardhan copper plates of the king Svamiraja, found in 1948. It mentions land donation made on the Chaitra

Figure 1. Virupaksha Temple, Pattadakal (Photo: Arian Zwegers; Wikimedia Commons).

amavasya in the Kalachuri year 322 that corresponds to the eclipse of 19 Mar. 573 CE (Mirashi 1949–50). It was a total over central India but mentioned nowhere so. Nagardhan narrowly missed the totality. The earliest Indian record specifically stating an eclipse as total is about that of 754 CE. It is on an inscription on a monolithic pillar of red sandstone in the Virupaksha Temple at Pattadakal (15.95°N, 75.82°E; Fig. 1) from the period of the Chalukyan king Kirtivarman II. Fleet 1894–95 suggested the total eclipse in the inscription as of 25 June 754 CE. A reference like 'total' can come from an actual viewing only. It turns out that Pattadakal actually witnessed the eclipse. It lay within the path of totality.

There is also a reference to an annular eclipse in an Indian literary classic of the 8th Century. The country in North India between the Ganges and Yamuna rivers had the prominent kingdoms of Thanesar (Kurukshetra) in the west and Kannauj to the east. The respective kings Lalitaditya and Yashovarman (r. 725−752 CE) were contemporary. Vakpatiraja, a poet in Yashovarman's court composed a historical poem Gaudvaho in Prakrit (735 CE). While eulogizing his king, the poet brings in eclipse of the Sun as a simile (Suru 1975: 123 / Eng. part: 92−93). The reference is not just literary but quite probably the record of a solar eclipse. What makes it exceptional is its being portrayed as an eclipse with the characteristics of an annular − "The sun's orb, pierced by (the black body of) Ketu, thus looking (from the earth) like a hole (vivarabha), drops down in the sky". Such a description of an eclipse in an Indian literary work is unprecedented that its author, quite likely, may have had a chance to see. (Jacobi 1888) identified it as the eclipse of 14 Oct. 733 CE. It was actually on 14 Aug. 733 CE. Kannauj (27.018°N, 79.912°E), 40 km off the northern fringe of the path saw the eclipse as partial, but with a large magnitude at 0.963.

3. Solar cult temples

The Sun−god has been worshipped in India since antiquity. Solar cult temples can be found all over India. Many are near the Tropic of Cancer. In some cases, the assigned

Figure 2. The mural at the Gurudwara Baba Atal Sahib Ji, Amritsar; Guru Nanak is at lower left and the eclipsed sun near top−left (Photo: R.C. Kapoor, March 2017).

dates coincide with or are close to the dates of solar eclipses of large magnitude. One such was at Dashapura (Mandsaur) in the Malva region, built under the king Bandhuvarman in 436−37 CE, the contemporary of the Gupta Emperor Kumaragupta I (r. 414−455 CE). A guild of silk weavers had migrated from the Lata district to Dashapura having been attracted by the virtues of the kings of the country. Here, with great feeling of reverence and gratitude, they built a 'House of the bright rayed Sun' in 436−37 CE (Fleet 1886). Interestingly, just a year before, an annular eclipse had occurred on 14 Feb. 435 CE. Mandsaur (24.037°N, 75.077°E) lay close to the central line of the path and saw annularity lasting over 8 minutes.

4. Kurukshetra: the eclipse capital

Traditionally, Kurukshetra is the most benefic place for ritualism during a solar eclipse. A dip here on the day is believed as a commemoration of the eclipse that occurred during the Mahabharata war. Beginning the 11th Century, Kurukshetra received important patronage and turned into a centre of pilgrimage and solar eclipse festivals.

Guru Nanak (1469−1539 CE) was a great poet-saint and the founder of Sikhism. His teachings are enshrined in the form of poetry and extensive dialogue with the learned in the Ādi Guru Granth Sāhib (ĀGGS). The account of his life and teachings is found in the later traditional works called the Janamsākhis. He was a reformer who sought to dispel superstition. The janamsaākhis refer to a solar eclipse that occasioned when Guru Nanak was visiting Kurukshetra. He used the occasion to address the pilgrims gathered there and sought to dispel their fears and superstitions associated with eclipses saying that it was a celestial phenomenon that had no influence on people's affairs on the Earth. Examining the eclipses in the relevant period of 1498−1521 CE with respect to other records, the one of 13 Jan. 1507 CE appears to be the most probable, with mag. 0.43 over Kurukshetra (Kapoor 2017). The incident is beautifully depicted in a 19th century mural in the Gurudwara Baba Atal Sahib Ji at Amritsar (Fig. 2).

5. The Mughal records

Political astronomy is about the role and the influence unexpected celestial phenomena make on the royalty or the people's life and their response thereto. These were regarded ill omens for the rulers. On the eclipse day, the Mughal Emperor would be weighed against grain, butter, etc. to be given in charity, right from Akbar's time until Bahadur Shah Zafar in the 1850s. The Emperor Akbar (r. 1556–1605 CE) abstained from meat on the first day of the Solar month, on Sundays and the days of lunar and solar eclipses, etc. to respect the Hindu sentiments. The Akbarnama (II: 422–23) speaks about a bloody conflict among two rival factions of Hindu ascetics over right to their space at Kurukshetra's holy tank, the occasion being the solar eclipse of 09 Apr 1567. Jahangir, the fourth Mughal Emperor (r. 1605–27 CE), had interest in astronomy. He has recorded in his journal Tuzuk−i Jahangiri a few solar and lunar eclipses and two bright comets of Nov 1618 that he observed. The solar eclipses are of 15 Dec. 1610 and 29 Mar. 1615. The latter he says lasted 3h 12 min, reaching a magnitude of four fifths of the Sun. His figures are very close to the modern computed value of 0.794 for the magnitude and a nearly 3h duration of the eclipse as at Agra.

6. War eclipses

Eclipses during wars have influenced the rulers, the warriors and lives of people. In India, historians have seldom associated eclipses with wars. A total solar eclipse of 17 Oct. 1762 CE that occasioned on the day of the festival of lights has been thought by some historians to have cast a decisive impact on the course of history in Punjab over which the path of totality passed. This surely was a War Eclipse. Ahmed Shah Abdali (1722–72 CE), acclaimed one of the greatest warriors of Asia, was in those times attempting to establish Afghan rule in Punjab. The incident in question happened while the invader came over to Amritsar with 60000 strong army to decimate the Sikhs who, numbering about 50000, had gathered there to face him. A fierce battle took place but an unexpected darkening noon forced an early retreat by the Afghans to Lahore. The eclipse at a magnitude 0.99 had its impact. All one can say is that the eclipse ended the war early which otherwise would have caused even more bloodshed. It possibly saved Abdali's life too (Kapoor 2010).

In relatively modern times, one such situation emerged during India's Uprising against the British in 1857 that shook the foundations of the colonial power in India. Delhi fell to the British on 21 Sept. 1857 CE. Days before, there occurred a solar eclipse on 18 September, an annular. Over Delhi it reached 90% obscuration just when the war against the British had reached a very critical stage. The eclipse had its impact on the morale of the Indian soldiers fighting a pitched battle in Delhi. The eclipse preceded Mughal Emperor Bahadur Shah Zafar's capture from Humayun's Tomb by Captain William Hodson by three days. Under the circumstances, the September 1857 CE eclipse may be called a War Eclipse (Kapoor 2018).

The eclipses continue to be interwoven with the life and culture in India.

This research has gratefully used the "Eclipse Predictions by Fred Espenak, NASA/GSFC Emeritus". I am thankful to the IAU, the Director, Indian Institute of Astrophysics, and to Mr Manu Kapoor and Dr. V. Muthu Priyal for help.

References

Bhat, M.R. 2010, *Brhat Samhita* (Delhi: Motilal Banarisidass)
Fleet, J.F. 1886, *The Indian Antiquary*, XV, 194
Fleet, J.F. 1894–95, *Epigraphia Indica*, III, 1
Hari, K.C. 2003, *Indian Journal of History of Science*, 38, 43
Jacobi, H. 1888, *Gottinger Gelehrte Anzeigen*, II, 67

Kapoor, R.C. 2010, *Indian Journal of History of Science*, 45, 489

Kapoor, R.C. 2016, *Journal of Astronomical History & Heritage*, 19, 264

Kapoor, R.C. 2017, *Current Science*, 113, 173

Kapoor, R.C. 2018, *Indian Journal of History of Science*, 53, 325

Mirashi, V.V. 1949–50, *Epigraphia Indica*, XXVIII, 1

Ramasubramanian, K., & Sriram, M.S. 2010, *Tantrasangraha of Nilakantha Somayaji*(New Delhi : Hindustan Book Agency), 331

Shukla, K.S. 1976, *Aryabhatiya of Aryabhata* (New Delhi: Indian National Science Academy)

Shylaja, B.S., & Ganesha, G.K. 2016, *History of the sky−on stones* (Bangalore : Infosys Foundation)

Suru, N.G. 1975, *Gaudavaho* (Ahmedabad : Prakrit Text Society)

Education and Heritage in the era of Big Data in Astronomy
Proceedings IAU Symposium No. 367, 2020
R. M. Ros, B. Garcia, S. R. Gullberg, J. Moldon & P. Rojo, eds.
doi:10.1017/S1743921321000703

Estrelleros: Astronomy in hospitals

Gloria Delgado-Inglada[1] (ORCID), **Diego López-Cámara**[2], **Alejandro Farah**[1],
Jorge Fuentes-Fernández[1], **Orlando García**[3], **Carolina Keiman**[1],
Tita L. Pacheco[4] **and Jaime Ruíz-Díaz-Soto**[1]

[1]Instituto de Astronomía, Universidad Nacional Autónoma de México, Apdo. Postal 70-264,
04510 Ciudad de México, Mexico
email: gdelgado@astro.unam.mx

[2]CONACyT – Instituto de Astronomía, Universidad Nacional Autónoma de México, Apdo.
Postal 70-264, 04510 Ciudad de México, Mexico

[3]Universidad del Valle de México, Campus Toluca, Apdo. Postal 52164 Metepec, México

[4]Dept. de Genética y Biología Molecular, Centro de Investigación y de Estudios Avanzados del
Instituto Politécnico Nacional, Apdo. Postal 14-740, 07000 Ciudad de México, México

Abstract. This project is designed for children under 18 years that have to frequently visit hospitals or that have to endure long-term hospital stays. The aims are to entertain these children and their families, to increase their scientific culture, and to promote scientific vocations. So far we have visited one hospital in Mexico City bringing astronomy to the patients and their families. We have developed five hands-on activities and one musical activity that ensures that all the children can participate independently of their conditions. We plan to expand this project to other hospitals and other cities in the country. Our next challenge is to start virtual visits to hospitals.

1. Introduction

Spending long-term stays in a hospital is an unpleasant experience for anyone but it is even worst for children that are hospitalized. They are bedridden while their friends are attending to school, playing in the park, or at the cinema.

Ludic activities in hospitals have proven to be very beneficial for children. Bermudez Rey (2009) listed an extensive list. According to her study these activities may contribute to: 1) Implement and develop the rights of hospitalized children, 2) Occupy their free time, 3) Improve their quality of life, 4) Counteract the problems derived from the monotonous life in hospitals, 5) Improve the integration of hospitalized children, 6) Reduce stress, anxiety, and isolation that generates hospitalization, 7) Contribute to normalize the situation, 8) Avoid the possible appearance of psychological problems, 9) Discover or affirm competences and abilities in the children, 10) Improve the self-esteem of the hospitalized children, 11) Develop creativity, 12) Restore the personal perception of control and competence, 13) Increase the cultural heritage, 14) Develop emotional bonds between participants, 15) Relieve parents of the physical and emotional burden.

Some of the benefits mentioned above are precisely some of the goals of outreach activities, in particular of our project. For example to enhance scientific culture, to develop creativity, and to discover or affirm competences and abilities. We want to use

Figure 1. A group of children and their families are building a space rocket following the instructions of the volunteers. As on that visit it rained, the activities were carried out in the hospital lobby.

astronomy as a tool to improve the stay of children in hospitals and we want to use hospital lobbies and rooms as places to promote science. Astronomy has always fascinated humans. It has many advantages over other scientific disciplines. We can use awesome images of galaxies, planets, and stars to generate curiosity and to attract people to science.

In the world there are some educational projects in hospitals that include astronomy among the subjects that are taught to the children. Astronomy activities are also included in the list of recreational programs in hospitals (along with clowns, storytellers, etc.). Two examples carried out in USA and Spain are described by Lubowich 2008; Alonso-Floriano, Cortés-Contreras, Pereira 2015. We are not aware of any astronomy outreach project in hospitals, besides ours, in Mexico (and probably in Latin America).

2. The project

The project "Estrelleros: Astronomy in hospitals" is funded by the Consejo Nacional de Ciencia y Tecnología (CONACyT) and by the Instituto de Astronomía of the Universidad Nacional Autónoma de México (UNAM). Our main goals are: 1) to entertain the hospitalized children (and their families), 2) to increase scientific culture, and 3) to encourage scientific vocations. More information about this project is provided in the webpage www.estrelleros.orgs.

In 2019 we visited four times the Hospital Shriners in Mexico City (see some photographs of the visits in Figures 1–4). This hospital provides care for children until age 18 with orthopaedic conditions and burn injuries. This forced us to create a variety of activities so that all the children with any condition can participate at least in a few of them. For example, all children (independently of their conditions) can participate in the activity "The music of the Universe" where music is the only tool we use to teach about chemical composition of the Universe.

Figure 2. A little girl is observing the Sun through a telescope with a special filter in the central garden of the hospital. This activity is one of the most successful ones together with the planetarium.

A typical visit included around 20 volunteers (students and academics) mainly from Instituto de Astronomía of the UNAM. Each visit lasted around four hours. The visit started with one brief conference without slides in the hospital waiting room. Then, the children chose among different activities: a mobile planetarium, a telescope to observe the Sun, and a few hands-on activities plus the musical activity mentioned above. Children learned about the components of spiral galaxies, our solar system, the chemical composition of the Universe, space exploration, and sundials. Occasionally we were able to visit children in their rooms. Around 120 people (between children and their families) participated in the different activities each day.

3. Next steps

Our next goal is to expand the project. On the one hand we want *Estrelleros* to visit other hospitals in Mexico City and in other cities of the country. To do this we will collaborate with astronomers in different locations. We will provide them all the material they need to carry out a typical visit. In 2020 we wrote a manual that includes the instructions for the six hands-on activities available at this moment (we have created

Figure 3. A contest to guess the scents of different objects in the Universe is taken place in the lobby.

Figure 4. The group of volunteer astronomers that visit the hospital.

one new activity about spectroscopy). We are also in the process of generating short videos with visual instructions. On the other hand, we have plans to start virtual visits. Due to the global pandemic we could not visit any hospital in 2020 and it is uncertain when we will visit hospitals in person again. This is a challenge for us and at this moment we are exploring possibilities. If we are able to carry out virtual visits we will be able to reach much more hospitals and children.

Acknowledgements

We want to thank all the volunteers that participated in the visits and all the staff from the Instituto de Astronomía that supported us. This project was funded in 2019 and 2020 by the Consejo Nacional de Ciencia y Tecnología (project number 297929 from the Convocatoria para Proyectos de Apropiación Social del Conocimiento de las Humanidades, Ciencias y Tecnologías 2019; and project number 849 from the Convocatoria para Fomentar y Fortalecer las Vocaciones Científicas 2020) and by the Instituto de Astronomía of the Universidad Nacional Autónoma de México.

References

Alonso-Floriano F. J., Cortés-Contreras M., Pereira V., 2015, hsa8.conf, 944
Bermúdez Rey, M. T., 2009, Actividades lúdicas en el hospital, *Padres y Maestros*, 327, 7
Lubowich D., 2008, ASPC, 400, 259

Posters

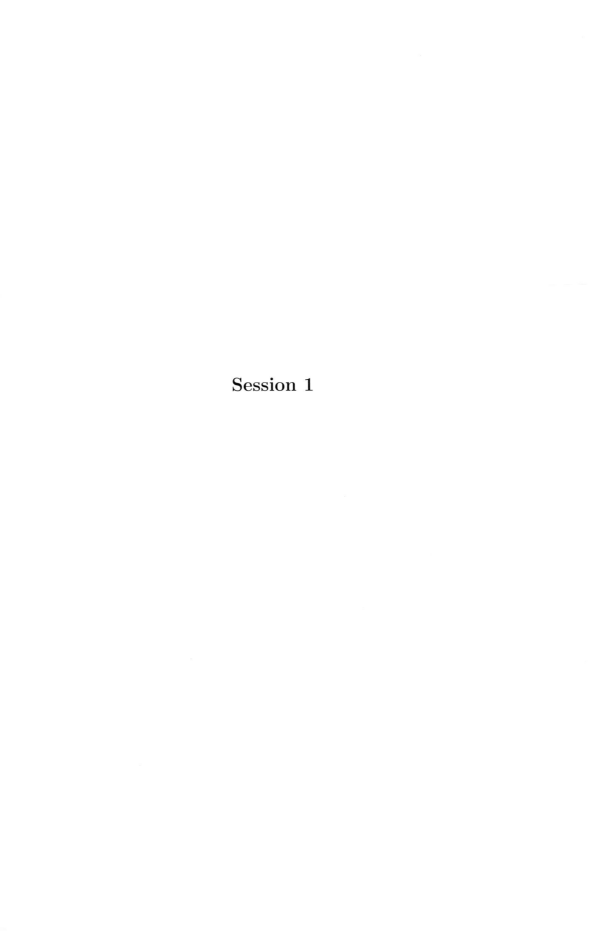

Session 1

Education and Heritage in the era of Big Data in Astronomy
Proceedings IAU Symposium No. 367, 2020
R. M. Ros, B. Garcia, S. R. Gullberg, J. Moldon & P. Rojo, eds.
doi:10.1017/S1743921321000247

Astronomy in The Malaysian National School Syllabus

Wan Mohd Aimran Wan Mohd Kamil[ID]

Dept. of Applied Physics, Faculty of Science and Technology
Universiti Kebangsaan Malaysia, 43600 Bangi, Selangor, Malaysia
email: `aimran@ukm.edu.my`

Abstract. This is a preliminary survey of astronomy topics incorporated into the Malaysian national school syllabus. Topics directly relevant to astronomy are situated within the general science subject from Year 4 (age 10 years old) until Year 9 (age 15 years old) and are grouped under four main themes: Earth-Sun-Moon system, the Solar System, Stars, and Exploration of Outer Space. Contemporary topics such as exoplanets and astrobiology are not explicitly mentioned, but students are required to engage in hypothetical thinking by speculating on planetary diversity and its implications for life in the Universe. We characterize the astronomy content in the Malaysian national school syllabus to be life-centric and relatively open-ended.

Keywords. Astronomy curricula, astronomy education, Malaysia

1. Introduction

A formal introduction to astronomy usually begins in a school classroom, whereby it is taught either as a standalone subject, or inserted discretely into other subjects such as science, geography and physics. Previous IAU Colloquiums devoted to the teaching and learning of astronomy have included reviews of astronomy topics contained in the school syllabi of China, Hungary, USA, Germany, Japan, Bulgaria, India, Egypt, and Thailand (Pasachoff & Percy 1990), as well as of France, Russia, UK, and Europe (Gouguenheim, McNally & Percy 1998). By presenting a brief overview of astronomy topics incorporated into the Malaysian national school syllabus, we intend to add Malaysia to the list of countries above, thus contributing to a more complete picture of the place of astronomy within the school syllabus across different countries, which facilitates comparative studies by future researchers.

2. Methodology

The syllabi for primary (Years 1 to 6) and lower secondary (Years 7 to 9) science subjects prepared by the Curriculum Development Division, Ministry of Education Malaysia were used as primary evidence to examine the incorporation of astronomy and astronomy-related topics in the national school syllabus (Ministry of Education Malaysia: Curriculum Development Division 2015–2020).

3. Discussion

Alternative ways of parsing the astronomy content in the Malaysian national school syllabus includes:

(a) *Spatially*. Beginning with Earth before 'moving' outwards step-by-step to encompass larger distances and 'space environments' such as such as the Earth → Moon

→ Sun → planets → minor bodies such as asteroids and comets → stars. The extension into new 'space environments' in a sequential fashion provides a 'big picture' that links and integrates individual topics on astronomy that spans several years of study (from Years 1 to 9).

(b) *The balance between pure and applied topics.* For instance, under Earth, mineral resources and geological disasters are discussed; under Earth-Moon-Sun system, the Islamic lunar calendar and space weather are discussed; under the Solar System, planetary habitability and the risks of collisions from asteroids are discussed; and under Stars, the utility of constellations as indicators of time and direction, and the importance of the Sun as proxy for stellar population are discussed.

(c) *The opportunity for practical observations.* Although practical observations are not explicitly stressed, students are encouraged to perform simple observations such as changes in shadows as proof of the rotation of the Earth, the changing phases of the Moon, the appearance of lunar and solar eclipses, and the identification of equatorial constellations.

(d) *Benchmarking with Big Ideas in Astronomy.* (Retrê *et al.* 2019) The greatest overlap occurs for Big idea 2 – Astronomical phenomena can be experienced in our daily lives, Big Idea 3 – The night sky is rich and dynamic, and Big Idea 7 – We all live on a small planet within the Solar System. Little or no overlap occurs for Big Idea 6 – Cosmology is the science of exploring the Universe as a whole, and Big Idea 10 – We may not be alone in the Universe.

4. Conclusion

Since astronomy content are inserted into the general science subject and since all students regardless of their subsequent specializations must take the general science subject, it is expected that a typical Malaysian student will have a basic familiarity with the Earth-Moon-Sun system, the Solar System, Stars and Exploration of Outer Space. Therefore, these topics represent the baseline for astronomical literacy of Malaysian students, specifically and the Malaysian public generally. Recognizing this baseline can help focus and streamline the planning, execution and evaluation of astronomy education efforts in Malaysia by various parties such as astronomy educators, amateur astronomers, and planetarium and observatory officials.

References

Gouguenheim, L., McNally, D., & Percy, J. R., eds. 1998, *New Trends in Astronomy Teaching: IAU Colloquium 162.* University College London and the Open University, July 8–12 1996 (Cambridge: Cambridge University Press)

Ministry of Education Malaysia: Curriculum Development Division. 2015–2019, *Kurikulum Standard Sekolah Rendah Sains Dokumen Standard Kurikulum dan Pentaksiran Tahun 1–5.* http://bpk.moe.gov.my/index.php/terbitan-bpk/kurikulum-sekolah-rendah (in Malay). 1st December 2020

Ministry of Education Malaysia: Curriculum Development Division. 2020, *Kurikulum Standard Sekolah Menengah Dokumen Penjajaran Kurikulum Tahun 2020 Sains Tingkatan 1–3.* http://bpk.moe.gov.my/index.php/terbitan-bpk/kurikulum-sekolah-menengah (in Malay). 1st December 2020

Pasachoff, J. M., & Percy, J. R., eds. 1990, *The Teaching of Astronomy: IAU Colloquium 105.* Proceedings of the 105th Colloquium of the International Astronomical Union, Williamstown, Massachusetts, 26–30 July 1988 (Cambridge: Cambridge University Press)

Retrê, J., Russo, P., Lee, H., Penteado, E., Salimpour, S., Fitzgerald, M., Ramchandani, J., Pössel, M., Scorza, C., Christensen, L., Arends, E., Pompea, S., & Schrier, W. 2019, *Big Ideas in Astronomy: A Proposed Definition of Astronomy Literacy.*

Education and Heritage in the era of Big Data in Astronomy
Proceedings IAU Symposium No. 367, 2020
R. M. Ros, B. Garcia, S. R. Gullberg, J. Moldon & P. Rojo, eds.
doi:10.1017/S1743921321000120

Astronomy through Continents

Joanna Molenda–Żakowicz[1] , Maciej Kokociński[2], Sylwester Kołomański[1] and Magdalena Ziółkowska–Kuflińska[2]

Instytut Astronomiczny, Uniwersytet Wrocławski,
ul. Kopernika 11, 51-622, Wrocław, Poland
email: joanna.molenda-zakowicz@uwr.edu.pl, sylwester.kolomanski@uwr.edu.pl

[2]Uniwersytet im. Adama Mickiewicza w Poznaniu,
ul. Wieniawskiego 1, 61-712 Poznań, Poland
email: kokot@amu.edu.pl, magdaz@amu.edu.pl

Abstract. We present recommendations for teachers and educators of science, based on the results of a survey carried out among secondary-school students from Poland, Australia, and the USA.

Keywords. Educational techniques, Project Method, sociology of science

The wealth, development, and security of each country depend to a large extent on their education systems. Those systems should be based on education curricula that are created in accordance with requirements of the present world, the changes of educational techniques, and the needs of a given society.

In order to shed light on educational needs of nowadays societies, we conducted a survey aiming at evaluation of the level of basic astronomical knowledge and understanding of fundamental astronomical concepts of secondary-school students (see Molenda-Żakowicz & Kołmański (2019)). We strove to formulate recommendations for teachers and educators which might be used for modifications of current educational curricula.

The survey was carried out in 2018 and 2019, among secondary-school students from Poland, Australia, and the USA. It had form of a questionnaire consisting of 20 test-sentences which covered basic astronomical knowledge that might be expected from a secondary-school alumnus. Those test-sentences had to be verified as true or false, however, it was possible to tick an 'I do not know' answer. Each correct verification was 1 point; the maximum score was 20 points. The questioned students provided also some additional information on their gender, the profile of education (STEM or no-STEM), the name of the city where they go to school, and their interest in astronomy.

Table 1 contains information on the number of the total (N_{tot}), male (N_{male}), and female (N_{female}) students surveyed in each country, as well as on their mean, median, minimum, and maximum score. Since not all students provided information on their gender, N_{male} and N_{female} do not necessarily sum up to N_{tot}.

We found that scores of Polish secondary-school students are noticeably worse than scores of their peers from the two other analysed countries. First, they are significantly lower, especially when compared to Australia. Second, in the Polish sample there is a much higher difference between scores gained by male and female or STEM and no-STEM students. The mean score of Polish male students is 2.4 points higher than the score of females while in Australia and the USA the respective differences are just 1.4 and 0.8 points. The mean score of Polish STEM students is 2.5 more points higher than the score of no-STEM peers while in Australia and the USA the respective differences are only 0.2 and 2.0 points. We observed also that Polish students from big cities gain significantly

Table 1. Basic statistics and scores.

Country	N_{tot} (N_{male}/N_{female})	Mean	Median	Minimum	Maximum
Australia	28 (21/6)	15.86	16	8	20
Poland	1212 (522/664)	11.76	12	3	20
USA	49 (31/18)	12.45	13	4	19

higher scores than those from medium-size and small cities (in Australia and the USA there were no comparison groups.) Finally, we found that Polish students declare much less interest in astronomy comparing to their peers from the other countries†.

The analysed data shows that the reported scores are closely related to the degree of prominence of astronomical knowledge in different societies. Since that prominence is being formed in the process of socialisation that starts in early childhood, the observed differences in scores reflect various social inequalities. Another issue worthy of consideration is the fact that the strong technological aspect of contemporary astronomical investigation disrupts links between the creator (an astronomer) and the piece of their work (a scientific discovery). That increases the divergence of the 'objective' and 'subjective' culture defined by Simmel (2005), lessens the degree of prominence of science, and enhances magical thinking (here: astronomy vs. astrology) in different societies. As a consequence, the current model of education, oriented on applied sciences and analysis of a process, and marginalising the basic sciences, may result in devaluation of fundamental scientific knowledge and predictions.

To remedy the situation, we recommend the teachers and educators to take into account in the process of education the differences in socialisation of the social roles of male and female scientists, and a generally lower degree of prominence of astronomy in small towns. In our opinion, the formal education needs to acknowledge the Project Method of teaching, in which observations should forerun analysis (see, e.g., Kołomański & Molenda–Żakowicz (2020)). Different results obtained in the framework of a project should be always assigned to respective individuals. Students should understand the whole process of the experiment and analysis, and see their own contribution. That formal education should be supported by pre-emptive teaching in a form of, for example, songs or educational games for kids, which can be started at the age of 2 or 4. Finally, the practical aspects of astronomical knowledge should not be unduly emphasised. Instead, astronomy should be presented as one of the essential sciences needed for proper understanding of fundamental physical concepts, including time, space, and the humanity's place in the Universe understood in the broadest possible sense.

References

Kołomański, S., & Molenda–Żakowicz, J. 2020, *Proceedings of the IAU Symposium 367/VIRTUAL Education and Heritage in the Era of Big Data in Astronomy, 8-12 December 2020*, poster 'School Workshops on Astronomy – an extracurricular activity for secondary-school pupils'

Molenda-Żakowicz, J. & Kołmański, S. 2019, *Urania – Postępy Astronomii*, vol 3, pp. 52–58

Simmel G. 2005, *The Philosophy of Money*, third edition, Edited by David Frisby, Translated by Tom Bottomore and David Frisbyfrom, Kaethe Mengelberg Taylor & Francis e-Library, pp. 450–473

† The reported differences in scores have not been affected by the differences between analysed samples, i.e., the sample size, the profile of education or the city size (Kruskal-Wallis test, $p < 0.01$).

Education and Heritage in the era of Big Data in Astronomy
Proceedings IAU Symposium No. 367, 2020
R. M. Ros, B. Garcia, S. R. Gullberg, J. Moldon & P. Rojo, eds.
doi:10.1017/S1743921321000144

Contribution of the student conferences "Physics of Space" to the continuing astronomy education system

Eduard Kuznetsov[1]📧, Dmitry Bisikalo[2], Olga Inisheva[3],
Konstantin Kholshevnikov[4], Boris Shustov[2] and Andrej Sobolev[5]

[1]Department of Astronomy, Geodesy, Ecology and Environmental Monitoring,
Ural Federal University, Lenina Avenue, 51, Yekaterinburg, 620000, Russia
email: eduard.kuznetsov@urfu.ru

[2]Institute of Astronomy of the Russian Academy of Sciences,
Pyatnitskaya Street, 48, Moscow, 119017, Russia
email: bisikalo@inasan.ru, bshustov@inasan.ru

[3]Lyceum, Ural Federal University, Danily Zvereva Street, 30, Yekaterinburg, Russia
email: o.v.inisheva@urfu.ru

[4]Department of Celestial Mechanics, St Petersburg University,
Universitetsky prospekt, 28, Stary Peterhof, St-Petersburg, 198504, Russia
email: kvk@astro.spbu.ru

[5]Kourovka Astronomical Observatory, Ural Federal University,
Lenina Avenue, 51, Yekaterinburg, 620000, Russia
email: andrej.sobolev@urfu.ru

Abstract. Ural Federal University is one of the main and most effective centers for educating young astronomers in Russia. The traditional student scientific conferences "Physics of Space" have been successfully held annually in Kourovka Astronomical Observatory of the Ural Federal University for 50 years and have gained recognition not only in Russia. The conference initiated many educational initiatives both in the Ural region and, in general, in Russia. The astronomy education system implemented by UrFU and partners includes the following activities: 1) education and career guidance of schoolchildren in the Lyceum of UrFU, 2) activities to attract applicants, 3) training at the speciality, undergraduate, and graduate level, 4) participation in the student conferences "Physics of Space", 5) postgraduate studies, 6) cooperation in the field of education. This activity ensures the attraction of promising youth to scientific research.

Keywords. continuing astronomy education system, student conferences

---

1. Introduction

Ural Federal University (UrFU) is one of the main and most effective centers for educating young astronomers in Russia. The traditional student scientific conferences "Physics of Space" have been successfully held annually in Kourovka Astronomical Observatory (KAO) of UrFU for 50 years and have gained recognition not only in Russia. Every winter, at the end of January, 40–50 students from Russia and 15–25 graduate students and young scientists from universities and academic astronomical organizations gather here. Foreign participants, as a rule, are fluent in Russian. The conference initiated many educational initiatives and, what is essential, allows them to maintain a high scientific and educational level. The astronomy education system implemented by UrFU and partners includes the following activities: 1) education and career guidance of schoolchildren in the Lyceum of UrFU, 2) activities to attract applicants, 3) training at the speciality, undergraduate, and graduate level, 4) participation in the student scientific conferences "Physics of Space", 5) postgraduate studies, 6) cooperation in the field of education.

2. Astronomy education system activities

1) *Education and career guidance of schoolchildren in the Lyceum of UrFU.* Special program for the graduates of Lyceum annually provides 4–6 graduates which enter UrFU and other universities that teach specialists in astronomy.

2) *Activities to attract applicants.* The thematic lecture hall on astronomy, implemented in KAO for schoolchildren of Yekaterinburg and Ural region and accompanied by excursions to the telescopes of the observatory, significantly expand the base of potential applicants for the speciality "Astronomy".

3) *Training at the speciality, undergraduate, and graduate level.* The UrFU has implemented training in the speciality "Astronomy" based on the best world experience. The programs are aimed at training specialists in digital technology in space sciences and big data mining. Practices in KAO and leading astronomical institutions are aimed at developing skills and experience in astronomy using various instruments.

4) *The student conferences "Physics of Space"* (https://astro.insma.urfu.ru/school). Conferences provide an opportunity to train skills, take part in the competition of scientific reports, get acquainted with leading scientists, and form cooperations for scientific work (Sobolev *et al.* 2003). Review lectures have been published since 1997. Conference site contains archive of review lectures and abstracts of the student and young scientist talks starting from 2002 (https://astro.insma.urfu.ru/school/archive). Starting from 2018 these publications became refereed, contain abstracts in English and have DOI numbers. Student conferences are a catalyst for how scientists of the future are formed from the students interacting with scientists of reknown.

5) *Postgraduate studies.* The contacts established at the conferences "Physics of Space" increase research efficiency by forming teams of researchers from potential graduate students and their leaders. Graduate students of UrFU, Institute of Astronomy of the Russian Academy of Sciences (INASAN), St. Petersburg University (more than 20 people) participate in joint projects of Russian and international research groups.

6) *Cooperation in the field of education.* The interaction of universities and scientific institutions during the organization and research, internships for students and graduate students provides access to modern equipment of the Terskol Observatory of INASAN, KAO UrFU, SAO RAS, Moscow State University and world facilities of the common use. Joint research involving students and graduate students in promising projects is crucial in the success of cooperation between research groups from various organizations.

3. Conclusions

Conferences "Physics of Space" play a significant role in the continuing astronomy education system, combining education and scientific research. At these conferences, employees of leading astronomical institutions in Russia and abroad present review lectures, students and graduate students test their work, and research groups are formed for new projects. Young scientists have the opportunity to get involved in the work of leading research groups, and the leaders of these groups participate in the preparation and selection of the most promising students and graduate students. Each year 30–40 people take part in this process as young researchers, ensuring involvement of the promising young people in the Russian scientific community.

Reference

Sobolev, A. M., Zakharova, P. E., Kuznetsov, E.D., Ferrini, F. 2003, *Astronomical and Astrophysical Transactions*, 22, 787

Education and Heritage in the era of Big Data in Astronomy
Proceedings IAU Symposium No. 367, 2020
R. M. Ros, B. Garcia, S. R. Gullberg, J. Moldon & P. Rojo, eds.
doi:10.1017/S1743921321000478

Dark sky tourism and sustainable development in Namibia

Hannah S. Dalgleish[1,2]⬤, Getachew M. Mengistie[1], Michael Backes[1,3]⬤, Garret Cotter[2] and Eli K. Kasai[1]

[1]Dept. of Physics, University of Namibia, Pionierspark, Windhoek, Namibia

[2]Dept. of Physics, University of Oxford, Keble Rd, Oxford, OX1 3RH, UK

[3]Centre for Space Research, North-West University, Potchefstroom, South Africa
email: hannah.dalgleish@physics.ox.ac.uk

Abstract. Namibia is world-renowned for its incredibly dark skies by the astronomy community, and yet, the country is not well recognised as a dark sky destination by tourists and travellers. Forged by a collaboration between the Universities of Oxford and Namibia, together we are using astronomy as a means for capacity-building and sustainable socio-economic growth via educating tour guides and promoting dark sky tourism to relevant stakeholders.

Keywords. dark sky tourism, astrotourism, Namibia, Africa, sustainable development, light pollution, capacity building

---

1. Introduction

Dark sky tourism (DST) attracts visitors to remote, unlit areas to observe celestial objects. Stargazing activities are carried out aided (with binoculars or telescopes) or unaided (with the naked-eye) and can be accompanied by other activities like astrophotography or storytelling. DST has been found to further many of the UN's seventeen Sustainable Development Goals (SDGs), which can help serve as a guide to implement dark sky experiences in a sustainable way (Dalgleish & Bjelajac 2021; Fig. 1).

DST contributes to the SDGs under all branches of sustainability. Economically, DST can generate significant income, providing jobs and extending tourism activity into off-peak times (Mitchell & Gallaway 2019; SDG 8—decent work and economic growth). Environmentally, the minimisation of artificial night at light prevents interference with freshwater, marine, and terrestrial wildlife (Davies & Smyth 2018; SDGs 14 and 15—life below water and life on land). Socially, DST presents educational opportunities for tourists and local residents, covering topics from astrophysics and light pollution to indigenous knowledge (Blundell *et al.* 2020; SDG 4—quality education); and empowers women in rural, underprivileged areas (see e.g. the `Astrostays` project; SDGs 5 and 10—gender equality and reduced inequalities). Stargazing also promotes health, well-being and connectedness with nature (Bell *et al.* 2014; SDG 3—good health and well-being).

2. Dark sky tourism in Namibia

The second least densely populated country in the world, Namibia has minimal light pollution and is therefore very well-suited to dark sky experiences. Namibia is vast with a wide array of climates; some areas (e.g. Sossusvlei) rarely experience cloud cover year-round, and thus, clear and dark skies can be found even during the wet (summer) season. A few lodges and "astrofarms" already take advantage of the country's pristine skies,

Figure 1. Dark sky tourism and its relationship with the SDGs (Dalgleish & Bjelajac 2021).

which are especially attractive to amateur astronomers and astrophotographers. Africa's first International Dark Sky Reserve can also be found in Namibia, at the NamibRand Nature Reserve. Thus, there is ample opportunity to extend and promote dark sky activities across the country, especially across wider tourist demographics.

In order to grow dark sky tourism sustainably, we have been working from both a bottom-up and top-down approach. For the former, we are developing a course comprising five main sections: (1) our place in the Universe, (2) astrophysics research in Namibia, (3) indigenous Namibian star lore, (4) practical astronomy, and (5) light pollution and sustainability. We will be delivering the course to Namibian tour guides in 2021, while ensuring that the content is adaptable for use by similar projects in other countries. We are also exploring options for delivering the course online. At the same time, we are working with the Ministry of Environment, Forestry, and Tourism, as well as tourism associations and other relevant stakeholders, in order to establish Namibia as a country at the forefront of DST.

In summary, astronomy provides a unique opportunity to build human capacity and diversify income generation in remote areas. Equally, dark sky tourism is well aligned with the 21st century ethos that tourism needs to be ecofriendly and sustainable. DST comes with many benefits, such as an increased awareness and understanding of science, environmental conservation (e.g. light pollution), and the celebration and preservation of indigenous heritage. These can all help to open up new avenues toward more meaningful and sustainable tourism practices.

Acknowledgements

This project is supported by the UKRI STFC Global Challenges Research Fund project ST/S002952/1 and Exeter College, Oxford.

References

Bell, R., Irvine, K.N., Wilson, C., & Warberd, S.C. 2014, *European Journal of Ecopsychology* 5

Blundell, E., Schaffer, V., & Moyle, B.D. 2020, *Tourism Recreation Research* 45, 4

Dalgleish, H., & Bjelajac, D. 2021, *in press*

Davies, T.W., & Smyth, T. 2018, *Glob Change Biol.* 24

Mitchell, D., & Gallaway, T. 2019, *Tourism Review* 74, 4

Education and Heritage in the era of Big Data in Astronomy
Proceedings IAU Symposium No. 367, 2020
R. M. Ros, B. Garcia, S. R. Gullberg, J. Moldon & P. Rojo, eds.
doi:10.1017/S1743921321000016

Informal astronomy education in Bulgaria at the beginning of the XXI century: organization, continuum, results

Penka Stoeva🄳 and Alexey Stoev

Space Research and Technology Institute, Bulgarian Academy of Sciences,
Stara Zagora Department, Bulgaria
email: `penm@abv.bg`

Abstract. The report shows the current opportunities for obtaining astronomical knowledge in school and outside it, through the use of non-formal education. These are school and extracurricular activities, schools, astronomical competitions and Olympiads, observation expeditions. For 25 years Bulgaria has been participating in the International Olympiads in Astronomy and Astronomy and Astrophysics with National Teams. The role and place of the system of Public Astronomical Observatories and Planetaria in the system of non-formal education in astronomy are discussed (In Bulgaria there are 7 Public astronomical observatories with a planetarium). Specialized activities in their school forms allow the formation of sustainable astronomical knowledge and observational habits.

Keywords. Sociology of Astronomy, History and philosophy of astronomy

---

Introduction: Informal education (IE) can be defined as an organized and systematic learning activity, outside the system of formal education (FE), oriented to the specific needs and interests of learners, regardless of their age, gender and level of education. Informal education has features such as flexibility, subjectivity and additionality in relation to FE, which can complement and develop in detail the educational activities in the school as a whole (http://omsu.ru/page.php?id=501). The International Standard Classification of Education defines the informal education as "any organized and continuing educational event that can take place both inside and outside educational institutions and covers people of all ages. Informal education programs do not have to be structured in stages and can have different durations (Rogers 2005).

Current state of astronomical education in Bulgaria (level of secondary education and informal educational institutions): Astronomy is one of the most ancient and socially significant sciences. The only place where the science of astronomy could be studied in its entirety – theoretically, practically and observationally, remained the astronomical observatories, planetariums, astronomical circles and local clubs. Their main resources and additional projects have made it possible to create and develop non-formal astronomy education, both in one-off events and in continuous multi-year courses and schools. In the context of developing science education, such centers allow the formation of a quality staff among young people who continue their education at the university. In the period 1961 – 1999 eleven Public Astronomical Observatories and Planetaria were established in Bulgaria – in the cities of Belogradchik, Varna, Gabrovo, Dimitrovgrad, Kardzhali, Silistra, Sliven, Smolyan, Sofia, Stara Zagora and Yambol, and in Haskovo and Troyan – astronomical centers. Some of them underwent various organizational and administrative transformations over time, and later an Astronomical

Association was established on the basis of the two Public Astronomical Observatories in Sofia. These astronomical complexes included observatories, some with planetarium, as well as sites for visual and telescopic observations of celestial objects and phenomena. Since 1997, a National Olympiad in astronomy has been held within the Ministry of Education and Science. To date, the Olympiads are held at three national and one international stage. For 25 years Bulgaria has been participating in the International Olympiads in Astronomy and Astronomy and Astrophysics with National Teams. During this period, 56 gold, 86 silver and 106 bronze medals were won at the Olympics in Brazil, Bulgaria, Greece, Russia, India, Indonesia, Italy, Kazakhstan, China, Poland, Romania, Taiwan, Sweden, etc. Modern information and communication technologies (ICT) also play an extremely important role in non-formal astronomical education in Bulgaria. Available global, European and national Internet resources in astronomy, space physics and astronautics, as well as social networks, allow students to receive up-to-date scientific information in real time. Significant role in drawing attention to astronomical knowledge is played by popular science literature. The books on astronomy and astronautics for children make a pleasant impression, with their well-illustrated pages and accessible language for the little ones. The main forms in the implementation of informal education in astronomy are courses, schools, seminars, project development, observation expeditions (Stoev & Stoeva 2008).

Perspectives: Non-formal education in astronomy in Bulgaria has been developing for more than 55 years. During this period, the individual forms, methods and programs for it were in the process of continuous improvement and modification: textbooks, presentations, manuals for teachers, astronomy tasks, amateur research observations etc.

Today, the development of informal education in astronomy as an additional opportunity to obtain astronomical knowledge is becoming extremely relevant. This is a pledge for the formation of well-prepared future students and doctoral students in physics, chemistry, mathematics, computer science at universities and research institutes.

Conclusion: The creation and implementation of adequate non-formal education in astronomy at the present stage allows to solve the following important tasks: – creating conditions for a comprehensive study of the kinematics and dynamics of celestial objects and phenomena, as well as physical processes in the near Earth cosmos, the Solar system, the Milky Way Galaxy; – creating preconditions for research activity during the whole training in astronomy; – formation of conceptual scientific notions about the recognizability of the Universe and space processes there; – acquisition of fundamental knowledge both in the field of astronomy and space physics, as well as in other sciences of the natural science educational cycle. Solving these tasks from the basic forms of non-formal education in astronomy makes them a connecting element between formal and non-formal education in other scientific disciplines, which include the acquisition of astronomical knowledge. In view of the above, the following conclusions can be drawn: First, astronomy as a subject of non-formal education contributes to the solution of general education and development tasks in secondary education in Bulgaria; Second, practical classes and research training in astronomy can play a central role in the training and development of students and out-of-school pedagogical institutions in astronomy, astronomical clubs and schools. The participation of students in various national and international astronomical projects and work with large databases has exceptional opportunities for this.

References

International Standard Classification of Education, URL: http://omsu.ru/page.php?id=501
Rogers A., 2005, Non-formal education: flexible schooling or participatory education. N.Y.
Stoev A.D., Stoeva P.V. 2008, *Adv. Sp. Res.*, 42, 1806

Education and Heritage in the era of Big Data in Astronomy
Proceedings IAU Symposium No. 367, 2020
R. M. Ros, B. Garcia, S. R. Gullberg, J. Moldon & P. Rojo, eds.
doi:10.1017/S1743921321000028

Processing methods and approaches for the analysis of images of the eclipsed solar corona taken during campaigns with the participation of amateur astronomers

A. Stoev[1], P. Stoeva[1][iD], S. Kuzin[2], M. Kostov[1] and A. Pertsov[2]

[1]Space Research and Technology Institute – Bulgarian Academy of Sciences, Stara Zagora Department, Stara Zagora, Bulgaria

[2]Lebedev Physical Institute, Russian Academy of Sciences, Russia, Moscow

Abstract. The increase in the amount of scientific information in heliophysics is related to both quantitative – increasing the number of high-power telescopes and the size of light receivers coupled to them, and qualitative reasons – new modes of observation, large-scale and multiple studies of the solar corona in different ranges, large-scale numerical experiments to simulate the evolution of various processes and formations, etc. The paper discusses the role and importance of methods for processing images of the solar corona, the store of obtained "raw" data and the need to access high-performance computing systems in order to obtain scientific results from the observational experiments, the need of international collaboration and access to the data in the era of increase in the amount of scientific information in heliophysics.

Keywords. Sun: corona, eclipses, techniques: image processing

---

Introduction: The brightness of the corona is a million times less than the brightness of the photosphere and is approximately equal to the surface brightness of the Moon. Because the Moon is not visible near the Sun, the solar corona can only be observed with the naked eye during the full phase of solar eclipses. Outside the eclipses, the corona is observed from the Earth's surface with the help of special telescopes – coronagraphs. The data obtained from observations of total solar eclipses (TSE) are usually "raw" observational materials, from various ground observations and experiments in the path of totality. Large amounts of data are also accumulating in the field of climatology, which studies the response of the Earth's atmosphere to a sudden deficit of solar radiation during the full phase. The size of astronomical radiation receivers – mainly CCDs – is growing significantly faster than the size and even the number of telescopes themselves. And the increase in the number of pixels in a CCD receiver is a proportional increase in the amount of information received from it. In heliophysics, scientific problems have arisen that require obtaining a large number of consecutive images of the same area of the corona with different exposures in search of rapid variability of its dynamic and kinematic characteristics. It is also important that all information obtained from such observations be processed immediately and added to the heliophysical database.

Big data in heliophysics – problems and approaches: The increase in the amount of scientific information in heliophysics is related to both quantitative – increasing the number of high-power telescopes and the size of light receivers coupled to them, and qualitative reasons – new modes of observation, large-scale and multiple studies of the solar corona in different ranges, large-scale numerical experiments to simulate the evolution of various processes and formations, etc. (Zhang & Zhao 2015). Typically, methods for processing images of the solar corona include the selection of characteristics and boundaries, as well as various types of filtering and conversion of histograms. In the case of images of the corona taken during the TSE, the task is to extract the characteristics and various structures of the corona, such as arcs, helmets, arches, loops and especially their anomalous development and connections with the underlying active photospheric and chromospheric formations. The task of determining their characteristics, as a rule, comes down to identifying the boundaries of these structures and their interaction. Another feature of modern heliophysical experiments is the need to store the obtained "raw" data and the need to access high-performance computing systems in order to obtain scientific results from the observational experiment and to analyze them. They require hundreds of gigabytes to preserve a "snapshot" of the solar corona. Such calculations are performed on distributed clusters with thousands of processors and the ability to work with such data in the database allows us to trace the history of the evolution of individual objects in the solar corona (Farivar *et al.* 2013). Heliophysical data obtained during expeditions for observations of total solar eclipses are completely heterogeneous – they are obtained with different tools using different methods, different observers and teams choose a completely different approach to different objects and processes in the solar corona.

Working with heterogeneous and distributed data is becoming an important problem of heliophysical science. The solution to such situations will require a global initiative and cooperation to develop universal standards for annotation, search and access to heliophysical data obtained during the TSE. Within this framework, centralized repositories are likely to be set up to store information for heliophysical data archives. With the help of standard protocols, it will then be possible to select from them those that contain information about the required coronal area, heliophysical event, observational experiment, exposures, filters, observation coordinates, etc. A characteristic feature of the current situation related to observations of the Sun during TSE is that the data become so diverse and so complex that they themselves cease to be valuable. The only important thing is to know how to analyze this data to get a scientifically significant result.

Stages of developmen of universal standards for annotation, search and access to heliophysical data obtained during TSE: Primary image processing, Astrometric image calibration, Photometric calibration of images, Match images, Creating a catalog. The ultimate goal of this international collaboration is to explore and develop a horizontally scalable communication network for corona imaging during total solar eclipses, which will be available to researchers around the world to process large amounts of raw data from current and future observations of TSE. The architecture of the communication network should provide ample opportunities for customization and modification of algorithms used at all stages of research. In this regard, the need to strengthen international cooperation of professionals and amateur astronomers is evident as well as the standardization of the observational experiments in the field of photometry and spectrometry of the solar corona during TSE, the creation of a database of observations for the purpose of further processing and interpretation. Training of amateur astronomers, organization of international schools and conferences, publication of specialized literature, Web-based teaching aids, instructions and consultations are also needed for the extensive research of this wonderful magnificent natural phenomenon – a total solar eclipse.

References

Zhang, Y. & Zhao, Y. 2015, Data Science Journal, 14, 11

Farivar, R., Brunner, R. J., Santucci, R., et al. 2013, Astronomical Data Analysis Software and Systems XXII, ASP Conf. 475, 91

Education and Heritage in the era of Big Data in Astronomy
Proceedings IAU Symposium No. 367, 2020
R. M. Ros, B. Garcia, S. R. Gullberg, J. Moldon & P. Rojo, eds.
doi:10.1017/S1743921321000880

Tanzanian experience of In-service Teacher Training in Astronomy through the NASE program

Noorali T. Jiwaji◉

Open University of Tanzania, Marian University College
email: `ntjiwaji@yahoo.com`

Abstract. We provide our first experience of Astronomy training as an in-service training of teachers of Science in Primary schools, and teachers of Geography, Physics and Mathematics in Secondary Schools necessitated due to lack of Astronomy specific training in their teacher training programs. The hands-on training was conducted in collaboration with the IAU Commission 46 Working Group program of Network of Astronomy Schools Education (NASE). Experiences from both face to face and virtual sessions conducted during the Covid19 period and in preparation of a major African solar eclipse, are discussed.

---

1. Introduction

Astronomy is taught only as the topic of Solar System in Primary schools and in the Geography subject in Secondary Schools; and as Astronomy topic in the Physics syllabus in Secondary schools, various (1996). During training, students are exposed to astronomy only through Physics courses such as Cosmology and Relativity and Astrophysics courses which are available in only a few Universities. Hence basic astronomy knowledge and concepts are not understood, while practical knowledge is lacking. These teachers have to teach Astronomy topics with a handicap with the content handled cursorily. More recent textbooks (B. McDowell (2010), Bwisa & Simiyu (2014)) have a more descriptive and illustrated Astronomy chapter, but it is used for memorization. The challenge of changing from Kiswahili to English during transition from Primary to Secondary schools adds to the problem, with only basic Astronomy lexicon being available in Kiswahili.

The advent into Tanzania of the IAU Commission 46 Working Group program of Network for Astronomy School Education (NASE) into this picture has been a most relevant to assist us to bring more effective hands-on training in Astronomy as an in-service training to help teachers to teach it better.

2. Astronomy training programs conducted in Tanzania

Several communications with NASE coordinator Dr Rosa M Ros resulted in the first NASE training in Tanzania with a two-week intensive in-service residential NASE teacher training, including night time sky observations, was conducted in December 2019, with permissions of the government authorities.

Teachers from within a semi-rural Ubungo suburb of Dar es Salaam city were selected purposively through recommendations of Heads of schools and acceptance by the teachers to show motivation. A total of 30 teachers balanced between Secondary schools and their respective nearby Primary schools were selected so as to provide mutual support when required after the training.

Figure 1. Need to innovate: Instead of a tripod, the telescope is taped to a football swiveled on a short cylinder

NASE facilitators from nighbouring Uganda and Zambia and NASE expert Rosa M Ros were assisted by local tutors from three Universities and neigbouring schools to initiate training of tutors (ToT) by assisting in teaching of the NASE the workshops. Local materials for hands-on demonstrations were gathered for hands-on activities and local language explanations were provided where needed.

The first NASE training snowballed into further Astronomy training, with the second NASE training held in June 2020 as a preparation of the Solar Eclipse of 21 June. This was conducted online due to Covid intrusion, but that also enabled additional participation of teachers from more distant schools. Practical demonstrations were shown online. Translation of the training material into Kiswahili was begun at this stage. An award of 10 telescopes to Tanzania by the Open Astronomy Schools (OAS) project provided another practical training opportunity in September 2020.

The OAS telescope kits were assembled by the teachers highlighting its optics concepts, followed by practical use to align with distant objects and finally observe the Moon. Lack of expensive telescope mounts such as a tripod resulted in an innovation where the small telescope was taped to a large ball om an open cylinder for viewing in all directions and hold the telescope steady (Fig. 1).

Enthusiastic comments were received from teachers with more than 90% of the teachers finding the training to be enlightening and beneficial. The number of topics and hands-on activities were found to be overburdening during the two-week program. Construction of models and finding alternative locally available material was challenging.

3. Conclusion

The IAU-NASE Astronomy training program has been extremely useful for introduce Astronomy directly to the teachers and get better understanding of its concepts and applications using hands-on training. Local resource people are available for further practice so as to continue with this training using guidance from NASE experts.

References

various *et al.* 1996, *Physics for Secondary Schools. Book Four.* Tanzania Institute of Education, 1995, DUP (1996) LTD, ISBN 9976-60-261-8

McDowell, B. 2010, *Physics Students' Book for Forms 3 and 4.* Tanzania Institute of Education, 2010, Pearson Education Limited, ISBN 978 1 4058 4211 2

Bwisa, J.-W., & Simiyu, S.-S. 2014, *Physics for Secondary Schools, Form 3.* Oxford University Press, 2014, ISBN 978 9976 4 0579 8

Education and Heritage in the era of Big Data in Astronomy
Proceedings IAU Symposium No. 367, 2020
R. M. Ros, B. Garcia, S. R. Gullberg, J. Moldon & P. Rojo, eds.
doi:10.1017/S1743921321001174

TUIMP: The Universe In My Pocket

Grażyna Stasińska[ORCID]

LUTH, Observatoire de Paris, PSL, CNRS 92190 Meudon, France
email: `grazyna.stasinska@obspm.fr`

Abstract. TUIMP is an outreach and educational project providing astronomical booklets that can be folded from one single sheet of paper. The booklets are written by professional astronomers and are intended for a broad audience. They can be downloaded for free from www.tuimp.org and used in classrooms, planetariums and in astronomy festivals. Presently they are available in 11 languages. Participation for writing new booklets or translating in any language is very welcome.

Keywords. Education and Outreach in Astronomy

1. Introduction

TUIMP, The Universe In My Pocket, is an outreach and educational project which started a few years ago. Its aim is to produce 16-page astronomy booklets that can be folded from just one sheet of paper. The booklets are designed for a broad audience: 'children from 9 to 99 years old curious about astronomy' as announced in our site. They can be downloaded for free from our webpage www.tuimp.org and used in schools, in planetariums or astronomy festivals. Each booklet ends with a small quiz that the reader can asnwer using the information contained in the booklet.

The project does not require any funding as all the participants give their time for free. One of its important characteristics is that it is truly international, as the authors of booklets originate from many countries. More importantly, the booklets are open to translation in any language. Figure 1 shows the folding of several booklets in Portuguese. Figure 2 shows a folded booklet in Polish.

2. Project evolution

Since its first presentation by Stasińska (2018), the TUIMP project has evolved considerably. Fifteen titles have already been published and others are in preparation. They go from purely descriptive such as "The nebular universe" or "The solar system" to more didactic ones such as "The sizes of celestial bodies" or "We come from the stars". Concepts such as spectroscopy, nucleosynthesis, or photoionization are introduced in a simple language. Sociological and historical aspects are not forgotten, such as the design of constellations in different cultures or the discovery of the 'guest star' in 1054 not only by Chinese astronomers but also in Europe, Arabia and Japan.

TUIMP booklets are presently available in eleven languages: Albanian, Arabic, Armenian, English, French, Greek, Italian, Polish, Portuguese, Russian, and Spanish. They are used in schools at various levels in different countries, as a support or a complement of physics or astronomy classes. The success is largest in Albania and Arabic countries, which do not benefit from a large corpus of astronomical literature in their own language. In Poland and France astronomical magazines include each month one page

Figure 1. Making booklets in Brazil.

Figure 2. A finished booklet in Polish.

with a TUIMP booklet to cut out and fold up by their readers, allowing a larger dissemination of the TUIMP project not only among amateur astronomers but also among educators.

While the TUIMP project was initiated by a small group of friends residing in different continents, it has expanded not only through a 'friends of friends' process but also thanks to the internet. For example, translations in Arabic language are done collaboratively by members of the Sirius Astronomy Association based in Constantine (Algeria) who contacted us on our site. This opened the way for TUIMP in Northern Africa and countries from the Middle East. In this case, even the translating activity itself has an educational aspect and could well be transposed to other countries.

3. The future

One of the important aims of the TUIMP project is to offer free outreach resources for populations that do not have an easy access to culture and education either because of language issues or because of their economical situation.

We are still far from reaching this goal. One obvious reason is that we are not able to reach these populations through internet precisely because of language issues. This is why the help of the international astronomical community is so important.

We openheartedly invite astronomers from all countries to join us for translating TUIMP booklets in their own language, especially in African or Asian languages but also in languages spoken by different communities in America. The small volume of each booklet ensures that the amount of work involved is not too heavy. Each collection of booklets in a given language can grow at its own pace.

Reference

Stasińska G., 2018, 'TUIMP: The Universe in My Pocket. Free Astronomy Booklets in All Languages', in 'Proceedings of the International Conference CAP2018', p.388, https://www.communicatingastronomy.org/cap2018news/ or arXiv:1806.02671

Session 2

Education and Heritage in the era of Big Data in Astronomy
Proceedings IAU Symposium No. 367, 2020
R. M. Ros, B. Garcia, S. R. Gullberg, J. Moldon & P. Rojo, eds.
doi:10.1017/S1743921321000296

Basics of Astrophysics in primary school: Marvel, play and learn

Ana Torres-Campos🆔

STEAM Department, Fundación Colegio Le Bret
Av. Antiguo Rancho a Morillotla No. 3014, San Andrés Cholula, Puebla, México, C.P. 72825
email: `astro@anatorrescampos.com`

Abstract. In this poster is presented the development and testing of a pilot project to teach basics of astronomy to primary school students. This is a learning program that bases on the interdisciplinary nature of astronomy with amusing and playful activities. The objective of the programme is to engage children in astronomy and make them aware of the importance of the development of science and technology for society. The program has been tested a small non-government-funded school in the state of Puebla, in Mexico. Due COVID-19 lockdown the classes had to switch from face-to-face to online. Over 80% of the students that completed the course had considerably increased their knowledge of astronomy and requested to continue with the classes on the next school term.

Keywords. Education, Astronomy teaching, Elementary school, etc.

---

1. Introduction

It is well known among the scientific community that Astronomy can be used as a tool for development in school and in life (van Dischoeck *et al.* 2020). Still, Astronomy is not included as one of the basic subjects in the elementary school plans worldwide. This project aims to create an elementary school astronomy program that allows children to practice the skills they need to perform successfully in their life for the coming years. In this part of the project only school children between 5 and 8 years old are targeted. The objective of the program is to marvel and inspire interest in science, while providing children with astronomical basic concepts.

2. Building up the astronomy teaching program

The project began with the study of Mexican government elementary education free textbooks (CONALITEG 2019) and analysis of outreach and education material from astronomy institutes, science museums and independent science communicators (e.g. REDPOP, Network for the Popularization of Science and Technology in Latin America and the Caribbean). Teachers and experts in education were also interviewed in order to identify children learning skills and knowledge at different ages (e.g. Allen and Kelly 2015; Norris *et al.* 2019).

The information collected guided the design of the programs to include the use of didactic material, games and hands-on activities to make the learning process fun and engaging, while developing children's motor and cognitive skills. The school-year program consisted in thirty-nine weekly one-hour classes. The classes consist of a short presentation or audiovisual exposition by the professor, followed by a hands-on activity or team game performed by the children in a 30-minute period, and a summary and closure. The designed activities are flexible enough as to be performed indoors, outdoors, individually or in teams. Finally, Kindergarten to third grade elementary school curricula, class activities (e.g. comparing the Sun's evolutionary stages and human's life cycle, giant planet's ring toss game, designing and drawing an imaginary monster from a rocky or gaseous planet) and low cost DIY didactic material (e.g. bottles filled with rise to compare the mass of the planets, scaled models of the Solar System) was created for each education level.

3. Implementation & testing

The pilot program was implemented and tested on the 2019-2020 school year in a Mexican non-government-funded small school, Fundación Colegio Le Bret, oriented towards families with lower-middle or middle socioeconomic level. The astronomy classes were given to the fifty students between Kindergarten to 3rd grade, distributed in four groups (one for each grade level). On the first class of the course an astronomy lottery game was played in the four groups. This activity revealed that twenty-six students were aware that astronomy studies objects in the sky and eighteen showed strong interest in the subject.

From august to march 2020 the classes were given weekly in the school classroom and playground. COVID-19 lockdown forced to continue the classes through zoom recurring group meetings and off-line video assisted activities. Thus needing to modify some of the previously planned activities. Nevertheless, children were always marvelled by Hubble Space Telescope images, the exploration of the Moon, properties of the stars and astronomy technology transfer. Even shy students reluctant to use zoom joined the class whenever they could to continue learning about the Universe with astronomical images, cartoons (e.g. Paxi cartoon from ESA Kids) and by drawing and creating astronomy themed hand-crafts.

At the end of the year trough an online multiple choice evaluation it was concluded that the thirty-nine students that could complete the course had achieved the expected knowledge. The student's general opinion was that they had enjoyed the course and were keen on taking more astronomy classes. The general comments from the student's parents were that their children had increased considerably their interest in astronomy and sciences in general.

4. Conclusions & Future work

Marvelling images, active participation and joyful activities lead to successful learning experience. Games, physical activities and handcrafts are eagerly performed by the students and raise student's focus on the class. The program has been updated to be fully virtual for 2020–2021 school-year. The next steps include video tutorials, student's workbook and learning log, teacher's guide and the 4th to 6th grades programs.

References

Allen L. R. and Kelly B. B., editors, Institute of Medicine; National Research Council, 2015, *National Academies Press (US)*, Jul 23, 4

Comisión, Nacional de Libros de Texto Gratuitos (CONALITEG); "Catálogo Digital de Libros de Texto Gratuito", 2019, *https://libros.conaliteg.gob.mx/catalogo.htm*

Norris, Emma; Steen, Tommy; Direito, Artur and Stamatakis, Emmanuel, 2019, *British Journal of Sports Medicine* 54

van Dishoeck, Ewine F. and Elmegreen, Debra Meloy, 2020, *The IAU Strategic Plan for 2020–2030: OAD*, 560, 562

Education and Heritage in the era of Big Data in Astronomy
Proceedings IAU Symposium No. 367, 2020
R. M. Ros, B. Garcia, S. R. Gullberg, J. Moldon & P. Rojo, eds.
doi:10.1017/S1743921321000533

With Covid-19:
Attempt of learning to observe the moon using a telescope at home

Hidehiko Agata (ORCID)

National Astronomical Observatory of Japan,
2-21-1 Osawa, Mitaka, Tokyo, Japan
email: h.agata@nao.ac.jp

Abstract. Students took an assembly-type telescope kit from a public elementary school and brought it to their homes in 2019. Three classes attempted to observe the Moon at home using the Kaifu-NAOJ Telescope Kit. As a result, all children observed the Moon at home using the kit. From their observations, around 90% identified the existence of craters and understood the reflection of sunlight on the Moon's surface. As Covid-19 prevention measures in education, we propose the introduction of at-home telescopic observations for STEAM activities.

Keywords. telescope, Moon, sociology of astronomy, Active Learning, STEAM

---

1. Introduction

Nowadays, STEAM education and Active Learning are in demand in schools. However, it is generally difficult to conduct evening and nighttime astronomical observation sessions at schools for the following reasons. (1) Ensuring the safety of children and students when they return home, (2) Faculty work time management, (3) How students spend their time after school, and (4) from the viewpoint of COVID-19 prophylaxis, there is a risk of infection at the viewing events, which are prone to contact infection and high concentrations of COVID-19. In this study, we investigate whether it is possible for each student to bring home an astronomical telescope kit developed by NAOJ and observe stars at home (Agata 2020).

2. Tools and methods

NAOJ has provided star observing workshop using Kifu-NAOJ telescope kit (NAOJ 2020) since 2019, which was developed as a ready-to-assembled telescope with reasonable price for an educational tool. (Fig. 1).

Active learning workshop was conducted in Shimauchi Elementary School with Mr. Teruyoshi Takizawa as a teacher in charge in the course of the unit "Moon and Sun" in 6th Grade Science using Kaifu-NAOJ telescope kit, which was rent out by NAOJ for 8 days. The unit includes 7 school hours (45min/school hour) and the workshop was held from October 29th to December 6 th, 2019 in three classes. The theme of the workshop was the "What's on the surface of the moon?" Each student was provided their own telescope kit, and a tripod if he/she needed. Students assembled the telescope and learned how to bring the telescope into focus in class (Fig. 2) and they took it home to observe the moon according to instructions on the worksheet. We investigated and evaluated its learning effectiveness using learning cards, questionnaires, and other tools.

Objective lens	Diameter 50 mm Focal length 399 mm Two achromatic lenses
Magnification	16x and 66x (replaceable eyepieces)
Full length	450 mm (about 490 mm at maximum extension)
Maximum diameter	67 mm (excluding protrusions)
Weight	Approx. 265 g
Eye pieces	25 mm (Huygens type) 6 mm (Plössl type)

Figure 1. Kaifu-NAOJ Telescope Kit before and after assembly and its specifications.

Figure 2. Students are assembling the telescope (left) and bringing it into focus (right).

70 students in class A and B were examined before and after observation at home to compare the proficiency while 35 students in class C as a reference group were examined only before observation.

3. Results and implication

All students in class A and B were able to observe the moon at home using the kit although some students (about 20%) needed support. After observation, more than 90% of them understood the existence of craters and nearly 90% of them understood that the moon shines due to the reflection of sunlight while less than 80% and 80% students did respectively in reference group without observation. In addition to achievement of proficiency, active attitudes to learning also tended to be nourished. Their impressions on observation at home excerpted from the post-questionnaire are as following.

- This was my first time using an astronomical telescope. I am glad that I was able to observe the moon thoroughly and ask many questions. It was a great memory for me to observe the moon in sixth grade science class.
- Observing the moon was the most fun I had in 6th grade science.
- It was fun! It was a rare experience, so it was good. I don't like science much, but it's something I enjoy and like to do, so it was good to see that it was fun. Looking at the moon is hard work, but when I saw it, I felt like I did my best.

Based on this positive results, we would like to propose the introduction of telescopic observation at home into problem-solving active learning in elementary and middle school science from the viewpoint of COVID-19 prevention.

References

Agata. H. 2020, JSEPA, 32(5), 4

NAOJ, 2020, *Kaifu-NAOJ telescope kit*, https://www.nao.ac.jp/study/naoj-tel-kit/en/

Education and Heritage in the era of Big Data in Astronomy
Proceedings IAU Symposium No. 367, 2020
R. M. Ros, B. Garcia, S. R. Gullberg, J. Moldon & P. Rojo, eds.
doi:10.1017/S1743921321000375

Reframing Pedagogy: Teaching Astronomy through STEAM Innovation

Exodus Chun-Long Sit[iD]

National Astronomy Education Coordinator (Chair of Hong Kong),
IAU Office of Astronomy for Education
email: `sitexodus@gmail.com`

Abstract. This contribution explores the reframing of promoting Astronomy as popular science, inspired by the COVID-19 pandemic. Through STEAM Innovation, integrating science and arts, such as Astro-Music and Space Art, would be a case in point of forced association. It redefines our methodology of Astronomy education and encourages the engagement of teachers from other disciplines. Supporting with user-centered design thinking, this pedagogy contributes effectively to the interactive teaching for solving real-life problems related to Astronomy.

Keywords. STEAM Education, Astro-Music, Innovation, Astropreneur

1. Introduction

The pandemic situation of COVID-19 had changed our daily lives, especially the forms of promoting astronomy and popular science. Many new events had been launched without borders, such as Astro At Home, virtual stargazing, conference live streaming, and online night sky observation. It then raised a follow-up question about the future of Astronomy Education, changing from sidewalk astronomy to interactive digitalization. For better implementation, it is crucial to consider the interdisciplinary approaches to effective teaching, based on the methodologies on Astronomy Education. Face-to-face lectures may possibly be replaced by convenient online platforms.

STEM Education has already been launched for several years in Hong Kong, which is mainly dominated by industries related to robotics, mechanical engineering, computer programming, and AI technology. It may raise a question on how could we can learn Astronomy or even Natural Science through STEM Education? How could we think out of the existing box, which are practically observational Astronomy and theoretically Astrophysics and Cosmology?

2. Overview

Based on observation and engagement in Astronomy Education and outreach programs in the past few years, I would like to propose a new idea called "STEM+A@Astronomy oriented". It was inspired by famous astronomer Kepler's famous piece of "The Harmony of the World" (Harmonices Mundi) (Gingras 2003). As he discovered the consonance between the planetary motions and music, he became the pioneer of Astro-Music and proposed his Third Law of Planetary Motion (Russell 1964).

One of the Astro-music pieces composed was "Valentine's Rosette" (Fig. 1) which was related to the digital composition of interactive music generated by the images of Rosette Nebula, as a well-known dark sky object. This music was innovatively composed, based on its stellar structures and nebulous features, such as NGC 2237, NGC 2238, NGC 2239, NGC 2244 and NGC 2246, celebrating Valentines Day in 2020. It would be my honour to

Figure 1. "Valentine's Rosette" as an example of Astro-Music composition.

cooperate with local astrophotographer with photos demonstrated in both Hubble Palette and Hydrogen-Alpha Image-processing. As the Prelude of "The Harmony of Mysterious Exoplanets" (Nebula), "Valentine's Rosette" had been showcased in TEDx seminars, Shaw-IAU workshop and IDA conference. It would be official performed in Hong Kong Space Museum in 2021, affected by the postponement of public lectures.

As for the instrumentation selected, it would be performed by Steel Tongue Drum. It is an unique musical instrument which can simulate the feeling and sound of Astro-Music. Inspired by the universal music (musica universalis) (Birat 2017), conceptual music could demonstrate the abstract connections and internal consonance of "Music of Sphere" model vocally. It would be a new interactive media to motivate people who are interested in Astronomy or Music.

3. Implications

Astro-Music is a flawless illustration of "STEM+A@Astronomy oriented". It is different from traditional STEAM Education, as we focus on Astronomy as the core subject on having connections with other potential fields of studies, no matter they are science-related or not. During the integration of Astronomy with other disciplines, we could generate more and more possibilities or even new Astronomy-related subjects in the future. Technically, it would be a progress of Forced Association (Mcfadzean 1999). By integrating Science and Arts, it could then form a new subject, such as Astro-Music.

For reframing astronomical pedagogy, we could try to think like an "Astropreneur" (Astronomer + Entrepreneur), who would be familiarized with future abilities, holistic design thinking, confident mindsets, and creative advocacy for preparing unexpected incidents in the future. To have an effective communication of popular science, it is suggested to consider learner-centered approaches (Michael 2005), focusing on what people wish to learn, rather than what we think they might be interested in. Music could be an incentive to motivate the general public to explore Astronomy.

References

Birat, J.P. 2017, *Musica Universalis or the Music of the Spheres*, SAM conference, p. 1–24

Gingras, B. 2003, *JRASC*, 97, 259G

Mcfadzean, E. 1999, *LODJ*, 20, p. 374–383

Michael, C.L. 2005, *AER*, 4, p. 83–89

Russell, J.L. 1964, *BJHS*, 2, p. 1–24

Education and Heritage in the era of Big Data in Astronomy
Proceedings IAU Symposium No. 367, 2020
R. M. Ros, B. Garcia, S. R. Gullberg, J. Moldon & P. Rojo, eds.
doi:10.1017/S1743921321000260

School Workshops on Astronomy – an extracurricular activity for secondary school pupils

Sylwester Kołomański[1] and Joanna Molenda–Żakowicz[1]

[1]Instytut Astronomiczny, Uniwersytet Wrocławski,
ul. Kopernika 11, 51-622, Wrocław, Poland
email: `sylwester.kolomanski@uwr.edu.pl`, `joanna.molenda-zakowicz@uwr.edu.pl`

Abstract. Interdisciplinary and egalitarian, the School Workshops on Astronomy have been being in their educational mission since 14 years. Here we present the concept, methods, and some example results of that educational technique.

Keywords. educational techniques, project method, hands-on activities

1. Hands-on astronomy

Hands-on activities are powerful educational tool based on a simple truth: if you do something yourself, you learn more and you learn faster. Keeping that in mind, we mixed the traditional approach to teaching with the project method to develop a genuine formula of teaching astronomy and science to secondary school pupils.

The School Workshops on Astronomy (SWA)† are a bi-annual event popularizing astronomy and other sciences, sparked off by Grzegorz Żakowicz, a teacher of physics at the XIII Lyceum in Wrocław (Poland). The SWA take place at the heart of the Izera Dark Sky Park in Poland, which is a perfect location for that kind of event. The staff members who are responsible for various educational activities that take place at SWA are astronomers from the University of Wrocław, teachers from high schools, and invited guest specialists. There have already been 23 editions of the School Workshops on Astronomy. The total number of pupils who took part in them exceeds 500.

The SWA focus on astronomical observations and hands-on activities during which pupils learn practical aspects of observing techniques which include orientation on the night sky with the help of sky maps and planispheres, assembling and using simple astronomical telescopes, performing visual or astrophotography observations, measuring brightness of the night sky to assess the level of light pollution, and presenting results of those observations in a form of reports.

Other workshops which build the SWA focus on providing pupils insight into fundamental physical laws, helping them to develop scientific thinking and the ability to formulate scientific predictions. A good example is an exercise in which the size of a crater made in sand by a small ball in a freefall can be extrapolated to the energies needed to form the real craters left by asteroids that hit our planet in the past (see Fig. 1). During that exercise, pupils learn that impact craters are the most common surface structure in the Solar System, and that they provide insights into the age and geology of the planetary body. Being able to see that a very simple experiment can lead

† School Workshops on Astronomy http://www.swa.edu.pl

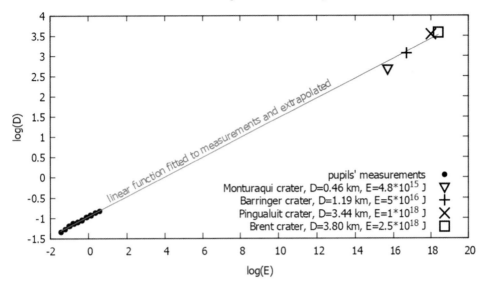

Figure 1. The relation between the diameter of the crater ($\log D$) produced as a result of an impact and the energy released in that event ($\log E$) obtained during one of the Workshops. Linear function was fitted to the measurements and extrapolated to real craters. The extrapolation fits well to a sample of four Earth craters which diameters and impact energy are known.

to valid scientific reasoning helps pupils build their self-esteem, while the new knowledge gained during the exercises and lectures widens pupils' intellectual horizons. All that may lead them to choosing science for the future career which is the ultimate purpose of the SWA. Even if they will not become scientists or engineers, they can possess skills that are very useful in everyday life: critical, analytical, and rational thinking, problem-solving skills, forming evidence-based opinions, curiosity and open-mindedness, awareness of existence of cognitive biases, ability to work in a team and to share knowledge with others.

Astronomy is the main content of the SWA, but other sciences (physics, biology, geography, meteorology, geology, computer science, ect.) are also present. This inclusion allows the pupils to see connections between ideas across different disciplinary boundaries.

2. Education beyond curricula

Over the last hundred years, we witnessed many astronomical discoveries that significantly changed our understanding of the Universe. The advancement in astronomy induced also technological progress and, as a result of it, influences our everyday life. Eventually, science and technology became foundation of modern society and an intrinsic part of modern civilisation. Therefore, it is obvious that science (including astronomy) has to be part of school education. Surprisingly enough, over the last 20–30 years astronomy has been significantly reduced in education curricula in Poland, resulting in unsatisfactory level of astronomical knowledge among pupils and a renaissance of unscientific concepts, like astrology, in the society (see, e.g., Molenda–Żakowicz, *et al.* 2020).

Such a trend is very difficult to counteract from the position of individual teachers and educators. However, even little help is important and therefore, extracurricular activities like the School Workshops on Astronomy play crucial role in filling gaps in knowledge and understanding among pupils.

Reference

Joanna Molenda–Żakowicz, Maciej Kokociński, Sylwester Kołomański & Magdalena Ziółkowska–Kuflińska 2020, *Proc. of the IAU Symp. 367/VIRTUAL Education and Heritage in the Era of Big Data in Astronomy, 8-12 December 2020, poster 'Astronomy through Continents'*

Education and Heritage in the era of Big Data in Astronomy
Proceedings IAU Symposium No. 367, 2020
R. M. Ros, B. Garcia, S. R. Gullberg, J. Moldon & P. Rojo, eds.
doi:10.1017/S1743921321001137

SeneSTEM: promoting STEM in Senegal

Salma S. Mbaye[1,2]🆔**, Modou Mbaye**[1] **and Katrien Kolenberg**[3,4,5,6]

[1]Institut de Technologie Nucléaire Appliquée, Université Cheikh Anta Diop,
Bp 5005 Dakar-Fann, Sénégal
email: `salma.sylla@ucad.edu.sn`

[2]Institut de Mécanique Céleste et de Calcul des Ephémérides, Observatoire de Paris,77 Avenue
Denfert-Rochereau
email: `salma.sylla@obspm.fr`

[3]Laboratoire de Télédétection Appliquée,Université Cheikh Anta Diop, Sénégal
email: `modou2812.mbaye@ucad.edu.sn`

[4]Department of Physics, Universiteit Antwerpen, Groenenborgerlaan 171, 2020 Antwerpen
email: `katrien.kolenberg@uantwerpen.be`

[5]Institute of Astronomy, KU Leuven, Celestijnenlaan 200D, 3001 Leuven, Belgium

[6]Physics and Astronomy Department, University of Brussels – VUB, Pleinlaan 2, Belgium

Abstract. SeneSTEM aims to bring Senegalese children and young people into contact with
science, and – by extension – with the STEM disciplines (Science Technology Engineering
Mathematics) in a very accessible and illustrative way. We do this with concrete workshops and
experiments, for both teachers and for groups of children and young people. In collaboration
with different educational organisations in Senegal, among which the Senegalese Association for
the Promotion of Astronomy, SeneSTEM ensures that all layers of the population are enthusiastic about science (education). Special attention is given to motivating girls for scientific careers.
SeneSTEM actions are based on an international collaboration partly supported by development
funds from the city of Antwerp and the University of Antwerp (Belgium).

Keywords. STEM, workshops, experiments, teachers, girls in STEM, international
collaboration

---

1. Introduction

SeneSTEM is born from the ideas of a long collaboration between Senegalese and
Belgian actors working in the fields of research in astronomy, physics, science education,
science promotion, and also art. It all started in 2010 with the organization of a seminar
at the University Cheikh Anta Diop (UCAD) as part of the Teaching Astronomy for
Development programs (TAD) of the International Astronomical Union. This made it
possible to bring together students from UCAD, secondary school teachers, and the
general public for five days to initiate them into basic astronomy via presentations but
also practical observation sessions in partnership with the Senegalese Association for the
Promotion of Astronomy (ASPA).

The great success motivated to organize other events the following years with specific
target audiences, with the goal of making our activities more concrete and sustainable.

With some seed funding from a development program of the city and University of
Antwerp (Belgium), we set up the SeneSTEM program in 2018. The program consists of
scientific activities intended for all levels of school education by putting astronomy at the

Visit to secondary schools with a telescope, thanks to SSVI (https://www.ssvi.be/)

Electricity workshop in Kolda, in the south of Senegal and at UVS (https://www.uvs.sn/)

Spacecraft materials' workshop, April 2019

Figure 1. Activities of SeneSTEM.

center thanks to its transversal character and its attractive power. In this framework we also take part in regular astronomical activities at the local level (Baratoux *et al.* 2017).

2. Mission and example activities

The SeneSTEM program responds to four Sustainable Development Goals (SDGs) namely:
– Quality education (SDG 4);
– Gender equality (SDG 5);
– Decent work and economic growth (SDG 8);
– Partnerships for the achievement of our objectives.
Some example activities (Fig. 1) include: (1) conferences in three secondary schools in Senegal in collaboration with the Association of Young Geologists and Environmentalists of Senegal, and donation of a telescope thanks to the SSVI program; (2) workshop on testing spacecraft materials to secondary school teachers thanks to the support of the Open Astronomy School, the Direction for the Promotion of Scientific Culture (DPCS) of the Ministry of Higher Education, Research and Innovation, and the Institute of Applied Nuclear Technologies (ITNA / UCAD); (3) participation in the science week organized

by the ambassador of the Next Einstein Forum (NEF) in Senegal; (4) participation in "Science Wednesdays" organized by the DPCS.

Reference

Baratoux, D., et al. 2017, *Eos*, 96, https://eos.org/features/the-state-of-planetary-and-space-space-sciences-in-africa; doi:10.1029/2017EO075833

Education and Heritage in the era of Big Data in Astronomy
Proceedings IAU Symposium No. 367, 2020
R. M. Ros, B. Garcia, S. R. Gullberg, J. Moldon & P. Rojo, eds.
doi:10.1017/S1743921321000284

Solar eclipses: A pump of curiosity for early humans?

Graham Jones[iD]

The University of Shiga Prefecture,
Hikone, JP 522-8533, Japan
email: graham@tensentences.com

Abstract. The dramatic nature and irregular frequency of solar eclipses may have helped trigger the development of human curiosity. If the kind of solar eclipses we experience on Earth are rare within the Universe, human-like curiosity may also be rare.

Keywords. eclipses, astrobiology, extraterrestrial intelligence, Earth, Moon

1. Introduction: Where is everybody?

According to the Copernican principle, there is nothing special about the Earth's place in the Universe. However, on the current evidence, human curiosity — defined here as our desire for reasons and explanations — is unique. No matter where we point our telescopes, we find no signature of extraterrestrial intelligence. This contradiction is an example of the Fermi paradox, also known as the Great Silence (Brin 1983), and described as "one of the most pressing problems in science" (Webb 2015).

A potential astrobiological solution is the Rare Earth hypothesis, which states that the development of complex life depended upon a combination of factors that may be vanishingly rare within the Universe. In short, humans may not be "so ordinary as Western science has made us out to be for two millennia" (Ward & Brownlee 2000).

2. A new question: Extending the Rare Earth hypothesis

Biological complexity does not automatically lead to curiosity. For instance, chimpanzees show no interest in magic tricks (Matsuzawa 2020). Nevertheless, human development arrived at a point where "we find it appropriate to describe the reasons why some things are arranged as they now are" (Dennett 2017). We can therefore ask: does the Earth have a special ingredient that led to the development of not only complex life, but also curious life?

The Earth's environment is notable for its regularity; even phenomena such as earthquakes are "part of life" in the areas where they tend to occur (Hinga 2015). Yet "too much regularity in the selective environment can be a trap" (Dennett 2017). Even though the wiring of the human brain evolved in an exceptional way (Ardesch *et al.* 2019), if novelty had remained below a certain threshold, early humans may not have received a sufficient trigger to begin forming the concept of reasons.

3. Solar eclipses: Off-the-scale novelty, by chance

The "dramatic nature" (Pasachoff 2017) of solar eclipses produces lasting effects on humans (ibid) and human culture (Blatchford 2016). Solar eclipses would have provided early humans with novelty on a scale unlike anything else within the environment. Francis

Baily, a pioneer of solar physics in the 19th century, noted that "I can readily imagine that uncivilised nations may occasionally have become alarmed and terrified at such an object" (Baily 1846).

Although solar eclipses are not rare in the Universe (Lazzoni 2020), the kind of eclipses that humans experience — end-of-the-world simulations that come, literally, out of a clear blue sky — may be exceedingly rare. Crucially, the nature and frequency of solar eclipses on Earth are the result of two chance circumstances: the Moon and the Sun have the same angular diameter, and the Moon's orbit is inclined to the ecliptic.

4. Conclusion: The pump of curiosity?

Solar eclipses, which "take place at very irregular intervals for a given place" (Meeus 1982), may have pumped novelty into the environment at an ideal rate to trigger the development of curiosity in early humans. If solar eclipses occurred more routinely, they may not have provided a sufficient dosage of novelty; if they occurred less frequently, they may not have delivered a sufficient number of injections of novelty, over time, into human communities.

Given that the nature and timing of solar eclipses is the result of chance, this pump of curiosity may be part of the solution to the Fermi paradox: even if complex life is common in the Universe, curious life may be rare.

Acknowledgements

I am grateful to the following for their comments and support: Ian Blatchford (Science Museum, London), John G. Cramer (University of Washington), Daniel C. Dennett (Tufts University), Cecilia Lazzoni (University of Padova), Tetsuro Matsuzawa (Kyoto University), Amelia Ortiz-Gil (University of Valencia), Jay M. Pasachoff (Williams College), Steffen Thorsen (timeanddate.com) and Stephen Webb (University of Portsmouth).

References

Ardesch, D.J., Scholtens, L.H., Li, L., Preuss, T.M., Rilling, J.K., & van den Heuvel, M.P. 2019, *PNAS*, 116(14), 7101
Baily, F. 1846, *MemRAS*, 15, 1
Blatchford, I. 2016, *Phil. Trans. R. Soc. A*, 374, 20150211
Brin, G.D. 1983, *QJRAS*, 24, 283
Dennett, D.C. 2017, *From bacteria to Bach and back: The evolution of minds* (WW Norton & Company)
Hinga, B.D.R. 2015, *Ring of fire: An encyclopedia of the Pacific Rim's earthquakes, tsunamis, and volcanoes* (ABC-CLIO)
Lazzoni, C. 2020, Personal communication, 20 August
Matsuzawa, T. 2020, Personal communication, 3 September
Meeus, J. 1982, *J. Brit. Astron. Assoc.*, 92, 124
Pasachoff, J.M. 2017, *Nat. Astron.*, 1(8), 1
Ward, P.D., & Brownlee, D. 2000, *Rare Earth: Why complex life is uncommon in the Universe* (Copernicus Books)
Webb, S. 2015, *If the Universe is teeming with aliens ... where is everybody?: Seventy-five solutions to the Fermi paradox and the problem of extraterrestrial life*, 2nd ed. (Springer)

Education and Heritage in the era of Big Data in Astronomy
Proceedings IAU Symposium No. 367, 2020
R. M. Ros, B. Garcia, S. R. Gullberg, J. Moldon & P. Rojo, eds.
doi:10.1017/S174392132100020X

The United Nations Open Universe Initiative

Ulisses Barres de Almeida[iD]

Brazilian Center for Physics Research (CBPF)
Rua Dr. Xavier Sigaud 150, 22290-180, Rio de Janeiro, Brazil
email: `ulisses@cbpf.br`

Abstract. In this contribution I will briefly introduce the concept and objectives of the Open Universe Initiative, as well as describe the first steps of its implementation by Brazil, in conjunction with the United Nations Office for Outer Space Affairs (UNOOSA), aiming to encourage new interested parties to join the Initiative.

Keywords. astronomical data bases, methods: data analysis, sociology of astronomy

1. Motivation and Concept

Much has been done in recent years, especially in space astronomy, to offer openly accessible data and user-friendly platforms that demonstrate a natural evolution in the field of space sciences towards a more transparent and inclusive ecosystem of services. However, despite the recent progress, there is still a considerable degree of unevenness in the resources available from various providers of space science data, and on the capabilities for exploitation of such information by the various potential groups of users across society and around the globe. Further efforts are therefore necessary to expand access to data and data services in space sciences, promoting a significant data-driven surge in training, education and discovery that may inspire the new generations. Such a process should be extended to non-scientific sectors of society.

The Open Universe Initiative, currently under implementation by the United Nations Office for Outer Space Affairs (UNOOSA), in cooperation with the Government of Brazil, is a project to foster the continued development of a culture of open shareable data in astronomy and spaces sciences, aiming to serve as a tool contributing to promote quality education and capacity building in the space sector.

Developed in the context of the Space 2030 Agenda and the United Nations Sustainable Development Goals, its principal objective is to impact education and development at a global scale. The original proposal of the Initiative – which was firstly presented to the Committee on the Peaceful Uses of Outer Space (COPUOS) in 2016 by the Government of Italy, with the support of Brazil (COPUOS Research Paper A/AC.105/2016/CRP.6) – stems from the view that scientific data, including space science and astronomical data, are a common good that should be freely available to the benefit of all people. In fact, the Initiative supports the idea that open data is today the primary entry door to an equitable access to space and to the democratic distribution of the benefits resulting from space exploration.

Under the leadership of the United Nations, Open Universe plans to achieve these goals by improving international cooperation and fostering partnerships to the development of innovative tools that improve data accessibility and usability at all levels, from the expert to the interested citizen, thus linking holders of astronomy and space science data with an ever-broader world community of users that can exploit those data for whatever scientific, educational or cultural purpose.

The United Nations Office for Outer Space Affairs (UNOOSA), as the UNs dedicated entity for space affairs, works on the legal, policy and capacity building aspects of international cooperation in the peaceful uses of outer space. For the Open Universe initiative, UNOOSA acts as the central node and executive secretariat. As the central operational node of the initiative, it will support the conduction of capacity building activities and foster best practices for open shareable data policies to contribute to the development of space sciences globally, helping to bridge the gap between developed and developing nations.

Having the UN as a central broker in the international cooperation, supported by Brazil and the other participating members, the Initiative is open to all types of public entities dealing with data in astronomy and space science, whether governmental institutions, non-governmental organisations or intergovernmental associations. Private entities accepting the licence conditions on their data are similarly welcome to take part in the Open Universe. Participant institutions will therefore be expected to contribute with the provision of data, documentation, software, training, and educational material.

2. Status Overview

Following the initial proposal at COPUOS in 2016, the initiative was included among the activities to be carried out in preparation of the fiftieth anniversary of the first United Nations Conference on the Exploration and Peaceful Uses of Outer Space (UNISPACE+50). A number of public discussions and international meetings were held, that defined in detail the structure and objectives of the Initiative, among which:

- the 2017 ASI-UNOOSA Preparatory Expert Meeting (A/AC.105/2017/CRP.22)
- the 2017 UN/Italy Workshop on the Open Universe Initiative (A/AC.105/1175)
- the participation in the UNISPACE+50 Conference, in 2018[†]
- and the participation at the UN/Austria World Space Forum, in 2019[‡]

As a conclusion of this extensive preparatory period, the Initiative has been finally welcomed to be developed under the leadership of UNOOSA as part of its Capacity Building Programme, and in the context of the activities to be conducted as part of the Space 2030 Agenda (COPUOS Working Paper A/AC.105/C.1/2020/CRP.16).

The formal start of the initiative is planned for 2021, in partnership with the Government of Brazil, through an implementation agreement between UNOOSA and the Ministry of Science, Technology and Innovation, after which official activities will commence, and additional partners are welcome to join.

Despite the initiative being still pending a formal foundation, pilot projects have been developed during the past several years by the various institutional partners in an effort to demonstrate the principles and feasibility of the proposal. A summary of all activities can be found in the recent reference by Barres de Almeida *et al.* (2020), the principal one being the Open Universe prototype portal developed by the Italian Space Agency[§].

Reference

Barres de Almeida, U., Giommi, P. & Pollock A. 2020, *arXiv:2006.09168*, to appear in "Annals of Brazilian Academy of Sciences", Proc. BRICS Astronomy Workshop (BAWG 2019).

† https://www.unoosa.org/oosa/en/ourwork/unispaceplus50/index.html
‡ See https://www.unoosa.org/oosa/en/ourwork/hlf/2019/world-space-forum-2019.html
§ https://openuniverse.asi.it

Session 3

Education and Heritage in the era of Big Data in Astronomy
Proceedings IAU Symposium No. 367, 2020
R. M. Ros, B. Garcia, S. R. Gullberg, J. Moldon & P. Rojo, eds.
doi:10.1017/S174392132100096X

An investigation on conceptual understanding about cosmology

Arturo Colantonio[1,2] 🄫, Irene Marzoli[1], Italo Testa[2,3] and Emanuella Puddu[2]

[1]School of Science and Technology, Physics Division, University of Camerino, Camerino, Italy
email: `arturo.colantonio@unicam.it`

[2]INAF – Astronomical Observatory of Capodimonte, Naples, Italy

[3]Department of Physics "E. Pancini", University Federico II, Naples, Italy

Abstract. In this study, we identify patterns among students beliefs and ideas in cosmology, in order to frame meaningful and more effective teaching activities in this amazing content area. We involve a convenience sample of 432 high school students. We analyze students' responses to an open-ended questionnaire with a non-hierarchical cluster analysis using the k-means algorithm.

1. Introduction and Aims

Cosmology is a meaningful context to teach, at high school level, contemporary physics topics, such as quantum mechanics, particles' standard model, nuclear reactions. The students beliefs and ideas about cosmology have been investigated in some aspects (e.g. the works of Prather *et al.* 2002, Wallace *et al.* 2012, Coble *et al.* 2018), but a coherent picture of students conceptual understanding is yet to be provided. We aims to identify patterns between such beliefs and pre-instructional ideas about cosmology, in order to frame meaningful and more effective teaching activities in this content area.

2. Methods

On the basis of previous studies (Wallace *et al.* 2012, Bailey *et al.* 2012, Trouille *et al.* 2013) we have identified two groups of conceptual dimension: "basic" and "advanced". The first group concerns fundamental astronomical entities (CO) such as stars, galaxies, constellations, nebulae, and time and length scales of typical astronomical events and objects (AD). The second group includes birth (BB) and age (Age) of Universe; temperature and chemical composition of the Universe changing over the time (T&C); space-time expansion (EX); hypothesis about the future of the Universe, black holes, dark matter and dark energy (BHDM). Then, we designed a questionnaire with 17 open-ended questions that addressed two or more aspects of the identified dimensions. We involved in this study a convenience sample of 432 high school students (17.9 ± 0.7 years old) attending extra-curricular activities about physics topics at the authors' Department. The collected data set was independently and completely analyzed by three researchers, who defined for each question five categories fitting the students' responses and ranging from "not given or unclear response" to "scientifically correct". To check students' responses categorization we used inter-rater reliability, obtaining at the end of process a satisfactory level of 0.82. We combined the students' answers on a given aspect by using the non-hierarchical cluster analysis and the k-means algorithm, with the aim to identify

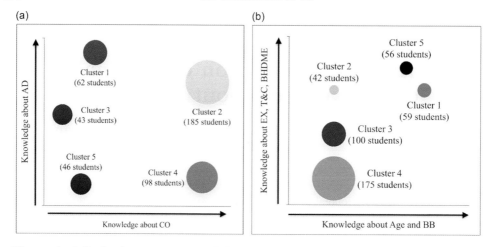

Figure 1. 2-D visual representation of the cluster distribution regarding basic dimensions (panel *a*) and advanced conceptual ones (panel *b*).

reasoning patterns corresponding to different levels of conceptual understanding about the targeted dimensions (Battaglia *et al.* 2019).

3. Results

For each dimension, we choose a five clusters solution, which reflect increasingly complex reasoning. The final interpretation of each cluster was validated by the same professional astrophysicists. In Fig. 1 we show our cluster solutions. In panel (a) we represent the position of each clusters with respect to the increasing levels of knowledge about the two basic dimensions. We note that there is no correlation between the knowledge of definitions of celestial objects and, between their mutual distance and the timeline of some events in the history of universe. Moreover, about 30% of them are unable to order from the nearest to the most distant one, compared to the Earth, a series of astronomical objects and fail to reconstruct the timeline of significant events in the universe history. In panel (b) we represent the position of each clusters with respect to the increasing levels of knowledge about Age and BB versus EX, T&C and BHDM. We note that the majority of students are in the lowest level of knowledge. Furthermore the students being in the higher knowledge levels about Age and BB show also an high levels of knowledge about expansion, temporal evolution and more advanced aspects of cosmology.

4. Conclusions

The collected students' responses suggest that cosmology is addressed during the curricular teaching and the dissemination activities in informal setting. Cluster analysis results point out that some relevant aspects are neglected, for instance how scientists support claims about theories of the Universe. Moreover, the curricular teaching seems to have a limited impact on students' ideas also about basic aspects, such as the role of gravity and does not allow a deep conceptual understanding about cosmology. To validate the identified clusters, we are in the process of administering a revised version of the questionnaire to a wider sample of students.

References

Wallace C., Prather E. & Duncan D. 2012 *Astron. Educ. Rev.*, 1, 1.
Coble K., Conlon M. & Bailey, J. M. 2012 *Phys. Rev. Phys. Educ. Res.*, 14, 010144.
Prather E., Slater T. & Offerdahl E. 2002 *Astron. Education Rev.* 1, 28.

Bailey J.M., Coble K., Cochran G., Larrieu D., Sanchez R. & Cominsky, L.R. 2012 *Astron. Educ. Rev.*, 11, 10.384.

Trouille L., Coble K. L., Cochran G., Bailey J., Camarillo T. C., Nickerson M. & Cominsky L. 2013 *Astron. Educ. Rev.*, 12. 10.3847.

Battaglia O., Di Paola B. & Fazio C. 2019 *Phys. Rev. Phys. Educ. Res.*, 2, 020112.

Education and Heritage in the era of Big Data in Astronomy
Proceedings IAU Symposium No.367, 2020
R. M. Ros, B. Garcia, S. R. Gullberg, J. Moldon & P. Rojo, eds.
doi:10.1017/S1743921321000338

Solar Eclipse: a didactic alternative for education in Astronomy

Viviana Sebben[1] and Claudia Romagnoli[2]

[1]Escuela de Normal Superior N° 34 "Nicolás Avellaneda" de Rosario, Argentina.
email: vrsebben@gmail.com

[2]Escuela de Posgrado. Faculad de Humanidades y Artes. Universidad Nacional de Rosario, Argentina.
emails: clauromag@gmail.com

Abstract. This article presents a didactic experience in teacher training carried out in the province of Santa Fe, Argentina. The training was carried out on the occasion of a total solar eclipse taking place in this region on July 2, 2019. Prior to this event, the authors, members of NASE, developed training meetings of the Ministry of Education of Santa Fe , on topics of Astronomy. From these workshops, participants of all educational levels with different specialties, carried out with their students school science activities where they applied Astronomy with an interdisciplinary perspective. In this way, the eclipse became an event that made possible an enrichment of astronomical education in the region.

Keywords. Solar Eclipse, Didactic experience, Teacher training, Astronomy.

1. Introduction

In 2019, a very important astronomical event took place in the southern hemisphere: a total solar eclipse, and it could be seen especially in a region of the Province of Santa Fe -Argentina. The authors of this article are ambassadors of NASE PG, IAU program for post-graduates, the main objective of NASE is to educate new generations of teachers and re-educate the current ones, in Astronomy topics.

This fact allowed teachers ambassadors of NASE, who were working within the framework of the Teacher Training Meetings developed by the Ministry of Education of Santa Fe, to consider an approach to topics related to Astronomy, with an interdisciplinary perspective. This decision was motivated by both the absence of this discipline at all educational levels as well as the lack of training of teachers in this field of knowledge. Six training meetings were held in different towns located in the south of the province of Santa Fe, where the solar eclipse could be observed. These towns are Rufino, Villa Cañás, Chañar Ladeado, Melincué, Pujato and Rosario.

2. Overview

2.1. *Purposes*

Training teachers to guide their students in the study of Astronomy through approaches to the study of the Universe, using activities which allowed the students to acquire concepts and link them with other disciplines. Guiding teachers to include topics related to Astronomy in the different curricular spaces they work in, favoring interdisciplinary school proposals.

2.2. *Methodology*

The training was presented through different actions, which include: Workshops inspired by NASE courses: practical constructions, modeling, observations and experimentations. The principal material support that used was Ros and García (2012) '14 steps to the Universe'. This book develops 14 sections, including conferences and workshops, that constitute an initial teacher training in Astronomy. Concret models -graphic, schemes, mockups and constructions- were used in the workshops with the purpose, as expressed Camino: 'to generate a dialogue between reality and the process of imagination and abstraction necessary for learning the concepts of the astronomical phenomena under study', (2004, p. 82) in this case the solar eclipse. Also, the Information about eclipses of Tignanelli and Feinstein (2005, pp. 108–122) has been considered as conceptual support for these teacher training. The training was presented through different actions, wich include: workshops, discussion groups, observations of the sky with the naked eye and with telescopes, day and – or night, discussions about the contributions of Astronomy – the Total Solar Eclipse 2019 – and its didactic approach to the school science.

3. Conclusions

Teachers from more than forty towns in the province of Santa Fe attended these meetings. In total, 114 teachers from different educational levels, modalities and skills training participated. The Interest in the extraordinary astronomical event that constitutes the solar eclipse, the construction of simple devices with accessible materials and astronomical literacy guidelines, observations and records, allowed the participants to investigate the near reality and to become involved in an autonomous way in the study of the environment, such as parts of this environment itself, creating special links that might improve their teaching and learning processes. The trainings achieved the motivation of the teachers who considered the shared resources as didactic tools to do school science where the eclipse was an unbeatable opportunity for the promotion of various student research. Ten of these studies were presented at different socialization and dissemination events of the classroom. Due to its spectacular nature and accessibility to be observed, this event was a meeting point for students, teachers and the community in general, where material of incalculable graphic and didactic value was shared through social networks. This first approach, made up of the training meetings and the didactic proposals that emerged from them, became an interdisciplinary vision of astronomical education in the province of Santa Fe. Finally, the didactic alternative for astronomy education presented, as expressed by Gangui and Iglesias (2015, p. 12) 'gives us guidelines to try to build knowledge in the most efficient and pleasant way that we can access. It takes us along the way, to generate learning that makes sense for people and, in general, for the whole community'.

References

Camino, N. 2004, Aprender a imaginar para comenzar a comprender: los modelos concretos como herramientas para el aprendizaje de la Astronomía. *Alambique*, 42, pp. 81–89

Feinstein, A. & Tignanelli, H. 2005, Objetivo: Universo, *Colihue*

Gangui, A. & Iglesias, M. 2015, Didáctica de la Astronomía, *Paidos*

Ros, R. M. & García, B. 2012, 14 pasos hacia el Universo. Curso de Astronomía para profesores y posgraduados de ciencias, NASE, UAI, *Antares*

Education and Heritage in the era of Big Data in Astronomy
Proceedings IAU Symposium No. 367, 2020
R. M. Ros, B. Garcia, S. R. Gullberg, J. Moldon & P. Rojo, eds.
doi:10.1017/S1743921321000193

The right to the night sky in punitive enclosure context

**Natalia Guevara[1]ⓘ, Rodrigo F. Haack[1], Victoria B. Acosta[1],
María A. Senn[1], Carolina A. Silva[1], Rocío Adamson[1], Jeremias Ruta[1],
Natalia M. Gómez[1], Karen A. Brellis[1], Marina S. Puga[1],
Bruno J. De Bortoli[1,2]ⓘ, Fiamma S. Pallazo[1] and Ayelén Lizzi[3]**

[1]Facultad de Ciencias Astronómicas y Geofísicas, Universidad Nacional de La Plata
Paseo del Bosque S/N, La Plata, Buenos Aires, Argentina.
contact email: guevaran@fcaglp.unlp.edu.ar

[2]Instituto de Astrofísica de La Plata, Universidad Nacional de La Plata,
Paseo del Bosque S/N, La Plata, Buenos Aires, Argentina.

[3]Facultad de Periodismo y Comunicación Social, Universidad Nacional de La Plata,
Diagonal 113 esquina 63 Nro 291, La Plata, Buenos Aires, Argentina

Abstract. The "Right to the night sky" outreach project holds astronomy workshops for children and teens deprived of their liberty in juvenile detention centers. It is carried out by an interdisciplinary group of students, graduates, and teachers of Astronomy, Geophysics, Educational Science, Law, Psychology, Social Work, and Social Communication. It's has been accredited and recognized by the Faculty of Astronomical and Geophysical Sciences, and the National University of La Plata (Argentina) since the year 2014. This work presents the diverse activities developed in the project, the methodologies used, and an analysis of how the project evolved, grew, and expanded over time, continuing what has already been presented by Charalambous *et al.* (2014) and Haack *et al.* (2019)

Keywords. astronomy education, human rights, outreach, enclosure context

1. Introduction

The project started as an informal proposal in 2013. Students and teachers of Astronomy, Law, Education Science and Social Work gave astronomy workshops to the teens living in Araoz Alfaro closed center, under the approval of the institution directors at the time. After this experience, a request was raised by several institutions in the area to repeat the experience. Therefore, in 2014, the education proposal evolved into an outreach project that could satisfy this demand and sustain itself in time. At present, it remains an outreach project of the Astronomy and Geophysics Faculty, part of the National University of La Plata. As team members change over time, and so the different disciplines involved, the educational proposal continues to improve and evolve, along with our formation as educators in the particular territory of punitive institutions.

2. Overview

The proposal is based on the human right to education and recreation, declared in various international, national, and regional legislations. These laws protect and promote their access, even for people deprived of their liberty, and emphasize their protection for children and teens. Imprisonment sentences should not affect these rights, especially when the local juvenile penal regime conceives prison confinement as a social-educative measure. Yet, these rights are not fully satisfied inside the punitive institutions for youths (see Daroqui *et al.* 2012).

The project intervenes in the Partido of La Plata, the epicenter of the juvenile detention institutes of the region. The access to the observation of the night sky it's almost null for the youths imprisoned there, as the cells on the vast majority of these establishments do not possess windows that allow the view of the sky and, even though in these institutions outdoor activities are allowed and nighttime activities are not explicitly forbidden, they are scarce, mainly because the night is often associated with "danger" or "insecurity".

On this basis, the project's main goal is to strengthen the access to the fundamental rights of education, recreation, and bonding with the night sky, of the children and teens deprived of their liberty. The proposal seeks to produce an educative and recreational environment where freedom of speech and creativity is encouraged; to reinforce the youths' astronomy knowledge, retrieving their previous education; to promote the youths' right to bond with the night sky, reinforcing the identity relationship person-landscape; and to reflect on the living regime of the institutions. To enhance the educational proposal communication channels are established with the teachers and psycho-social team of the institutions, also helping on the project's continuity over time.

3. Methods

To produce a working environment that supports the goals of the projects, we utilize hands-on workshops. This method favors the active participation of the subjects, conjugating the context and knowledge involved in a learning process that results from the exchange between the participants towards the collective construction of wisdom. The hands-on approach creates a relaxed setting, that enables not only participation but also dialog. We understand dialog as an "encounter" where knowledge and action are combined, contributing to -in our case- the democratization of astronomical knowledge. Moreover, the dialog produces a trusting climate between the subjects, which is encouraged by the conformation of a horizontal relationship between the participants and the members of the project, making it possible to bring out personal experiences, thoughts, and ideas of the subjects involved.

At present time, there are four thematic workshops (Cardinal Points, Moon, Solar System, and Space Technology) grouped under a cycle, that we offer for a group of 15 to 20 participants. Each themed workshop has its corresponding planning, they are linked with each other and follow an order. Each one has a "classroom moment" of indoor activities, and an outdoor night sky astronomical observation, where we use a telescope, binoculars or just observe with the naked eye.

The workshops take place every 15 days thus, we work for two months with the same group of youths. After that, for closure, we organize along with the directors of the institution a daytime visit to La Plata Observatory and Planetarium, where we have a guided tour, watch a function, have outdoor activities in the green area of the Observatory and, if the weather allows it, we conclude with a solar observation.

References

Charalambous C., Sáez M., Scalia M.C., Fusé M.D., Godoy J.P., Giménez D. L., San Sebastián I. L., Fasciolo M. I., García F., Gargiulo I., Kornecki M.P., Montúfar Codoñer S.E. , Novarino M. & Zeballos M. L. 2014, *CRAAA*, 57

Daroqui, A., López, A., Cipriano García, R. 2012, *Sujetos de Castigo* Ed: Homosapiens.

Haack, R. F., De Bortoli, B. J., & Pessi, P. J. 2020, *BAAA*, 61B, 242

Education and Heritage in the era of Big Data in Astronomy
Proceedings IAU Symposium No. 367, 2020
R. M. Ros, B. Garcia, S. R. Gullberg, J. Moldon & P. Rojo, eds.
doi:10.1017/S1743921321000168

The sample of eight LLAGNs: X-ray properties

Nadiia G. Pulatova[1], **Anatoliy V. Tugay[2]** and **Lidiia V. Zadorozhna[2]**

[1] Main Astronomical Observatory of the National Academy of Sciences of Ukraine,
Akademika Zabolotnoho str. 27, Kyiv, Ukraine, 03143
email: nadya@mao.kiev.ua

[2]Taras Shevchenko National University of Kyiv, Physical faculty,
ave. Glushkova 2, building 1, Kyiv, Ukraine, 03680
emails: tugay.anatoliy@gmail.com, zadorozhna_lida@ukr.net

Abstract. LLAGN are very important objects for studying as they are found in a large fraction of all massive galaxies. Nevertheless this topic needs more investigation as fraction of LLAGN in all AGN are much more higher than fraction of researches dedicated to LLAGN among all AGN studies. The goal of our work is checking out X-ray properties of LLAGN. For this purpose we created a sample of LLAGN by selecting most prominent LLAGN from literature and analyzed their X-ray spectral properties. As a result, we obtained 12 LLAGN and for 8 of them XMM X-ray observations are available. The spectra from one XMM camera, PN, were fitted with power law + absorption of neutral hydrogen. In the current report we present the previous results of this study. We plan to increase numbers of objects in our future studies.

Keywords. LLAGNs, X-ray spectra, XMM Newton.

1. Introduction

For nowadays numerous observations in all electromagnetic spectrum are available for public using and searching for a new ideas with a purpose to clarify the nature of AGN (active galactic nuclei). X-ray variability of different types of AGNs has beed studied for many years (Hernández-García (2016), Hernández-García (2017)). It is generally accepted that most time of its "life" galaxy spend in a quiescent mode. Therefore all studies connected with LLAGN plays very important role in understanding of the nature and physical mechanisms that take place in a galactic nuclei. The intrinsic faintness, i.e. $L_{Bol} < 10^{44} erg \cdot s^{-1}$, and the low level of activity are the distinctive characteristics of LLAGNs, which mainly consists of local Seyfert galaxies and Low-Ionization Nuclear Emission-line Regions (LINERs).

2. Overview

The common properties of LLAGNs were studied on the base of different samples of LLAGNs. Minfeng Gu & Xinwu Cao (2009) found an interesting difference between Low-Ionization Nuclear Emission-line Regions (LINERs) and luminous active galactic nuclei (AGNs). A significant anti-correlation was revealed between the hard X-ray photon index and the Eddington ratioL_{Bol}/L_{Edd} for a sample of LINERs and local Seyfert galaxies compiled from literature with Chandra or XMM-Newton observations. To the most brightest LLAGNs were dedicated numerous studies by many scientific groups from different countries. For example, M87 that is included to our sample of LLAGNs. In Jolley, E.J.D., Kuncic Z. (2007) to the well known LLAGN M87 was applied an accretion model that attributes the low radiative output to a low mass accretion rate, rather than a low radiative efficiency. Authors calculated the combined disk-jet steady-state broadband spectrum. A comparison between predicted and observed spectra indicates that M87 may be a maximally spinning black hole accreting at a rate of $\sim 10^{-3} M_{sun} yr^{-1}$ In the resent report made by Younes *et al.* (2019) is presented the analysis of simultaneous XMM-Newton+NuSTAR observations of two low-luminosity Active Galactic Nuclei (LLAGN), NGC 3998 and NGC 4579. In the paper were calculated parameters of X-ray emmision of these LLAGNs, and were discussed these results in the context of hard X-ray emission from bright AGN, other LLAGN, and hot accretion flow models.

The goal of our study was to obtain XMM spectra for LLAGNs and to search for the common behavior of their spectra. Therefore we performed facial analysis of XMM observations for every source. To find the best spectral fit to the data, several models were tested in this work. For the beginning we fitted every obtained spectra with a simple phabs·powerlaw model that shows that X-ray emission is driven by non-thermal processes. We found that X-ray spectra of only 3 sources from our sample can be fitted with simple phabs·powerlaw model. Another 5 LLAGNs have very complicated spectra and fitting with phabs·powerlaw model doesn't provide good χ^2 statistic.

The data extraction was performed with the Science Analysis System (SAS) version 16.0.0, The PN data are selected using event patterns 0–4 and 0–12, respectively. We extracted source events for all observations from a circle that should hold 90% of X-ray emission of the source and the center should be equidistant from the ring.

We used archived XMM-Newton observations for 8 from 12 selected LLAGNs. (see Table 1 in the full version of the article, available here: arXiv:2012.13518) NGC3642 is not bright in X-ray range and for 3 LLAGNs, namely NGC404 NGC3368 and NGC4203 there is no XMM observations. For sources with more than 5 XMM observations we chose observations distributed in time with the reason that we can observe some changes in their spectral properties, if they exist (see Table 2 in the full version of the article, available here: arXiv:2012.13518).

3. Implications

Conclusions. We created the sample of 12 LLAGNs for investigation of their X-ray properties. Redshift z in our sample varies from −0.000140 in M81 to 0.00562 in NGC 4579. It means that all objects are relatively close to us galaxies. From Table 1 (in the full version of the article, available here: arXiv:2012.13518) we can see that 75% (9 from 12) of the host galaxies of selected LLAGNs are spiral and only 25% (3 from 12) are elliptical. We obtained XMM spectra for LLAGNs and tried to find common behaviors in their spectra. Nevertheless their relatively low brightness (if compare with AGNs), we found that X-ray spectra of only 3 sources from our sample can be fitted with simple phabs·powerlaw model. X-ray spectra of another 5 X-ray sources required models with multiple components and shows complicated features. Obtained model's parameters are: Photon Index varies from 1.2 in NGC 1052 to 2.90 in NGC 3486; absorption by neutral hydrogen $N_H = 2.22 \cdot 10^{20} cm^{-2}$ in NGC 3998 to $N_H = 7.5 \cdot 10^{22} cm^{-2}$.

References

Hernández-García L., Masegosa J. et al 2016, *ApJ*, 824, 7
Hernández-García L., Masegosa J. et al 2017, *A&A*, 602, A65
Jolley, E.J.D., Kuncic, Z. 2007, *Ap&SS* 311, 257
Minfeng Gu, Xinwu Cao 2009, *MNRAS*, 399, 349
Younes, George; Ptak, Andrew; Ho, Luis C.; Xie, Fu-Guo; et al. 2019, *ApJ*, 870, 73

Session 4

Education and Heritage in the era of Big Data in Astronomy
Proceedings IAU Symposium No. 367, 2020
R. M. Ros, B. Garcia, S. R. Gullberg, J. Moldon & P. Rojo, eds.
doi:10.1017/S1743921321000090

Astronomy Education in Covid Times

Priya Hasan⑩ and S N Hasan

Maulana Azad National Urdu University, Hyderabad, India
email: `priya.hasan@gmail.com`

Abstract. We shall describe the various activities done by us in Covid Times including outreach and educational workshops in Physics and Astronomy. We shall discuss the caveats in virtual teaching of Astronomy and the lessons learnt in the process.

Keywords. astronomical data bases, sociology of astronomy

1. Introduction

The pandemic of 2020 came as a shock and a challenge to the teaching community all over the world. It meant adapting to a new system of communication as the sole source with its limits on interaction and reach. It was a learning experience for teachers as well as students because of its strong dependence on pedagogical methodologies and technologies.

2. Activities

We collaborated with Jana Vigyan Vedika (JVV) Telangana which is a member organisation of the All Indian People Science Network (AIPSN) and a member of the National Council for Science and Technology Communication (NCSTC) in India. The NCSTC is a scientific programme of the Government of India for the popularisation of science, dissemination of scientific knowledge and inculcation of scientific temper. JVV is an effective Science Forum in the state and has a very good reach in rural and urban Telangana.

We started off with a Six-Day Online Astronomy Class from the 18-23 May 2020 from 4:30–6:30 pm IST. We conducted a quiz at the end of the event and gave certificates to participants who got the minimum mark. We had 475 registrations and many were given certificates with JVV.

Encouraged by the success of this, we conducted a Six-Day Workshop 'Joy of Learning Physics@ Home' from 15-20 June 2020 for school students. We showed them live demonstratons on Newton's Laws, Gravity, Simple Machines, Pressure, Optics and Electricity & Magnetism. At the end of each session, we demonstrated an experiment with questions to be answered by students the next day. It was very interactive and students were very enthusiastic about it. We had an International-Astronomical Union-Office of Astronomy for Development (IAU-OAD) Project 'Clear Skies' for school children and hence alot of the material for activities were available to us.

In the meanwhile, the IAU-OAD announced a special call for Covid Times Projects. We proposed a project 'Astronomy from Archival Data' which involved Educational Activities for Under-Graduate and Post-Graduate Students (Possel 2020). The project is close to completion. For the four months of August-November 2020, we had sessions every weekend on Saturdays and Sundays where students learnt python, astropy, virtual observatory tools like Topcat (Taylor 2003), Aladin, ESASky and sources of archival data. Resource persons from all over the world contributed in guiding students in 10

Research Projects. The Internal Virtual Observatory Alliance (IVOA) supports this project. Participants will present their projects in the end of January. A more detailed report of this is presented in this series in the form of a talk.

Prior to the pandemic, we had received funding from the US Consulate, Hyderabad to conduct a Hands-on Astronomy Workshop for School Teachers. Initially we were waiting for the situation to improve, but when it did not, we got 68 astronomy kits delivered to School Teachers all over the country. We then had a Two-Day Hands-on Astronomy Workshop online for almost 100 teahers on the 30-31 October 2020. This Workshop covered topics like: Day-time Astronomy activitiies, Sun-Dials, Optics, Telescope Malking and a basic introduction to Astronomy. We also had a Panel Discussion on application of Mathematics and Physics concepts in Astronomy.

To add on to the recent activity, we had another Hands-on Astronomy Workshop on the 2,9 and 16 December with sessions by Prof Daniel Barth on Low-cost Activities in Astronomy†.

As a team with Prof S N Hasan, we run Shristi Astronomy (https://shristiastro.com/). Details of our activities are on the website and the reader is encouraged to have a look. All the sessions have been recorded and are available on youtube‡.

3. Lessons Learnt

In the course of our sessions we learned the following:

• Planning of the events has to be done carefully since most potential participants had their own online sessions on. We selected weekends, evenings or holidays for all our events.

• There are various issues like internet connectivity problems, power as well as overlapping events. Hence all our activities were recorded and made available on youtube so that participants could attend at their convenience.

• A few of the participants were from non-English speaking countries, hence we enabled subtitles on our videos for their benefit.

• It is good to have a variety of speakers ranging from students, research scholars, post-doctoral fellows and junior and senior faculty. The level needs to be matched to the average participant.

We hope that these resources we created would be useful even when better times return.

References

Possel, M. 2020, *The Open Journal of Astrophysics*, 3, 1
Taylor M. B. 2005, *ASPC*, 347, 29

---

† Barth, D. E. (2019). Astronomy for Educators. Open Educational Resources. Available at: https://scholarworks.uark.edu/oer/2

‡ https://www.youtube.com/c/shristiastronomy

Education and Heritage in the era of Big Data in Astronomy
Proceedings IAU Symposium No. 367, 2020
R. M. Ros, B. Garcia, S. R. Gullberg, J. Moldon & P. Rojo, eds.
doi:10.1017/S1743921321000077

Classical Astronomy as an educational resource in a Faculty of Education

Antonio Eff-Darwich[1] iD **, Erik Stengler**[2]**, Gad El-Qady**[3]**,**
Usama Rahona[3]**, Ashraf Shaker**[3]**, Makram Ibrahim**[3] iD **,**
Manuel Núñez[4]**, Victor Medina**[5]**, Adán Yanes**[6]**, Pere Ll. Pallé**[7]
and Jesús Martínez-Frías[8]

[1]Departamento de Didácticas Específicas, University of La Laguna, Tenerife, Spain
email: adarwich@ull.edu.es

[2]Museum Studies, SUNY Oneonta, USA

[3]National Research Institute of Astronomy and Geophysics, Egypt

[4]IES Alcalde Bernabé Rodríguez, Tenerife, Spain

[5]Colegio Virgen del Mar, Tenerife, Spain

[6]Colegio Salesiano San Isidro, Tenerife, Spain

[7]Instituto de Astrofísica de Canarias,Tenerife, Spain

[8]Instituto de Geociencias, IGEO (CSIC-UCM)

Abstract. History, Maths and Astronomy are all mixed up in an innovative educational project that is being carried out in the Faculty of Education of the Universidad de La Laguna, in Spain. Students learn how to teach (to primary school students) about the shape of the Earth, the distances to the Moon, the Sun and other planets, collecting their own data with simple instrumentation and, most important, to connect ideas and different disciplines. The structure and contents of this project are presented, as well as examples of the activities that are carried out.

Keywords. Astronomy, pre-service teachers, primary school

1. Introduction

'Primary science learning resources' is a core project-based subject of the degree in primary education at the Universidad de La Laguna, Spain. This subject is divided into three projects: one devoted to life sciences, the second to energy and matter and the third one consists in an educational multilevel (from elementary school to research institutions) science project about Classical Astronomy. In this sense, our purpose is to expose pre-service education teachers to the potential of Astronomy as a multidisciplinary and engaging educational resource.

The work of Eratosthenes was chosen as the topic for the astronomy project that was carried out during the last quarter of 2020. Eratosthenes was able to measure the radius of the Earth, by analyzing the shadow cast by the sun during the summer solstice in two Egyptian cities, present-day Alexandria and Aswan. It has been widely used for educational purposes (e.g., Décamp & de Hosson (2010), Božić & Ducloy (2008) and Camino & Gangui (2012)), being an ideal experiment to learn and teach about triangulation, to make measurements, to control variables and to analyze data. Moreover, it is a good opportunity to understand the historical, cultural and social context behind scientific discoveries. Our work consisted of a set of activities covering contents from maths, social and natural sciences and museum studies. It was designed as an international collaboration

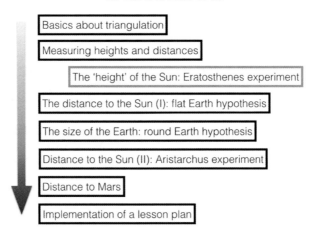

Figure 1. Set of activities contained within the Eratosthenes project.

between Spain (Tenerife), the USA (Cooperstown, NY) and Egypt, the country where Eratosthenes carried out his famous experiment. Besides the 120 pre-service teachers, the project involved students from the master's degree of museum studies (State University of New York), primary and secondary students from three schools in Tenerife and scientists from the Spanish Institute of Geosciences, Instituto de Astrofísica de Canarias and the Egyptian National Research Institute of Astronomy and Geophysics. In total, 300 primary, secondary, undergraduate and master's students participated in this initiative.

2. Implementation

The sequence of activities (two-week-long project) for pre-service teachers (Fig. 1) contains an introduction to triangulation and how to measure distances and heights to closely objects and to distant bodies (sun, moon, planets, etc...). The project also contains activities to expose the students to the general ideas of the scientific method by working on the flat and round Earth hypothesis. In this sense, the solar altitude might be used to infer the distance to the Sun (under the flat Earth and non-parallel sun rays hypothesis) or to derive the radius of the Earth (under the round Earth and parallel sun rays hypothesis).

Students from the different schools, pre- and in-service teachers, museum studies graduate students and scientist carried out a version of the famous Eratosthenes experiment by measuring the shadow cast by a rod at solar noon at different locations (latitudes) in the world (22 and 23 of october 2020). One of the most valuable aspects of this project is that we also include the original sites where Eratosthenes carried out the experiment, Alexandria-Cairo and Aswan. From all the data obtained at the different locations, it was obtained, under the Round Earth hypothesis, that the radius of the Earth ranges from 6200 to 6455 km, whereas, under the Flat Earth hypothesis, the distance from the Earth to the Sun ranges from 3200 to 3860 km. The pre-service teachers use the information and experience obtained with all these activities to design a final project for primary school students that conforms 35% of the final grading of the subject.

During the implementation of the project, the pre-service teachers found extremely motivating and educational these three aspects: to design activities that mix science, maths and history; the confrontation of hypothesis, in this case flat vs round Earth; the importance of error propagation in scientific measurements

References

Božić, M. & Ducloy, M. 2008, *Physics Education*, 43, 165
Camino, N. & Gangui, A. 2012, *The Physics Teacher*, 50, 40
Décamp, N. & de Hosson, C. 2010, *Sci. & Educ.*, 21, 911

Education and Heritage in the era of Big Data in Astronomy
Proceedings IAU Symposium No. 367, 2020
R. M. Ros, B. Garcia, S. R. Gullberg, J. Moldon & P. Rojo, eds.
doi:10.1017/S174392132100003X

Penn State Summer Schools

Gutti Jogesh Babu🆔

Dept. of Statistics
Dept. of Astronomy & Astrophysics, Penn State University,
University Park, USA
email: babu@psu.edu

Abstract. Intensive week-long Summer Schools in Statistics for Astronomers were initiated at Penn State in 2005 and have been continued annually. Due to their popularity and high demand, additional full summer schools have been organized in India, Brazil, Space Telescope Science Institute.

The Summer Schools seek to give a broad exposure to fundamental concepts and a wide range of resulting methods across many fields of statistics. The Summer Schools in statistics and data analysis for young astronomers present concepts and methodologies with hands on tutorials using the data from astronomical surveys.

Keywords. statistical inference, classification, clustering, high-performance computing.

---

1. Introduction

Astronomical research today often involves organizing vast surveys of imaging, photometric and spectroscopy data producing tera/petabyte databases and billion-object catalogs. Many time-domain surveys in visible light are underway: DES, SDSS IV-V, ZTF, SNF, Pan-STARRS, and others leading to ever increasing databases. Radio interferometric telescopes like the NRAO Jansky Very Large Array and Atacama Large Millimeter Array (ALMA) are producing enormous datasets. The forthcoming Vera C. Rubin Observatory project's 10-year Legacy Survey of Space and Time (LSST) is expected to deliver 500 petabyte set of images and data products. While the scientific promise of these surveys is great, the scientific goals cannot be achieved with a narrow suite of statistical methods and old-fashioned labor-intensive approaches in common use by the astronomical community.

Astronomers can be enormously assisted by improved knowledge of fundamental statistical inference and of modern application methods. The procedures may be buried in the standard reduction of data from a detector, in the exploration of a megascale multivariate sample, in the analysis of a stochastic time series, and in the fitting of a nonlinear astrophysical model to observational results. Model fitting methods such as maximum likelihood estimation, Bayesian inference, and nonparametric goodness-of-fit tests are needed. Due to the structure of undergraduate and graduate curricula, astronomers are not well trained in statistics or informatics, receive inadequate conceptual foundations in mathematical statistics, and little training in advanced algorithms and computational techniques.

2. Week-long summer schools

Intensive week-long Summer Schools in Statistics for Astronomers were initiated at Penn State in 2005 and have been continued annually. Additional full summer schools have been organized in India, Brazil, Taiwan, and Space Telescope Science Institute, with shorter roving tutorials in a dozen countries on four continents. These schools are organized by the Center for Astrostatistics at the Pennsylvania State University.

The curriculum emphasizes principles and methods of statistical inference. Lectures are accompanied by hands-on software tutorials with applications to contemporary astronomical datasets. Ambitious in scope for a short course with about 40 contact hours, it is taught roughly at the level of senior classes for undergraduate statistics majors (for example at the level of the textbook by Ross (2019), but requires coverage of material normally distributed over many courses. A typical School has 8-10 different instructors, most experienced Professors of Statistics. Some lectures are given by statistically-expert astronomers.

The topics covered during the very popular Penn State schools include: probability, inference, regression, model selection & validation, maximum likelihood methods, non-parametric methods, multivariate methods, clustering & classification, Bayesian inference, MCMC, spatial statistics, Time series, Fundamentals of scientific computing, high-performance computing, Bayesian computation, Machine Learning algorithms, optimization, Gaussian process regression, neural networks & deep learning.

3. Statistical issues arising in astronomical research

The following are some of the questions often arise in astronomical research.

How can one reliably extract a population of supernovae from a huge multi-epoch sky survey? How do we model the vast range of variable objects arising from accretion onto compact objects? How can we find faint emission line galaxies or diffuse structures in from large spectro-imaging datacubes? How do we address data limitations such as heteroscedastic measurement errors, censoring and truncation (nondetections), sparse data, and irregular time sampling? How can uncertainties in astrophysical models be incorporated into the inferential process? How do we classify time series expected to be generated by LSST type surveys, that are generally sparse, irregular, heteroscedastic, and describe a large set of different transient or variable objects (AGNs, binary stars, supernovae, etc.)? The summer school curriculum is designed to train the participants to address these issues.

4. Conclusion

Penn State Summer Schools at the Center for Astrostatistics have trained about 950 young astronomical researchers in statistical inference and in the use of the R statistical computing environment. They consistently attract large number of women who are underrepresented in the physical sciences. In addition to US, participants arrived from Australia, Belgium, Canada, Chile, China, Czech Republic, Denmark, France, Germany, Hungary, Indonesia, Italy, Iran, Israel, Japan, Mexico, Morocco, Netherlands, Poland, Russia, South Africa, Spain, Sweden, Switzerland, Taiwan, UK. In 2018, a full week of astroinformatics is added. This new component gave training in high performance computing with an emphasis on Bayesian computing. Participants used the hybrid CPU/GPU cluster at Penn State. The lectures and tutorials of these summer schools are archived at the Center for Astrostatistics's website https://astrostatistics.psu.edu/.

References

Ross, S. 2019, *A First Course in Probability* (10th Edition), New York: Pearson.
Center for Astrostatistics's Website. *Lecture notes and tutorials of recent summer schools.*

Education and Heritage in the era of Big Data in Astronomy
Proceedings IAU Symposium No. 367, 2020
R. M. Ros, B. Garcia, S. R. Gullberg, J. Moldon & P. Rojo, eds.
doi:10.1017/S1743921321000065

The Case for Coordinating Earth & Space Science Education Worldwide

William H. Waller🆔

Rockport Public Schools, Endicott College, and *The Galactic Inquirer*,
IAU/OAE US National Astronomy Education Coordinator
243 Granite Street, Rockport, MA 01966 USA
email: williamhwaller@gmail.com

Abstract. Despite the many amazing advances that have occurred in the space sciences (planetary science, heliophysics, astronomy, and cosmology) these subjects continue to play minor roles in pre-collegiate science education. Similarly, the Earth sciences are woefully under-represented in most school science programs – despite their vital relevance to our physical well-being. Some countries have educational standards that formally prioritize the Earth & space sciences as much as the physical and life sciences, but even they fail to actualize their mandated priorities. I contend that better coordination and advancement of Earth & space science education at the national, state, society, and educator levels would lead to better educational outcomes worldwide.

Keywords. astronomy education, Earth and space science education

1. Introduction

The Earth sciences address processes within and among the rocky Earth, its ice caps, oceans and atmosphere. The space sciences consider Earth as a planet among other planets in the Solar System and the greater cosmos - what is commonly called astronomy. Together, the Earth and space sciences span what we know - and what we would like to know - about our place in space and moment in time. The Earth & space sciences, in concert with the life sciences, comprise what is commonly known as the natural sciences. Herein, I argue in support of teaching the Earth & space sciences together, so that students can attain a more holistic understanding of their planetary environment, how it came to be, and where it is headed. Such teaching (and teachers) should receive the same priority as in the teaching of physics, chemistry, and biology. My reasoning for bundling and advancing Earth & space science education has pedagogical, scientific, and cultural underpinnings. These will be discussed along with ideas for enhancing the interaction, cooperation, and coordination of Earth & space science educators worldwide.

2. Pedagogical Underpinnings

Some countries have educational standards that prioritize the Earth & space sciences as much as the physical and life sciences. These include the U.S. Next Generation Science Standards (NGSS) which have a separate Earth and Space Science track that spans the K-12 grades. Other countries don't explicitly feature Earth & space science education but rather include it as small parts of courses in physics and geography, if at all. Even where Earth & space science education is formally prioritized, relevant courses and teaching at the high-school level are often under-represented. This shirking of educational mandates is apparent as a clear deficit of teaching opportunities in the Earth & space sciences, amounting to only 8% of available science teaching openings in a recent survey.

At the collegiate level, astronomy is strongly linked to physics. Indeed about 83% of all astronomy programs listed in the AAS's directory are parts of physics departments, or physics & astronomy departments. This vital link can explain why astronomy and the space sciences are typically taught in many high schools as small parts of physics courses, if at all. Similarly, the Earth sciences often appear as small parts of high school geography courses, if ever. Clearly, we could and should do better to advance both Earth & space science education at both the collegiate and secondary school levels.

3. Scientific Underpinnings

Earth & space science education is vital to our collective well-being, as it is the best way towards understanding the photochemical process of atmospheric warming by greenhouse gases – and for tracking our industrialized society's increasing contributions to this warming. Proper coordination of Earth and space science education also helps students better understand their place in space and moment in time as beneficiaries of incredible cosmic transformations that have transpired over the past 14 billion years. From fundamental particles to atoms, stars, planets, complex molecules, life, and spacefaring cultures, the Earth & space sciences tell a compelling tale of being and becoming. What better way to understand our complete natural history and to ponder our possible future as emergent citizens of the Milky Way galaxy??

4. From Awareness to Action

How best to coordinate and advance Earth and space science education worldwide?? Astronomy on its own is too vulnerable to being neglected as an essential topic of study!! The same goes for the Earth sciences, especially at the high-school level.

• We need to communicate with and learn from our Earth Science education counterparts, perhaps through an IAU working group on Coordinating Earth & Space Science Education. This endeavor will involve the IAU partnering with the AGU, AAS, ASP, AMS, NAGT, NSTA, and their international peers.

• We need to advocate for state and national departments of education to make Earth & space science education a greater priority at the high-school level – and to actualize this priority!!

• At the collegiate level, we need to advocate for integrated degree programs in the Earth & space sciences that will be attractive to pre- and post-service science educators.

• We need to promote the development of coordinated curricula, textbooks, and other resources in Earth & space science education.

• We need to do a better job of communicating our coordinated assets for widespread use by educators. Hopefully, the IAU's Office of Astronomy Education (OAE) will make significant progress addressing this goal. No doubt, professional certificate courses and workshops in the Earth & space sciences will be necessary to engage and coach educators in effective use of these resources.

⇒ So, let's get going!! To provide input and track further progress in these regards, please contact me at williamhwaller@gmail.com.

5. Related Article

Waller, W. (2020), "The Case for Coordinating Earth & Space Science Education," The Galactic Inquirer, http://galacticinquirer.net/2020/09/the-case-for- coordinating-earth-space-science-education/ (accessed January 2021)

Session 5

Education and Heritage in the era of Big Data in Astronomy
Proceedings IAU Symposium No. 367, 2020
R. M. Ros, B. Garcia, S. R. Gullberg, J. Moldon & P. Rojo, eds.
doi:10.1017/S174392132100051X

Antofagasta: Astronomy education on the shoulders of giants

Eduardo Unda-Sanzana[ID]

Centro de Astronomía (CITEVA), Universidad de Antofagasta,
Avda. U. de Antofagasta 02800, Antofagasta, Chile
email: `eduardo.unda@uantof.cl`

Abstract. This work presents the motivation, history and current status of the *Primera Luz* initiative, a long-term educational project run by researchers and professionals of the Astronomy Center (CITEVA) at Universidad de Antofagasta (UA, Chile), primarily focused on connecting Astronomy with the people living in the Region of Antofagasta (Chile) but producing results designed to be shared with international audiences.

Keywords. education, schoolchildren, space

1. Introduction

It is expected that in 2025 the European Southern Observatory (ESO)'s ELT, the largest optical astronomical facility in the world, will attain first light. ELT will be hosted by the Region of Antofagasta in the North of Chile. This is the same region which currently hosts ESO's Very Large Telescope (VLT); also the Atacama Large Millimiter/submillimeter Array (ALMA) one of the largest radiotelescopes worlwide; the highest telescope worldwide, MiniTao, and soon the Tokyo Atacama Telescope (TAO); and it is getting ready to host the largest gamma-ray observatory worlwide, the Cherenkov Telescope Array (CTA). However, in the past, the local public reaction about these developments has been mild and sometimes critical, perceiving the astronomical observatories as disengaged from the life of the local community. As an effort to build a bridge between the local community and the many opportunities provided by the presence of these large facilities the author proposed in 2014 to the Regional Government of Antofagasta (GORE) a long-term program called *Primera Luz* (First Light, PL from now on). PL is esentially a set of educational and outreach activities aimed to strengthen the foundations for the teaching and communication of Astronomy in the Region of Antofagasta. However, several of its products have been developed with an international audience in mind, doubling as promotional materials acting in support of the development of special interest tourism in this region.

During the last decade GORE decided to strengthen the public funding devoted to Astronomy, using it as a tool for cultural change, funding its own education and astro-engineering projects. The author has been the PI of several of these projects, joined by several researchers and professionals of CITEVA, and has thus made progress in the construction of PL.

2. Main work

Given the lack of funding opportunities for long-term educational projects, PL has been built on the basis of a series of short-term projects or portions of projects. The first years of PL were spent in producing educational materials and infrastructure. The main landmarks of this period are:

- 2015: Opening of Ckoirama, the first Chilean-State owned observatory under the skies of the north of Chile, including one 14-inch telescope devoted to educational activities (see Char *et al.* (2016)).
- 2016: First collection of educational videos explaining why Astronomy is important from a regional perspective.
- 2018: Creation of "Region of Antofagasta, Gateway to the Universe", an astrophotography exhibition of 30 panels depicting regional sites by night, updated in 2020 to encompass all the counties of the Region of Antofagasta.
- 2020: Second collection of educational videos, portraying sites of value for astronomical educational, as well as a series of Astronomy lessons and accompanying activity boxes and videos to be used by local schools.

In the development of the educational products an inclusive and international approach has been used, often providing English subtitles or translations when applicable. In 2016 an agreement was signed between ESO and UA, formalizing ESO's support to the efforts of UA in the context of this program of activities. This has helped to make the materials of PL reach an international audience. In 2020 UA took steps to formalize PL as an "emblematic outreach program" of CITEVA.

The ongoing pandemic of COVID-19 has affected progress in the next step of PL, which aims to connect the regional community with its products through workshops and other interactive activities. A fraction of these activities had been been executed already thanks to a program of public observations carried out monthly since 2013, and through a program of yearly teacher schools organized since 2014. In 2020 a first series of online workshops was offered to schoolchildren from 4 regional counties (out of a total of 9 counties) in the context of a regional astroengineering project. Given the constraints set by the pandemic the focus of these workshops has been programming as a tool for working with astronomical technologies. In 2021 the author expects to run a pilot of a massive open online course (MOOC) using the educational materials developed so far.

3. Related developments

One regional project of astroengineering is connected to PL, aiming to develop modern portable telescopes to strengthen local education and tourism. Also, the increasing concern over light pollution from ground and space sources has led the team to engage in the scientific and technical study of the problem by using Ckoirama (see Char *et al.* (2016) and by participating in projects attempting to measure it by using drones and new satellites. The output of these efforts will be integrated in the educational materials of PL. All the PL products are available to the public in http://www.astro.uantof.cl/primeraluz.

Acknowledgement

The author acknowledges Juan Pablo Colque's help to develop to PL, as well as other researchers and professionals' contributions throughout the years.

References

Char, F., Unda-Sanzana, E., Colque, J., Fossey, S. & Rocchetto, M. 2016, *Boletín de la Asociación Argentina de Astronomía*, 58, 200

Tregloan-Reed, J., Otarola, A., Ortiz, E., Molina, V., Anais, J., Gonzlez, R., Colque, J. P. & Unda-Sanzana, E. 2020, *Astronomy & Astrophysics*, 637, L1

Education and Heritage in the era of Big Data in Astronomy
Proceedings IAU Symposium No. 367, 2020
R. M. Ros, B. Garcia, S. R. Gullberg, J. Moldon & P. Rojo, eds.
doi:10.1017/S1743921321000272

A User Centred Inclusive Web Framework for Astronomy

Wanda Díaz-Merced[1], Mizzani Walker-Holmes[2], Christian Hettlage[3], Paul Green[4], Johanna Casado[5,6], and Beatriz García[5]

[1]email: wanda.diaz.merced@gmail.com,

[2]Northwestern Mutual Chicago, United States

[3]South African Astronomical Observatory, Observatory Cape Town, South Africa

[4]Smithsonian Astrophysical Observatory, United States

[5]Inst. de Tecnol. en Detec. y Astropart. (CNEA, CONICET, UNSAM), Argentina.

[6]Instituto de Bioingeniería, Univ. Mendoza, Argentina

Abstract. This contribution presents the ongoing work to develop the Sensing the Dynamic Universe (SDU) webpage elements, based on the sonoUno framework, a user centered software for sonification of astronomical data. The SDU provides an inclusive experience that transcends the approach of assuming and limiting design based on the Web Content Accessibility Guidelines (WCAG) and ISO recommendations.

Keywords. digital accessibility, user center design, virtual assistants.

1. Introduction

Digital access to information is a human right, and effective and individualised support measures are needed to allow People with Disabilities (PwD) to fully enjoy this right. However, despite many initiatives, we are far from achieving that goal. Focusing on Astronomy, the Sensing the Dynamic Universe (SDU) team proposes an approach to web access for inclusion targeting the comorbid aspect of 5 groups of diverse functionalities, namely: a) mastoid problems/hearing impaired, b) skin disorders, c) orthopedically impairment, d) Neurodiversity, and e) B/VI (Blind/Visually Impaired), using as a framework the sonoUno interface (Garcia 2017) which permits people with other sensory styles to explore scientific data and make science by sonorization. To support these target audiences, SDU reduces complexity in navigation, offers selection and saving of user preferences, and generates strategies to dynamically identify and utilize at will desired web elements, which may communicate dynamically with each other. Users can decide freely what path to take for task performance, error prevention and error recovery. It also attempts to embed the framework of a virtual assistant: Stephanie (Gupta 2017).

2. Methodology

SDU uses JavaScript local storage method, which does not communicate with servers, does not use cookies and gives the option to add an expiry date. While this method has rarely been used for accessibility purposes so far, it permits to create web interfaces that adapt to the user, goes beyond the subjectivity of the WCAG (https://www.w3.org/TR/WCAG20/), ISO *Guidance on software accessibility* (https://www.iso.org/standard/39080.html), etc., aiming for high granularity, usability,

W. Díaz-Merced et al.

efficiency, effectiveness, performance and user satisfaction. The webpage aims to embody the mindset of "how to create an interface in which the user may work at their own maximum without submitting them to transactions that may further exclusion?". The SDU team began by identifying, and experimenting with, structural factors that support PwD. The list will grow and change over time, as it is multi-axial and it includes:

(a) Choices. There is a lack of intuitive options to alter choices in digital systems, rendering the individual excluded or dependent. Many PwD become inured to the frustrations of inaccessibility or break down. Choices may be grouped as:
- Dynamic choices: In digital interfaces dynamically let the user focus their attention as desired, and let them activate options and freely switch I/O options.
- Choices that cause or diminish attention demands: SDU tries to preserve structure and consistency, for example employing contrast options, saving user choices and reducing language, without overloading the user. The combination of words and actions does not jeopardise the user action-word processing and movement, considering the relation between the verbs and the actions to be performed.
- Choices to meet the needs of cognitive processing: The user is not overloaded with information which is not needed or distracting.
- Choices about information needs: what does the user want to learn multisensorially about a subject. SDU provides a sensorial display and expansion of specific examples, and provides a direct link to the sonoUno to display visually and auditorily and explore data using their own settings.
- Navigation choices. The options to reduce the learning curve are linearised according to user preferences and allow the user to select how to start the interaction and change I/O modalities without exhaustion. SDU employs the alt-text functionality of html to provide a comprehensive description of the content and options, attempting to assist users in the process of identification.
(b) Interruptions. As interruptions undermine the user's ability to engage in decision making and integrative thinking, they are minimised. SDU offers saving preferences and optional alerts, allows changes of text size or font, and avoids crowded I/O management.
(c) Body physical transactions. The location of displayed information seeks to minimise the number of physical transactions the user has to carry out to access a functionality, to deliberate or to experience a display.

The website will soon be available at https://www.cfa.harvard.edu/sdu.

3. Final remarks

While some people with disabilities/impairments express that traditional barrier removal is enough, the multifactorial and multidimensional nature of disabilities imply that traditional barrier removal may make interactions more difficult for some users or impose a very high learning curve not related to their academic abilities. Given that the features required for more inclusive access were easy, one has to ask why it is hard to find a wider engagement on allowing PwD to bring their diverse functionalities on board. This work has been performed partially under the Project REINFORCE (GA 872859) with the support of the EC Research Innovation Action under the H2020 Programme SwafS-2019-1.

References

García, B., Díaz-Merced, W., Casado, J., Cancio, A. 2017, *Evolving from xSonify: a new digital platform for sonorization, ISE2A 2017, EPJ Web Conferences,* DOI:10.1051/epjconf/201920001013
Gupta, U. 2017, *Stephanie virtual assitant, https://github.com/SlapBot/stephanie-va*

Education and Heritage in the era of Big Data in Astronomy
Proceedings IAU Symposium No. 367, 2020
R. M. Ros, B. Garcia, S. R. Gullberg, J. Moldon & P. Rojo, eds.
doi:10.1017/S1743921321000119

Integrated Active Learning Utilizing the Stories of Tomorrow at Elementary School in Japan

Akihiko Tomita[1]⓪, Fumihito Kubo[2]⓪, Masashi Maeda[3]⓪ and Rosa Doran[4]⓪

[1]Graduate School of Teacher Education, Course Specializing in Professional Development in Education, Wakayama University, Wakayama City 640-8510, Japan
email: `atomita@wakayama-u.ac.jp`

[2]Attached Elementary School, Faculty of Education, Wakayama University,
1-4-1 Fukiage, Wakayama City 640-8137, Japan
email: `kubo1214@wakayama-u.ac.jp`

[3]Attached Elementary School, Faculty of Education, Mie University,
359 Kan-nonji-cho, Tsu City 514-0062, Japan
email: `mmaeda@fuzoku.edu.mie-u.ac.jp`

[4]NUCLIO – Interactive Astronomy Nucleus,
Largo dos Topázios, 48, 3rd Front, 2785-817 São Domingos de Rana, Portugal
email: `rosa.doran@nuclio.org`

Abstract. In 2018, two schools from Japan participated in Stories of Tomorrow, a computer-based STEAM educational practice for primary school students. We were able to learn from the students' feedback that through problem-solving, a spirit of collaboration, a spirit of overcoming failure, and a deep understanding of scientific research and technology development have been nurtured. We also confirmed the importance of translation and coordination to cross over the language barrier.

Keywords. Problem-solving learning, international collaboration, Stories of Tomorrow

---

1. Introduction

The Stories of Tomorrow is a problem-solving project based on the hypothetical challenge of traveling and migrating to Mars. This is a good example of STEAM education. The project began in 2017 as an experiment in Europe with funding from the European Commission (http://www.storiesoftomorrow.eu/). Several schools from each of the five European countries, Portugal, France, Germany, Greece, and Finland have participated in the project, and in 2018, two elementary schools attached to Faculty of Education of Wakayama University and Faculty of Education of Mie University were invited to participate. In the two schools in Japan, the two teachers, the authors, Kubo at Wakayama and Maeda at Mie, implemented the project in their classrooms. Tomita served as the contact person in Japan. Doran coordinated the international network of the Stories of Tomorrow and was in frequent contact with Tomita.

2. Practice at Mie

The students practiced making a simulated Martian environment made from cardboard, placing a piece of metal on it, and creating a simulated Mars rover made from

Lego bricks to search for the piece of metal. In considering the creation of a "class that ignites students' minds," Maeda has felt that themes that aim at the boundary between "I know" and "I don't know yet" was quite effective for students to pursue tasks proactively and freely. We got feedbacks, such as "The research activity was full of failures, disagreements, and quarrels during the activity, but we got the light on at last." Some of the students returned feedbacks with a proactive mind in discovering new ideas and admiring the device development, such as "We have made a rover that lights up when it finds a metal object, but it can only find it, not take it back to Earth, so I want to make something that can take it back to Earth." There are feedbacks full of challenging spirit and a deep understanding of scientific research and technological development, such as "I was amazed that researchers continue to study Mars, despite repeated failures and revising the programs."

3. Practice at Wakayama

The class was divided into seven groups to create the e-picture book. Kubo asked the students to think about the structure of the book in relation to the Japanese language classes' storytelling. After that, each group came up with a draft, and then they worked proactively and collaboratively to revise the draft. We got feedbacks, such as "Though the task was difficult for me, I felt a sense of accomplishment when it was completed." It was the appropriate difficulty of the task that led to such collaborative feedback from the students. This is similar to the strategy of aiming for the boundary between "I know" and "I don't know yet" in the Maeda class.

4. Crossing over the language barrier

In Japan, the language is all Japanese in the classroom. To implement an international project such as the Stories of Tomorrow, translating the contents into Japanese is definitely crucial. All contents of the Stories of Tomorrow were translated into Japanese carefully in advance by many volunteers. The web-based platform for the e-book creation adopted all languages of participating countries including Japanese, which was prepared through the great effort by technical staff of the project. The summer school was held as an exchange and training session. It was a quite valuable opportunity for the participating Japanese teachers to interact directly with teachers from many other countries. With the careful and well-considered preparation above, Japanese teachers could incorporate the project into their classes through the curriculum design.

5. Conclusions

(a) We confirmed that the key to success in problem-solving learning is an appropriately difficult task that lies on the border between "I know" and "I don't know yet."

(b) We confirmed that through problem-solving, we can see from the feedbacks that a spirit of collaboration, a challenging spirit, and a deep understanding of scientific research and technological development are fostered.

(c) We confirmed the importance of translation and coordination to cross over the language barrier in implementing an international project.

A detailed explanation of the activity was published in Tomita *et al.* (2021).

Reference

Tomita, A., Kubo, F., Maeda, M.& Doran, R. 2021, *Bulletin of Course Specializing in Professional Development in Education, Wakayama University*, 5, in press

Education and Heritage in the era of Big Data in Astronomy
Proceedings IAU Symposium No. 367, 2020
R. M. Ros, B. Garcia, S. R. Gullberg, J. Moldon & P. Rojo, eds.
doi:10.1017/S1743921321000557

Making sciences with smartphones: The universe in your pocket

Martin Monteiro[1] and Arturo C. Marti[2]

[1]Universidad ORT Uruguay, Uruguay
email: monteiro@ort.edu.uy

[2]Facultad de Ciencias, Universidad de la República, Uruguay
email: marti@fisica.edu.uy

Abstract. We propose a set of modern and stimulating activities related to the teaching of Astronomy orientated to high school or university students using smartphones. The activities are: a) the experimental simulation of asteroid light curves including the determination of the period of rotation of asteroids, b) the experimental simulation of exoplanet detection by transit method, c) the experimental simulation of stellar distances using parallax and d) the use of virtual and augmented reality.

Keywords. smartphones, asteroid light curves, stellar distances, virtual and augmented reality

---

Smartphones are pocket computers with many sensors that can be used as portable laboratories for a wide variety of scientific and educational activities, Vieyra et at. (2015). Here, we show some activities useful in basic courses of astronomy or geosciences.

Asteroid light curves. The vast majority of asteroids are irregular objects, which due to their small size and great distance can hardly be appreciated as points of light from the best astronomical observatories. However, a careful analysis of the sunlight reflected by an asteroid allows in some cases to determine some characteristics of its shape, as well as its period of rotation around its axis. This photometric method is called the *light curve*. If asteroids were spherical and uniform in color, their light curves would be flat. The more irregular its more complex shape the curve and the greater the changes in brightness. We propose to simulate this photometric technique using a scale asteroid which can simply be a bit of molding mass or a real asteroid, from real data, made with 3D printer. Data files can be obtained freely at https://nasa3d.arc.nasa.gov/models/printable. The scale asteroid is rotated on a circular platform, a turntable or by the axis of a slow motor. While it is rotating the light intensity is measured with a smartphone's light sensor by means of an app like Physics Toolbox or PhyPhox (left panel of Fig. 1).

Exoplanets and planetary transits. In recent years the discovery of planets orbiting stars other than the Sun, called extrasolar planets or simply exoplanets, has become one of the most innovative and fastest growing themes of modern astronomy. Today more than 3,500 exoplanets have been confirmed. The great distance to which the stars makes it impossible to observe them directly with the necessary resolution, even for the largest telescopes. Thus, several techniques have been developed to discover exoplanets by indirect detection methods. One of these methods is planetary transit, a photometric method that involves observing the light of a star and measuring the small change in brightness that occurs when an exoplanet passes in front of the star. Several characteristics of exoplanets can be determined from light curves as size (estimated from the reduction in the brightness of the star) or the orbital period (determined from the periodicity of the

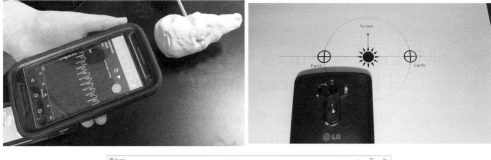

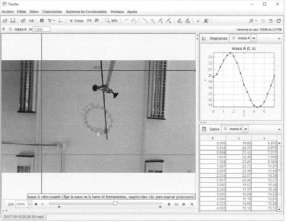

Figure 1. Some examples of the activities discussed in this presentation.

light curve). To simulate the transit method a lamp can be used as a star, spheres of different sizes as planets (which are rotated around the star) and the light sensor of a smartphone to record the light curve using a suitable app as mentioned in the previous activity (Barrera-Garrido 2015).

Parallax and parallactic ellipse. Usually, astronomers employ different methods to determine the distance to celestial objects according to the scale of astronomical distances in question. To measure the distances of the stars, the method par excellence and the oldest is that of parallax. Parallax is the change in the angular position of an object when the point from which it is observed is changed. Here we model the effect of parallax and show how can be used to measure stellar distances.

Annual parallax can be used to determine the distance to the stars thanks to the translational movement of the Earth around the Sun. To observe the parallax effect, we just take two photographs from different positions and observe the change of position with respect to the background as proposed in Fitzgeral et al. (2011). To observe the parallactic ellipse, a scaled star is placed on top of a scaled solar system (center panel of Fig. 1). The experiment consists of recording a video with the smartphone, while it moves as if the camera lens were the Earth above the paper with the scaled solar system and later analyze the video, to determine the parallactic ellipse and calculate the distance at which the star is located (right panel of Fig. 1).

Virtual reality consists of the generation of three-dimensional images, which provide a sense of depth and spatiality. For this reason there are some applications for smartphones that what they do is generate two images on the phone screen, left and right, one for each eye. In this case, it is needed the help of special lenses such as Google Cardboards, which are made of cardboard or some other more comfortable model that today are very economical (of the order of a few dollars). The smartphone is placed inside these special lenses and thus each image is sent to each eye separately, generating the sensation of *spatiality*. Virtual Reality *apps* take advantage of the smartphone sensors to recognize perspective changes when we turn our heads or tilt them up or down. For this reason, acceleration, gyroscope and magnetic field sensors are required.

Augmented Reality consists of overlapping several real images with virtual images. This technique is increasingly used in several industrial contexts, in games and, of course, also in education. A very interesting example to get closer to the world of space exploration is the NASA Spacecraft 3D app, which allows you to visualize several spacecraft, robots and other devices linked to space. Augmented reality not only creates the illusion that an object is in our environment but can also be observed from different perspectives and distances as if it were a real object that we are seeing through the camera.

As a conclusion, we remark that smartphones and mobile-devices in general offer a very wide variety of possibilities for experimentation in Astronomy teaching.

References

Vieyra, R., Vieyra, C., Jeanjacquot, P., Marti, A. & Monteiro, M. 2015, *Sci. Teacher*, 82(9), 32

Barrera-Garrido, A. 2015, *Phys. Teach*, 53, 179

Fitzgerald, M.T., McKinnon, D. H. , Danaia, L., & Woodward, S. (2011), *Astron. Educ. Rev.*, 10, 010108-1

Education and Heritage in the era of Big Data in Astronomy
Proceedings IAU Symposium No. 367, 2020
R. M. Ros, B. Garcia, S. R. Gullberg, J. Moldon & P. Rojo, eds.
doi:10.1017/S1743921321000545

Outreach in the era of big data with the Pierre Auger Observatory

Karen S. Caballero-Mora[1]🄳 for the Pierre Auger Collaboration[2]

[1]Universidad Autónoma de Chiapas,
Boulevard Belisario Domínguez, Km. 1081, Sin Número, Terán, Tuxtla Gutiérrez, Chiapas,
México, 29050. Tel:52 (961) 617-8000, Mexico
email: `karen.mora@unach.mx`

[2]Observatorio Pierre Auger,
Av. San Martín Norte 304 Malargüe, Mendoza, Argentina
Full author list: `http://www.auger.org/archive/authors_2020_12.html`
email: `auger_spokespersons@fnal.gov`

Abstract. The Pierre Auger Observatory, built to study the physics of astroparticles, has expanded its endeavours in outreach and education. Since its inception, the collaboration has been informing the general public about its discoveries. From creating a visitor center in Malargüe, Argentina, to providing talks for different audiences, science fairs for students, international measurement workshops, and making compilations of the scientific contributions of women in different countries throughout history, we share the passion for science with the population. The collaboration processes a huge amount of data and requires a large storage capacity. However, 10% of this data is made available to the general public in a useable format. The collaboration developed special masterclasses aimed at high school students, providing the software that can be used to analyse the public data set. In addition, a summary of the different outreach activities performed by the members of the Pierre Auger Collaboration will be presented.

Keywords. Outreach, Education, Astro-particles, Pierre Auger Observatory

1. Introduction

Astroparticles are particles from the universe arriving on Earth. The Pierre Auger Observatory studies those particles having ultra-high energies. Their study can give information on their origin, of acceleration and propagation mechanisms, and also clues on the laws governing interactions between particles at the highest energies. This study area has helped to develop the physics and also the technology applied in the detectors used. Outreach and education has been a very important component of the work of the Pierre Auger Collaboration, informing several kinds of audiences about our discoveries. The Collaboration is formed by researchers from 17 countries and 81 institutions, who prepare different outreach and educational activities to share the science discoveries with high-school and university students, as well as other members of their communities. In the following sections, some of them are described.

2. Outreach and education at the Auger Observatory

The Pierre Auger Observatory has an instrumented area of $3000 \, \text{km}^2$ with different kinds of detectors: water-Cherenkov detectors, fluorescence telescopes, scintillators and radio antennas. The technology used for these detectors is explained to visitors

Figure 1. Visitor Center, Permanent exhibition, Auditorium and some visitors.

Figure 2. Science Fair in Malargüe and the Collaboration in the Parade.

so that they can understand the physics used to measure astroparticles (Castellina A 2019). As early as 2001, a visitor center has been created (see Figure 1), which welcomes around 9000 visitors a year. Nowadays, a visitor tour is also available, which can be found in the outreach web page of the Collaboration at: https://www.auger.org/index.php/edu-outreach. In addition, the bond between the city of Malargüe and the Collaboration is expressed by our participation in the annual parade of the city. Every year during the Collaboration meeting in November, a Science Fair is organised at which high-school students, mostly from the Mendoza province, present projects to the researchers of the Collaboration (Figure 2). Students show what they have learned and they discuss with scientists who may enrich and inspire them. The Collaboration promotes diversity in science. An example thereof is a virtual exhibition, created by members of the Collaboration, called "Women hold up half the sky", which can be found in English at https://izi.travel/en/c824-women-hold-up-half-the-sky/en and in Spanish at https://izi.travel/en/c824-las-mujeres-sostienen-la-mitad-del-cielo/es.

3. Masterclasses, Data and Software release

Ten percent of the data collected by the Observatory is made available to the general public in a useable format for educational purposes. A Public Event Browser is available at https://labdpr.cab.cnea.gov.ar/ED/index.php?lang=en. Students and teachers are aided through the masterclasses developed by members of the Collaboration, consisting of lectures, notes and hands-on data analyses through online tools or tools installed on a local PC. More information can be found at: https://www.lip.pt/ https://www.lip.pt/experiments/auger/?p=public-data. These master class tools are used for instance when members of the collaboration participate in the International Cosmic Day, organized annually by DESY (https://icd.desy.de/).

Reference

Castellina, A., for the Pierre Auger Collaboration 2019, *EPJ Web Conf., Volume 210, Ultra High Energy Cosmic Rays 2018, Article Number 06002, 2019*

Education and Heritage in the era of Big Data in Astronomy
Proceedings IAU Symposium No. 367, 2020
R. M. Ros, B. Garcia, S. R. Gullberg, J. Moldon & P. Rojo, eds.
doi:10.1017/S1743921321000235

Recognising 'academic expert' does not equal 'education expert' and the need to facilitate a dynamic, two-way flow of expertise for the future of communicating astronomy

Elizabeth J. Iles ⓘ

Department of Physics, Faculty of Science, Hokkaido University,
Sapporo 060-0810, Japan
email: iles@astro1.sci.hokudai.ac.jp

1. Introduction

In the era of community science, big-data surveys and public outreach, it is unavoidable that astronomy discourses may be guided by a professional who is not trained in astronomy but is considered an expert in their own field (e.g. general education teachers in classrooms). Despite this, the importance of contact with experts who are fully immersed in the discourse (e.g. professional scientists and academics) has not been neglected. In recent years, there has been a clear movement within the discipline to increase the frequency of interaction between academics and the general public. However, this is accompanied by what seems to be a widespread failure to recognise that, although these academics may be domain specific experts in astronomy, they often lack relevant training in education or effective communication, thus hindering the results.

The number of resources available to train and support education professionals in the discourse of astronomy are increasing rapidly, while the same cannot be said for resources to support the increasing public engagement requirements on academic professionals, particularly in the context of sharing their own research. In many academic institutions, there is an endemic deficiency in the provision of guidelines, resources or advice on how to translate the knowledge of astronomy into effective teaching practices with which students can engage and learn. This means that we rely primarily on our experts intuitive knowledge, personal experience and the individual drive of the audience we hope to reach to support engagement and learning. Additionally, there is often the expectation that it is an individual researcher's own choice whether to develop [waste time on] an understanding of educational theories and practice. Under the increasing pressure for institutions to engage, not just in token outreach programs but, in the opening of effective discourses with the community at large, this practice is no longer proving sustainable. We are now expecting our academics to successfully engage with the full spectrum of possible audiences in a dizzying array of contexts which they may only experience once or twice a year, with little training or support.

2. An Interdisciplinary Approach

The problems of effective outreach and effective tertiary education should not be treated as separate issues. Both arise from a misconception that the years of study required to become a scientific expert will also somehow produce expertise in communicating that knowledge. While expertise in education, communication and astronomy are not mutually exclusive, neither is one a by-product of the other. We would not expect a senior member of the Faculty of Education to be able to derive any fundamental astrophysical principles. So, why then do we expect senior astronomers to be able to immediately put into practice a full spectrum of educational theories in all manner of contexts?

Sustainable long-term engagement with astronomy is driven both by the opinion of the wider community towards astronomy and the situation within the academic field. However, these are not independent but rather, interdependent. This interdependency can be seen clearly by considering linguistic choices in the discourse of astronomy. There exists a traditional bias in long-respected fields of science, that true understanding is exclusive to experts who have devoted many years to the study of the field. This is the 'expert' discourse and is perpetuated unconsciously by the way we speak and present ideas, both among ourselves as academics and with the 'uninitiated' general public. Conversely, to more effectively communicate ideas and develop true understanding, it is necessary to consider the state, needs and best methods of approach for each audience. In terms of educational theory, this requires considering age, context and pedagogy: what is the developmental age/stage, existing knowledge and experience of the audience; what is the format of the interaction and expected content; and how can this be delivered to promote effective engagement. The intersection between the existing expert discourse and an audience-sensitive discourse is the most effective method for communicating astronomy and consequently, driving the development of astronomy for the future.

3. Immediate & Practical Actions

In an ideal world, we would simply train astronomy academics in the appropriate educational theories and all educators in the discourse of astronomy. In reality, this is unachievable. We need to work as a joint astronomy-education community with recognition that the expertise on both sides is of equal worth. Current imbalances in the perception of the skills and roles of academic experts and their education counterparts is not new and has been limiting the development of astronomy, as an academic discipline as much as a community resource, for many years. With the advent of this era of citizen science and public engagement with astronomy, it can no longer be ignored. Just as it is important to support, train and develop tools for the non-experts of astronomy to engage with astronomical discourse and pass on the knowledge of astronomy in their communities, it is necessary to acknowledge that experts in astronomy require similar support, training and tools to supplement their understanding of effective communication skills which are not inherent in the learning science itself.

We recommend the following immediate and practical actions. Develop an instrument or toolset which is globally relevant so non-experts in education can easily identify and use appropriate educational theories in context. Encourage the academic community to consider that the way they learn may not be the only way to learn, specifically in the context of outreach where the audience's age, context, cognitive capability, experiences and existing knowledge may be in extreme opposition to our own. Address the preconception that it is optional to actively engage students in classes, removing such statements as: "if they are not learning, they are clearly not motivated enough". Finally, we must change the expectation that becoming an academic expert automatically makes someone equipped to be an effective educator or communicator. It is necessary for the future of astronomy to develop a dynamic, two-way flow of expertise with a range of other fields.

Education and Heritage in the era of Big Data in Astronomy
Proceedings IAU Symposium No. 367, 2020
R. M. Ros, B. Garcia, S. R. Gullberg, J. Moldon & P. Rojo, eds.
doi:10.1017/S1743921321000053

The AstroCamp Project

Carlos J. A. P. Martins[iD]

Centro de Astrofísica, Universidade do Porto, Rua das Estrelas, 4150-762 Porto, Portugal
email: `Carlos.Martins@astro.up.pt`

Abstract. This contribution describes the concept, main structure and goals, and some highlighted outcomes, of the AstroCamp—an international academic excellence program in the field of astronomy and physics created in 2012 and organized by Centro de Astrofísica da Universidade do Porto (CAUP) together with the Paredes de Coura municipality and several national and international partners.

Keywords. Pre-university education, International programs, Academic excellence programs

---

1. Introduction

The AstroCamp is an academic excellence program in astronomy and physics created by the author in 2012 and organized by CAUP with the Paredes de Coura municipality and other partners—including, since 2017, the European Southern Observatory (ESO). It is intended for students in the last 3 years of pre-university education. Initially it was restricted to students living and studying in Portugal, but it now accepts applications from (in the 2020 and 2021 editions) 42 eligible countries, in Europe and the Americas.

The three main Astrocamp goals are to promote scientific knowledge, with high-quality training in a secluded and tranquil setting, to stimulate student curiosity and skills of critical thinking, team work and group responsibility, and to stimulate interactions between students with different backgrounds and life experiences but common interests.

AstroCampers are selected according to their motivation, academic merit and potential. As a point of principle the camp is free for students living in Portugal, and for foreign students the costs (if they exist) are a maximum of 400 Euro. In recent years about half of the applicants have been foreign. In common with other academic excellence programs, AstroCamp typically attracts more applications from girls than from boys.

2. Structure and logistics

The core of the AstroCamp is a two-week residential camp, held in mid-August at the Centre for Environmental Education of the Corno de Bico protected landscape, in the Paredes de Coura municipality (in the north of Portugal). Due to COVID-19 the 2020 edition was a hybrid one, with students living in Portugal in residence as usual and students living in other countries joining through a set on online collaboration tools.

Participation by invitation, after an application (in April) and a selection phase including an interview in English (at the end of May). There are no quotas of any kind, so academic merit and potential are the only selection criteria. In the 9 editions from 2012 to 2020 a total of 118 students (67 girls and 51 boys, from 14 countries) have been selected to participate. Basic statistics of the camp participants are in Table 1.

The core AstroCamp scientific activity are two courses, each with 10x90min classes and a two-hour written exam. Course lecturers must have a PhD and be currently active in research. (A pre-approved list of courses is proposed to the students, and they can then

Table 1. Basic statistics on 2012-2020 AstroCamp participants.

Accepted Students	Years 2012-2015	Years 2016-2020
Portuguese	100%	54%
Foreign	0%	46%
10th Grade	24%	34%
11th Grade	52%	45%
12th Grade	24%	21%
Boys	47%	41%
Girls	53%	59%

Notes: In the first four editions, only Portuguese students were eligible. The middle part uses the Portuguese names for the last three years of pre-university education.

choose the ones they prefer to take.) Other activities are observational and high-level scientific programming projects (in Python or Matlab), stargazing sessions, Zoom chats with foreign researchers and selected documentaries. There are also community service projects and public talks, hiking (including an overnight one) and other recreational activities. Finally there are post-camp research and mentorship projects.

An example of a community service activity is the Solar System Hiking Trail. This is one of only 16 (in Europe) scales model of the Solar System, that is accurate both in terms of sizes and distances of the objects, which was developed during the first four AstroCamps, and officially opened on 13 August 2016.

The camp is fully residential: students, teachers and camp monitors (former AstroCamp students now doing university degrees) work and live together for 14 days. By choice the students have no internet access (and only have their mobile phones for very limited periods) to maximize their interactions with their peers and to enable them to focus on learning at a much faster pace than in their normal school classes.

3. Highlights of Outcomes

The scientific excellence of the camp can be illustrated by the fact that work done in the camp computational project has been included in several peer-reviewed publications, three recent examples being Alves *et al.* (2017), Alves *et al.* (2018) and Faria *et al.* (2019).

Post-camp student debriefings and follow-up activities (including an annual alumni lunch and an alumni weekend during the camp) demonstrate its impact. As an example, at the time of writing, 5 of the 10 students of the 2012 editions are doing PhDs in science topics, in the universities of Aveiro, Cambridge (two of them), Edinburgh and Minho.

Acknowledgments

This work was funded by FEDER-COMPETE2020, and Portuguese FCT funds, under project POCI-01-0145-FEDER-028987 and PTDC/FIS-AST/28987/2017. The AstroCamp is supported by Fundação Millenium bcp, Ciência Viva, U.Porto and ESO.

References

Alves, C.S., Silva, T.A., Martins, C.J.A.P., & Leite, A.C.O. 2017, *Phys. Lett. B*, 770, 93
Alves, C.S., Leite, A.C.O., Martins, C.J.A.P., Silva, T.A., Berge, S.A. & Silva, B.S.A. 2018, *Phys. Rev. D*, 97, 023522
Faria, M.C.F., Martins, C.J.A.P., Chiti, F., & Silva, B.S.A. 2019, *Astron. Astrophys.*, 625, A127

Education and Heritage in the era of Big Data in Astronomy
Proceedings IAU Symposium No. 367, 2020
R. M. Ros, B. Garcia, S. R. Gullberg, J. Moldon & P. Rojo, eds.
doi:10.1017/S1743921321000302

The first scientific role-playing game

Nicolás P. Maffione[1,2]🆔

[1]Universidad Nacional de Río Negro. Río Negro, Argentina

[2]Consejo Nacional de Investigaciones Científicas y Técnicas. Argentina
email: npmaffione@unrn.edu.ar

Abstract. We present here an edutainment strategy to communicate science and technology which is strongly based on personal motivation of the learner candidate: participants/players learn because they find it useful/interesting in order to achieve their own goals in some unique game. Our own goal is to capture young people's attention in an immersive and collective storytelling experience within the framework of a role-playing game specifically designed for scientific literacy. The first experiences in public high schools of Argentina are reported here.

Keywords. Role-Playing Games, Edutainment, Scientific Literacy

---

1. Introduction

A usual choice for the public communication of science and technology (PCST) are popular science talks which success depends not only on how much engaging is the topic for the people attending but also on the ability of the speaker to present such contents. The following informal edutainment (educational entertainment) strategy changes this approach: it is strongly based on personal motivation. In other words, participants/players do not learn because somebody try to engage them with an interesting topic, they learn because they need to in order to achieve their own goals in an immersive and collective storytelling (shared imagination) experience framed within a well-known game format: role-playing games (RPGs for short; Bowman 2018). RPGs are games where the players assume some fictional role in an also fictional setting. Tabletop RPGs (being Dungeons & Dragons the most popular) focus on personal interaction between players who work as a party (or may be not) in some adventure proposed by the game master (GM). The GM is also responsible for all the setting, including the role of all non-player characters. We present here a RPG specifically designed for scientific literacy that offers an informal framework for the PCST by exploring its unique hard sci-fi universe. We also report our results using the aforementioned RPG-based edutainment strategy that had been offered as an extracurricular activity in public high schools of Argentina.

2. A serious game-changer: scientific role-playing games

"Chameleon 792" (the title of our game, 792 for short) is a hard sci-fi tabletop RPG given that: (1) its carefully designed fictional universe is completely built on scientific concepts; (2) it needs facetoface interaction in order to achieve a more personal experience. Then, it is a serious game that offers the perfect environment to apply our three-steps edutainment strategy: (1-Setting) the GM (world)builds an adventure with the scientific concepts that want the party to become aware of; (2-Mission) the party plays the adventure and finds the concepts previously planted in it by the GM; (3-Discussion) the party and the GM have a brief closure talk about the most appealing scientific content faced by the players during the mission. Then, participants undergo self-paced learning, and those contents that the party finds most interesting are used by the GM to provoke them intellectually. In conclusion, 792 might be a serious game-changer candidate, being the first title of a new class of RPGs that can be used as tools for scientific literacy and that we have collectively termed as scientific roleplaying games (SciRPGs for short).

3. The first Sci-RPG: a complete hard sci-fi worldbuilding set

The present open playtest version of the Sci-RPG Chameleon 792 starter book has more than 300 pages to play the 792 right out of the box. It includes a last appendix, "Scientia", where we describe, however briefly, more than 160 scientific concepts from all sciences that were used to build up from the scratch the embryonic version of the 792 hard sci-fi universe. The "Mission Ground Zero: 6EQUJ5" (also included in the starter book) is our first full Sci-RPG mission. It consists of an astronomy-based example that includes all the material to simulate an open-world Sci-RPG experience. 6EQUJ5 extends for more than 25 pages: two self-awareness (human-like intelligence) and four instinctive (non-human-like animals) species are described, an extrasolar binary system composed of nine planets is introduced with two of those planets that are inhabited by the self-awareness species, tokens for three megacities and a space station are also included. Some of the scientific contents planted in the mission are: extrasolar systems and the Solar System, propulsion mechanisms for interplanetary exploration and sustainability.

4. The first Sci-RPG school workshops: a novel edutainment strategy

Based on the 6EQUJ5 mission we organized four workshops on three different public high schools of San Carlos de Bariloche (Argentina) during 2018 and 2019 (due to the COVID-19 outbreak the programmed workshops for 2020 were cancelled). The workshops consisted of 3 to 4 sessions of 2 to 4 hours each and for groups of 5 students (girls and boys). We show here a few answers collected in an anonymous poll as part of the positive and encouraging feedback received from the participants, notice that motivation was key to foster their curiosity about astronomy related content:

(1) *Yes, I would recommend the game because one can have fun while solving situations that the mission forces you to face, and not only using the laws of physics but also breaking them. (2) Yes, totally help me to socialize. (3) Time flies because the game is so interesting. (4) I would kept playing more sessions if I could. (5) Yes! It would have been awesome to explore further the fictional world. (6) Actually, I did not expect a game like this. The idea is very original. (7) I had the feeling while we were playing that the story changed with our decisions and actions. (8) After the sessions I started to be curious about black holes. (9) I would like to learn more about the planets. (10) I want to know more about astronomy, dark matter, time and space and the different dimensions.*

5. Final remarks

Find the starter book at https://bit.ly/39F5M3Z (in spanish) and https://bit.ly/3qtxPu1 (english sneak peek). In 2021 we will present the 792 in a dedicated website.

Reference

Bowman, S. 2018, in: J.P. Zaga &, S. Deterding (eds.), *Immersion and Shared Imagination in Role-Playing Games* (Role-Playing Game Studies: Transmedia Foundations. New York: Routledge), p. 379–394

Session 6

Education and Heritage in the era of Big Data in Astronomy
Proceedings IAU Symposium No. 367, 2020
R. M. Ros, B. Garcia, S. R. Gullberg, J. Moldon & P. Rojo, eds.
doi:10.1017/S1743921321000156

Astronomy, an Amazing First Contact with Science for High School Students

Vinicius de A. Oliveira[iD]

Space Geoscience and Astrophysics Laboratory (LaGEA), UNIPAMPA,
Campus Caçapava do Sul, Brazil
email: viniciusoliveira@unipampa.edu.br

Abstract. The contact with science is very important to the development of a conscious citizen, even if he, or she, never work with. Astronomy is an excellent way to do this approximation, simply because everyone can see the sky and has questions about it. This project aims to use observational night to discuss the classroom situations. We have made interventions in a rural local school in Caçapava do Sul (Brazil) during five years which we explain what we will see and, after that, we observe it. Since 2018, we have used the Caçapava do Sul Meteorite, that available a direct contact with a real space rock, when everyone can test the resistance and the weight of this meteorite. During this time it was almost 2 000 people has been attending by us and we noted that the students had increased their notion and comprehension about science in general.

Keywords. Astronomy, Education, Scientific Divulgation

---

1. Introduction

This work is an experience report about interventions in a local rural school (E.M.M. Antônio José Lopez Jardim) carried out by the 'Laboratório de Geociências Espaciais e Astrofísica' (LaGEA), at 'Universidade Federal do Pampa' (UNIPAMPA), Campus Caçapava do Sul, Brazil. It had been active during from 2015 to 2020, having carried out two interventions per year.

According to Lachel & Nardi (2011) the teachers do not known well the issues of Astronomy and, almost ever, then during the classes they present alternative conception and popular knowledge about this theme. This problem arises from the recent obligation, by law, that they must explain the basic concepts about Astronomy but they have not studied these during their graduation. In fact, Oliveira (2016) shows when teachers, and their students, participate in events using planetarium and/or night observation they appropriate these concepts and apply them in the classroom.

In order to attack this problem, this project aims to bring a basic knowledge of Astronomy to the teachers. At the same time, it intends to show how this can be applied in the regular courses of high school.

2. Methodology

To support the observational night we have taken talks about what we are seeing (Fig. 1A). The presentations made necessary for us found how to introduce knowledge about Astronomy to the public that is not familiar with this subject or the usual terms used, and how to make this interesting for them. For example, when the planets Venus and Jupiter are visible at night, we explain the main difference between them before

Figure 1. (A) An theorical explaination before the observational night; (B) the CSM; (C) the telescopes used in this project; (D) a photo took by us during an observational night in 2019.

observe. At the same time, we usually did round tables to discuss how that information could be used in formal learning.

In the matter, since 2018 we have taken an exposition of the Caçapava do Sul Meteorite (CSM), an octahedrite siderite found in the early 1900s (see Fig. 1B). The CSM is presented with the purpose of showing what the rock bodies in the Solar System are composed. We used this concept to explain the importance of understanding the meteorites since they keep records of the composition of the celestial bodies that constitute the Solar System.

3. Observations

We used a 12″ and a 6″ telescope to realize astronomical observational (Fig. 1C). Of course the main target is the Moon (see an example in Fig. 1D), however, according to the epoch of the year we added some near planets (Venus, Mars and Jupiter) to the target list. Naked eye observations, using a green laser to point, was a great success too. During these activities the participants had the opportunity to take a good look at the sky and learn how identify the main constellations and stars.

Unfortunately, once or another the meteorology did not cooperate, then we used a mobile planetarium to show the night sky. A planetarium section is a completely different event, however the sensation is similar to the public. The curiosity about science arise naturally, interesting question comes up after each presentation.

4. Results and discussion

During these five years, the project had seen in almost 2 000 people from the local community of this high school (students, their teachers and parents). It was observed that in each new interaction with the same students, as some followed the project throughout their high school, a deepening the questions and solutions presented by themselves. This shows a strong interaction with school subjects and the growing interest in science.

5. Conclusion

In summary, the teachers told us that this experience was fundamental for the students. They noted strengthening what was seen in class and opening up to new questions.

By the same way, some of these students are at the university now, studying Geology, Geophysics, Physics or Maths, much because projects like this one. This assures us that we are forming citizens concerned with Science, Technology and Society.

References

Achel, G.; Nardi, R. 2011, Análise do impacto de um curso de astronomia na formação continuada de professores da educação básica. In: *VIII ENPEC*, v. 01. p. 1–12

Oliveira, V.A. 2016, Astronomy like the First Contact with Sciences, *IESRJ*, v.2, issue 8

Education and Heritage in the era of Big Data in Astronomy
Proceedings IAU Symposium No. 367, 2020
R. M. Ros, B. Garcia, S. R. Gullberg, J. Moldon & P. Rojo, eds.
doi:10.1017/S1743921321000508

Considerations on the importance of building a national astronomical glossary

Hidehiko Agata[iD]

National Astronomical Observatory of Japan,
2-21-1 Osawa, Mitaka, Tokyo, Japan
email: h.agata@nao.ac.jp

Abstract. Children in elementary and middle school learn fundamental concepts of science in their local language. There seems to be some discrepancies between what they learn in school and up-to-date terminologies because academic terminologies are usually updated and shared only in English by each specific academic society. We introduce the online dictionary called as "The Internet Encyclopedia of Astronomy" compiled and provided by the Astronomical Society of Japan as an efficient solution for such terminology problem.

Keywords. sociology of astronomy, glossary, technical terms

1. Introduction

When we visited developing countries such as Myanmar, Cambodia, Mongolia, Nigeria, Saudi Arabia and so on, we found out that essential terms for understanding astronomy may not exist in the local language. In some cases, basic terms were misunderstood from the original concepts as long as we noticed on local school textbooks and descriptions about displays in science museum, which is also the case in Japan. The IAU plans to introduce only elementary terms online, based on the list of national terms compiled by the OAE. Not only this activity, each academic society in each country should take the responsibility to update concepts and terminologies and make them common in public for local science education. Aiming to provide dictionary rich enough to understand modern astronomy in local language, the Astronomical Society of Japan published "The Internet Encyclopedia of Astronomy" in Japanese in April 2018 (Okamura *et al.* 2018) so as to make everyone including children available to access the up-to-date astronomical concepts and terminologies.

2. "The Internet Encyclopedia of Astronomy" and Overall Evaluaton

The Internet Encyclopedia of Astronomy (ASJ 2018) is available online and the sample pages are shown in Fig. 1 and 2. The dictionary includes:

Total number of terms:	~ 3,000	
Elementary School Category Terms:	~ 40	(1%)
Middle School Category Terms:	~ 80	(3%)
High School Category Terms:	~ 800	(25%)
For undergraduate and graduate students:	~ 2200	(71%)

Unlike books, online terminology databases are free of charge and open to all users, and offer many advantages, such as the addition of images and videos, advanced search functions, and the ability to update the content at any time.

Figure 1. The top page design of "The Internet Encyclopedia of Astronomy".

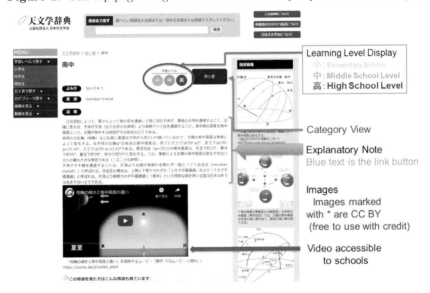

Figure 2. An example page of a term.

The online dictionary got more than 90% of positive feedback. According to feedback, it is helpful for undergraduate students who have just started an astronomical study under the circumstances that they were unable to access the reference materials in the library due to COVID-19. It is also useful for faculty staff to catch up with the latest developments. From this successful result so far, we recommend each country provide astronomical dictionary in local languages. We would also like to mention that it is only a recommendation for the use of terms in the encyclopedia and should not be enforced because language is subject to change.

References

Okamura, S., Agata, H., & Handa, T., 2018, *The Astronomical Herald*, 111, 601
ASJ, 2018, *The Internet Encyclopedia of Astronomy*, http://astro-dic.jp/

Education and Heritage in the era of Big Data in Astronomy
Proceedings IAU Symposium No. 367, 2020
R. M. Ros, B. Garcia, S. R. Gullberg, J. Moldon & P. Rojo, eds.
doi:10.1017/S1743921321000326

Construction of correspondences between the sky we see and the heliocentric model: problem-based learning in 7th grade of elementary school

Fernando Ariel Karaseur[1]🄳 and Alejandro Gangui[2]

[1]Universidad Nacional del Centro de la Provincia de Buenos Aires, Facultad de Ciencias Exactas, B7000 Tandil, Buenos Aires, Argentina
email: fkaraseur@hotmail.com

[2]CONICET – Universidad de Buenos Aires,
Instituto de Astronomía y Física del Espacio (IAFE),
C1428ZAA Ciudad Autónoma de Buenos Aires, Argentina

Abstract. We present the results of the implementation of a didactic sequence based on the formulation and resolution of astronomical problems by seventh grade elementary school students from the Autonomous City of Buenos Aires, Argentina. Its objective is to generate a meaningful understanding of the heliocentric model of the Solar System from the systematization of topocentric observations of the sky, either direct or mediated by resources such as diagrams, Stellarium software and tables, which we correlate with the parallel globe, other models with specific material and the Solar System Scope software. Throughout the sequence we address topics such as the diurnal and annual movement of the Sun, the night sky, astronomical ephemeris, Moon phases and eclipses. These are developed in parallel to the sphericity of the Earth and the concept of motion in science. For each of these topics we start from its recognition. We then implement strategies to guide students towards a possible description from the local point of view, and then extend it to other locations on the surface of the Earth. We encourage them to explain their ideas about the possible links between these topocentric observations and the corresponding relative positions of the celestial objects as seen from an external point of view to the Earth. These ideas are then contrasted with geocentric and heliocentric models. Here we highlight the integrative instances in which the students formulated problems in small groups and shared them for their resolution. Thus, motivated and challenged by the collaboration between peers, they became the protagonists of their learning.

Keywords. problem-based learning, elementary school, topocentric and external models.

---

1. Formulating and solving problems

The transversal axis of the teaching sequence proposed by the teacher (first author) is based on the construction of spatial skills that give relevance to the position of the observer (Gangui & Iglesias 2015; Longhini 2014). A problem is a new situation that a group needs to solve and for which it does not have a direct or routine strategy that leads to the solution. However, the techniques and skills necessary to solve it are previously exercised (Kaufman & Fumagalli 1999). Each problem and solution are the students' own elaborations. For this short presentation we selected and translated three examples:

Problem 1. Imagine that we take two photos of the same object throughout a day. In the first the shadow was short while in the second it was longer than the first one. The

(a)   (b)   (c)

Figure 1. (a) Screenshot of the Solar System Scope software for Problem 3. (b and c) Photographs taken by students while explaining their answers.

next day I showed them to my friends Florencia and Martin. Florencia told me that the first photo had been taken at 5:00 p.m. and the second at 11:00 a.m. On the other hand, Martín told me that the first photo had been taken at 1:00 p.m. and the second could have been at 9:00 a.m. or 6:00 p.m. Who is right? How did you find it out? *Martin is right because when the shadow is shorter, the Sun is higher and at 1 p.m. the Sun is high. On the other hand, when it starts to go down at around 6 o'clock, it is lower, so the shadow is longer. It could also have been at 9:00 because it is the other extreme.*

Problem 2. If I am in China and my flight leaves at 12:50 p.m. on December 1 (the flight lasts 13 hours and there are 14 hours difference), it arrives in Mexico City at 11:50 a.m. The flight is from west to east, in the same direction as the planet's rotation.
When did I arrive in Mexico? *December 1st.*
How could the plane land before the day and time it took off? *Because the two locations are in different time zones. When the plane leaves China, in Mexico it is 10:50 p.m. November 30. After a 13-hour flight, he arrives in Mexico at 11:50 a.m.*

Problem 3. In this photo (Figure 1.a) we see that the Sun is in the center and the planets around it. It is June 10, 2046.
What planets could be seen by day from Earth? *Venus and Neptune could be seen.*
And at night? *Mars could be seen.*
If the Moon is smaller than Mars, why do we see it larger in the sky? *The smallest sphere is the one in front, the Moon, but since it is much closer than the sphere that symbolizes Mars, it looks bigger* (Figures 1.b and 1.c).

2. Comments

The implementation of this teaching sequence allowed the students to approach astronomical phenomena that were initially unknown to them. In a challenging and motivating context, they managed to construct knowledge that progressed from their initial natural conceptions towards models closer to those scientifically valid. With the appropriate modifications, we think this experience could be applied to other educational contexts.

References

Gangui, A. & Iglesias, M. 2015, *Didáctica de la astronomía: actualización disciplinar en Ciencias Naturales: propuestas para el aula.*, Paidós.

Longhini, M. 2014, *Ensino de astronomía na escola: concepções, ideias e práticas*, Átomo.

Kaufman, M. & Fumagalli, L. 1999, *Enseñar Ciencias Naturales. Reflexiones y propuestas didácticas*, Paidós.

Education and Heritage in the era of Big Data in Astronomy
Proceedings IAU Symposium No. 367, 2020
R. M. Ros, B. Garcia, S. R. Gullberg, J. Moldon & P. Rojo, eds.
doi:10.1017/S1743921321000983

Discovering exoplanets in the classroom

Margarita Metaxa[1,2] and Anastasios Dapergolas[2]

[1]Arsakeio High School, 1 L. Marathonos, 14565, Athens, Greece
email: `marmetaxa@gmail.com`

[2]National Observatory of Athens, Institute for Astronomy, Astrophysics,
Space Applications and Remote Sensing,
Ioannou Metaxa and Vasileos Pavlou, GR-15236, Athens, Greece
email: `adaperg@noa.gr`

Abstract. This resource was developed to help bring this exciting area of research into the classroom. It consists of two practical activities appropriate for the *K12 curriculum*.
Each of the activities is standalone, takes around 60 minutes to complete and can be used either during lessons or as part of a science club. Each offers plenty of opportunity for extension work and includes a taking it further section to allow students to build on what they have learnt through independent research. The activities can be used individually, or in combination. We have already implemented them during the astrophysics summer courses we offer to school students at the National Observatory of Athens, with great success.

Keywords. Pbl exercise, exoplanets, transit method, spectral type.

1. Introduction

The discovery of planets outside our solar system, has been at the forefront of astronomical research since 1995, when Michel Mayor and Didier Queloz announced the first discovery of a planet outside our Solar System. Since then over 4600 exoplanets have been found in our home galaxy. More exoplanets are still being discovered, with different sizes, types and orbits that are piecing together the puzzle that can teach us about planet formation and our own origins. But the biggest payoff is yet to come: capturing evidence of a distant world hospitable to life. To find another planet like Earth, astronomers are focusing on the 'habitable zone' Heller (2015) around stars–where it's not too hot and not too cold for liquid water to exist on the surface. All these information and discussions are worth to come to education1 in a figurative way since as groundbreaking as exoplanet science is, the basic physics is quite accessible to K12 level physics students. To further illustrate this point, we developed this exercise that generates real exoplanet data taken from a simple telescope2, to provide students and teachers with interactive learning activities. Using introductory physics concepts, with the data taken we check:

a. the existence of an exoplanet with the transit method
b. the spectral type of the host star, so that the hosted exoplanet to have chances to belong to the habitable zone. We have created a student's fact sheet, task sheet, and an information sheet about exoplanets.

1 https://www.stem.org.uk/resources/elibrary/resource/31030/exoplanets
2 https://www.cfa.harvard.edu/smgphp/otherworlds/ExoLab/teachers/pdfs/Activity1 WelcomeV5.pdf

HD189733b 18 July 2020

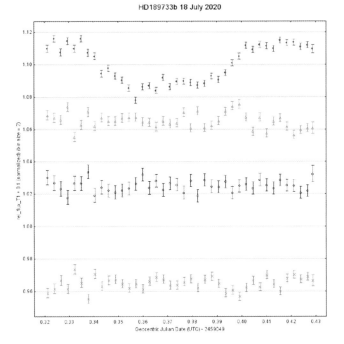

Figure 1. The uppermost light curve refer to the transit of the exoplanet, while the other curves refer to the comparison stars.

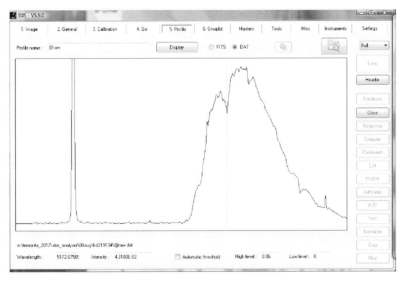

Figure 2. The trace of the star HD219134. We see on the left the trace of the undispersed star (order 0) and right the dispersed spectrum. The indicated strong absorbtion is a blend of MgI, FeI etc.

2. The case of exoplanets of the star HD189733b

We used images of the exoplanet HD189733b, taken with a 8inch f/4 newtonian tele-scope with an detector DSLR canon 40D. In order to increase the signal the images were defocused and the exposure time was increased to 60sec. Then with the software *a. Muniwin* we extracted and used the green (G) colour of the RGB images and reduced the data for the dark and flat corrections and *b*. with the software Astroimagej we conducted the photometry (Fig. 1). In the diagram orizontal axis is Julian Date in UTC and the vertical relative flux. The plotted points has bin size equal to 2.

3. The case of the star HD219134

HD 219134, is a main-sequence star in the constellation of Cassiopeia with at least five exoplanets orbiting around. Using the same telescope we took the spectra images of the mother star with the star analyser 100 transmission grating and a CMOS camera ASI120M. We analysed the spectra using the ISIS software to extract the spectral trace were we can identify the metal lines eg of Mg which is a strong indicator that the star is of spectral type between F to K, and thus the star can host exoplanets at the hapitable zone (Fig. 2).

4. Conclusion

It is a totally hands-on exercise, with an additional observational experience for the students. Through the proposed exercise on "discovering exoplanets" in the classroom, we effectively teach students to develop the skills of Problem based Learning ie Critical thinking, Team Work, Independent thinking, Communication and Digital Literacy. This method of teaching requires a high level of thinking. The students analyse, create and evaluate a challenging problem posed. This problem is open-ended and might not have solutions. Additionally the use of a telescope, as a mean for the students to obtain their own data, is an extremely challenging opportunity beyond the ordinary process, leading them to exciting exploration.

Reference

Heller, R. 2015, "Better than Earth", *Scientific American* (January 2015), 32–39.

Education and Heritage in the era of Big Data in Astronomy
Proceedings IAU Symposium No. 367, 2020
R. M. Ros, B. Garcia, S. R. Gullberg, J. Moldon & P. Rojo, eds.
doi:10.1017/S1743921321000181

Mars – a target for teachers and science students

Maria Sundin[1] [ID], Peter Bernhardt[2], Peter Ekberg[3], Jonas Enger[4] and Andreas Johnsson[5]

[1]Department of Physics, University of Gothenburg,
412 96 Gothenburg, Sweden
email: `maria.sundin@physics.gu.se`

[2]Department of Radiation Physics, University of Gothenburg,
Sahlgrenska Universitetssjukhuset, 41345 Gothenburg, Sweden
email: `peter.bernhardt@gu.se`

[3]Larkverket Kvarnfallsvagen 7, 42353 Gothenburg, Sweden
email: `peter@peterekberg.se`

[4]Department of Physics, University of Gothenburg,
412 96 Gothenburg, Sweden
email: `jonas.enger@physics.gu.se`

[5]Department of Earth Sciences, University of Gothenburg,
Box 460, 40530 Gothenburg, Sweden
email: `andreasj@gvc.gu.se`

Abstract. An interdisciplinary course about Mars for teachers and science students is presented. The focus of the course is on planning for a journey with humans to the planet Mars. Issues in ethics, morality, rights and obligations, conflict management and human psychology as well as rocket orbits, fuel economy, radiation hazards and knowledge of the solar system are included. Examination of the teacher students include interpretation of the course material for future pedagogical usage.

Keywords. Mars, ethics, didactics, philosophy of space exploration, radiation, Martian landscape, environmental ethics

1. Introduction

A case study is here presented of an interdisciplinary course about Mars for teachers and science students undertaken at the Department of Physics, Gothenburg University. The course in question is titled: Mars- a return trip. The course is given through lectures and teaching in base groups with group supervision. We aim to share the experience of creating an interdisciplinary approach with lecturers spanning physics, geology, radiation physics and philosophy. Issues in ethics, morality, rights and obligations, conflict management and human psychology as well as rocket orbits, fuel economy, radiation hazards and knowledge of the solar system have proven to be a valued and successful initiative for the further training of teachers and science students.

The focus of the course is on planning for a journey with humans to the planet Mars. This provides a great opportunity to package complex societal problems in a physics context. The course is offered with a special sustainability content mark. Future research could be done on the impact of this course on the education in different levels.

2. Landscape, Radiation, Ethics, Didactics, Learning Goals and Examination

Mankind has always had a strong and dependent relationship with the physical landscape. Understanding the physical environment has been crucial for our survival and development. The same will be equally, or more important for Mars where life conditions are much more extreme. We highlight similarities and differences in the geologic processes that have shaped Earth and Mars (Garry & Bleacher 2011). What conditions do the future explorers on Mars have to manage? We then enter the modern era and explore the dynamic Martian landscape of today. Also, by learning to read the landscape we may find locations of shelter such as vast systems of lava tubes, or locations of essential resources such as preserved glacial ice etc (Starr & Muscatello 2020).

A journey to Mars will cause substantially higher personal irradiation than obtained on Earth. The radiation part of the course lectures starts with defining the different radiation types and the biological effects these different types of radiation will cause. Then, the difference between the irradiation on Earth to the elevated irradiation in space and on Mars is described (Hassler et al. 2014). Thereafter, it is discussed if this elevated radiation burden can cause acute biological effects, e.g. fatigue, vomiting and death, and late biological effects as cancer induction. Last, possible radiation protection strategies are described and discussed. The philosophy of space exploration consists of philosophical approaches to ethics, presently applied to the topic of Mars exploration and colonization, with environmental ethics (anthropocentric vs ecocentric) and value theory at its core (e.g. Klein 2007). Four main uses of philosophy are distinguished: ethics, aesthetics, cognition and existentialism.

Research has shown that visual representation is an important part for students to be able to create a deeper understanding of concepts as well as context about the material that is taught. Interdisciplinary and complex societal problems have also been shown to be important in science teaching.

After taking the course the students are expected to have gained:
- In-depth knowledge and understanding of Mars as it is today and how the planet has changed during its history.
- Basic knowledge of how a trip to Mars is achieved and general knowledge of how a colonization might be made possible.
- Knowledge concerning ethical questions and challenges regarding planetary colonization and humans in the space environment.

Apart from these skills we require written and oral participation from the student. Since a fraction of the students are teachers looking for further education, they must develop and interpret the course material for own future application. The other students are required to use the course material to write non-fiction and be able to popularize the material for a general audience.

References

Garry, W.B & Bleacher, J.E. 2011, *Analogs for Planetary Exploration. Geological Society of America Special Paper*, Vol. 483

Hassler, D.M., Zeitlin, C., Wimmer-Schweingruber, R.F., Ehresmann, B., Rafkin, S., Eigenbrode, J.L., Brinza, D.E., Weigle, G., Böttcher, S., Böhm, E., Burmeister, S., Guo, J., Köhler, J., Martin, C, Reitz, G., Cucinotta, F.A., Kim, M.H., Grinspoon, D., Bullock, M.A., Posner, A., Gómez-Elvira, J., Vasavada, A., Grotzinger, J.P. 2011, *Science*, 343, 6169

Klein, E.R. 2007, *Space Exploration: Humanity's single most important moral imperative*, Philosophy Now, Issue 61

Starr, S.O. & Muscatello, A.C. 2020, *Planetary and Space Science*, Vol. 182, 104824

Education and Heritage in the era of Big Data in Astronomy
Proceedings IAU Symposium No. 367, 2020
R. M. Ros, B. Garcia, S. R. Gullberg, J. Moldon & P. Rojo, eds.
doi:10.1017/S1743921321000818

Preparation to Brazilian Astronomical Olympiad on a public middle school

Rafael R. Ferreira[1] [ID], Isabelly M.C. Teixeira[2], Eliane T. de Melo[2] and Marília T. de Melo[3]

[1]Federal University of Rio Grande do Norte, Physics Department, 59072-970, Natal, RN, Brazil
email: rafaelferreira@fisica.ufrn.br

[2]Municipal Secretariat for Education and Culture, Parnamirim, RN, Brazil

[3]Federal University of Rio Grande do Norte, Social Anthropology Department, 59072-970, Natal, RN, Brazil

Abstract. This investigation was derived from the observation and teaching process carried out in elementary classes at the Brigadeiro Eduardo Gomes Municipal Middle School (Natal, Brazil) at the end of 2019. The sciences classes taught were the following: the development of astronomy and astronautics from antiquity to the modern era and, introduction to stellar evolution. The goal was to focus on preparing for the Brazilian Astronomical Olympics (OBA) for students in these classes. The results, both quantitative and qualitative, were quite satisfactory. Taking into account the region's socio-economic vulnerability profile. In this context, the study of Astronomy in the classroom in search of creating new mechanisms to improve pedagogical activities encourages the critical and creative development of students, showing that they have the potential and want to make use of it.

Keywords. astronomy literacy, public middle school, basic astronomy

1. Introduction and school description

The Brigadeiro Eduardo Gomes Municipal School is a public middle school, located in the city of Parnamirim, RN, Brazil. The community in which the school is inserted in the socioeconomic profile of the middle-low class. The students are from several social classes, most of them are from more distant communities and quite lacking in basic public services, which is why they need school transport to school.

The school has a good physical structure, its classrooms were equipped in good condition and fans in full working order, its facilities were well painted and clean, its sports court has recently undergone a renovation. With regard to the physical space, infrastructure, and organization, it consists of 9 classrooms, laboratory, library, bathroom for teachers and students, warehouse, pedagogical technical team room, director's office, management office, teacher office, kitchen, 2 computer rooms and multifunctional room dedicated to special students.

2. The goal and methodology

This pedagogical intervention project had the purpose of experiencing the teaching practice, not only in the theoretical readings and discussions but also to develop methodologies and themes that can be addressed in the daily routine as a teacher, which later could be used for the development of the future interventions. Basically, the methodological procedures during these classes consisted of making several observations both in the school, in the teachers' room, and finally in the science classes of the three classes of the 6th year of elementary school. Right after the observation period, expository classes were held for 6th-grade students, on relevant themes of stellar astronomy. These classes focus on preparing these students for the Brazilian Astronomical Olympics (OBA).

3. Discussion

A possible analysis of the set of observations and rules in the classroom, it is evident the good preparation of the teacher and students. Within the limitations of basic education in that community, the school has a well-developed physical structure and management for the task they propose to perform, We really appreciate it.

The interventions, i.e., the classes taught focused on the historical development of astronomy and astronautics from ancient times to the modern era, and it was also a preparation for the OBA for students in these classes. The basic material used were: Oliveira Filho (2017), Harwit (2000) and Steiner (2018). The activities were carried out between March 25 and April 25, 2019, the first days with the observation of some classrooms, as well as the teacher and the subjects covered and methodology used by him, to later teach in the three classes observed to teach science/astronomy classes. It is important to note that the structure of the classrooms is common to the school standard, only having simple desks and whiteboards. Multimedia projectors and whiteboards were used in the classes, due to the fact that the methodology used by the professor of the discipline is valued.

During the interventions, We tried not to run away from the teaching structure already used by the Teacher. We learned numerous techniques, methodologies for teaching-learning. Thus, We used an expository class model, in order to work on basic concepts for an understanding of astronomical science. Within this proposal, We tried to start from the students' daily lives, asking questions based on examples that are present in their own daily lives, on television, in the films and programs they watch. We also sought to dialogue with interdisciplinarity, whereas the contemporary teaching-learning practice invites us to work with interconnection and with a network of knowledge. For such an action, We sought to dialogue with History and other disciplines of the natural sciences, such as Biology, in order to attract the attention of students, with the intention of embracing the various languages present in the room, facilitating the absorption of content for most of the class, respecting and challenging each student based on their own language and ability/assimilation capacity and cognition, as learned in the authors Vygotsky and Piaget from Perrenoud (2000), so used in our educational and teaching-learning model. In the approaches in the classroom, We used exercises as didactic resources, with oral and written questions, precisely so that those students who have more resourcefulness and understanding through orality and listening can get good assimilation, as for students who need writing for fixing the contents. The projector was also used to better visualization, through images, for detailed explanations and calculations.

References

Oliveira Filho, K. de S. Astronomia e Astrofísica. São Paulo: Livraria da Física, 2017. 640 p.

Harwit, M. Astrophysical Concepts. 3. ed. New York: Springer, 2000. 651 p. (Astronomy and Astrophysics Library).

Steiner, J. Astronomia Uma Visão Geral. In: Astronomia Uma Visão Geral I., 2018. Available on Youtube: LINK

Perrenoud, P. Dez novas competências para ensinar. Porto Alegre: Artmed, 2000.

Session 7

Education and Heritage in the era of Big Data in Astronomy
Proceedings IAU Symposium No. 367, 2020
R. M. Ros, B. Garcia, S. R. Gullberg, J. Moldon & P. Rojo, eds.
doi:10.1017/S174392132100017X

Archaeoastronomy in the Big Data Age: Origin and Peculiarities of Obtaining Data on Objects and Artifacts

Mina Spasova[1], P. Stoeva[2]ⓘ and A. Stoev[2]

[1]Institute of Philosophy and Sociology – Bulgarian Academy of Sciences, Sofia, Bulgaria

[2]Space Research and Technology Institute – Bulgarian Academy of Sciences, Stara Zagora Department, Stara Zagora,Bulgaria

Abstract. The impressive transition from an era of scientific data scarcity to an era of overproduction has become particularly noticeable in archaeoastronomy. The collection of astronomical information about prehistoric societies allows the accumulation of global data on: – the oldest traces of astronomical activity on Earth, emotional and rational display of celestial phenomena in astronomical folklore, "folk astronomy" and timekeeping, in fine arts and architecture, in everyday life; – the most ancient applied "astronomy" – counting the time by lunar phases, accumulation and storage of ancient databases in drawings and pictographic compositions in caves and artificially constructed objects; – "horizon" astronomy as an initial form of observational cult astronomy, preserved only in characteristic material monuments (the oldest cult observatories) with indisputable astronomical orientations. The report shows the importance of collecting the maximum number of artifacts and monuments from prehistory associated with the early emergence of interest in celestial phenomena. Spiritual, emotional and rational (including practical) needs that have aroused interest in Heaven are discussed. The huge variety of activities in realizing the regularity (cyclicity) of celestial phenomena as a stimulus for their use for orientation in space and time is shown.

Keywords. Sociology of Astronomy, History and philosophy of astronomy

1. Introduction

Archaeoastronomy (AA) is a scientific discipline that has a methodological connection with many sciences, but mostly with natural sciences. It unites the efforts of astronomers, archaeologists, architects, surveyors, geologists, epigraphists, ethnographers, mathematicians. AA is part of an interdisciplinary scientific approach that provides solutions to various problems related to the study of astronomical knowledge of ancient societies. An artifact in AA is a man-made object that has certain physical characteristics, astronomical orientations, and signs containing symbolic astronomical information; bearer of socio-cultural information, means of communication and is a subject of prehistoric culture in three main areas of its existence: material culture, spiritual culture, human relationships.

2. Archaeoastronomy as a science

AA is a branch of science that is formed in the contact zone between the humanitarian and natural sciences. The subject of its research are monuments from the ancient illiterate epoch of Mankind, studied by the methods of archeology, astronomy and other basic or applied sciences. The aim of the research is to clearly restore the astronomical knowledge and culture of ancient societies, taking into account the evolutionary factors in the surrounding natural environment. Astronomical knowledge in antiquity built any concept of nature, encouraged the creation of myths, presupposed relationships between mythological characters and events. What does archaeoastronomy study ? – (1) Objects that show a simple interest in the sky, celestial objects and phenomena and their cyclical appearance: decorations of domestic ceramics, jewelry, tools; (2) Objects that testify to astronomical knowledge and skills and their practical use: megalithic and rock – cut monuments, ancient sanctuaries, masonry and rock tombs, astronomical images and calendar ornaments, ancient calendars and calendar records; (3) Ethnoastronomy: myths, legends.

3. Digitization of archaeoastronomy

A problem with the documentation of archaeoastronomical monuments requires the appropriate levels of accuracy and adequacy of the methods. This suggests that researchers scan ancient megalithic and rock-cut monuments with an astronomical orientation, obtaining accurate 3D images with high resolution. The creation of digital libraries containing hundreds of archaeoastronomical artifacts will require the creation of specialized digital laboratories. They will provide experimental astronomical azimuths of orientation, connection with sunrises, sunsets and culminations of bright celestial bodies, GIS modeling, traceological methods and mathematical statistics for studying the archaeoastronomic culture of ancient societies. The finished 3D model of the archaeoastronomical object can become part of a collection, documented in such a way that it can then be shown to all interested parties without violating the physical original. The very possibility of creating and transmitting electronic copies of archaeoastronomical objects will be a huge step towards worldwide collective research. The digital model of the stone astronomical instrument can be emailed to colleagues for joint research or consultation (Beale & Reilly 2014). One of the areas of application of 3D reconstruction will be the study of petroglyphs and monochrome paintings with astronomical and calendar content. With the help of photogrammetry and 3D scanning it will be possible to obtain a digitally scaled copy of the drawings' surface with a high resolution. Information on the technology of making and morphology of petroglyphs and monochrome paintings can be used for their relative dating. It will be possible to automatically classify traces of different types of tools on the surface of petroglyphs, as well as the evolution of the trace during the aging of the rock substrate (Hurcombe 2014). With the help of a special program it will be possible to automatically restore the 3D relief of the drawing, made with a stone or metal tool, with fingers or a brush.

4. Conclusion

The main goal of creating a national, regional and global digital database of 3D models of archaeoastronomical objects is to move to solving research problems using volumetric models with high resolution (Forte & Campana 2017). Researchers will be able to analyze the shape and orientation of stone astronomical instruments, comparing them on a number of metric parameters, basic characteristics and landscape relationships. The virtual three-dimensional model created during the excavations will allow to record and save all the information about the spatial characteristics of the studied archaeoastronomical object much more accurately than text, depictions, exact drawings and photographs. Laser scanning, structured lighting technology and photogrammetry can be used in field 3D fixation. Accurate measurements allow high-level research and, in combination with trasological research, to restore the function of the archaeoastronomical artifact. Comparing the shape and orientation of megalithic and rock-cut monuments can provide researchers with information on the most likely migration routes and possible cultural interactions in the field of observational technologies of ancient people.

References

Beale, G., Reilly P., 2014, Virtual archeology in a material world: new technologies enabling novel perspectives // CAA 2014 Paris: 21st Century Archeology

Forte, M., Campana, S. (eds.) 2017, Digital Methods and Remote Sensing in Archeology. The Age of Sensing. New York: Springer, p. 496

Hurcombe, L. 2014, Archaeological Artefacts as Material Culture. Routledge

Education and Heritage in the era of Big Data in Astronomy
Proceedings IAU Symposium No. 367, 2020
R. M. Ros, B. Garcia, S. R. Gullberg, J. Moldon & P. Rojo, eds.
doi:10.1017/S1743921321000107

La Serena School for Data Science: multidisciplinary hands-on education in the era of big data

A. Bayo[1,2] , **M. J. Graham**[3] , **D. Norman**[4], **M. Cerda**[5,6],
G. Damke[7,8], **A. Zenteno**[8], **and C. Ibarlucea**[8]

[1]Inst. de Física y Astronomía, Universidad de Valparaíso, Chile
email: `amelia.bayo@uv.cl`

[2]Núcleo Milenio de Formación Planetaria (NPF)

[3]California Institute of Technology.

[4]NSF's OIR Lab, Tucson, AZ

[5]Inst. of Biomedical Sciences & Center for Medical Informatics and
Telemedicine, Universidad de Chile.

[6]Biomedical Neuroscience Institute, Santiago, Chile.

[7]Instituto de Investigación Multidisciplinar en Ciencia y Tecnología,
Universidad de La Serena

[8]Association of Universities for Research in Astronomy (AURA).

Abstract. La Serena School for Data Science is a multidisciplinary program with six editions so far and a constant format: during 10-14 days, a group of ∼30 students (15 from the US, 15 from Chile and 1-3 from Caribbean countries) and ∼9 faculty gather in La Serena (Chile) to complete an intensive program in Data Science with emphasis in applications to astronomy and bio-sciences.

The students attend theoretical and hands-on sessions, and, since early on, they work in multidisciplinary groups with their "mentors" (from the faculty) on real data science problems. The SOC and LOC of the school have developed student selection guidelines to maximize diversity.

The program is very successful as proven by the high over-subscription rate (factor 5-8) and the plethora of positive testimony, not only from alumni, but also from current and former faculty that keep in contact with them.

Keywords. astroinformatics, statistics, data bases, surveys, machine learning, big data.

1. Introduction: the challenge

The volume and complexity of astronomical data continue to grow as the current generation of surveys come online (Gaia, SDSS / APOGEE, etc). Beyond these challenges, astronomers will need to work with giga-, tera-, and even peta-bytes of data in real time in the era of LSST. Large data-sets pose the challenge of developing and using new tools for data discovery, interoperability and access, and analysis.

This framework brings also new opportunities for interdisciplinary research in applied mathematics, statistics, machine learning, and other areas under active development. Astronomy provides a sandbox where scientists can come together from diverse fields to address common challenges within the "Big Data" paradigm.

But of course, astronomy is not alone. Society's inexorable digitization of data and the rapidly evolving Internet are driving the need for global transformation of data intensive science in many fields. Indeed, "Big Data" now impacts nearly every aspect of our modern society, including retail, manufacturing, financial services, communications & mobile services, health care, life sciences, engineering, natural sciences, art & humanities.

Clearly, our research leadership hangs on whether the next generation can be productive within the petabyte-sized data volumes generated in different domains. Unfortunately, the development of "Big Data" and "Artificial Intelligence" (AI) related skills (including, for example machine learning) is not present commonly enough in the University curricula worldwide. This is particularly true in Chile (with a few examples that have emerged in the last years like the astroinformatics initiative from Universidad de Chile), and also in the US outside of the main / top Universities.

For instance, reports in Chile yield a deficit in highly trained AI related professionals in the thousands per year (as claimed in the "Política Nacional de Inteligencia Artificial" draft presented by the Chilean government in December 2020).

La Serena School for Data Science (LSSDS) emerged in 2013 with the leadership of AURA, aiming at covering part of the gap in training via a combine effort of Chilean and US individuals and institutions.

2. A diverse school with a rich history

LSSDS targeted since the beginning students either in their last years of undergraduate school, or the first years of graduate school (with a new pilot program involving high-school students). The school has welcomed students with majors or minors in either mathematics, statistics, physics, computer sciences, astronomy, and more recently, bio-related subjects.

The program is very intense, and lasts between 10 and 14 days with constant inter-actions between the core faculty and the students. The teaching philosophy is project oriented with ~34% of the time spent on lectures (covering basic to intermediate level statistics, basics of Data Science, Machine Learning, Distributed computing and data-bases), ~ 23% of the time spent on hands-on labs (that settle the content of the lectures), and the remaining ~43% of the time is devoted to work on the group's project with one of the faculty acting as the mentor. The transition between the three types of activities are gradual through the school, with the first days being more "lecture" heavy, and the last days of the school being focused solely on project time (Cabrera-Vives *et al.* (2017)).

The school also offers opportunities for social interactions, with the "top" activity being the visit to several of the AURA telescopes stargazing close to the summit.

The groups for the projects, with typically four students, are purposely design to maximise the diversity among the students. The projects themselves are proposed by the core faculty and tend to have different steps in difficulty with a possible open-ended final goal. Some of these projects have in fact resulted in successful observing proposals, continued collaboration, and even the replication of parts of the school by some of the students in their home institutions.

Diversity in the school is pursued since the very beginning, starting with the student selection process. When sorting the students applications, we try to keep a balance between the different minors and majors, we look for diversity of institutions (keeping a balance between students from big / well-known schools and those less renowned), gender balance, and nationality.

We believe that most of the success of the program can be grouped in two aspects: the inherent richness obtained from the multidisciplinary and diversity conscious student (and faculty) selection process, and the strong commitment from the faculty. Regarding the latter, more than half of the professors stay through the whole school serving as lecturers, hands-on instructors, and as mentors for the student group projects.

Another very relevant factor is the very strong commitment and support from AURA, NOIRLab, the LSST Corporation, NSF, REUNA, CONICYT, CORFO and other Chilean institutions that, since the very beginning understood the need to train the new generation of, not only astronomers, but scientists, in data driven problems, which particularly benefit from diversity and multidisciplinarity.

Reference

Cabrera-Vives, G. Reyes, I., Förster, F., Estévez, P. A., Maureira, JC.. 2017, *Apj*, 836, 97

Education and Heritage in the era of Big Data in Astronomy
Proceedings IAU Symposium No. 367, 2020
R. M. Ros, B. Garcia, S. R. Gullberg, J. Moldon & P. Rojo, eds.
doi:10.1017/S1743921321000132

Machine learning for the extragalactic astronomy educational manual

Maksym Vasylenko📵 and Daria Dobrycheva

Main Astronomical Observatory of the National Academy of Sciences of Ukraine
27 Akademik Zabolotnyi St., Kyiv, 03143 Ukraine
emails: `daria@mao.kiev.ua`, `vasmax@mao.kiev.ua`

Abstract. We evaluated a new approach to the automated morphological classification of large galaxy samples based on the supervised machine learning techniques (Naive Bayes, Random Forest, Support Vector Machine, Logistic Regression, and k-Nearest Neighbours) and Deep Learning using the Python programming language. A representative sample of $\sim 315\,000$ SDSS DR9 galaxies at $z < 0.1$ and stellar magnitudes $r < 17.7^m$ was considered as a target sample of galaxies with indeterminate morphological types. Classical machine learning methods were used to binary morphologically classification of galaxies into early and late types (96.4% with Support Vector Machine). Deep machine learning methods were used to classify images of galaxies into five visual types (completely rounded, rounded in-between, smooth cigar-shaped, edge-on, and spiral) with the Xception architecture (94% accuracy for four classes and 88% for cigar-like galaxies). These results created a basis for educational manual on the processing of large data sets in the Python programming language, which is intended for students of the Ukrainian universities.

Keywords. machine learning, morphological galaxy classification, education

1. Introduction

Morphological classification of galaxies is one of the basic aim for extragalactic astrophysics and observational cosmology because the morphology of galaxies plays a vital role in reflecting the evolutionary history and large-scale structure growing in the Universe. The visual labeling is the most precise method of galaxy morphological classification, which is successfully used to within 2 T-type units of the Vaucouleurs scale before era of big data surveys. Modern galaxy sky survey DES generated 50 TB of data over its six observation seasons (average 2 TB of data each night); SDSS detected 116 TB at all; the forthcoming LSST will detect 30 TB per night. It means that these hundreds of millions of galaxies could be impossible to classify manually, and the human mind is not able to comprehend complex correlations in the diverse space of parameters.

So, the multidimensional analysis is the best tool for determining the various features between different types of galaxies. All that exaggerates the interest to use the alternatives in the form of machine learning (ML), including deep learning (DL), for the automated classification of galaxies by their features (Ball & Brunner (2010), Conselice et al. 2014, Vavilova et al. (2020b,c), Elyiv et al. (2020)). Students as the future professional astronomers should be familiar with modern methods of big data processing. In this sense, the educational manual teaching the ML and DL methods for galaxy morphological classification is a good introduction into the modern course of extragalactic astronomy. Our textbook was prepared taking into account the experience with another manual on the multi-wavelength properties of galaxies and galaxy clusters, first of all, in optical and X-ray spectral ranges (Chesnok et al. (2009), Vol'Vach et al. (2011), Babyk & Vavilova (2014), Pulatova et al. (2015), Vasylenko et al. (2020)).

2. Experimental results for the educational manual

We have successfully applied a supervised machine learning methods to the automated morphological classification of large galaxy samples. Unlike the most other authors, we paid attention to the visual cleaning of the dataset (Dobrycheva (2013)). The studied sample contains of $\sim 315\,000$ SDSS DR9 galaxies at $z < 0.1$ and stellar magnitudes $r < 17.7^m$ with unknown morphological types (Dobrycheva et al. (2018)).

We apply classical ML methods (Naive Bayes, Random Forest, Support Vector Machine, Logistic Regression, and k-Nearest Neighbours) to binary morphologically classification of galaxies into early E (from elliptical to lenticular) and late L (from spiral $S0a$ to irregular Im/BCG) types. The training sample consisted of 6 163 galaxies, which were randomly selected at different redshifts and with different luminosity. For training the classifier, we used the absolute magnitudes, color indices, and inverse concentration index (Melnyk et al. (2012)). As a result, the Support Vector Machine provides the highest accuracy – 96.4% correctly classified (96.1% early E and 96.9% late L) types. It allowed us to create the Catalogue of morphological types of $\sim 315\,000$ SDSS-galaxies at $z < 0.1$ applying the Support Vector Machine: $\sim 141\,000$ E-type and $\sim 174\,000$ L-type galaxies among them (Vavilova et al. (2017), Vasylenko et al. (2019), Khramtsov et al. (2019)).

Deep learning methods were applied to classify images of SDSS-galaxies into five visual types (completely rounded, rounded in-between, smooth cigar-shaped, edge-on, and spiral) We have retrieved target sample RGB images, composed of gri bands colour scaling, each of $100\times100\times3$ pixels. The convolutional neural network classifier (Vasylenko (2020)), namely Xception, was trained on the images of galaxies from the target sample, matched in the Galaxy Zoo 2 dataset. We provided the data augmentation that was randomly applied to the images during learning. Our method shows the state-of-art performance of morphological classification, attaining $> 94\%$ of accuracy for all classes, except cigar-shaped galaxies $\sim 88\%$ (Khramtsov et al. (2019), Khramtsov et al. (2020)).

It is very important to share the accumulated knowledge with the next generation. Educational manual based on these our results in the Python programming language is intended for students of the Ukrainian universities.

References

Babyk, I., & Vavilova, I. 2014, *Ap&SS*, 349, 415
Ball, N.M., & Brunner, R.J. 2010, *Int. J. Modern Phys. D*, 19, 1049
Chesnok, N.G., Sergeev, S.G., Vavilova, I.B. 2009, *Kinemat. Phys. Cel. Bodies*, 25, 107
Conselice, C.J., Bluck, A.F.L., Mortlock, A. et al. 2014, *MNRAS*, 444, 1125
Dobrycheva, D.V. 2013, *Odessa Astron. Publ.*, 26, 187
Dobrycheva, D.V., Vavilova, I.B., Melnyk, O.V. et al. 2018, *Kinemat. Phys. Cel. Bodies*, 34, 290
Elyiv, A.A., Melnyk, O.V., Vavilova, I.B. et al. 2020, *AA*, 635, A124
Khramtsov, V., Dobrycheva, D.V., Vasylenko, M.Y. et al. 2019, *Odessa Astron. Publ.*, 32, 21
Khramtsov, V., Dobrycheva, D.V., Vasylenko, M.Yu. et al. 2020, *AA* (submitted for review)
Melnyk, O.V., Dobrycheva, D.V., & Vavilova, I.B. 2012, *Astrophysics*, 55, 293
Pulatova, N.G., Vavilova, I.B., Sawangwit, U. et al. 2015, *MNRAS*, 447, 2209
Vasylenko, M. Y., Dobrycheva, D. V., Vavilova, I. B. et al. 2019, *Odessa Astron. Publ.*, 32, 46
Vasylenko, M., Dobrycheva, D., Khramtsov, V., et al. 2020, *Communications of BAO*, 67, 354
Vasylenko, A.A., Vavilova, I.B., & Pulatova, N.G. 2020, *Astron. Nachr.*, 341, 801
Vavilova, I.B., Dobrycheva, D.V., Vasylenko, M.Y. et al. 2020, *arXiv eprints* 1712.08955v2

Vavilova, I., Dobrycheva, D., Vasylenko, M. et al. 2020, in: P. Skoda & F. Adam, *Knowledge Discovery in Big Data from Astronomy and Earth Observation* (Elsevier), p. 307

Vavilova, I., Pakuliak, L., Babyk, I. et al. 2020, in: P. Skoda & F. Adam, *Knowledge Discovery in Big Data from Astronomy and Earth Observation* (Elsevier), p. 57

Vol'Vach, A.E., Vol'Vach, L.N., Kut'kin, A.M., et al. 2011, *ARep*, 55, 608

Education and Heritage in the era of Big Data in Astronomy
Proceedings IAU Symposium No. 367, 2020
R. M. Ros, B. Garcia, S. R. Gullberg, J. Moldon & P. Rojo, eds.
doi:10.1017/S1743921321000387

Using Hadoop Distributed and Deduplicated File System (HD2FS) in Astronomy

Paul Bartus[iD]

School of Computer Science and Mathematics
Lake Superior State University
Sault Ste. Marie, Michigan, USA, 49783
email: `pbartus@lssu.edu`

Abstract. During the last years, the amount of data has skyrocketed. As a consequence, the data has become more expensive to store than to generate. The storage needs for astronomical data are also following this trend. Storage systems in Astronomy contain redundant copies of data such as identical files or within sub-file regions. We propose the use of the Hadoop Distributed and Deduplicated File System (HD2FS) in Astronomy. HD2FS is a deduplication storage system that was created to improve data storage capacity and efficiency in distributed file systems without compromising Input/Output performance. HD2FS can be developed by modifying existing storage system environments such as the Hadoop Distributed File System. By taking advantage of deduplication technology, we can better manage the underlying redundancy of data in astronomy and reduce the space needed to store these files in the file systems, thus allowing for more capacity per volume.

Keywords. Astronomy, Big Data, Deduplication, Education, File System, Hadoop, Storage System

---

1. Introduction

The use of deduplication has shown potential to remove storage redundancy in similar files across file systems. The concept of a file can be adapted to refer to chunks (data blocks) and file recipes. The file recipe for a file is a synopsis that contains a list of chunk identifiers (fingerprints) that comprise the file. Each chunk identifier can be created using a collision resistant hash over the contents of the block. Once the chunk identifiers in a file recipe have been obtained, they can be combined as prescribed in the file recipe to reconstruct the file. Hadoop Distributed File System (HDFS) is used to solve the storage problem of huge data, but does not provide a handling mechanism of duplicate files. HDFS is based on Google File System (GFS) and it operates on top of the operating system. HDFS has a name node, an optional secondary name node, and several data nodes. The name node is managing access and storing all metadata, such as file names, file attributes, and block locations.

2. Deduplication

Deduplication systems divide files into chunks (data blocks) and identify redundant chunks by comparing their identifiers (fingerprints). The chunk index contains the chunk identifiers of the stored chunks. Every deduplication system has an additional persistent index to store the information that is necessary to rebuild file contents based on file recipes. Chunk index (fingerprint) is a unique chunk identifier for each stored chunk and can be created using a collision resistant hash function such as SHA-256 over the chunk

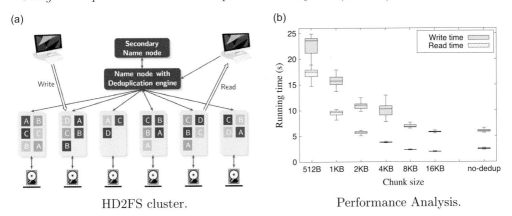

(a) HD2FS cluster.

(b) Performance Analysis.

Figure 1. Comparative box plots for Write and Read times for 100 MB files using HD2FS cluster for different chunk sizes.

contents. File recipe is a list of chunk indexes and it will be the new file. The deduplication ratio is defined as the ratio between the original data size and the non redundant data size. A higher *DedupRatio* value shows a high redundancy in the file content while a lower ratio shows a high number of unique chunks. The deduplication ratio is given by the following formula:

$$DedupRatio = \frac{Total\,chunks}{Distinct\,chunks}$$

Bartus & Arzuaga (2017) presented the concept of a file-aware deduplication storage system and provided a study on the relation between the percentage of duplicate content and the percentage of duplicate chunks for the most common file types.

3. Hadoop distributed and deduplicated file system (HD2FS)

To improve data storage capacity and efficiency in distributed file systems, the use of Hadoop Distributed and Deduplicated File System (HD2FS) is proposed. Instead of storing multiple copies of the same astronomical data, HD2FS (Figure 1a) will store only the data that is different along with a map to reconstruct all data files. Bartus (2018); Bartus & Arzuaga (2018) tested HD2FS performance. A set of user files of size 100 MB each was written and then read from HD2FS to determine how the deduplication chunk size influence the overall performance. As Figure 1b shows, for very small chunk sizes such as 512 B, or 1 KB, the write times are relatively high compared to chunk size values over 2 KB. The no-dedup experiment was done with the original unmodified HDFS cluster, therefore without deduplication. HD2FS performance is very similar to the performance of the original HDFS, but with the big advantage of improving storage space requirements.

4. Conclusion

HD2FS was designed and implemented using Hadoop Distributed File System (HDFS) by adding deduplication. The advantages of using HD2FS have been presented. The results show that the obtained deduplication values are superior in the case of using HD2FS when compared to HDFS. This means that there is potential for HD2FS to be effectively integrated to improve storage management of big data in Astronomy.

References

Bartus, P. 2018, *Using Deduplication to Improve Storage Efficiency in Distributed File Systems*, PhD Dissertation, University of Puerto Rico, Mayaguez Campus

Bartus, P., & Arzuaga, E. 2018, *Gdedup: Distributed file system level deduplication for genomic big data.*, IEEE International Congress on Big Data, July 2-7, 2018, San Francisco, CA, USA.

Bartus, P., & Arzuaga, E. 2017, *Using file-aware deduplication to Improve capacity in storage systems.*, IEEE Colombian Conference on Communications and Computing (COLCOM), pages 1–6.

Session 8

Education and Heritage in the era of Big Data in Astronomy
Proceedings IAU Symposium No. 367, 2020
R. M. Ros, B. Garcia, S. R. Gullberg, J. Moldon & P. Rojo, eds.
doi:10.1017/S1743921321000211

Astronomy education and culture

Magda Stavinschi[1] and Alexandra Corina Stavinschi[2]

[1]Astronomical Institute of the Romanian Academy
email: magda_stavinschi@yahoo.fr

[2]Independent researcher
email: alecorina@yahoo.it

Abstract. The 2020 pandemics has brought about a revolution in education, thanks to the pervasiveness of online teaching. Contents, methods and techniques can now be rapidly shared across the globe. On the downside, a number of disciplines have been neglected or dropped altogether. Our paper aims to address the following questions: How has Astronomy in culture been affected? Why is it important to keep it alive? What are the solutions? We suggest that it has been dismissed for two reasons: first, it is perceived as a niche topic – some sort of erudite chatter about non-essential curiosities – that can be sacrificed in favour of more practical information; second, it is heavily culture-specific, meaning that it requires extra effort from the teachers, as it cannot be easily copied or translated from other sources.

Astronomy is first and foremost a science of nature, although astronomical elements can also be found in other sciences, arts, folklore and religion. It used to be a subject in the school curriculum. However, it has become increasingly obvious that education through clear-cut disciplines is not always the most effective strategy. This is particularly true today, when everything around us is changing faster than ever. Many students will soon choose professions that do not even exist yet. In response, teachers have switched to inter-, multi- and transdisciplinary methods, which provide more flexibility and a better perspective.

Furthermore, 2020 has brought about a revolution in education, thanks to the pervasiveness of online teaching. Contents, methods and techniques can now be rapidly shared across the globe. Not all teachers were prepared for this; here is where astronomy and astronomers can help. On the one hand, those accustomed to observing the sky at a distance have adapted faster to online teaching, on the other hand, astronomy is found both in the classical scientific education (mathematics, physics, chemistry, biology, geography, etc.) and in the arts (literature, music, history, religion, etc.).

We will first look back at the history of astronomy in order to draw important points that need to be addressed in education; secondly, we will look ahead at widely available technological resources (websites, online videos, podcasts, etc.) to show how we can easily and efficiently provide empowering resources to open the mind of the new generation. It is important to show them that the recent discoveries about the universe are not only the product of top technologies, but also the outcome of research and observations that have been carried out around the world for millennia.

Astronomy teaches to draw connections not only between different branches of science, but also between science and arts, science and religion, science and local traditions, between the universal and the particular. In fact, one of the cornerstones projects of International Year of Astronomy 2009 that has been adopted by IAU and UNESCO is

"Astronomy and World Heritage". Its main goal is to establish a link between Science and Culture on the basis of research aimed at acknowledging the local cultural and scientific values connected with Astronomy. Just as the sky belongs to all of us, so does the cultural heritage on Earth. Ignoring or neglecting it does not only threaten its survival, but also risks altering future humanist thinking. Therefore, it is crucial to raise awareness about the topic, which is now an underrated aspect of education.

The internet generation has been using computers to capitalize and store the information concerning the past of world Astronomy, to come to know better the history of their native places through the history of Astronomy and to get a glimpse into the future of the society they belong to.

If we look at the relationship between Astronomy and History, we have a very wide margin to "travel" with the students both in time and in space. Not only does the history of astronomy go back right to the beginning of civilization, but it can also be encountered everywhere on Earth. Teachers can select the best examples from their own country or region. In this way, the knowledge of astronomy is intertwined with that of the history of one's own people. To take one example, related to the Romanian history: if we talk about the prehistoric monument Stonehenge, we must also mention a similar monument in Sarmizegetusa Regia, the capital and the most important military, religious and political centre of the Dacians prior to the wars with the Roman Empire.

The same method can be used for the relation between astronomy and art. Teachers can give many examples from the arts, such as Van Gogh's famous painting The Starry Night or Beethoven's Moonlight Sonata, and find examples from the culture of their own people. They can easily illustrate them online, drawing parallels between the knowledge of the sky and the artwork they picked out. For many nations, folk art can also be used as a relevant source. Architectural or ornamental details, sometimes even oral folklore, provide evidence not only of a solid knowledge of the sky but also of the deep interest in the universe, from the closest elements to infinity.

A delicate issue is the relationship between science and religion. In many countries, religion is taught in schools as a discipline from the very first years. Specific training is crucial, both for teachers of religion and for those who teach scientific issues related to the origin of the universe. The Book of Genesis can easily clash with the scientific information that students receive about the birth of the universe.

Finally, there is one more astronomy-related area that needs to be addressed: astrology. Dismissing the topic altogether does not seem to be a good idea. The pseudo-scientific information pouring in through the media turns out to be completely counterproductive from an educational point of view. Teachers have an excellent opportunity to discuss the fine line between science and pseudo-science, between real and fake news, between science fiction and conspiracy theories.

Of course, any universe-related subject requires well-trained teachers who understands the topic, are able to select the most appealing resources and motivate students to read, to travel and, last but not least, to look up at the stars. No matter how many powerful instruments one might have, nothing compares to the emotion that the sky can give you.

1. Conclusion

Astronomy is the science of the Universe but also a universal science. In the times of globalization, of online teaching, when school curricula are severely simplified, when science is advancing at an unprecedented rate, astronomy can offer the bridge between disciplines, the way to understand each one of them and the inspiration to build the (wo)man of the future.

Education and Heritage in the era of Big Data in Astronomy
Proceedings IAU Symposium No. 367, 2020
R. M. Ros, B. Garcia, S. R. Gullberg, J. Moldon & P. Rojo, eds.
doi:10.1017/S1743921321000363

Astronomy in romanic churches

Ederlinda Viñuales Gavín ⓘD

Astronomía y Astrofísica, Universidad de Zaragoza
C/ Pedro Cerbuna 12, Ed. Matemáticas
50009-ZARAGOZA-SPAIN
emails: ederlinda.vinuales@gmail.com and naseprogram.treasurer@gmail.com

Abstract. In this poster we present a study of the orientation of the church of San Adrián de Sasabé in Borau, Huesca (Spain) in a practical way. This church is a characteristic Romanesque construction, predominant in the High Middle Ages, mainly in southwestern Europe.

The apse of Romanesque churches are oriented towards the east. But, in some churches, the apse has three windows and these are oriented in the direction of the sunrises on the days of the solstices and equinoxes. **But sunrises and sunsets depend on the latitude of the place**.

The church of San Adrián de Sasabé, the object of our study, has three windows in the apse, which allows us to carry out the necessary calculations to determine its orientation with precision outside the church.

Keywords. Heritage, Education, Orientation and Astronomy.

---

1. Introduction

At present, in the construction of temples and public buildings their orientation is not taken into account, mainly due to the value of the land and urban planning laws. They are built where and how is convenient. But it was not always like this. All religious buildings for centuries have been oriented. From the pyramids of Egypt and Mexico to the Hindu and Chinese temples, all its builders sought to erect their temples in a specific position and orientation. Astronomical or religious reasons?

Culturally, Romanesque art was the first great purely Christian and European style that unified the different currents that had been used in the High Middle Ages and managed to formulate a specific and coherent language applied to all artistic manifestations. It emerged gradually and almost simultaneously in Spain, France, Italy and Germany but with its own characteristics, although with enough unity to be considered the first international style, within the European scope.

In Spain the Romanesque is introduced through the Camino de Santiago due to the numerous pilgrimages that come from all over Europe. Thus, the north of Spain, see reference Canellas (1992), is crowded with Romanic buildings.

But Romanesque is a style that is not only used for the construction of churches. We can find castles, palaces, cathedrals, bridges, etc. Also painting and sculpture reached great relevance (Fig. 1).

2. San Adrían de Sasabé church

We will start by saying that the orientation of the Romanesque churches is due to a Christian and not astronomical symbolism, although astronomy is so present. See Hani (1997).

In some well-oriented Romanesque churches with three windows in their apse it is observed that approximately, these three windows are located in the direction of sunrise,

Figure 1. Examples of Romanesque art: a castle, a cathedral and a painting.

Figure 2. Front door and apse of San Adrián de Sasabé.

on the horizon, on the first day of each of the seasons: winter, spring and summer (since that the point of the autumn and spring dawns coincide). But not always the location of the windows in the apse meets this astronomical characteristic. In some cases due to the peculiarities of the land on which the church was built and in others with the purpose of celebrating the saint to whom the church was dedicated.

San Adrián de Sasabé has three apses and is well oriented. But, how can we check the angle that the sunrise forms between a solstice and an equinox (or between the two solstices) in a church?

We know that this angle depends on the latitude of the place.

By a theoretical method, using the *AstroMath* following the program of Abad (1998) and knowing the latitude and longitude of the place: Latitude: $42^o40^m33^s$ N and Longitude: $0^o35^m26^s$ W and the declination of the Sun δ, of the Sun at the solstices and equinoxes, we calculate the difference in azimuths between the sunrises of the summer and winter solstices days that turns out to be around $65^o31^m50^s$.

But it is much more interesting and educational to do this calculation in the church itself and with students. Read the article Viñuales (2012).

How can we proceed? Look at the Fig. 2 (right).

Working outside of the church, we project the sill of the central window on the floor and that of one of the side windows. Next, we draw the perpendicular through the midpoint in the projection of the windowsills. The angle formed by the two red lines in the image of San Adrián de Sasabé will give us the angular displacement of the sunrise from a solstice to the equinoxes. So, the angle obtained by measuring it *in situ* was about 65°, approximately that obtained theoretically.

As a consequence all of the above, *we can conclude that the three windows of the apse of San Adrián de Sasabé was oriented astronomically with all intention, so that the first rays of the Sun entered through the three windows of the apse the solstices and equinoxes days.*

References

Abad Medina, Alberto. 1998, in University of Zaragoza, *AstroMath*, program for *Mathematica*.

Hani, Jean. *El simbolismo del templo cristiano*. Sophia Perennis. Barcelona 1997.

Canellas Lopez, Angel 1992, *Aragón, La España románica*, v.4. Eds Encuentro. Madrid 1992.

Viñuales Gavín, E. *La orientación en las iglesias románimas*. Web de NASE, 2012.

Education and Heritage in the era of Big Data in Astronomy
Proceedings IAU Symposium No. 367, 2020
R. M. Ros, B. Garcia, S. R. Gullberg, J. Moldon & P. Rojo, eds.
doi:10.1017/S1743921321000041

Historical constellations in the planetarium

Susanne M Hoffmann[iD]

Michael Stifel Center of the Friedrich Schiller University and Planetarium Jena,
Ernst-Abbe-Platz 2, 07743, Jena, Germany
email: susanne.hoffmann@uni-jena.de

Abstract. This contribution summarizes the reconstruction of historical constellations. It is based on studies of classics, philologies, history of science and history of art. In the given brevity, I can only sketch the strategic, scientific and educational reasons.

Keywords. reference systems, astronomical data bases: miscellaneous, history and philosophy of astronomy

1. Introduction: Why should we care?

Indigenous constellations are a cultural heritage of people all over the world. Before the decisions in the 1920s (i) to define a set with international validity, (ii) to base this set of constellations on the 48 ancient Greek constellations and, thus, the Euro-Arabic tradition, (iii) to assign a Belgian priest with new definitions according to equatorial longitudes and latitudes, and (iv) to accept his suggestion of 88 areas by the IAU and make it canonical (Delporte, E. (1930)), each culture used its own set of constellations.

The reconstruction of the indigenous constellations is therefore a tribute to the cultural history and helps peoples and tribes all over the world to self-identify with research on the sky. If it is possible to present the constellations in the local planetarium dome, it is certainly much more immersive but also the visualisation on flat screens of desktop computers has an appeal. The latter has the advantage that the learners – children, youngsters or adults – can play around and really use it, create their own visualisations and study the sky. Either way, the local history and intellectual heritage supports the teaching and learning process.

With a global community of astronomers in mind, this is a very attractive way to introduce the IAU constellations as official frame of reference without loosing the own roots by showing the indigenous constellations of many cultures. This also demonstrates that constellations are human conventions and not given by nature which is why we must not derive anything from them. It is a rather obvious way to visually demonstrate (without arguing) the absurdity of astrology.

Constellations have always been a frame of reference that kept existing although mathematical frames of reference were introduced in the 1st millennium BCE and in Ptolemy's Almagest from the 2nd century CE, even a coordinate system was provided but people kept using constellations additionally. Thus, roughly 2000 years of history show without doubt that the patterns of gestalt-seeing with a local culural background are important for many people: They help memorizing the patterns and connect to the human cultures.

2. Making Off

Currently, Zeiss Planetarium software does not offer the possibility to display anything else but the IAU constellations – as area borders, as skickfigure and as drawing. Only the open source software Stellarium allows to choose and change between roughly thirty different sky cultures, most of them incomplete. In the era of big data and software development, knowledge is never fixed or final and in the age of databases, sorted lists can always easily be extended.

For the Zeiss planetaria Jena and Berlin, I created and installed the almost completely preserved Babylonian constellations based on MUL.APIN (−2nd millennium), the ancient Greek (around year "0") and the medieval set of constellations (∼ 850 CE): Hoffmann, S.M. (2017b). In Stellarium, I contributed the original Greek sky culture in shape of the Almagest stickfigures (a work of my students) and two Babylonian sky cultures of the −1st and −2nd millennium: Hoffmann, S.M. (2020).

The mathematical reconstruction of the area of ancient constellations is based on coordinates and dates of rising, setting, culmination or whatever is available: Hoffmann, S.M (2016); Hoffmann, S.M. (2017a). The figures are reproduced from historical depictions. This creates an image that is as realistic as possible and displays figures in a way that is known to the people (the ancestors as well as us because objects of their art are preserved). The uncertainties and ambiguities of identifications are documented in scientific papers.

3. Outlook

Applications of these visualisations are not only the usage in daily public presentations and narrating local history. Constellations are a frame of reference and used to identify historical transients for research in astrophysics. However, their high cultural value helps people all over the world to find access to all types of research concerning astronomy and astrophysics. It will be necessary to build a knowledgebase for native constellations.

Currently, I am setting up a structure for storing the data for my already reconstructed sky cultures. I already talked to the producers of planetaria at Zeiss about the requirements for better performance in implementing further cultures and I keep contributing to Stellarium.

My goal is mainly to provide the structure for a collection and, then, invite colleagues to contribute their work on this topic. This will enable a learning and research process that is data-driven instead of professor-driven, i. e. it is unbiased and nonauthoritarian.

Acknowledgements

I thank Prof. Dr. Birgitta König-Ries (Computer Science, FSU Jena), Dr. Björn Voss (LWL Planetarium Münster, Germany), Jürgen Neye and Dr. Monika Staesche (Planetarium am Insulaner, Berlin, Germany), Stefan Harnisch (Planetarium Jena) and my students Alina Schmidt, Marie von Seggern and Lea Jabschinski for the collaboration.

References

Delporte, E. 1930, *Atlas Cèleste*, (Cambridge University Press)

Hoffmann, S.M. 2016, *Orbis Terrarum*, 14, 33–49

Hoffmann, S.M. 2017, *Hipparchs Himmelsglobus*, (Wiesbaden, Ney York: Springer)

Hoffmann, S.M. 2017, in: Wolfschmidt, G. (ed.), *Nuncius Hamburgensis*, 41, 135–157

Hoffmann, S.M. 2020, in: Draxler, S., Lippitsch, M. and Wolfschmidt, G. (eds.), *Harmony and Symmetry, Proceedings of the SEAC 2018 Conference in Graz*, (Hamburg: SEAC Publications Vol. 01)

Zotti, G., Hoffmann, S.M., Wolf, A., Chéreau, F., Chéreau, G. *Journal for Skyscape Archaeology*, accepted, forthcoming

Education and Heritage in the era of Big Data in Astronomy
Proceedings IAU Symposium No. 367, 2020
R. M. Ros, B. Garcia, S. R. Gullberg, J. Moldon & P. Rojo, eds.
doi:10.1017/S1743921321000442

Japan Astronomical Heritage:
The First Two Years

Toshihiro Horaguchi [ID]

National Museum of Nature and Science, Japan,
on behalf of the Committee of Japan Astronomical Heritage,
4-1-1 Amakubo, Tsukuba, Ibaraki 305-0005, Japan
email: horaguti@kahaku.go.jp

Abstract. The Astronomical Society of Japan has started authorization of Japan Astronomical Heritage since 2018. The society certificates two or three sites/materials/literature every year not only for preservation but also for utilization. The certification influences citizens and local governments, and stimulates various movements. The idea of national astronomical heritage will help to preserve valuable properties of each country and to promote the utilization.

Keywords. astronomical heritage

1. ASJ and the Japan Astronomical Heritage

The Astronomical Society of Japan (ASJ, https://www.asj.or.jp/en/) has started authorization of Japan Astronomical Heritage, recognizing that the preservation, inheriting, notification and utilization is one of the most important missions of the society. ASJ, which is founded in 1908, has about 2,200 regular members, and is supported by many associate members, such as amateur astronomers, educators, etc., of about 1,100. Every year, ASJ calls to the members for candidates of heritage. The committee of Japan Astronomical Heritage selects a few from them, and the representatives of the society finally approve the Japan Astronomical Heritages of the year.

Japan Astronomical Heritage has three categories, which are 1) historic sites and buildings, including observational facilities, 2) materials, such as telescopes and instruments involved in an important astronomical discovery, and 3) literature, such as astronomy-related documents of historical significance. Those include National Treasures of historic era but also local heritages which are not well known nationwide and modern equipment that should be preserved.

The certification of heritage is presented to the owner or the managing organization at the next spring meeting of the society, though the latest was postponed to September due to COVID-19. The press release of heritages is made public by not only national media but also many local media such as newspapers and broadcasters.

2. The first two years of Japan Astronomical Heritage

The first call for candidate was August 2018, and two heritages were selected: 明月記 ("Meigetsuki"), a diary of a famous medieval poet which includes description of astronomical phenomena such as the supernova explosion, SN1054, now known as crab nebula (Duyvendak (1942), Fig. 1), and 会津日新館天文台跡("Aizu Nisshinkan Tenmondai-Ato", Fig. 2), a historic site of an early 19th-century observatory at which observation for making luni-solar calendar was done (Watanabe (2016)). The historic site is being prepared by the local government to construct information center for visitors.

Figure 1. HST image of the supernova remnant of SN1054 (Crab Nebula) by courtesy of NASA, ESA, and J. Hester (Arizona State University).

Figure 2. Historic site of early 19th-century observatory at Aizu. Image courtesy of Aizuwakamatsu City Board of Education.

Figure 3. Star chart drawn in late 7th–early 8th century. Image courtesy of Nara National Research Institute for Cultural Properties.

Figure 4. Observation site of 1887 total solar eclipse at Niigata. Image courtesy of Committee of Japan Astronomical Heritage.

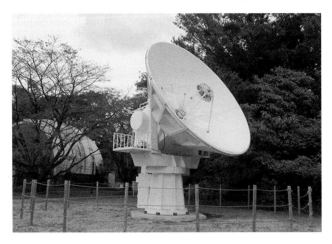

Figure 5. 6m millimeter radio telescope made in 1970. Image courtesy of Committee of Japan Astronomical Heritage.

The second call of 2019 results in three heritages: one is キトラ古墳天井壁画 ("Kitora Kofun Tenjo Hekiga", Fig. 3), a fine mural of star chart drawn on the ceiling of late 7th- or early 8th-century tomb (Sôma (2016)), another is 明治 20 年皆既日食観測地及び観測日食碑 ("Meiji-20nen Kaiki-Nisshoku Kansokuchi oyobi Kansoku-Nisshoku-Hi", Fig. 4), a historic site where the first modern observation of total solar eclipse was done in 1887 (Arai (1888)), and the other is 6m ミリ波電波望遠鏡 ("6m Mili-Ha Denpa Bouenkyou", Fig. 5), the cornerstone telescope of Japanese radio astronomy made in 1970 (Akabane et al. (1974)). The news of the historic sites has reminded citizens of the value of the heritages and the pioneering astronomers.

3. The goals of Japan Astronomical Heritage

Japan Astronomical Heritage advances the recognition of historically important properties and promotes the utilization such as in the astronomical community and the local community. ASJ will continue to authorize Japan Astronomical Heritage in order to make citizens and professionals re-recognize the value of the sites/materials/literature not only for preservation but also for utilization such as education of astronomy, regional history, etc. The activity will also contribute to the project of World Astronomical Heritage.

We believe that the idea of national astronomical heritage helps to preserve valuable properties of each country and to promote the utilization.

References

Akabane, K. et al. 1974, *Publ. Astron. Soc. Jpn.*, 26, 1

Arai, J. 1888, *Mem. R. Astron. Soc.*, 49, 271

Duyvendak, J.J.L. 1942, *PASP*, 54, 91

Sôma, M. 2016, *Rep. Natl. Astron. Obs. Jpn.*, 18, 1 (in Japanese)

Watanabe, J. 2016, *Nikkei Science*, 46(11), 62 (in Japanese)

Education and Heritage in the era of Big Data in Astronomy
Proceedings IAU Symposium No. 367, 2020
R. M. Ros, B. Garcia, S. R. Gullberg, J. Moldon & P. Rojo, eds.
doi:10.1017/S1743921321000892

Telescopic sunspot observations during the last four centuries: a forgotten world heritage

José M. Vaquero🅾

Departamento de Física, Universidad de Extremadura,
Avda. Santa Teresa de Jornet, 38, 06800, Mérida, Badajoz, Spain
email: `jvaquero@unex.es`

Abstract. Sunspot observations, normally made with small telescopes during the last four centuries, should be considered a world heritage for many reasons including purely scientific aspects and other cultural and social reasons. Here, these aspects are briefly reviewed.

Keywords. Sunspots, Heritage, Literature, and Cultural Astronomy

1. Introduction

Sunspot observations, normally made with small telescopes during the last four centuries, should be considered a world heritage for many reasons including purely scientific aspects and other cultural and social reasons.

2. Science, literature, and society

The sunspot observations made by amateur and professional astronomers over the past four centuries are of crucial importance to today scientific community (Clette *et al.* 2014; Muñoz-Jaramillo & Vaquero 2019; Arlt & Vaquero 2020). They are the basis for reconstructions of solar activity. In particular, the *sunspot number* is constructed with the accumulation of these observations (Vaquero *et al.* 2016). Furthermore, the sunspot number has become the most famous time series in astronomy and statistics.

Sunspots have become an element of the culture of humanity including, for example, literature. Weiss & Weiss (1979), mention as an example the Andrew Marvel (1621–1678) satirical poem *The instructions to a painter* (published in 1689), which contains apparently the first mention in English language of the disappearance of the sunspots in the seventeen century. Moreover, García Santo-Tomás (2014) has shown that the sunspots appear in other famous literary texts such as the satire *La hora de todos y la Fortuna con seso* by Francisco de Quevedo (1580–1645) or the moralistic novel *El diablo cojuelo* by Luis Vélez de Guevara (1579–1644). In 1641, Guevara quotes in his novel, first of all, Copernicus and Galileo, talking about the *Galileo's telescope* (the lens that Quevedo had already discussed without citing Galileo and presenting it as a Dutch invention). He also mention the *charms of Copernicus*. Quevedo only talks about the *optical tube*, an invention according him also Dutch, a long-sight telescope *that finds a spot in the Sun* (Hora de todos, 293. Ed. Felicidad Buendía: Obras Completas, Vol. I).

These works show how sunspots were quickly incorporated into European culture after their discovery with the telescope in 1610. A captivating example is the engraving (Fig. 1) that serves as the frontispiece to the work titled *Il Cannocchiale aristotelico* (The Aristotle's spyglass) by Emanuele Tesauro (1592–1675). It shows a paradoxical situation: Aristotle helps a woman to observe sunspots with a telescope.

Figure 1. Frontispiece of *Il Cannocchiale aristotelico* (The Aristotle's telescope) by Emanuele Tesauro published in 1670 (and an enlarged detail). It is the second edition of this work.

Finally, these observations are essential for scientists to solve problems of greatest current interest to modern society such as space weather (Baker 2002) or climatic change (IPCC 2013) because the long-term solar activity is crucial to a better understanding our planet and the interplanetary medium (Hanslmeier 2007).

3. Conclusion

The telescopic sunspot records preserved in archives and libraries are a forgotten world heritage that the astronomical community must preserve for the future generations. Furthermore, these observations are a clear example of how astronomy influences the culture of societies (López & Giménez Benítez 2010). Finally, this heritage can also be used in the field of general education and in the teaching of astronomy in particular.

References

Arlt, R. & Vaquero, J.M. 2020, *Living Rev. Solar Phys.* 17, 1
Baker, D.N. 2002, *Sci.* 297, 1486
Clette, F., Svalgaard, L., Vaquero, J.M. & Cliver, E.W. 2014, *Spa. Sci. Rev.* 186, 35
Hanslmeier, A. 2007, *The sun and space weather*, second edition (Dordrecht: Springer), p. 315
IPCC 2013, *Climate Change 2013: The Physical Science Basis. Contribution of Working Group I to the Fifth Assessment Report of the Intergovernmental Panel on Climate Change* (Cambridge, United Kingdom and New York: Cambridge University Press), p. 1535
López, A. & Giménez Benítez, S. 2010, *Ciencia hoy* 20, 17
Muñoz-Jaramillo, A. & Vaquero, J.M. 2019, *Nature Astron.* 3, 205
García Santo-Tomás, E. 2014, *The Refracted Muse. Literature and optics in early modern Spain* (Chicago: The University of Chicago Press), p. 320
Vaquero, J.M., Svalgaard, L., Carrasco, V.M.S., Clette, F., Lefevre, L., Gallego, M.C., Arlt, R., Aparicio, A.J.P., Richards, J.-G. & Howe, R. 2016, *Solar Phys.*, 291, 3061
Weiss, J.E. & Weiss, N.O. 1979, *Q. Jl. R. astr. Soc.*, 20, 115

Education and Heritage in the era of Big Data in Astronomy
Proceedings IAU Symposium No. 367, 2020
R. M. Ros, B. Garcia, S. R. Gullberg, J. Moldon & P. Rojo, eds.
doi:10.1017/S1743921321000806

Solar Eclipse: a didactic alternative for education in Astronomy

Paula Chis🄳

Colegiul National George Baritiu, Cluj-Napoca, Romania
email: `paulaonica@yahoo.com`

Abstract. Influence of astronomy education, in other disciplines, can give some possible explanation even in the history field. We propose to link astronomy, history and heritage through Big Data, without a telescope.

If we look to the Dacian Draco flag we can find a similitude with Draco constellation and that was a little bit intrigue. In the era of digitalization we can use computers to see back in time.

Keywords. Didactic experience, Teacher training, Astronomy.

This extraordinary symposia "Education and Heritage in the era of Big Data in Astronomy" hosted by IAU was a big opportunity for each of us to contribute with different and original materials in this interesting field of study. My students also are very interested in astronomy and the links with other sciences like history, archeology or geography but also computer science are a benefit. They were curious about the historical and archeological heritage since the era of dacians and the new modern tools we have today like Stellarium application.

We found that astronomy education used in other disciplines, can give some possible explanation even in the history field.

Stellarium as an educational instrument easy to use in the classroom, represents a planetarium software that shows exactly what you can see when you look up to the sky. But the most interesting quality is that you can change the place and time of observation. Going back in time using Stellarium, for the period of ancient ancestors you can contribute with a small part in historical research, in the explanation of some symbols or some legends. The intrigue was about the shape of Dragon constellation and the same style for the battle banner for our ancestries living here in Dacia, Sarmizegetusa Regia.(Dacian Kingdom, 168 BC-106 AD).

Therefore we were wondering, the people living here in those times what they saw when they were looking up to the sky. Now, using stellarium we can answer this question. Mostly at their latitude they saw the Dragon constellation above their heads. If we look to the Dacian Draco flag we can find a similitude with Draco constellation and that was a little bit intrigue.

Among the Dacians, the draco was undoubtedly seen by the army as a special protective symbol, while it also played an important role in the religious life of the people (Oltean 2007), maybe like the Draco constellation which guided them from the sky.

One of the most important evidence is that on the Trajan's Column (113 AD), Dacian soldiers are represented carrying a draco in 20 scenes. Another evidence we can find on the Roman coins of Dacia.The draco appears on coins of Roman Emperor Antoninus Pius (r.138–161 AD), indicating that it was still the characteristic emblem in the 2nd century. On Arch of Galerius in Thessaloniki: the characteristic Dacian dragon emblem

Figure 1. Draco constellation, sculpture from center of Orastie city.

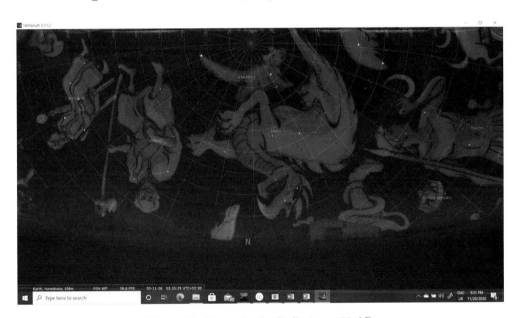

Figure 2. Draco in the Stellarium 100 AD.

is carried by a group of Dacian horsemen depicted on the Arch of Galerius and Rotunda in Thessaloniki, Greece.

Our ancestries carried with them the symbol of Draco constellation from zenith to the ground. During Dacian era the constellation above that latitude was Draco and the traditional name of Alpha Draconis, Thuban, means "head of the serpent", all these might be just a legend, or an impressive history story but this is what our students need to be curious, to research and find answers, more than that to make connections between the areas of study.

Reference

Oltean, D., 2007, Burebista si Sarmizegetusa. Ed. SAECULUM I.O., ISBN/COD: 978-973-642-125-9

Education and Heritage in the era of Big Data in Astronomy
Proceedings IAU Symposium No. 367, 2020
R. M. Ros, B. Garcia, S. R. Gullberg, J. Moldon & P. Rojo, eds.
doi:10.1017/S1743921321000089

The oldest astronomical observatories in Ukraine

I.B. Vavilova[ID] and T.G. Artemenko

Main Astronomical Observatory of the National Academy of Sciences of Ukraine
27 Akademik Zabolotnyi St., Kyiv, 03143 Ukraine
email: irivav@mao.kiev.ua

Abstract. We present the data related to the revealed astronomical oldest observatories at the territory of Ukraine and describe briefly the principles of observations which could be realized at these sites with usage of megalithic stones. Among these oldest observatories are as follows: the stone complex at the Lysyna Kosmatska mountain (Charpatian region); Bakhchysaray Menhir in Crimea; a complex of shafts at the Mavrin Maidan near Pavlograd city.

Keywords. archeoastronomy, oldest observatories, cultural astronomy

---

The Mavrin Maidan (48°33'34.92" N, 35°48'19.08" E) is located in the Dnipro region near Pavlograd city. It is a complex of shafts having the correct relief form. The Mavrin Maidan reminds the shape of a spider from the height of the bird's eye (Fig. 1, left). It is really a mystical place with remarkable acoustics, where trees do not grow and a water does not stay in the rainy season. The sizes are as follows: diameter of the central depression is 54-60 m, depth is 10 m; length of whiskers is 65-200 m; area is 5.5 hectares. In the center of this construction is a bulk circle, in the middle of which there are three "entrances" and long wavy shafts on the outer edge of the circle. On the day of spring equinox, the Sun goes beyond the horizon exactly in the center of one of the valleys.

In 1987, the Mavrin Maidan was investigated by the archaeologist Churilova L.M. who dated it to the II Millennium BC. The Mavrin Maidan under No.6348 is included into the list of regional archeological sites by the Resolution of the Dnipropetrovsk Executive Committee No.424 on Nov 19, 1990. The "Kosmopoisk" expedition (headed by Bondarenko K.) studied this place during the sunset on March 19-20, 2011, with compass, GPS Garmin, Belwar dosimeter, quartz watch, dowsing frame (Fig. 1, right) and reported: good acoustics; the word spoken at any point is well audible around the perimeter; radiation background is normal; difference between the readings of the clock in the center and outside is not detected; magnetic anomalies are not detected; openings in the circle can indicate "the House of the Sun on the longest and shortest days".

The Bakhchysarai Menhir. Among the most interesting astronomical places in the Crimea, pointing to the existence of ancient observatories of the Eneolithic era, are the Crimean menhirs (rocky ceilings, II Millennium BC) near the Bakhchysaray city (Fig. 2). The Bakhchysaray Menhir (44°46'06.5" N, 33°54'57.2" E) is not as large as the Stonehenge (Ruggles (2015)), but also belongs to the archeo-observational sites of the solar calendar type (Vavilova & Artemenko (2011, 2014), Gullberg (2019)). The oldest sky images can be reconstructed with methods by Savanevych et al. (2015, 2018).

In the beginning of 2000s this archeoobservatory was studied by the Crimean astronomer A. Lagutin. Menhir and the window could serve as two diopters of a huge visual instrument that fixed an east-west visor axis, on which the Sun could be observed

Figure 1. The Mavrin Maidan, Dnipro region, Ukraine (II Millennium BC).

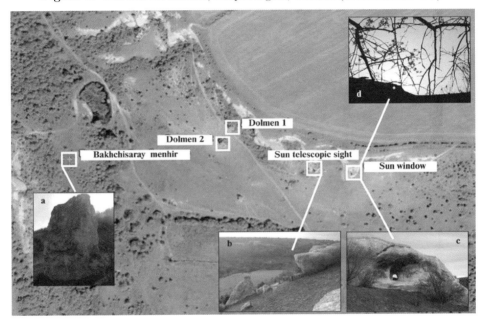

Figure 2. The Bakhchysaray Menhir, Crimea, Ukraine (II Millennium BC) at the Google Earth map. Photos by A. Terebizh

only on certain days near spring and autumn Equinox. Namely, if anybody stays near the menhir and looks east, one can see the hole (window) in the rock on the opposite side of the beam (at a distance of about 300 m). It took several years to confirm or disprove this hypothesis, as weather conditions (fog and clouds) obstructed observations. But the success was exciting: the descending Sun through this window in the rock rays really fell into the eye observer near menhir (Vavilova (2008)).

The megalithic stone complex at the southern slope of the Lysyna Kosmatska mountain (Ukrainian part of the Carpathian mountains) covers area of about 400 sq. m. The sanctuary is an elongated (10×40 m) east-west slope of gigantic stone slabs. The top plates of enormous size form the 6-m long tunnel in the "north-east – south-west" direction. The astronomical character of the sanctuary is confirmed by the observations of the ethnological-archaeological expedition "Carpathians – Dniester" in 2006–2011 (Kohutyak (2011)). The scheme of observations coincides with the Stonehenge plan, where, for example, the stone tunnel is analogue of the "Stonehenge alley", a summer solstice sunset event is related to the sacrificial calendar pit at the Lysyna Kosmatska (see, video, www.youtube.com/watch?v=CTJp96HpcVs, in Ukrainian).

The more information is available at the UkrVO web-site (Vavilova et al. (2017)).

References

Gullberg, S.R. 2019, *Astron. Nachr.*, 340, 23
Kohutyak, M. 2011, Antiquities of the Hutsul region, (Lviv, Manuscript), vol. 1, 447 p.
Ruggles, C.L.N. 2015, *Handbook of Archaeoastronomy and Ethnoastronomy*, 1223
Savanevych, V.E., Briukhovetskyi, O.B., Sokovikova, N.S. et al. 2015, *MNRAS*, 451, 3287
Savanevych, V.E., Khlamov, S.V., Vavilova, I.B. et al. 2018, *AA*, 609, A54
Vavilova, I.B. 2008, in: *Innovation in Astronomy Education*, Cambridge Univ. Press, p. 321
Vavilova, I.B., Artemenko, T.G. 2011, *IAU Proceedings*, 260, E7
Vavilova, I.B., Artemenko, T.G. 2014, *J. Astron. History and Heritage*, 17, 29
Vavilova, I.B., Yatskiv, Y.S., Pakuliak, L.K. et al. 2017, *IAU Proceedings*, 325, 361

Education and Heritage in the era of Big Data in Astronomy
Proceedings IAU Symposium No. 367, 2020
R. M. Ros, B. Garcia, S. R. Gullberg, J. Moldon & P. Rojo, eds.
doi:10.1017/S1743921321000314

The uniqueness of astronomical observatory publications

Ole Ellegaard[1] and Bertil F. Dorch[1,2]

[1]University Library of Southern Denmark,
SDU, Campusvej 55, 5230 Odense, Denmark
email: `oleell@bib.sdu.dk`

[2]Department of Physics, Chemistry and Pharmacy,
SDU, Campusvej 55, 5230 Odense, Denmark

Abstract. Astronomical observatory publications include the work of local astronomers from observatories around the world and are traditionally exchanged between observatories through their libraries. However, large collections of these publications appear to be rare and are often incomplete. In order to assess the unique properties of the collections, we compare observatories present in our own collection from the university at Copenhagen, Denmark with two collections from the USA: one at the Woodman Library at Wisconsin-Madison and another at the Dudley Observatory in Loudonville, New York.

Keywords. Observatory publications, Historical collections

---

1. Introduction

Astronomical observatory publications (AOPs) are considered to be of a high scientific standard and is guaranteed through the reputation of the individual observatory and in their modern form dates back to the middle of the 18th century Holl & Vargha 2003. Initially, AOPs present the results of the institutions' own observations obtained with local equipment and issued with or without peer review. AOPs have been a cheap and very popular form of knowledge exchange, and for many low-budget observatories an indispensable source of information. Today, almost no observatories exchange physical material. Instead, information is published in international journal articles or exchanged either via institutional repositories or websites, but problems remain in terms of storage, registration and retrieval of the material. Older AOPs can still be relevant and valuable. An example is photometry of variable stars. Many AOPs are listed in NASA's ADS database, and many documents have been scanned (cf. adsabs.harvard.edu/historical.html) but the coverage is far from complete. Large physical collections of AOPs are rare and e.g. special examples are found in the United States at the Woodman Library at Wisconsin-Madison (W-M) (cf. www.library.wisc.edu/astronomy) and the Dudley Observatory (Dud) (cf. dudleyobservatory.org/ collections-overview). The library at the University of Southern Denmark (SDU) has recently acquired a large collection of observatory publications from the now discontinued, historical library collections at the Niels Bohr Institute, University of Copenhagen. The collection's material dates from a period of several centuries and consists of tables of observations, annual reports, bulletins, circulars, newsletters, reprints etc. It has been collected for a period of more than a hundred years at the observatory in Copenhagen. The oldest material in the collection dates back to the 1700s. To understand the importance

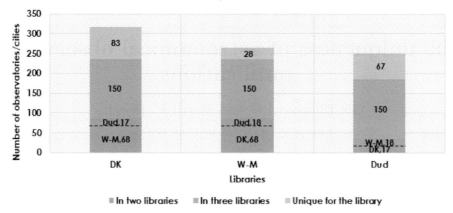

Figure 1. Combination of cities/observatories in the various collections.

and coverage, we examine the difference between the material found in the various collections. In practice, we compare the number of cities/observatories present.

2. Results

Publications of all three library collections are registered under the city where the observatory is located. For some of the largest cities are more than one observatory registered.

318 observatories are found in the danish collection, 264 in the W-M collection and 252 in the Dudley collection (Fig. 1). There is a overlap between the observatories present in the three collections. 83 or (26%) are unique and can only be found in the Danish collection, 28 (11%) in the W-M collection and 67 (27%) in the Dudley collection. The Dudley collection is the smallest but with the relatively largest number of unique observatories.

3. Discussion

The traditional AOP has now been replaced by the more widespread publication in traditional journals. A few libraries around the world still have quite large collections of older AOPs in paper format. Some of this material is included in the ADS, and is based on the collection at the astronomical institution at Wolbach Library, Harvard and supplemented by other collections, but other publications can only be found by searching the more or less complete registers of the relevant libraries. For example, the register in the Danish collection is present as a card catalog, and the Dudley collection is registered as incomplete items that are transferred directly from the cards. Not all cities/observatories are represented in the individual collections, as shown by the current study. The Danish collection represents the largest number of unique cities/observatories, followed by the Dudley Collection. We encourage astronomers looking for different types of older, rare materials that are hard to find, to be aware of and to search in ADS or in at least one major library that has a large collection of observatory publications.

Reference

Holl. A., & Vargha, M. 2003, In: B. Corbin, E. Bryson, & M. Wolf (eds.), *Observatory Publications-Quo Vadis?*, Library and Information Services in Astronomy IV (LISA IV), Prague, Czech Republic, p.109–116

Education and Heritage in the era of Big Data in Astronomy
Proceedings IAU Symposium No. 367, 2020
R. M. Ros, B. Garcia, S. R. Gullberg, J. Moldon & P. Rojo, eds.
doi:10.1017/S1743921321000223

Written on stone: Engraved European medieval solar eclipses

María J. Martínez[1]🄳 and Francisco J. Marco[2]

[1]Dept. de matemática Aplicada, IUMPA, Universitat Politècnica de València,
Camino de Vera, s/n 46022 Valencia, Spain
email: mjmartin@mat.upv.es

[2]Dept. de Matemáticas, IMAC, Universidad Jaime I,
Avinguda de Vicent Sos Baynat, s/n, 12071 Castelló de la Plana, Castelló
email: marco@mat.uji.es

Abstract. In the context of the European Middle Ages as the period roughly covering from the 5th to the 15th centuries, the astronomical records are rarery found in scientific treatises. At least, not until the 15th century. A few surprinsing examples in which the observations were recorded in a particularly original way on stone are found. In this poster we will shortly review the only four cases in which this occurred in Europe.

Keywords. Solar Eclipses, Middle Age, History of Astronomy

1. Introduction

One of the greatest natural spectacles that can be seen throughout a human life are solar eclipses, especially those in which the star is totally or largely hidden. In times when science was not developed and the only explanation for these phenomena was the will of the gods, these events were often viewed as omens of good or bad fortune, reinforcements of recent decisions, or ominous signs of future happenings. For this reason, numerous records of solar eclipses are found throughout history, from the first and second millennium B.C., up to the present day.

Over the years there have been numerous authors who have studied and compiled records of ancient total and annular solar eclipses (see, for example, Stephenson F.R. (1997)). This has provided a large number of publications that have proven useful for research both from the historical and astronomical point of view. Most of the observations come from the Far East, especially China and Japan. In this case, the observer is usually a professional astronomer who provides relevant data other than the date, such as the duration of the eclipse, the precise place of the observation, the time and even the magnitude.

A smaller number of observations come from European sources, although in these cases, and at least until the middle of the fifteenth century, the records do not usually come from professionals, but are found in narrative sources mixed with historical facts to which they usually provide some kind of reinforcement. The causes of this shortage are very numerous, highlighting the cultural ones, with the generalized loss of knowledge of classical culture that only survived in very limited areas, and the historical vicissitudes that were experienced in these centuries, including the Arab occupation of southwestern Europe and the border tensions both in the East and the West, the fragmentation of the European territory first in two and later in multiple kingdoms and the constant wars between them. Thus, we find in Europe many accounts of solar eclipses but most of them

Figure 1. Engraved fragment found in the ruins of St Nicholas church in Soria (Spain).
CVRAT(us) EST SOL(is) ER(a) MCXXLX(xvii) (in the stript at the bottom).

Figure 2. Inscription B from the portal of the market square of Sos del Rey Católico. It was
found upside down due to the successive reconstructions of the portal.

are found in written chronicles, almanacs and diaries. It comes as a surprise the finding
of a few of these register engraved on stone. In fact, to our knowledge, only four stone
engravings have been found in Europe, three of them corresponding to the well-known
AD1239 eclipse and the other to the one in AD1354.

2. Total Eclipse of June 3, 1239

It is very unusual to find records of European eclipses engraved in rock from medieval
times. until recently, the only well-known case is the one from Marola recorded by
Stephenson for the A.D.1239 solar eclipse (Stephenson F.R. (2008)). But for this eclipse
two more records were found in Soria (Spain) recently. See Figure 1 and (Martínez *et al.*
2016). for a further description.

3. Hybrid Solar Eclipse of September 17, 1354

For the AD1354 eclipse a much more remarkable inscription was found in an ashlar
stone of the portal of the market square of Sos del Rey Católico (Zaragoza, Spain). In fact,
not one but two engravings were found in two different stones. The so called Inscription
A states "Anno domini MCCC XXXIX" and possibly relates to the date of construction
of the arch. Inscription B (see Figure 2) was hard to read. It is longer and more crudely
engraved and the space between the letters and the size of the engraving are irregular.
The emblem of Aragon is clearly visible below on the right of the image. The engraving
states: Anno domini MCCC: L: IIII XVII die septembris: hora prima: Obscura uit sol.
The translation will be: In the first hour (hora prima) of September 17 of the year of Our
Lord Jesus Christ, 1354, the sun was darkened. The maximum of the eclipse occurred

around 9 in the morning. The "hora prima" was a canonical hour that corresponded to the first hour after sunrise, which that day occurred around 8 in the morning.

References

Martínez, MJ.; Marco, FJ. & Ibáñez, L. 2016, *Journal for the History of Astronomy*, 47, 61

Stephenson F.R. 1997, *Historical Eclipses and Earth's Rotation* (Cambridge University Press, Cambridge)

Stephenson F.R. 2008, in: Selin H. (eds) *Encyclopaedia of the History of Science, Technology, and Medicine in Non-Western Cultures.* (Springer, Dordrecht)

Session 9

Education and Heritage in the era of Big Data in Astronomy
Proceedings IAU Symposium No. 367, 2020
R. M. Ros, B. Garcia, S. R. Gullberg, J. Moldon & P. Rojo, eds.
doi:10.1017/S1743921321001022

Astronomy outreach for hospitalized children at Kyoto University Hospital

Y. Tampo[1] , S. Mineshige[1], J. Arimoto[2], Nico-Nico Tomato[3],
T. Hayakawa[1], S. Kawamura[1], R. Ohuchi[1], K. Kihara[1],
and S. Okamoto[1]

[1]Department of Astronomy, Kyoto University,
Kitashirakawa-Oiwake-cho, Sakyo-ku, Kyoto, Japan
email: tampo@kusastro.kyoto-u.ac.jp

[2]Ohka-do, Kyoto, Japan

[3]Kyoto University Hospital, Shogoin-Kawaramachi, Sakyo-ku, Kyoto, Japan

Abstract. Department of Astronomy, Kyoto University has been conducting astronomy outreach programs in Kyoto University Hospital since 2006. In this proceeding, we report our activities in the hospital, survey results from graduate students, and discuss future directions.

1. Organizing team and objective

This program has been carried out by the graduated students of Department of Astronomy, Kyoto University, under the supervision of Prof. Mineshige and Mr. Arimoto. This is also a part of the activities of "Nico-Nico Tomato" †, volunteer group acting in the pediatrics department, Kyoto University Hospital, and collaborating with the "Ohka-do" ‡, astronomical volunteer team based in Kyoto.

The purpose of our activity and Nico-Nico Tomato is to provide a "fun and joy" moment to children even in the hospital, where is the place for medical treatments. Most of the participating children in this activity need long-term treatment and should obey strict restrictions; e.g., they are not allowed to go out. Therefore, they have really few chances to look up night skies. Our events offers opportunities for such children to feel and enjoy the starry sky and universe even in the hospital and to stimulate their curiosity. Also from the perspective of inclusive astronomy, since some of these children are hard to go to public museums or planetariums, our activity is a few of their chances to enjoy Astronomy.

2. Our activity

We hold our events roughly every three months at the hospital. Our activity is divided into two parts, the afternoon and evening session. In the afternoon session, we have storytelling and quizzes related to astronomy and space (left panel of Figure 1). We often select the seasonal astronomical events as the theme, so that they can feel the change of the seasons and enjoy the real sky more if they had chance to look up night sky. The evening session, which starts from around 6pm, is held at an open space in the hospital. In this space, there is a large window so that we can see moons and stars even from

† Facebook page of Nico-Nico Tomato in Japanese: https://www.facebook.com/nicotomakyoto
‡ website of Ohkado in Japanese: http://www.oukado.org/

Figure 1. Scenes of our activities in Kyoto University Hospital. Left: storytelling in the afternoon session. The theme of this talk was about moon at this time, as this was in autumn and a time for moon watching. Right: gazing the moon and stars from inside the hospital.

inside of the building. With small telescopes, we mainly watch the moon and sometimes the planets (right panel of Figure 1). Another popular content for the evening session is universe travel using MITAKA, which is a universe-visualizing software developed by NAOJ §.

3. Benefit and take away for participated graduate students

Until now, over 30 graduate students have participated in this activity. From the standpoint of graduate students, this is more than a normal outreach event of astronomy; this program is a good chance to learn and understand inclusive astronomy and socially vulnerable people through interacting with hospitalized children. Each student needs to attend a mini-lecture by our supervisor Mineshige beforehand to learn what to beware of for our activity with hospitalized children, so as to avoid any inappropriate actions, even though they are unintentional, from the viewpoints both of medical care and protection of privacy.

We conducted a survey to the participated graduate students. All they answered that they are satisfied with the activity, mainly because they are happy to join the outreach program and to see the smile faces of children. One suggestion for the future is that we should refresh some part of our contents, which are less attractive to children.

4. Future direction

Today, all voluntary activities in Kyoto University Hospital including ours are limited since March 2020 due to COVID-19. The play room, where the children can play with toys and where Nico-Nico Tomato performs its activities, is closed as well. To develop something kids can enjoy in their own rooms and to keep our member active, we create a mobile craft kit, which is shaped as our solar system. As nobody knows when we can re-enter the hospital for our activity, we will keep small activities to pass on our knowledge and experiences to our younger students.

Acknowledgement

We appreciate all support from Nico-Nico Tomato, Ohkado, Kyoto University Hospital, Department of Astronomy, Kyoto University and participated graduated students. We also thank to IAU S367 SOC, LOC and all the participants to give us this opportunity to present our activity.

§ MITAKA can be downloaded from https://4d2u.nao.ac.jp/html/program/mitaka/index_E.html

Education and Heritage in the era of Big Data in Astronomy
Proceedings IAU Symposium No. 367, 2020
R. M. Ros, B. Garcia, S. R. Gullberg, J. Moldon & P. Rojo, eds.
doi:10.1017/S1743921321000995

Inclusive education and research through African Network of Women in Astronomy and STEM for GIRLS in Ethiopia initiatives

Mirjana Pović[1,2], Vanessa McBride[3], Priscilla Muheki[4], Carolina Ödman-Govender[5], Somaya Saad[6], Nana Ama Brown Klutse[7], Aster Tsegaye[8], Tigist Getachew[9], Melody Kelemu[10], Hanna Kibret[1], Jerusalem Tamirat[1], Deborah Telahun-Teka[9], Beza Tesfaye[11], and Feven Tigistu-Sahle[9]

[1] Ethiopian Space Science and Technology Institute, Ethiopia,
email: mpovic@iaa.es.

[2] Instituto de Astrofísica de Andalucía (CSIC), Spain.

[3] Office of Astronomy for Development-IAU, South Africa.

[4] Mbarara University of Science and Technology, Uganda.

[5] Inter-University Institute for Data Intensive Astronomy, University of the Western Cape, South Africa.

[6] National Research Institute of Astronomy and Geophysics, Egypt.

[7] University of Ghana, Ghana.

[8] Addis Ababa University, Ethiopia.

[9] Ethiopian Biotechnology Institute, Ethiopia.

[10] International Institute for Primary Health Care, Ethiopia.

[11] Ethiopian Space Science Society, Ethiopia.

Abstract. The African Network of Women in Astronomy and STEM for GIRLS in Ethiopia initiatives have been established with aim to strengthen the participation of girls and women in astronomy and science in Africa and Ethiopia. We will not be able to achieve the UN Sustainable Development Goals without full participation of women and girls in all aspects of our society and without giving in future the same opportunity to all children to access education independently on their socio-economical status. In this paper both initiatives are briefly introduced.

Keywords. Astronomy; women and girls in science

1. African network of women in Astronomy- AfNWA

Considering the latest report of the UNESCO and UN-WOMEN, the number of female researchers in the world (both part- and full-time) is on average $< 30\%$ (UNESCO 2019). For most of countries this number becomes even lower when STEM (Science, Technology, Engineering and Mathematics) fields are considered. Therefore, globally we are facing significant gender gap in science. In Africa, most of countries have a number of female scientists below 25%. Many factors may be responsible for the low number of female scientists (e.g., poverty and lack of access to education, social constraints, cultural biases and beliefs, lack of female mentors and role models, etc.), but the final result is that these difficulties mean we are losing huge potential that could benefit our society. We will never be able to reach the UN Sustainable Development Goals (SDGs) without giving our best in empowering girls and women who make $\sim 50\%$ of world population. Astronomy and space sciences are currently experiencing significant growth in Africa (Pović et al. (2018)). For the benefit of all society we would like to guarantee future participation of girls and women at all levels in astronomy and science developments in Africa.

The African Network of Women in Astronomy (AfNWA)† is an initiative established in September 2020 under the African Astronomical Society (AfAS)‡ that aims to connect women working in astronomy and related fields in Africa. Our main objectives are improving the status of women in science in Africa, and using astronomy to inspire more girls to do STEM. These objectives are planned to be achieved through different activities such as: the creation of AfNWA network, organisation of needed courses and trainings for improving research and leadership skills of women in astronomy in Africa, giving visibility to women in astronomy and related fields (through yearly published reports, newsletters, website, public talks, public communications, given awards, etc.), organisation of outreach activities given by women astronomers, understanding of the main factors responsible for the lack of women in astronomy and science in different African countries, and development of efficient methods of retaining women in astronomy.

Since AfNWA objectives are there for the benefit of the whole society, we strongly encourage participation and support of the whole population in our activities, both women and men.

2. STEM for GIRLS in Ethiopia

In the last Ethiopian Growth and Transformation Plan (GTP) II (2016 - 2020) it has been raised that women in Ethiopia face multiple challenges, including illiteracy and inequality in education, unequal division of labour, unequal power relationships, and limited participation in leadership and decision-making (GTP II, 2016). In addition, participation of young girls in STEM and women in science is still far from reaching gender balance due to different social and economical challenges. STEM for GIRLS in Ethiopia initiative was established in 2019 in collaboration with the Society of Ethiopian Women in Science and Technology (SEWiST) with aim to help to achieve the goals of the GTP II and to improve in future the participation of women and girls in STEM. It is based on creating a strong connection between SEWiST members and the rest of society, in particular grade 9 and 10 girls and their teachers, and use SEWiST's strength and long experience to promote women in science and technology as role models. Beside inspiring and encouraging more girls to do STEM, we also aim in understanding what are the main factors responsible for the lack of girls and women in STEM in Ethiopia and how can we improve it in future. To reach proposed objectives, we organised different activities with girls along 2019 and beginning of 2020. Almost all activities were carried out in Addis Ababa, in average once per month. We organised interactions between girls and women scientists, so that women can share their professional experiences together with their life story on how they became scientists and therefore serve as role models, and that girls can share their life stories as well, their future plans, challenges they are facing, etc. Up to now the interaction has been made with almost 1000 girls, with many activities being focused on astronomy. We also developed a questionnaire that girls can feel voluntarily, for understanding better the reality and main factors behind the lack of girls in STEM, seeing clearly the tendency of girls to continue with care-type professions (up to 70%). In addition, in 2019 we organised the very first workshop for teachers to discuss why

† https://www.africanastronomicalsociety.org/afnwa/
‡ https://www.africanastronomicalsociety.org/

it is important to bring more girls into STEM. Our activities were interrupted during 2020 due to the COVID-19 pandemic and closure of schools in Ethiopia, but we are now re-starting the activities.

Acknowledgements

AfNWA and STEM for GIRLS in Ethiopia have been supported by the Nature Research and Estée Lauder through the MP 2018 Inspiring Science Award. Support of the South African DSI to AfAS-AfNWA is also highly acknowledged.

References

Growth and Transformation Plan II (GTP II), 2016, Federal Democratic Republic of Ethiopia, National Planing Commission

Pović, M., et al., 2018, Nature Astronomy, 2, 507

UNESCO 'Women in Science' report, 2019, Fact Sheet No. 55 (FS/2019/SCI/55)

Education and Heritage in the era of Big Data in Astronomy
Proceedings IAU Symposium No. 367, 2020
R. M. Ros, B. Garcia, S. R. Gullberg, J. Moldon & P. Rojo, eds.
doi:10.1017/S1743921321001149

Inclusion of deaf and hearing-impaired people in a planetarium show

Guillermo M. García[iD], Juan I. Gerini and Andrea F. Costa

Planetarium – Complejo Astronómico Municipal de Rosario, Santa Fe, Argentina
email: `planetario@rosario.gob.ar`

Abstract. In the Planetarium of the city of Rosario the deaf and the hearing-impaired people could only perceive the visual elements of the show, such as the starry sky and the video images, which are meaningless without the narration. The technical team detected the problem and implemented a solution, adapting a particular show with a team of interpreters and a deaf person, with a low cost of development and arriving to great results.

Keywords. inclusion, planetarium show, deaf people, hearing-impaired people

1. Introduction

The Planetarium of the city of Rosario has an analogic Carl Zeiss Model IV projector and 3 regular size (non-full dome) video projectors functioning as a digital complement. For an adequate analogic projection, the dome requires absolute darkness.

The deaf and the hearing-impaired could only understand, with the assistance of a sign language interpreter, the welcome and introductory speech which are delivered with the lights on. Once the show started, the darkness eliminated the possibility of interpretation, and the deaf inevitably missed the explanations, concepts and metaphors from the voice-over. They could only perceive the visual elements of the show, such as the starry sky and the video images, which are meaningless without the narration.

2. Approach to the problem

Detect the problem. The customer service staff should always attempt to identify the problems the audience might have. Planetarium staff set the objective of making the shows accessible to that group.

Propose experimental solution. Consisting of two steps: 1- The interpretation is inserted in the video displayed by the three projectors. Subtitles can be added if necessary. 2- Auxiliary screen: its goal is to show the sign language interpretation and subtitles whenever there's a voice-over accompanying the sky projection with the analogic planetarium in total darkness.

Apply to one show. "A Tour Through The Solar System" was selected, because it is the show educational institutions demand the most.

3. Development

Team. A multidisciplinary team composed of interpreters from the Argentinian Sign Language association (Lengua de Señas Argentina – LSA), a deaf person, and technicians.

Script interpretation. The interpretation of the script was the first – and the most extensive – step. Interpreters observed that many words of the astronomical glossary do not have a translation. At present time, there are groups of people working in this task in our country. The script was interpreted almost entirely and the film duration is only 30 seconds longer.

Recording. The second step was to mount a film set to record the interpreter on a chroma background to allow for a better insertion on the screen. D'Angelo *et al.* (2011).

Post production. Subtitles and interpreter were inserted on videos.

New screen. The video shown on the auxiliary screen is displayed by an old projector contained in a black box to avoid light pollution on the dome.

 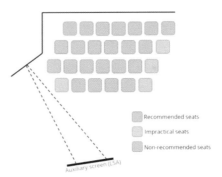

4. Results and conclusions

• **Concrete implementation**: low cost, reuse of old resources and quick configuration whenever switching to the adapted version is required.

• **Positive social impact**: the deaf and the hearing-impaired community showed excitement and gratitude. Likewise, the involvement of a deaf person in the production process was appreciated.

• **The show is not exclusive** to the deaf and the hearing-impaired. Both the hearing and the deaf enjoy the show equally at the same time, which makes the show truly inclusive.

• **Possibility of conceiving new productions** including an interpreted version from the very beginning.

Reference

D'Angelo C. & Massone M. 2011, *La accesibilidad a los medios audiovisuales: la narración en Lengua de Señas Argentina y el subtitulado para personas sordas.*, INCAA.

Session 10

Education and Heritage in the era of Big Data in Astronomy
Proceedings IAU Symposium No. 367, 2020
R. M. Ros, B. Garcia, S. R. Gullberg, J. Moldon & P. Rojo, eds.
doi:10.1017/S1743921321000259

Astronomical collections in Observatories: promoting preventive conservation

Maria Rosalia Carotenuto, Ileana Chinnici⬚ and Donatella Randazzo

INAF – Astronomical Observatory "G. S. Vaiana",
Piazza del Parlamento, 1, Palermo, Italy
email: carotenutomr@gmail.com, ileana.chinnici@inaf.it,
donatella.randazzo@inaf.it

Abstract. The practice of preventive conservation of cultural heritage consists of "all measures and actions aimed at avoiding and minimizing future deterioration or loss" of it (ICOM-CC, 2008). Unlike conservation treatments, preventive conservation deals with entire collections and their surrounding environment. It is known that exposing historical objects to the environment has a significant impact on their degradation process. Studying and managing risk factors is an indispensable practice within the management policies of any cultural institution. The National Institute for Astrophysics (INAF) holds some valuable historical collections, heritage derived from the Italian contribution to astronomy over the centuries. The management and protection of these collections faces many challenges. A preventive conservation plan, aimed at assessing and managing risks that threaten the collections, may offer many long-term benefits, allowing us to use available resources in the best possible way. In the past few years INAF-Astronomical Observatory "G.S.Vaiana" of Palermo has been working on the development of preventive conservation projects for its archival, bibliographic and scientific heritage. The present contribution reports on these ongoing experiences and intends to stimulate a discussion within the scientific community in order to individuate the problems we are called to respond to in Astronomical Observatories.

Keywords. Preventive conservation, museum, archive, scientific instruments, astronomical observatory, historical library

---

1. What does "preventive conservation" mean?

At the 15th Triennial Conference held in New Delhi in 2008, ICOM-CC adopted a resolution on a terminology for conservation to facilitate communication among the international communities. The following terms have been adopted: "preventive conservation", "remedial conservation", and "restoration" which together constitute "conservation" of the tangible cultural heritage. These terms are distinguished according to the aims of the measures and actions they encompass. Specifically, preventive conservation was defined as "all measures and actions aimed at avoiding and minimizing future deterioration or loss. They are carried out within the context or on the surroundings of an item, but more often a group of items, whatever their age and condition. These measures and actions are indirect – they do not interfere with the materials and structures of the items".

2. The scientific heritage held by INAF

The National Institute for Astrophysics holds remarkable historical collections, preserved in the italian Observatories. The collections include instruments, books, and historical archives, mostly still *in situ*. The management and protection of these collections face many challenges, including limited resources, presence of heterogeneous materials (with different conservation requirements), location inside historical buildings, lack of room for appropriate storage of the collections, use of historical showcases. A preventive conservation plan, aimed at assessing and managing risks that threaten the collections, would help in establishing priorities of intervention and adopting timely and sustainable strategies. It may offer many long-term benefits, allowing available resources to be used in the best possible way. Many Observatories throughout the world face the same problems. We would like to share our experiences.

2.1. *A case study: preventive conservation project for the collections of Palermo Observatory, Museo della Specola*

The project starts in 2018 with the aim of identifying and managing the "risks" to which objects are exposed in the museum in order to establish intervention priorities and identify where improvements can be made. The "risk" is defined as 'the chance of something happening that will have a negative impact on our objectives'. The impact of risks is expressed as the expected loss of value to the heritage asset It does not involve just any type of material damage to the heritage asset, but rather the loss of information about it, or the inability to access heritage items.

At the moment, we are working on: 1) identifying the relevant aspects of the context in which the collections are situated; 2) identifying the risks that threaten our collection, building and site, through: a) elaboration and compilation of condition reports of some selected objects on display; b) thermo-hygrometric monitoring of the museum rooms; c) sampling campaign to test air quality in the museum and inside the showcases.

2.2. *Preventive conservation interventions in the historical library and archives of Palermo Observatory*

Since 2016, in collaboration with Palermo University, preventive conservation interventions were made on the historical archives papers, on a small collections of 19th-century portrait photographs and on many books of the antique collections (printed before 1830). A preliminary survey was made to establish general conditions and evaluate the risks for the collections through use. A conservation assessment allowed us to ascertain the absence of environmental factors that could adversely affect collections. No complex care strategies were involved. The storage folders were replaced with boxes or folders made from acid-free buffered board. Many damaged books, that could not undergo restoration due to lack of funds, have been provided with appropiate boxes to prevent further deterioration. Whenever possible, safe storage of photographs or documents is guaranteed by appropiate enclosures.

3. Why promote preventive conservation?

Preventive conservation is the most efficient form of conservation as it allows you to: focus on the entire collection of a museum; reduce, over time, the need to act directly on assets with costly and invasive interventions, acting on the causes of degradation processes and not on the effects; guarantee a suitable conservation environment for objects that have already undergone restoration treatments in the past. The resulting measures are not always expensive or particularly complex to implement. On the contrary, knowing the real needs of the museum and collections and having established the risks and priorities, targeted strategies can be planned guaranteeing a more efficient management of the available resources.

References

Antomarchi, C., Pedersoli Jr., J.L., & Michalski, S. 2016, *Guide to Risk Management* (ICCROM)
Chinnici, I. 2009, in *Proceedings of ICOMOS International Symposium, 2008*, p. 227–231

Education and Heritage in the era of Big Data in Astronomy
Proceedings IAU Symposium No. 367, 2020
R. M. Ros, B. Garcia, S. R. Gullberg, J. Moldon & P. Rojo, eds.
doi:10.1017/S1743921321001162

Solar eclipses as a chance for professional-amateur scientific collaboration

Yoichiro Hanaoka[iD]

National Astronomical Observatory of Japan,
2-21-1 Osawa, Mitaka, Tokyo 181-8588, Japan
email: yoichiro.hanaoka@nao.ac.jp

Abstract. Total solar eclipses are popular targets for amateur astronomers. At the same time, the eclipses are still scientifically important to observe the solar corona. Therefore, the eclipses are a good chance for amateurs to participate in scientific observations. In fact, some of amateur astronomers in Japan have been carried out scientific observations at the total solar eclipses collaborating with professional solar scientists over more than ten years. Some scientific results have been produced from the collaboration. We present here our collaborative activities as a practical example of citizen science.

Keywords. solar eclipses, solar corona, citizen science

1. Advantages of the Pro-Ama collaboration in the eclipse observations

Energetic phenomena on the Sun, which occasionally affect the electromagnetic environment of the Earth, occur in the solar corona. Therefore, the solar corona is an important scientific target, and total solar eclipses enable us to observe the corona from just above the limb to several solar radii. That is one of the reasons why the solar eclipse observations are still carried out in the space age. However, only a snapshot of the corona can be observed from a site. Therefore, to track the evolution of the corona, multi-site observations of the solar eclipses are necessary, and they can be realized by amateur observers, who widely spread along the total eclipse path. Furthermore, the multi-site observations help to mitigate the risk of the weather conditions. Thus, the participation of amateurs in the scientific observations of solar eclipses is very valuable (for an extreme example, see Peticolas et al. 2019).

2. Scientific results produced by the Pro-Ama collaboration and more advanced observations

The professional and amateur collaboration for the observation of the solar eclipses in Japan, which has been continued for more than ten years, have really produced some scientific results (Hanaoka et al. 2012; Hanaoka et al. 2014; Hanaoka et al. 2018). Most of the cutting-edge astronomical observations require expensive, sophisticated instruments, but, on the contrary, imaging observations of the solar eclipses can be carried out with small telescopes. There are many amateur observers who take photos (white-light images) of the solar corona. Just an addition of the acquisition of calibration data (dark and flat-field) makes their photos scientifically valuable data; the above-mentioned results have been obtained with such observations. In the case of solar eclipse observations, amateur observers sacrifice nothing to participating citizen science; they can enjoy the eclipse as usual.

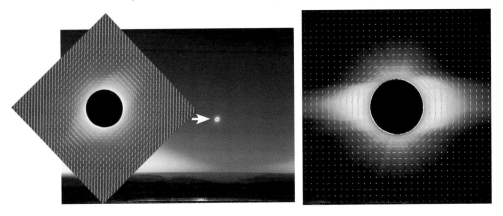

Figure 1. Images taken during the eclipse on July 2, 2019. Left: Raw polarization signals of the corona shown with green ticks and a wide-field view of the total eclipse. The zenith is to the top. The vertical components correspond to the polarization of the sky, which is remarkable at the low elevation (13 deg in this case). Left: Polarization data from which the sky components were removed.

Furthermore, some of amateur astronomers are eager to carry out more sophisticated observations. The polarimetry of the corona is a good target for them. The information of polarization is the key to distinguish the K-corona (hot plasma belonged to the Sun) and the F-corona (interplanetary dust). Some amateur astronomers, who are collaborating with us, started to the polarimetry observation of the eclipses from 2016. They successfully obtained the polarization data in the 2017 and 2019 eclipses as well. Figure 1 shows an example of the results of the polarimetry of the solar corona taken during the 2019 solar eclipse. It is not very easy to set up the polarimetry instrument or the observation procedure, but its accomplishment is more rewarding.

3. Conclusion

In this way, the solar eclipses are a good chance for the citizen science based on the professional-amateur collaboration involving a broad range of amateur astronomers. As Haklay (2013) discusses, data collection by amateurs accompanied by devising its method by themselves can be considered as a high-level participation of amateurs in citizen science.

Although we could not carry out coordinated observations at the eclipse in December 2020, we will continue to the eclipse observations including the polarimetry. Furthermore, international professional-amateur collaboration will be a course of action in future.

References

Haklay, M. 2013, in: D. Sui, S. Elwood, M. Goodchild (eds.), *Crowdsourcing Geographic Knowledge* (Dordrecht: Springer), p. 105

Hanaoka, Y., Hasuo, R., Hirose, T., Ikeda, A.C., Ishibashi, T., Manago, N., Masuda, Y., Morita, S., Nakazawa, J., Osamu Ohgoe, O., Sakai, Y., Sasaki, K., Takahashi, K., & Toi, T. 2018, *ApJ*, 860, 142

Hanaoka, Y., Kikuta, Y., Nakazawa, J., Ohnishi, K., & Shiota, K. 2012, *Solar Phys.*, 279, 75

Hanaoka, Y., Nakazawa, J., Ohgoe, O., Sakai, Y.,& Shiota, K. 2014, *Solar Phys.*, 289, 2587

Peticolas, L., Hudson, H., Johnson, C., Zevin, D., White, V., Oliveros, J. C. M., Ruderman, I., Koh, J., Konerding, D., Bender, M., Cable,, C., Kruse, B., Yan, D., Krista, L., Collier, B., Fraknoi, A., Pasachoff, J. M., Filippenko, A. V., Mendez, B., McIntosh, S. W., Filippenko, N. L. 2019, *ASP-CS* 516, 337

Education and Heritage in the era of Big Data in Astronomy
Proceedings IAU Symposium No. 367, 2020
R. M. Ros, B. Garcia, S. R. Gullberg, J. Moldon & P. Rojo, eds.
doi:10.1017/S1743921321000454

The sky is the limit
fifteen years of literacy in astronomy

Daniel Cabezas, Omar Curcio, Joaquín De La Rosa, Jorge Escudero[ID], José Fabbro, Daniel Flores, Rafael Girola, Patricia Iglesias, Carlos Lucarelli, Verónica Pernicone, José Plem, Norma Racchiusa and Néstor Vinet

EnDiAs, Teaching and Public Education of Astronomy,
San Miguel and Moreno, Buenos Aires province, Argentina
email: info@endias.com.ar

Abstract. EnDiAs is an argentine group of astronomy fans founded fifteen years ago, whose goal is the public dissemination of Astronomy through Non-Formal Education. We consider that the sky is a World Heritage and everyone should have access to its observation and study. This paper shows the activities carried out by EnDiAs to fulfill its goal.

Keywords. Non-Formal Education, Astronomy Fans

1. Our history

EnDiAs (Teaching and Public Communication of Astronomy) was founded in 2005 by Lic. Hipólito Falcoz, at the National Observatory of Cosmic Physics of San Miguel, in order to communicate astronomy within the field of non-formal education. Non-Formal Education provides the possibility of continuing to build knowledge, throughout life, by personal choice and self motivation.

This framework implies the design of activities, processes and means, based on explicit training goals, although not linked to the granting of degrees typical of the formal educational system.

In order to fulfill its goals, EnDiAs has developed a lot of activities in different contexts, among which we can mention the frequent observational outings to different locations in the province of Buenos Aires, the delivery of open courses for the entire community at San Miguel and Moreno (two towns in Buenos Aires province where our educational centers are located), the teacher training courses in General Roca, Río Negro (Escudero *et al.* 2009). and San Salvador de Jujuy, Jujuy (Escudero *et al.* 2014), the telescope construction workshops, the stands and the talks at several Book Fairs, the participation in numerous congresses, seminars, workshops, etc.

The main goal of EnDiAs is astronomy literacy, hence the emphasis on teaching courses. Currently, the courses offered to the community are: Basic Astronomy (for those who are starting on the path of discovering the universe and have the vocation of Astronomy fans), Modern Astronomy (in-depth development of topics related to the evolution of chemical elements in stars and the structure of the universe), Children's Astronomy (for Primary Education children, between 8 and 11 years old, who have a literacy base) and Cultural Astronomy (for those who are interested in perceptions, conceptions and practices related to the firmament that were and are manifested in the symbolism and in the material culture of different societies. Topics covered include positional astronomy, archaeaostronomy, ethnoastronomy, astronomy in the visual arts, philately, vexillology, numismatics, and science fiction literature and cinema).

Although the covid-19 pandemic prevented the realization of the observational outings and face-to-face courses, EnDiAs continued its activities with the community through radio interviews and online talks, and the Basic Astronomy course was given through YouTube. Only the Telescope Construction Workshop started in 2019 was suspended, because it must be carried out in person. We hope to restart it soon.

We considered it of great importance to continue our activities for the communication of Astronomy during 2020 and not abandon our bond with the community in such difficult times, because the observation and knowledge of the stars made it possible for us to free ourselves from worries for a few moments and to find ourselves a space of freedom watching the sky, despite the forced confinement. In addition, it kept us active and prepared to continue with normal activities when the pandemic ends, which we hope will be as soon as possible.

Acknowledgment

We do not want to miss the opportunity to thank these people and institutions, who make our work possible:

Domingo F. Sarmiento Municipal Popular Library of San Miguel, for granting a centennial space dedicated to education, to develop our courses.

National University of General Sarmiento, for prioritizing our work with its permanent support.

Municipal Observatory of Mercedes. To its Director, Miguel de Laurenti, always ready to receive us. Copernicus Institute. To its Director, Dr. Jaime R. García, for giving us a place in his Star Party.

Astronomical Park Cielos del Sur, Chivilcoy. To its Director, Armando Zandanel, for always involving us in his Astronomy Conference.

To Dr. Néstor Camino, because participating in the Analema project was a great enriching challenge (Vinet *et al.* 2016).

To Dr. Leonardo Pellizza, President of the Argentine Astronomy Association, for supporting our defense of the Historical Heritage, the Observatory of Cosmic Physics of San Miguel.

We proudly celebrate 15 years working on Scientific Literacy in Astronomy through Non-Formal Education, in the communities of San Miguel and Moreno, to move towards a better world.

References

Escudero J., Fabbro J. Flores D., Girola R., Racchiusa N., Santos M. 2009, in: Dissemination Conference and Teacher Training in Astronomy. Physical Phenomena and Energy, Horacio Pedro Fernández), *Training in didactics of astronomy*, CEM 1 J. M. García, Gral. Roca, Río Negro

Escudero J., Fabbro J. Girola R., Lucarelli C. Racchiusa N., Santos M. 2014, in: Astrojujuy 2014, Provincial Ministerial Resolution No. 0821, 19/08/2014), *Training in didactics of astronomy*, Provincial Baccalaureate No. 18 Hugo Cazón at Purmamarca. School No. 38 Juanita Stevens at San Salvador de Jujuy

Vinet N., Pernicone V., Racchiusa N., Escudero J., Rinaudo S., Girolar R. 2016, in: Bulletin of the 59th Annual Meeting of the Argentine Astronomy Association), *EnDiAs contribution to the Observational Determination of Analemma Project. South American Common Observation Project*

Education and Heritage in the era of Big Data in Astronomy
Proceedings IAU Symposium No. 367, 2020
R. M. Ros, B. Garcia, S. R. Gullberg, J. Moldon & P. Rojo, eds.
doi:10.1017/S1743921321000661

21 Years Activity of the ITAU Activities

Hasan Baghbani[1], Mahdi Rokni[2], Akihiko Tomita[3] and Maryam Hadizadeh[4]

[1]Graduate in Geology, President of Iranian Teachers Astronomy Union, Bushehr, Iran
email: hasan.baghbani1971@gmail.com

[2]Undergraduate Student in Software Engineering, Astronomy Instructor and International Coordinator of Iranian Teachers Astronomy Union, Bushehr, Iran
email: mahdirokni75@yahoo.com

[3]Graduate School of Teacher Education, Course Specializing in Professional Development in Education, Wakayama University, Wakayama City 640-8510, Japan
email: atomita@wakayama-u.ac.jp

[4]Undergraduate student of English Teaching, Astronomy Teacher, Farhangian University, Fasa, Iran
email: Maryam.hadizadeh79@gmail.com

Abstract. ITAU, the Iranian teacher's astronomy union, has attempted for many years to introduce astronomy to the teachers and students not only in Iran but also around the world by providing various materials and projects. The primary goal of ITAU is also to protect the environment with a help of students; therefore, first, try to change the attitude of students. SINA, Student's International Network for Astronomy, will follow the goals.

Keywords. Iranian teacher's astronomy union, student's international network for Astronomy, international collaboration, environment education

---

1. Introduction

SINA is the Student's International Network for Astronomy in which students interested in astronomy from all around the world connect to each other and practice in groups. The primary goal of SINA is to create an easy access platform for students to learn astronomy practically, and also create a situation for various students to know each other, acknowledge world peace and importance of the environment, and strengthen their sense of altruism.

In recent decades, the education system and schools are not synced to the technological advances. Modern schools have started their way to educate and prepare students for modern life and industrial developments, however, they could not be updated with the rate of development in technology, leading to inefficiency. One of the most important needs in society is understanding environmental situations and world-wide peace. Attitudes towards the development and the current situation need to be changed, therefore, a method should be taken to bring students' attitude from local to global. They must put aside their desire for unbridled developments and become acquainted with sustainable developments.

2. History and Development of ITAU

Astronomy has been taught to students in Bushehr, the south of Iran, aiming to create a non-political and geographical perspective for more than 20 years using the sky without borders (ITAU web site: http://itaubu.com/). ITAU, the Iranian teacher's astronomy union, has provided various projects besides educational courses to governmental education using world-wide knowledge and teachers' experiences. One of the most successful projects is the Sky Explorer festival which is held for more than 9 years by ITAU with aim of challenging students in environmental challenges, managing efficient usage of water and food, creating a world-wide perspective for students using the sky without borders, and increasing the spirit of teamwork among the students and giving students opportunity to be creative in the face of challenges. The poster of the Sky Explorers festival in which students can learn, enjoy, and become a leader in new courses, was accepted by the 2019 Munich Conference, which indicates that this project not only improves individuals' abilities but also helps educate new instructors and leaders. In the festival, first, students learn the instructions of the festival and then get involved in the project.

In this project, students are asked to mention their local environmental challenges, as environmentalists to demand the government to solve the environmental problems. This plan is one of the primary plans to control global warming and protect species in danger of extinction. Teachers and students in the south of Iran tried to raise awareness about environmental challenges. Expressing such experiences would help the education and would be promising.

One popular part of the Sky Explorer festival is expressing the experiences. This part is led by a student, and teachers express their experiences and ideas and also analyze the students' experiences. Another interesting part of the Sky Explorer festival is cultural, sports, and artistic programs in which different styles of music and different traditions are introduced to the students, and not only they would have physical movements but also it would be amusing and joyful.

The teacher-training program of NASE, Network for Astronomy School Education (www.naseprogram.org), is incorporated into the activity in Iran. The first Iranian NASE course was held in 2019 (Ros and Rokni 2019). The special on-line Iranian NASE course was held in 2020 (Ros and Rokni 2020).

3. Future perspective

Holding this festival for 9 years and its popularity among students and teachers indicates this project would get great attention and feedback from all around the world. It is suggested that UNESCO help in holding this project in the least developed countries that students and teachers meet their needs. Teachers could not contact other teachers around the world, however, now in the pandemic of Covide-19, they catch the opportunity to express their ideas to other teachers using virtual platforms. ITAU will share its experiences with teachers and students in SINA with online workshops and international astronomical projects for all of the students around the world.

References

Ros, R. & Rokni, M 2019, *First Iranian Course on Didactics of Astronomy, NASE-137, 9-12 March 2019, Bushehr, Iran*, URL: http://sac.csic.es/astrosecundaria/en/cursos/realizados/reglados/132_iran_2019/ListaDocs.php

Ros, R. & Rokni, M 2020, *Special On-Line Iranian Course on Didactics of Astronomy, NASE-187, 17-18 August 2020, Tabriz, Iran*, URL: http://sac.csic.es/astrosecundaria/en/cursos/realizados/reglados/186_busher_2020/ListaDocs.php

Education and Heritage in the era of Big Data in Astronomy
Proceedings IAU Symposium No. 367, 2020
R. M. Ros, B. Garcia, S. R. Gullberg, J. Moldon & P. Rojo, eds.
doi:

Closing Session

Introduction

Insightful closing remarks were given by IAU President-Elect Debra Elmegreen. Debra began by sharing that this virtual symposium became a positive experience during the pandemic. She emphasized the much larger virtual audience than would have been the case face-to-face. Debra recalled the part of the IAU's mission statement that says "promote and safeguard astronomy in all its aspects (including research, communication, education and development) through international cooperation," and commended IAU global initiatives that have now "engaged several hundred thousand people in education, heritage, and outreach activities. She then shared her observations of symposium highlights and mentioned key messages such as how astronomy inspires interests in STEM and STEAM. She ended by expressing how all were enriched by the many contributions and she thanked the organizers.

Education and Heritage in the era of Big Data in Astronomy
Proceedings IAU Symposium No. 367, 2020
R. M. Ros, B. Garcia, S. R. Gullberg, J. Moldon & P. Rojo, eds.
doi:10.1017/S1743921321000958

Closing Remarks

Debra Meloy Elmegreen

IAU President-elect
Vassar College, Department of Physics & Astronomy, Poughkeepsie, NY USA 12604
email: `elmegreen@vassar.edu`

Abstract. This symposium has highlighted key first steps made in addressing many goals of the IAU Strategic Plan for 2020–2030. Presentations on initiatives regarding education, with applications to development, outreach, equity, inclusion, big data, and heritage, are briefly summarized here. The many projects underway for the public, for students, for teachers, and for astronomers doing astronomy education research provide a foundation for future collaborative efforts, both regionally and globally.

1. Introduction

This symposium, the first fully virtual one of the International Astronomical Union, provided a positive experience in a year otherwise marred by the global pandemic. While we all missed a face-to-face gathering, this online meeting allowed a much larger audience to share ideas about education, astronomical heritage, and big data than would have been possible in person, with more than 600 participants from around the globe.

The IAU, initially founded a century ago to share scientific ideas among astronomers, has evolved to include a much broader perspective. This is reflected in its expanded mission statement to "promote and safeguard astronomy in all its aspects (including research, communication, education and development) through international cooperation," as stated in the Strategic Plan for 2020–2030†. In her opening remarks, President van Dishoeck noted that our research-based efforts represented by nine scientific Divisions and dozens of Commissions and Working Groups are linked to four outward-looking essential and interconnected cornerstones of the IAU, which are the Offices of Astronomy for Development, for Education, Outreach, and Young Astronomers. Global initiatives such as IAU100 have engaged several hundred thousand people in education, heritage, and outreach activities.

IAU Vice President Hearnshaw reflected on the deliberate evolution and transformation that the IAU has undergone, particularly over the last half century. The five main strategic goals of the IAU include worldwide coordination, fostering communication, dissemination of astronomical knowledge, inclusive advancement of astronomy, using astronomy as a tool for development, engaging public through access to astronomical information, and using astronomy for teaching and education. A new Code of Conduct includes an ethics policy and an anti-harassment policy, applicable to all IAU functions so that everyone feels welcome. The Strategic Plan provides a blueprint both for social revolution in astronomy, and for using astronomy to make progress in society. This symposium has taken the first steps in addressing many of these goals, and in paving the way for future collaborative efforts both regionally and globally.

† https://www.iau.org/administration/about/strategic_plan/

2. Highlights

Highlights of the symposium included discussions about astronomy as a tool for development through human capacity-building, astronomical heritage and culture, databases and online resources, teacher training and student workshops, women, diversity, and inclusion in astronomy, OAE strategic plans, astronomy education research, engaging through museums, planetaria, astro-tourism, eclipses and citizen science, and astronomy as a STEM gateway. Brief summaries are listed below, with invited speakers noted whose papers appear in this Proceedings.

OAE Director Pössel described the newly formed Office of Astronomy for Education, headquartered at the Haus der Astronomie, and National Astronomy Education Coordinators (NAECs; already 300 in 80 countries), along with OAE centers and nodes. Efforts are underway to provide teachers with access to high-quality resources, hold schools for astronomy education, define standards and create databases for resources and disseminate best practices for teaching and evaluation, as well as foster connections among teachers, astronomers, and the astronomy education research community. Division C President Deutsua described the IAU strategic plan for educational efforts, and the reorganization of the division (which includes education, outreach, and heritage) and its Commissions and Working Groups to complement OAE efforts. The Network for Astronomy School Education (NASE) teacher training schools and newly formed Astronomy Education Journal are housed in the division.

A comprehensive review of astronomy teaching over the past century points to avenues for future astronomy education research, such as methods to develop cognitive and spatial skills in addition to knowledge content (Fitzgerald). An AER database has been developed as a resource to the community. In teaching astronomy and engaging with the public, it is important to respect and be aware of cultural and social diversity, non-Western as well as Western heritage, and archaeoastronomy to emphasize that astronomy connects all of humanity (Camino, López, Gangui, González-García). Recognition of cultural and social roots of astronomy is a key for engaging local communities in development. IAU-UNESCO efforts are underway for the designation of astronomy heritage sites.

Research and public engagement in solar eclipses were discussed, with a view towards the eclipse at the conclusion of this symposium (Pasachoff, Young). Posters and oral contributions highlighted pandemic-driven efforts to develop online learning methods even in remote regions. STEAM activities, multidisciplinary and interdisciplinary approaches to teaching astronomy, robotic telescope observations, museums and planetaria (some online), virtual observatories, heritage sites, astro-tourism, hackathons, and the IAU International Schools for Young Astronomers (ISYA) are all areas of ongoing activities.

Data-driven activities play important roles in astronomy education, outreach, and public science literacy (Soonthornthum, Li). Open access to data is vital for public communication and education as well as for astronomical research. The free and open source platform desktop planetarium program Stellarium (Zotti) provides multi-language support and multi-cultural constellations and mythological figures as well as vantages from archaeoastronomy sites. For teaching and outreach there is a wide variety of online resources, including free online textbooks, lab activities, videos, sky-viewing tools, interactive simulations, image collections, citizen science projects, podcasts, and virtual worlds (Impey).

Discussion of diversity and inclusion in STEM activities included the fact that less than 30% of scientists worldwide are women (and the percentage of astronomers who are women is half this value), so a global cultural change is needed to achieve equity (Hallberg). It is also important to have diverse role models so that everyone can see themselves reflected in real scientists. This is critical, since a diverse astronomical community fosters ideas and advancement. The future is in the hands of young astronomers, who connect with subsequent generations through their outreach efforts (Ödman).

One impediment in science involves career moves, and ASTROMOVES is a project that examines the associated issues and impacts (Holbrook). Posters and discussions on these topics included STEM for girls, and girls & women networks to provide role models and foster connections. There are many activities and exhibits designed for the hearing-impaired and visually-impaired to make astronomy more accessible to all. There are also outreach activities in such diverse settings as children's hospitals, homes for the elderly, and institutions for youths. Furthermore, lifelong learning is a key to addressing many challenges that communities face, so ongoing accessibility to knowledge is crucial (Hammer).

Astronomy for development touches on many of the United Nations Sustainable Development Goals. Because astronomy is a source of inspiration, astronomy education is a primary contributor to development by leading to many other technological and scientific interests (Soonthornthum, Ödman). New ideas have arisen at the boundaries between astronomy and other fields, and engagement in research develops skills that lead to progress through interdisciplinary approaches to problems. When people are better educated about science, they can make informed decisions about policies, which can have positive impacts on their communities.

3. Key Messages

The activities and efforts presented at this symposium had many commonalities. The inspiration of astronomy leads to further STEM and STEAM interests because it is multidisciplinary. Astronomy is a tool for capacity-building through training and development of skill sets applicable to many problems. Teacher effectiveness requires training, databases, education research, and resources. Teaching the use of data and data analysis develops critical thinking, which in turn leads to progress on many of the UN Sustainable Development Goals. Engaging in astronomy improves scientific literacy, which leads to better-informed global citizens who can make a positive impact in their communities. Internet activities and data connect us globally and encourage collaborative efforts, which is at the heart of the IAU. Spreading the message of astronomy requires inclusive, accessible, and multi-lingual resources, activities, and opportunities. Finally, linking astronomy to our various heritages and cultures emphasizes that astronomy unites all humankind.

We leave the symposium enriched by the contributions of all the symposium speakers and participants, and look forward to further progress and collaborations as we employ the message and methods of astronomy for the betterment of society.

4. Acknowledgments

We are very grateful to the SOC, chaired by Beatriz Garcia and Rosa Ros, for planning this informative and enlightening symposium. We also thank the LOC for making all the arrangements for this virtual symposium, and the session chairs for coordinating the meeting. We appreciate the important contributions by the invited speakers and oral and poster presenters, as well as the participants whose questions stimulated discussions. Finally, we thank the many sponsors who made this free symposium possible.

Education and Heritage in the era of Big Data in Astronomy
Proceedings IAU Symposium No. 367, 2020
R. M. Ros, B. Garcia, S. R. Gullberg, J. Moldon & P. Rojo, eds.
doi:

Conclusions

The symposium's aim to present a global vision of astronomical education and heritage was a resounding success. The many great talks, papers, posters, and workshops made this a reality. These trying times with the Covid-19 pandemic led to this being the first all-virtual symposium of the IAU. The pandemic has affected everyone around the globe – limiting travel, limiting interpersonal relationships, and has caused much suffering. Considering such circumstances, the results of IAUS 367 were remarkable with 635 attendees spread across six continents. There were 23 invited talks, 50 oral presentations, 110 poster presentations, four public conferences, three workshops, and one roundtable discussion. The Local Organizing Committee did an amazing job, first in organizing a superb presencial symposium, but then even more so when they took what had been created and reworked the event to be entirely virtual. These proceedings represent this great achievement.

The virtual symposium left us without informal discussions and kept us from renewing relationships with our colleagues. We were not able to experience the wonderful local culture in Bariloche and we did not get to see the 2020 total eclipse of the Sun! On the other hand, being virtual made the symposium available to so many around the world who otherwise might not have been able to attend. Attendance exceeded expectations and there were great contributions of talks and posters from scholars representing countries seldom represented in the past. Becoming better acquainted with the state of global astronomy education initiatives at these locations was a tremendous benefit that resulted from being online. Being virtual also allowed us the privilege of having the current IAU president and the IAU president-elect as our opening and closing speakers. At traditional face-to-face events it is possible that one or both of them would be unable to attend in-person. Another positive factor of being online was in the way that time management was enhanced in the virtual format. This worked very well and the talks of the symposium stayed on time!

We now know that IAU symposiums can be very successful online, and this bodes well for the future. It is our desire to reach out to and include as many as we can from anywhere in the world, and the virtual format of the symposium proved that this can be done most effectively. Our results underscore that all such IAU events in the future should include a virtual element in a hybrid format. We certainly want to meet with each other in person again, but we also want to continue to draw collaborations and contributions from the many colleagues we otherwise might not come to know because they are unable to attend in person. We will take the lessons learned here and apply them in the future as we refine the format to work both onsite and online simultaneously. Hybrid events are the future!

The global initiatives in astronomy education highlighted in this symposium are exciting! Restrictions due to the pandemic will eventually end and when they do it will allow us to share our research directly with each other once again. Including virtual participation as well will enable greater collaboration with additional colleagues around the globe.

Rosa M. Ros
Beatriz Garcia
Steven R. Gullberg
Javier Moldon
Patricio Rojo

Author index